INTRODUCTION TO

COASTAL PROCESSES & GEOMORPHOLOGY

SECOND EDITION

GERHARD MASSELINK • MICHAEL HUGHES • JASPER KNIGHT

INTRODUCTION TO

COASTAL PROCESSES & GEOMORPHOLOGY

SECOND EDITION

GERHARD MASSELINK • MICHAEL HUGHES • JASPER KNIGHT

First edition published 2003
Second edition published 2011 by
Hodder Education, a member of the Hodder Headline Group

Published 2014 by Routledge
2 Park Square, Milton Park, Abingdon, Oxon OX14 4RN
711 Third Avenue, New York, NY, 10017, USA

Routledge is an imprint of the Taylor & Francis Group, an informa business

British Library Cataloguing in Publication Data
A catalogue record for this book is available from the British Library

Library of Congress Cataloging-in-Publication Data
A catalog record for this book is available from the Library of Congress

ISBN 13: 978-1-444-12240-4 (pbk)

Typeset in Palatino Light by Dorchester Typesetting Group Ltd

Cover image: © Gerhard Masselink

CONTENTS

PREFACE

This textbook has been written for intermediate-level undergraduate students. We have attempted a level of explanation of geomorphological processes in relation to the formation of coastal landforms that is introductory, but that will also extend students' analytical skills. We draw on examples from around the world, and many of these have a geographical bias favouring the British Isles and Australia because this is where we have undertaken much of our own research.

This second edition builds on the strengths of the first (published in 2003), but we have corrected a few errors, removed out-dated material and updated with new material, in particular related to sea level. We have also added new chapters on climate and sand dunes. New boxes and case studies in each chapter expand upon elements of the text. Also shown in the book are locations where supporting online materials are available on the publisher's website. A headphones symbol indicates that answers to the reflective questions are available as mp3 files. A mouse symbol shows where expanded case studies and worked examples are available online. These additional resources are designed to increase students' understanding and hopefully their interest in coastal morphodynamics.

A number of friends and colleagues have commented on aspects of the first edition of this book: Kitty Bos, Rob Brander, Jo Bullard, Bruce Hegge, Paul Kench, Gui Lessa and Wayne Stephenson. Their comments have been very useful and have greatly improved the readability of this book. We are, of course, responsible for any remaining errors, omissions, inappropriate embellishments and bad prose. We also thank the following colleagues for providing original figures for the second edition of this book: Martin Austin for Figure 8.8; Helene Burningham for Figure 12.1a; Giovanni Coco for Figure 1.14; Peter Cowell for Figure 8.23a; Simon Haslett for Figure 8.35; Peter Keene for Figure 10.7c; Aart Kroon for Figure 8.28b; Julian Orford for Figures 4.13b and 4.16; Andy Short for Figures 8.9a, 8.23e and 8.23f; Matt Tomczak for Figure 7.12; and Wayne Stephenson for Figure 10.13. Thanks also to Mark Szegner for cartographic assistance.

We thank the following for permission to reproduce figures: Academic Press for Figures 2.8, 8.36, 8.37, 9.10, 9.17 and 11.9; American Geographical Society for Figure 11.6; American Geophysical Union for Figures 7.21b and 10.25; Blackwell Publishers for Figures 2.2, 5.5, 5.7, 5.19a, 5.22a, 5.23b, 6.5, 6.8, 6.11b and 7.18; Butterworth Heinemann for Figure 11.3; Cambridge University Press for Figures 1.3, 1.12b, 1.15, 2.1, 7.8, 8.26a, 8.29b, 8.32, 8.33, 8.39, 9.2, 9.6, 9.7, 10.28, 11.1 and 11.5; Coastal Education and Research Foundation for Figures 4.13, 6.16, 7.9, 8.19b, 8.20, 8.27, 8.31, 10.1 and 10.12; Edward Arnold for Figures 3.12 and 8.14; Elsevier Science for Figures 3.11, 3.14, 9.15, 10.3, 10.14, 11.8 and 12.5; Geographical Association for Figure 10.5; Geological Association of Canada for Figure 8.29a; Geological Society of America for Figure 6.7, 6.9 and 11.11; Geological Society of London for Figures 9.19 and 11.4; George Allen & Unwin for Figure 5.11; Geoscience Australia for Figure 6.12; International Association of Sedimentologists for Figures 5.21 and 6.3; Irish Marine Institute for Figure 7.3; John Wiley & Sons for Figures 1.12a, 4.4, 4.20, 8.1, 8.4, 8.9a, 8.13, 8.18, 8.19a, 8.21, 8.22, 9.1, 9.14, 10.2, 10.16, 10.21, 10.23, 10.27 and 11.2b; Joint Nature Conservation Committee for Figure 10.6b; Macmillan Magazines for Figures 6.14 and 10.18; NASA for Figures 3.16, 10.19 and 12.3; Methuen for Figure 11.10; Natural Resources Canada for Figure 4.16; Oxford University Press for Figure 7.10; Parabolic Press for Figures 5.10a and 6.11a; Pearson Education for Figures 8.10 and 8.12; Routledge Education for Figure 11.2a; Springer-Verlag for Figure 5.19b; Society for Sedimentary Geology for Figures 7.2 and 7.4; Society for Sedimentary Petrology for Figure 5.23a; Taylor & Francis Books for Figures 6.2, 8.35 and 12.4; Thomas Telford for Figure 7.21a; The University of Chicago Press for Figure 1.13; University of Colorado for Figure 2.11; and US Army Corps of Engineers for Figure 4.5. In a few cases it was not possible to trace the copyright holders of figures used in this text. We apologise therefore for any copyright infringements that may have occurred.

LIST OF SYMBOLS

A	constant; dimensional parameter; delta plain aggradation rate (m)
a	least squares regression coefficient; height of saltation (m); amplitude of tide (m); longest axis of clast (m)
a_i	amplitude of partial tide (m)
B	height of the subaerial beach (m)
b	spacing of wave rays (m); least squares regression coefficient; intermediate axis of clast (m)
b_o	spacing of wave rays in deep water (m)
C	constant; wave velocity or wave celerity (m s^{-1}); sediment concentration (kg m^{-3})
C_d	drag coefficient (-)
C_g	wave group velocity (m s^{-1})
C_A	accelerated sediment compaction (m)
C_N	natural sediment compaction (m)
C_o	deep water wave velocity (m s^{-1})
C_s	shallow water wave velocity (m s^{-1})
c	shortest axis of clast (m)
CSF	Corey shape factor (-)
D	fetch duration (minutes or hours); grain size (m); fractal dimension (-)
D_r	reference grain diameter (0.25 mm)
D_*	non-dimensional grain diameter
D_{50}	median sediment size (m)
d	wave orbital diameter (m)
d_0	wave orbital diameter at the sea bed (m)
E	vertical shoreline displacement caused by eustatic change (m); wave energy density (N m^{-2})
e_b	bedload efficiency factor (-)
e_s	suspended load efficiency factor (-)
F	fetch length (km); tidal form factor (-)
F_c	centripetal force (N)
F_g	gravitational attractive force (N)

F_{lg} local gravitational force (N)
F_T tractive force (N)
F_t tide-generating force (N)
F_r Froude number (-)
F'_r densimetric Froude number (-)
f constant; frequency (Hz)
f_i frequency of partial tide (hr^{-1}); percentage of grains in grain size class interval
f_w wave friction factor (-)
G universal gravitational constant ($N\ m^2\ kg^{-2}$)
G_i phase of partial tide (radians)
g gravitational acceleration ($m\ s^{-2}$)
H wave height (m)
H_b breaker height (m)
H_l wave height of the largest waves (m)
H_o deep water wave height (m)
H_{rms} root mean square wave height (m)
H_s significant wave height (m)
$\bar{H}_s$ mean annual significant wave height (m)
$H_{1/3}$ significant wave height (m)
H_{sx} nearshore storm wave height that is exceeded only 12 hours each year (m)
h bed elevation (m); water depth (m)
$\bar{h}$ local tidally-averaged water depth (m)
h' thickness of river plume (m)
h_b breaker depth (m)
h_c thickness of the Earth's crust (m); closure depth (m)
h_l limiting depth for significant cross-shore sediment transport of sand by waves (m)
h_i thickness of the ice sheet (m)
h_* depth of the shoreface (m)
I alongshore length of island (m)
J distance between island and mainland (m)
K constant
K_r wave refraction coefficient (-)
K_s wave shoaling coefficient (-)
k wave number ($2\pi/L$)
k_s Nikuradse roughness length (m)
k' skin friction roughness length (m)
k'' bedform roughness length (m)

L	wave length (m); tidal wavelength (m)
L_o	deep water wave length (m)
L_s	shallow water wave length (m)
M	tectonic and lithospheric changes in Earth's mass (kg)
$M_{\phi i}$	midpoint of grain-size class interval on phi scale (phi)
MSR	mean spring tide range (m)
m	dimensionless exponent; mass (kg)
m_E	mass of the Earth (kg)
m_M	mass of the Moon (kg)
n	wave transformation parameter (-)
P	elevation of the raised feature above MSL (m); pressure (kg m^{-1} s^{-2}); wave energy flux (N m^{-1} s^{-1} = J s^{-1}); coastline length (km)
P_l	longshore component of the wave energy flux (N m^{-1} s^{-1} = J s^{-1})
P_o	deep-water wave energy flux (N m^{-1} s^{-1} = J s^{-1})
Q_l	volumetric longshore sediment transport rate (m^3 day^{-1})
q	sediment transport rate (kg m^{-1} s^{-1})
q_b	bedload sediment transport rate (kg m^{-1} s^{-1})
q_s	suspended load sediment transport rate (kg m^{-1} s^{-1})
R	distance between centres of mass of two bodies (m); wave runup height (m)
R_e	Reynolds number (-)
R^*_e	boundary Reynolds number (-)
R_g	grain Reynolds number (-)
R	wave runup height exceeded by 2% of the runup events (m)
RTR	relative tide range (-)
r	radius of the Earth (m); radius of a sediment grain (m); rod length (km)
S	rise in sea level (m); horizontal extent of the swash motion or swash length (m)
S_c	compressive strength of the rock (N m^{-3})
S_{xx}	cross-shore component of radiation stress (N m^{-1} s^{-1} = J s^{-1})
T	wave period (s)
T_p	peak spectral wave period (s)
T_s	significant wave period (s)
$\overline{T}_s$	mean annual significant wave period (s)
T_{sx}	wave period associated with H_{sx} (s)
T_z	zero-crossing or mean wave period (s)
$T_{1/3}$	significant wave period (s)
t	time (s)
$\tan\beta$	beach gradient (-)

U	amount of uplift (m); wind speed at 10 m height (m s^{-1})
U_A	wind stress factor (m s^{-1})
u	horizontal current velocity (m s^{-1})
u_g	velocity of sediment grain (m s^{-1})
u_m	maximum orbital wave velocity (m s^{-1})
u_0	maximum orbital wave velocity at the sea bed (m s^{-1})
u_z	flow velocity at elevation z above the surface (m s^{-1})
u_{100}	velocity measured 100 cm above the bed (m s^{-1})
u_*	shear velocity (m s^{-1})
u_{*t}	threshold shear velocity (m s^{-1})
$\overline{v}_l$	longshore current velocity at the mid-surf zone position (m s^{-1})
w_0	maximum deflection of the land surface under the ice sheet (m)
w_s	sediment fall velocity (m s^{-1})
w_*	width of the shoreface (m)
x	cross-shore direction (m), distance (m)
$\overline{x}$	mean (first-moment) of grain size distribution (phi)
y	longshore direction (m)
Δy	shoreline retreat due to sea-level rise (m)
Z_{berm}	height of berm (m)
Z_{step}	height of beach step (m)
z	height above the bed (m)
z_0	hydraulic bed roughness length (m)
α	wave angle (^{o})
α_b	breaker angle (^{o})
α_o	deep-water wave angle (^{o})
β	bed slope-angle (^{o})
γ	breaker index or breaking criterion (-)
$\langle\gamma\rangle$	relative wave height (-)
$\delta^{18}O$	oxygen isotope ratio (-)
δy	shoreline retreat (m)
ε	factor accounting for sediment density and porosity (-); surf scaling parameter (-)
ε_b	bedload efficiency factor (-)
ε_s	suspended load efficiency factor (-)
η	water surface elevation (m); bedform height (m)
$\overline{\eta}$	mean sea level (m); departure of the water level from still water level (m)
η_r	water level departure from predicted tide (m)
$\overline{\eta}_s$	wave set-up at the shoreline (m)
θ	angle (^{o})

θ_c	critical Shields parameter required for sediment motion (-)
κ	von-Karman constant (-)
λ	bedform spacing (m)
λ_{cusp}	spacing of beach cusp morphology (m)
μ	molecular viscosity of fluid (N s m^{-2})
ξ	Irribarren number (-); eddy viscosity (N s m^{-2})
ρ	water density (kg m^{-3})
ρ_c	density of the Earth's crust (kg m^{-3}); density of coastal waters (kg m^{-3})
ρ_i	density of ice (kg m^{-3})
ρ_m	density of the Earth's mantle (kg m^{-3})
ρ_r	density of river water (kg m^{-3})
ρ_s	sediment density (kg m^{-3})
σ	wave radian or angular frequency ($2\pi/T$); standard deviation (2nd-moment) or sorting of grain size distribution (phi)
τ	fluid shear stress (N m^{-2})
τ_0	bed shear stress (N m^{-2})
τ_w	bed shear stress under waves (N m^{-2})
ϕ	friction angle of sediment (^{o})
ψ	transport stage (-)
Ω	dimensionless fall velocity (-)

CHAPTER 1

COASTAL SYSTEMS

AIMS

This chapter outlines how the coastal zone can be defined and discusses the properties and characteristics of systems, highlighting the importance of sedimentary processes and concepts such as self-organisation. Consideration of systems is useful because it provides a framework by which the dynamic behaviour of geomorphological processes and landforms can be understood.

1.1 Introduction

This book is about coastal processes and geomorphology. The spatial boundaries of the coastal zone as considered in this book are defined in Figure 1.1 and follow the definitions set out by Inman and Brush (1973). The upper and lower boundaries correspond to the elevational range over which coastal processes have operated during the Quaternary period, and include the coastal plain, the shoreface and the continental shelf. During the Quaternary (from 2.6 million years ago until present), sea level fluctuated over up to 135 m vertically due to expansion and contraction of ice sheets and warming and cooling of the oceans. The landward limit of the coastal system therefore includes the coastal depositional and marine erosion surfaces formed when sea level was high (at

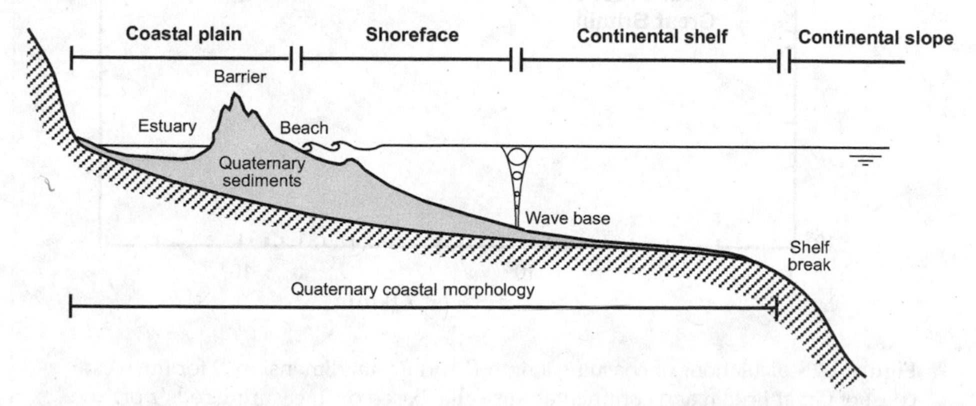

Figure 1.1 Spatial boundaries of the coastal zone considered in this book.

or just above its present-day position) during global warm periods. The lowest sea levels, during global cold periods, placed coastal processes close to the edge of the continental shelf. The seaward limit of the coastal system is therefore defined by the continental shelf break, which typically occurs in water depths of 100–200 m. Changes in global climate during the Quaternary markedly shifted the position of the coastal zone, changing coastal geography particularly in areas with extensive shelves (Case Study 1.1). Many of today's coastal landscapes reflect their Quaternary (and older) geological inheritance, and are responding dynamically to human activity in the coastal zone and ongoing global warming.

Case Study 1.1 How long are the world's coastlines?

In 1967 the mathematician Benoit Mandelbrot (1924–2010) posed the question: How long is the coastline of Britain? There are several reasons why we need to know the answer to this question. First, coastline length has

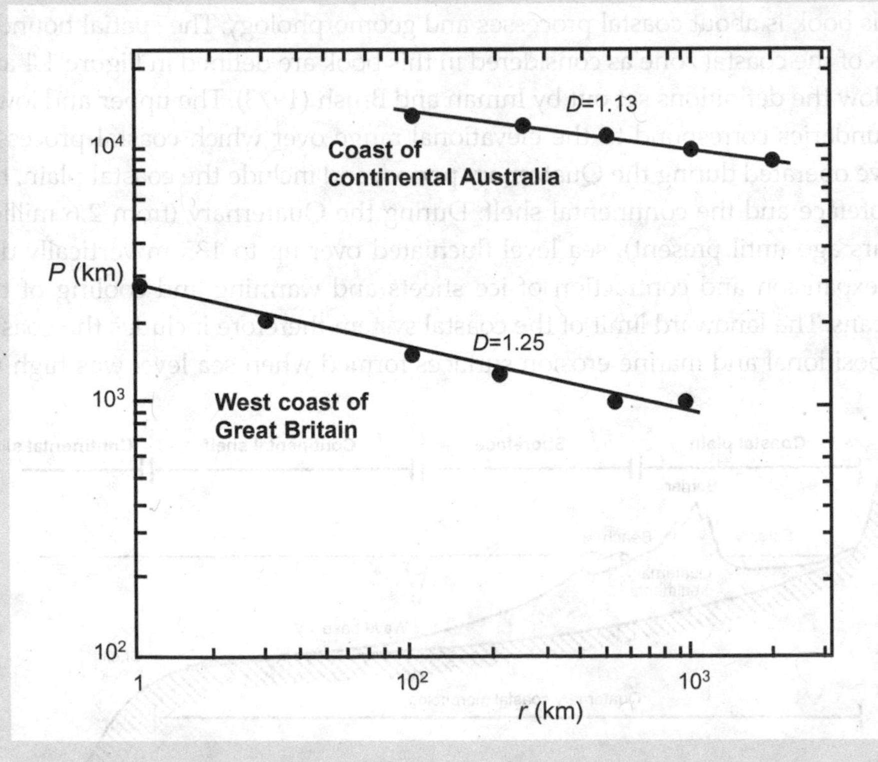

Figure 1.2 Calculations of coastline length P and fractal dimension D for the west coast of Great Britain and continental Australia, based on 'measuring rods' r of different length. (Data from Mandelbrot, 1967.)

implications for calculation of length (and therefore cost) of coastal protection such as sea walls, and the length (and area) of protected areas. Second, coastline length has implications for calculation of coastal resources such as length of rocky or sandy coasts, productivity of the intertidal zone, or volume of offshore aggregate. Calculating coastline length, however, is not straightforward, because all coasts exhibit complex outlines over different scales. For example, the position of the water–land boundary varies due to tidal state, and by coastal erosion and deposition. Highly indented rocky coastlines are also longer than smooth, sandy coastlines.

An important property of almost all coastline types and coastal locations is that they exhibit **scale invariance** (Mandelbrot, 1967). This means that there is the same degree of complexity in the coast's outline shape irrespective of the scale at which it is observed. Figure 1.2 shows the length of a mapped coastline P using rods r of different lengths: the shorter the rod the longer the measured coastline. What this means in practice is that very accurate measuring devices give higher coastline lengths. The results plot in a straight line on a log-log scale, which means that the coastline is scale invariant over that scale range. The slope of the line gives the **fractal dimension** D. The coastlines of Great Britain and Australia exhibit a similar fractal dimension, but Australia is larger than Britain so both P and r are larger.

Having defined the spatial boundaries of the topic of this book, we now need to indicate the time scale we are interested in. Cowell and Thom (1994) group the time scales over which coastal processes operate into four overlapping classes (Figure 1.3):

- **Instantaneous time scales** – These involve the evolution of morphology during a single cycle, for example, the formation and destruction of wave ripples by large waves, and onshore migration of an intertidal bar over a single tidal cycle.
- **Event time scales** – These are concerned with coastal evolution by processes on time spans from individual wave/tide events to seasons. Examples include the erosion of coastal barriers by a major storm or seasonal closure of an estuary by a sand bar.
- **Engineering time scales** – These are time scales of years to centuries and include the migration of tidal inlets and buildup of sediment in beaches or foredunes. Coastal engineers and managers are most concerned with processes that operate on these time scales.
- **Geological time scales** – These time scales operate over decades to millennia and correspond to directional trends in the driving forces of sea

level, climate and tectonics. Examples include the infilling of a tidal basin or estuary, onshore migration of a barrier system, and switching of delta lobes.

This book will deal with coastal processes and geomorphology on all four time scales.

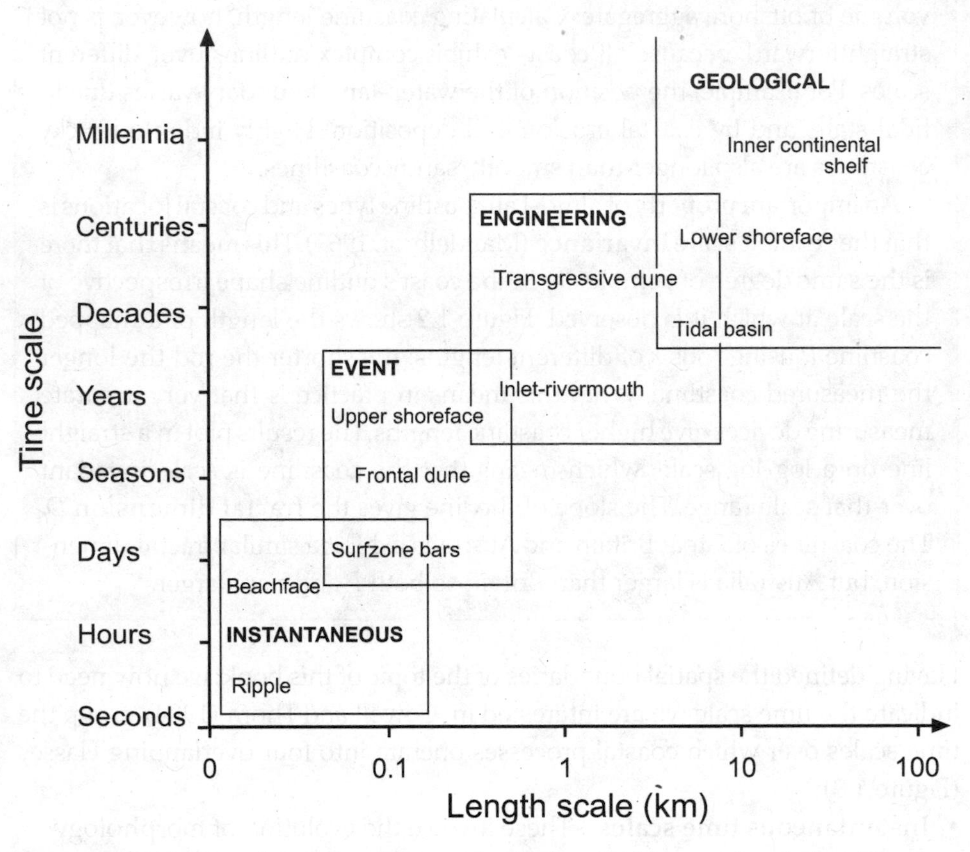

Figure 1.3 Definition of spatial and temporal scales involved in coastal evolution. (From Cowell and Thom, 1994.) (Copyright © 1994 Cambridge University Press, reproduced with permission.)

1.2 Coastal classification

A variety of coastal landforms and coastline types exists in nature and coastal geomorphologists are concerned with describing and classifying them. **Coastal classification** schemes can be useful from a conceptual point of view and help to assess the different forcing factors and controls (e.g. sea-level history, geology, climate, waves, tides) that give rise to coastal landforms. Most early classification schemes are based on the role of sea-level variations, and so distinguish between submerged and emerged coasts (Johnson, 1919). **Submerged coasts** include drowned river and glacial valleys, termed rias/fjards and fjords, respectively. Coastal plains are characteristic of **emerged coasts** where sea level has fallen. Shepard (1963) identified primary and secondary coasts. **Primary coasts** result mainly from non-marine processes and include drowned river valleys, rocky and deltaic coasts. **Secondary coasts** result mainly from marine processes or organisms and include barrier coasts, coral reefs and mangroves. Coasts can also be classified with respect to their tectonic position (Inman and Nordstrom, 1971). **Leading edge coasts**, also termed collision coasts, are located adjacent to subducting plate margins such as along the Pacific coasts of South America, Japan and New Zealand. Here, tectonic processes have formed mountain belts that have steep, erosive and rocky coastlines, boulder beaches, and falling relative sea levels. **Trailing edge coasts**, however, are located away from subducting plate margins, are tectonically benign, older and of lower elevation. Examples are coasts of Africa, Australia and Atlantic coasts of North and South America. These coasts are typically sediment-rich, progradational, with large deltas and sandy beaches.

The main shortcoming of these classifications is that they emphasise geological inheritance rather than hydrodynamic processes that shape coastal landforms. Davies (1980) identified coastal types based solely on wave height and tidal range. Because waves are generated by wind, the **distribution of wave environments** varies by latitude, reflecting global climate zones (Figure 1.4). Coastlines dominated by storm waves are located in higher temperate and arctic latitudes, whereas swell-dominated coasts are located in lower temperate and tropical latitudes where cyclones (hurricanes) are also important. The tidal range

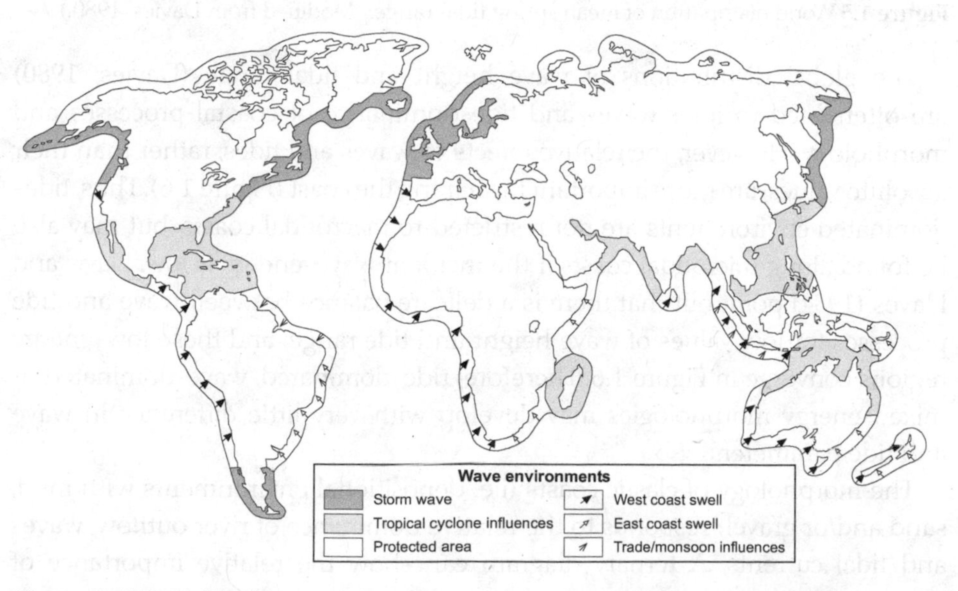

Figure 1.4 World distribution of wave environments. (Modified from Davies, 1980.)

in the middle of oceans is quite small (less than 1 m), but increases towards the coast and may reach in excess of 10 m. The amplification of tides depends on the gradient and width of the continental shelf, the location and shape of continents, and the presence of large embayments. The global **distribution of tidal range** is therefore controlled by large-scale coastal configuration (Figure 1.5). Macrotidal ranges exceeding 4 m are mostly observed in semi-enclosed seas and funnel-shaped entrances of estuaries. Microtidal ranges below 2 m occur along open ocean coasts and almost fully enclosed seas. Small but noticeable tidal fluctuations also occur in large lakes.

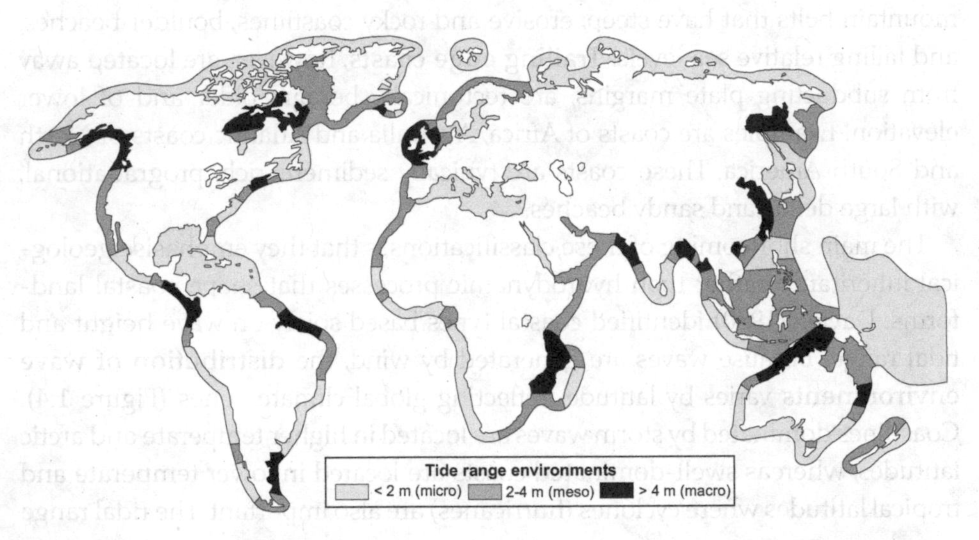

Figure 1.5 World distribution of mean spring tidal range. (Modified from Davies, 1980.)

The global distributions of wave height and tidal range (Davies, 1980) are often used to infer wave- and tide-dominance of coastal processes and morphology. However, the relative effects of waves and tides, rather than their absolute values, are more important in shaping the coast (Figure 1.6). Thus, tide-dominated environments are not restricted to macrotidal coasts, but may also be found along microtidal coasts if the incident wave-energy is low. Davis and Hayes (1984) point out that there is a delicate balance between wave and tide processes for low values of wave height and tide range, and these low-energy regions converge in Figure 1.6. Therefore, tide-dominated, wave-dominated or mixed-energy morphologies may develop with very little difference in wave and tide parameters.

The morphology of clastic coasts (i.e. depositional environments with mud, sand and/or gravel) responds to the relative dominance of river outflow, waves and tidal currents. A ternary diagram can show the relative importance of

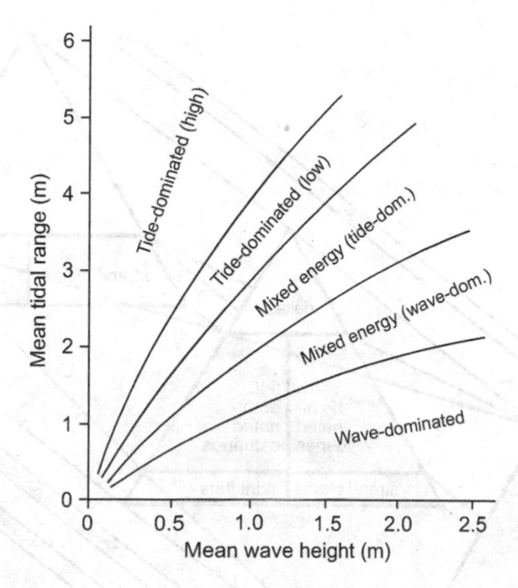

Figure 1.6 Relationship between mean tidal range and wave height delineating different fields of wave/tide dominance. A particular coastal stretch may span several fields. (Modified from Davis and Hayes, 1984.)

these three factors. Deltas are positioned at the fluvial apex because a fluvial sediment source dominates, while prograding, non-deltaic coasts are on the opposite wave–tide side, because in these environments sediment is moved onshore by waves and tides. Estuaries occupy an intermediate position because they have a mixed sediment source and are affected by river, wave and tidal factors. Dalrymple *et al.* (1992) provide an evolutionary aspect to the ternary classification by including time as an additional component (Figure 1.7). According to this **classification of clastic coastal environments**, time is expressed in terms of coastal accretion (or progradation) and coastal inundation (or transgression). Progradation of the coastline generally occurs with a relatively constant sea level and a large sediment supply, whereas transgression takes place when sea level is rising across the land, pushing the coastline farther back. Progradation entails the infilling of estuaries and their conversion into deltas, strand plains or tidal flats. This is shown in Figure 1.7 by movement to the back of the triangular prism. Changes associated with transgression, such as the flooding of river valleys and the creation of estuaries, are represented by movement towards the front of the prism. Dalrymple *et al.*'s (1992) model also acknowledges the dynamic nature of coasts, because changes in sea level that lead to progradation and transgression are caused primarily by changes in climate.

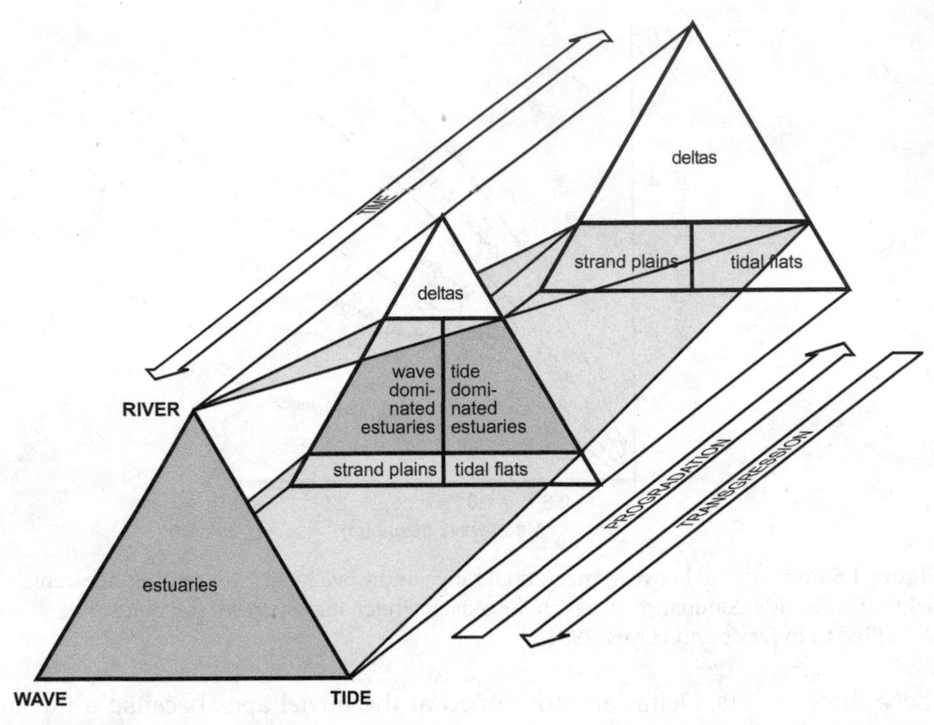

Figure 1.7 Evolutionary classification of coastal depositional environments. The long axis of the three-dimensional prism represents relative time with reference to changes in relative sea level and sediment supply (i.e. transgression and progradation). The three edges of the prism correspond to fluvial, wave and tidal dominance. (Modified from Dalrymple *et al.*, 1992.)

1.3 Morphodynamic approaches to coastal systems

The coastal classifications discussed above indicate which environmental factors are most important in shaping the coast, but to understand their interactions a more in-depth approach is required. Wright and Thom (1977) applied a **systems approach** to coastal morphology and evolution by viewing the coast as a **geomorphic system** with inputs and outputs of energy and material, influenced by surrounding environmental conditions (Figure 1.8). Wright and Thom (1977) introduced the term 'coastal morphodynamics' for their approach, defined as 'the mutual adjustment of topography and fluid dynamics involving sediment transport'. The morphodynamic approach has since become the paradigm for studying coastal evolution. We now discuss the main environmental conditions and properties of coastal morphodynamic systems.

1.3.1 Environmental conditions

Environmental conditions are the 'set of static and dynamic factors that drive and control coastal systems' (Wright and Thom, 1977). They are not affected by

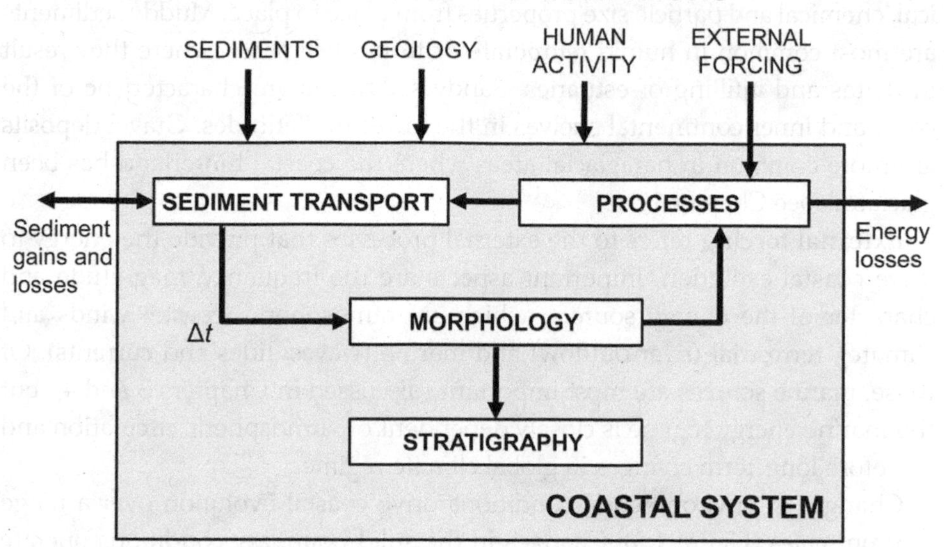

Figure 1.8 Primary components involved in coastal morphodynamics. Δt signifies the time dependence inherent in coastal morphodynamic evolution. (Modified from Cowell and Thom, 1994.)

the coastal system itself but form the boundary conditions of the system, and are responsible for geographical variations in coastal geomorphology. The main types of environmental conditions are geology, sediments and external forcing (Figure 1.8).

Geology describes the initial state and properties of the solid boundaries, including the surrounding geology and pre-existing morphologic state (shelf and shoreline configuration, lithology). Globally, tectonics controls the width and slope of the continental shelf. Wide and flat continental shelves (e.g. east coast of the USA) permit more rapid coastal progradation for a given rate of sediment supply than steep and narrow shelves (e.g. southeast coast of Australia). In addition, wide shelves lead to greater reduction in wave height by frictional dissipation and are also responsible for amplification of tides. Regionally, coastline configuration can be important by controlling wave processes such as reduced wave energy levels in the lee of offshore islands and within coastal embayments. Lithology is a significant factor on the recession rate and cliff profile development along eroding rocky coasts.

Unconsolidated sediments are essential for coastal evolution because they build into landforms such as beaches. The nature and abundance of the sediment is examined in Chapter 5. Sediment availability depends on the location and volume of sediment sources, and transport processes between the sources and the location where sediments are deposited. Sediments may have a marine, fluvial/deltaic, terrestrial or biological origin, and so they may vary in their physical, chemical and particle size properties from place to place. Muddy sediments are most common in humid temperate or tropical climates, where they result in deltas and infilling of estuaries. Sandy sediments are characteristic of the coast and inner continental shelves in the lower midlatitudes. Gravel deposits are more common in paraglacial areas where the coastal hinterland has been glaciated (see Chapter 11).

External forcing refers to the external processes that provide the energy to drive coastal evolution. Important aspects are the frequency, magnitude and character of the energy sources, which are atmospheric (coastal winds and climate), terrestrial (river outflow) and marine (waves, tides and currents). Of these, marine sources are most important (discussed in Chapters 3 and 4), but the marine energy regime is closely dependent on atmospheric circulation and therefore long-term changes in global climate regime.

Changes in environmental conditions drive coastal evolution over a range of time scales (Figure 1.3). Changes in the solid boundary conditions operate on geological time scales and are mainly related to vertical land movements resulting in coastal emergence or submergence by falling and rising relative sea levels, respectively. Changes in sediment type and abundance operate mainly on geological and engineering time scales. On geological time scales, changing sea levels during the Quaternary affected sediment availability across the continental shelf. On engineering time scales, human intervention in coastal catchments and drainage basins, and their ability to deliver sediment to the coast, has most significantly affected coastal sediment availability. Greatest temporal changes in external forcing, however, occur on instantaneous and event time scales, such as seasonal or storm-related changes in weather and waves.

1.3.2 The mechanics of systems

The term 'system' describes the organisational framework through which energy and matter are moved around on different scales within and upon the Earth. In **open systems** there is a free exchange (termed flux) of energy and matter between different parts of the system, or between the system and its surroundings. Typical open systems on the Earth's surface include river basins and beaches. Figure 1.8 describes a simple coastal system. The movement of energy and matter is manifested as geomorphological processes which represent the fluxes of energy from one part of the system to another. These processes create or destroy the landforms that represent the dynamic state of the system.

All systems have a number of shared characteristic properties. For example, systems are found on different scales, from atomic to global; they have limits or boundaries; they are associated with processes, which are the means by which changes to the system are made; they have structural organisation, which refers to the pathways through which energy and mass are moved within the system; and the component parts of a system tend towards a steady or stable state over time, which is referred to as the system being in **dynamic-** or **quasi-equilibrium**.

The directional trajectory of processes that drive a system towards equilibrium is regulated within the system by a mechanism that is termed feedback. There are two types of feedback. **Negative feedback** refers to those processes which act to maintain a steady state (quasi-equilibrium) of the system. Negative feedback dampens down the system by internal self-regulation. **Positive feedback** refers to those processes that lead to a directional change in the system that has cumulative effect over time. Under positive feedback ('self-forcing') a system follows a directional trajectory of change away from one quasi-stable state. Negative feedback is then required in order to bring the system back under control and to stabilise it around a different quasi-stable state (Figure 1.9). Under some circumstances positive feedback works so effectively that a system can accelerate out of control, such as during the Neoproterozoic period (*c.* 540–1000 million years ago) when positive climate feedback led to near-total global glaciation.

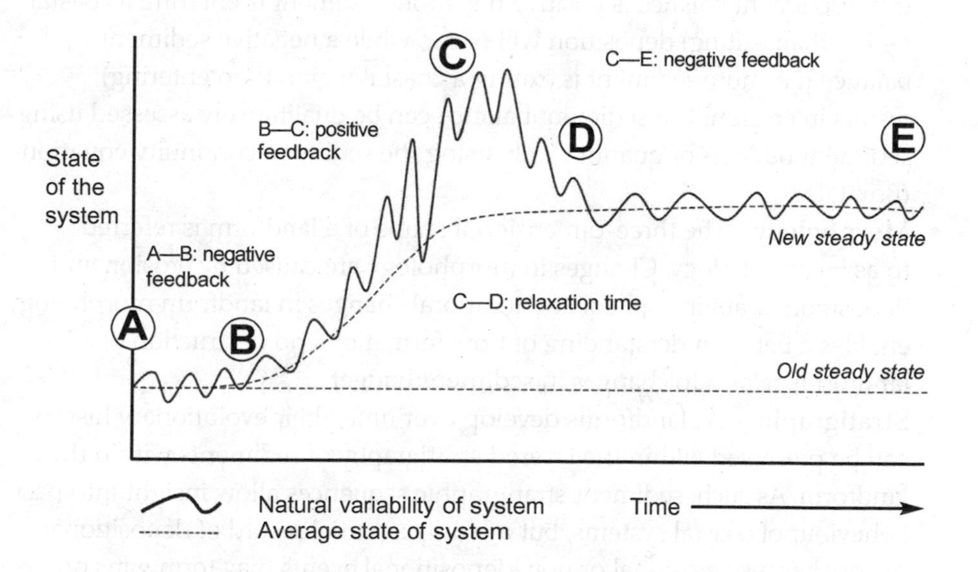

Figure 1.9 Illustration of periods of negative and positive feedback within a single system over time. Negative feedback in periods A–B and C–E allows the system to persist at equasi-equilibrium around a steady state. Positive feedback in period B–C allows the system to follow a trajectory of change. The time period required for the system to attain a steady state following initial disturbance (period C–D) is the relaxation time.

1.3.3 Properties of coastal systems

A coastline is a typical open system because its component parts are linked by the flow of energy and material but are also influenced by environmental conditions outside of its system boundaries (Figure 1.8). These external factors include climate, sea level, sediment supply and human activity. The coastal system itself has four main properties that correspond to the flow of energy and material through the system (Cowell and Thom, 1994):

- **Processes** – These represent energy and material in transit from one part of the system to another, which in coastal environments take place by sediment erosion, transport and deposition. The most important coastal processes are driven by hydrodynamic (waves, tides and currents) and aerodynamic (wind) external forcing. Weathering also contributes to sediment generation on rocky coasts, and biological processes are important in coral reef, salt marsh and mangrove environments.
- **Sediment transport** – The interaction between a moving fluid and an unconsolidated bed induces shear stresses that may result in the entrainment and transport of sediment (see Chapter 5). Patterns of erosion and deposition can be assessed using the concept of net sediment balance. If the sediment balance is positive (i.e. more sediment is entering a coastal region than exiting) deposition will occur, while a negative sediment balance (i.e. more sediment is exiting a coastal region than entering) results in erosion. The sediment balance can be qualitatively assessed using sediment budgets or quantitatively using the sediment continuity equation (Box 1.1).
- **Morphology** – The three-dimensional shape of a landform is referred to as its morphology. Changes in morphology are caused by erosion and deposition. Mapping spatial and temporal changes in landform morphology enables a better understanding of how formation and destruction of landforms relates to changes in sediment budget.
- **Stratigraphy** – As landforms develop over time, their evolutionary history can be preserved within the layered stratigraphy of sediments within the landform. As such, sediment stratigraphic sequences allow insight into past behaviour of coastal systems, but are only a partial record of depositional history because erosional or non-depositional events may form gaps or truncations in the record.

Box 1.1 Sediment budgets and sediment balance

Morphological change is the net result of sediment transport processes. **Sediment budgets** can be used to gain an understanding of the sediment inputs (sources) and outputs (sinks) involved, which are measured in sediment volume (Figure 1.10). Key components of the sediment budget are sediment fluxes, which are vectors representing the direction and amount of sediment transport, expressed as the quantity of sediment moved per unit of time (e.g. kg s^{-1} or m^3 yr^{-1}, etc). Whether a sediment flux is considered input or output depends on the point of view. For example, the transport of sediment from beach to dunes entails a sediment loss (output) for the beach, but a sediment gain (input) for the dunes. If the sediment fluxes are known, sediment budgets can be used to quantitatively predict morphological change over time.

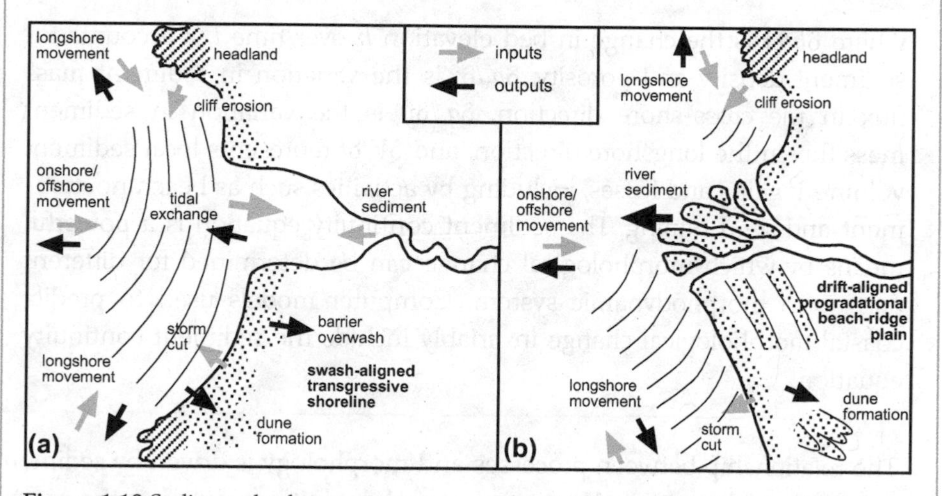

Figure 1.10 Sediment budgets on estuarine and deltaic coasts. (a) Fluvial sediment may be deposited into an estuary, which may also receive sediment from the seaward direction through tidal processes. (b) On a deltaic coast, riverine sediment contributes to the coastal sediment budget, and is moved down the drift-aligned coast, giving rise to a beach-ridge plain. (Modified from Carter and Woodroffe, 1994.)

For example, consider an estuary with a surface area of 1 km^2 that receives an annual input of sediment of 10,000 m^3 yr^{-1}. If we assume that this sediment is evenly spread over the estuary floor, then the depth of the estuary will decrease by

$$\frac{\text{sediment input}}{\text{surface area}} = \frac{10{,}000\ \text{m}^3\ \text{yr}^{-1}}{1{,}000{,}000\ \text{m}^2} = 0.01\ \text{m yr}^{-1}$$

If the average depth of the estuary is 10 m, then the estuary will be infilled in

$$\frac{\text{depth}}{\text{accretion rate}} = \frac{10 \text{ m}}{0.01 \text{ m yr}^{-1}} = 1000 \text{ yr}$$

This simple example assumes that the amount of sediment entering the estuary does not change while the estuary is infilling, and thus ignores one of the main principles of morphodynamic systems – that there is feedback between morphology and process.

A more rigorous analysis of morphological change is through the **continuity equation** for sediment transport

$$\frac{\delta h}{\delta t} = \varepsilon \left(\frac{\delta q_x}{\delta x} + \frac{\delta q_y}{\delta y} \right) - \frac{\delta V}{\delta t}$$

where $\delta h/\delta t$ is the change in bed elevation h over time t, ε accounts for sediment density and porosity, $\delta q_x/\delta x$ is the variation in sediment mass flux in the cross-shore direction, $\delta q_y/\delta y$ is the variation in sediment mass flux in the longshore direction, and $\delta V/\delta t$ represents local sediment volume V gains and losses, including by activities such as beach nourishment and sand mining. The sediment continuity equation is a powerful means by which morphological change can be determined for different parts of a morphodynamic system. Computer models used to predict coastal morphological change invariably include the sediment continuity equation.

The relationship between processes and morphology is driven by sediment transport. As sediment erosion, transport and deposition continues the conditions encountered by the hydrodynamic processes are progressively modified. For example, sand is transported to a beach by waves under calm weather conditions, resulting in beach accretion. As the beach builds up, its seaward slope progressively steepens, which affects the wave breaking processes and sediment transport. At some point during beach steepening, the hydrodynamic conditions may be sufficiently altered to stop further onshore sediment transport. The feedback between morphology and processes can be either negative or positive and is fundamental to controlling coastal morphodynamic behaviour. This relationship also means that cause and effect of morphodynamic change is not always readily apparent. This makes it difficult to predict coastal development, especially over long time scales.

1.4 Morphodynamic behaviour of coastal systems

The fundamental properties of all systems, including feedback and relaxation time, can be demonstrated in the morphodynamic behaviour of coastal systems. For a more detailed discussion of coastal systems the reader is referred to Cowell and Thom (1994).

An example of negative feedback leading to morphological change is shown in Figure 1.11. Here, the refracted waves approach the initially-straight coastline at an angle, and therefore drive longshore sediment transport (Figure 1.11a). This leads to accretion at both ends of the embayment and erosion in the centre. This change in beach planform entails negative feedback, because the wave angle and longshore transport decreases over time, and as wave approach angle and coastline shape reach a state of equilibrium with each other (Figure 1.11b). Under these conditions, longshore currents and sediment transport become insignificant. Development of equilibrium conditions along coasts have also been identified for the shoreface profile, tidal inlets and tidal basins.

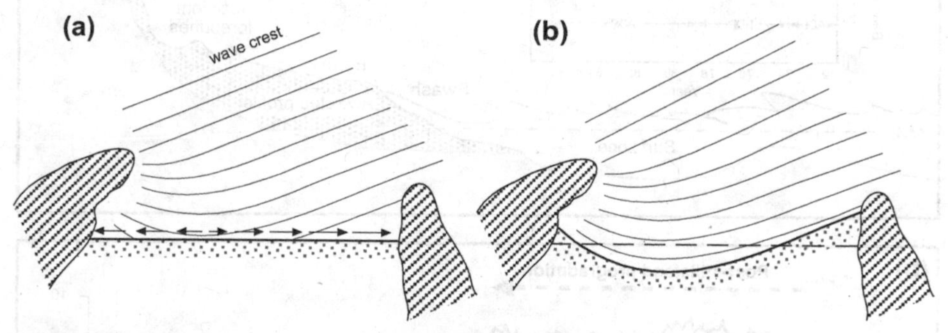

Figure 1.11 (a) Disequilibrium and (b) equilibrium beach planforms. The arrows in (a) represent the direction of longshore sediment transport.

An example of positive feedback leading to morphological change is the infilling of deep estuaries by marine sediments due to asymmetry in tidal flow. In a deep estuary, flood currents are stronger than ebb currents, resulting in a net influx of sediment and infilling of the estuary. As infilling takes place, the tidal asymmetry decreases as friction and shoaling effects are enhanced by reduced water depth, which in turn increases the rate of infilling. Eventually, tidal flats develop, which reduce estuary size and asymmetry of the flood tide so that the estuarine morphology approaches steady state as sediment imports and exports equilibrate.

1.4.1 Relaxation time

The time required for a system to reach equilibrium following disturbance is referred to as its **relaxation time** and is a measure of 'inertia' within the system. The bigger and more complex the system, the longer the relaxation time. In coastal environments, relaxation time depends on:

- **Energy level** – The higher the energy level, the larger the sediment transport rates and the shorter the relaxation time (i.e. the system responds quickly). Therefore, coastal morphological change takes place most dramatically under higher energy (e.g. large storms) than lower energy conditions (post-storm recovery).
- **Sediment mobility** – Relaxation time increases with decreasing sediment mobility. Rocky cliffs and shore platforms have a large resistance to change

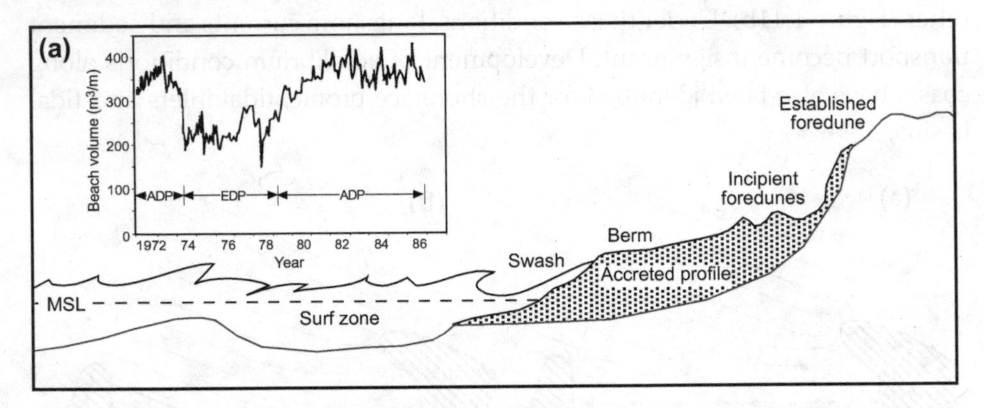

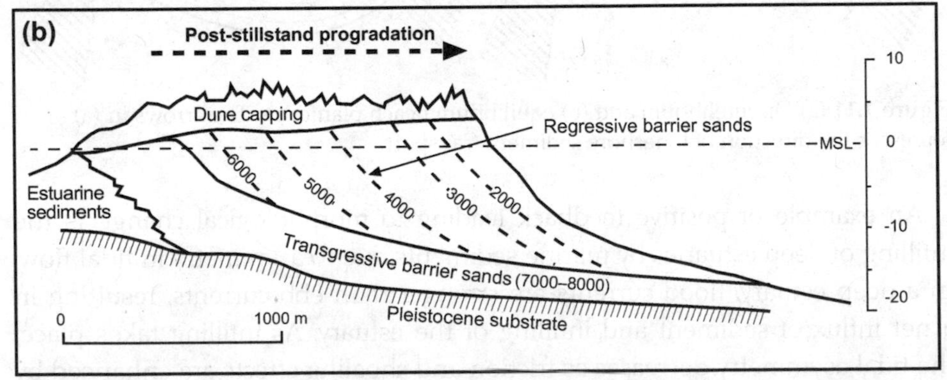

Figure 1.12 Examples from southeast Australia illustrating the presence of relaxation times. (a) Decadal scale relaxation time is apparent in the time series of beach volume change on Moruya beach, New South Wales, Australia, which are the accretion-dominated period (ADP) and erosion-dominated period (EDP). (From Thom and Hall, 1991.) (Copyright © 1991 John Wiley & Sons, reproduced with permission.) (b) Millennium time scale relaxtion time is shown in the chronostratigraphy of a prograded barrier system. The isochrons (dashed lines) are in years BP. (From Cowell and Thom, 1994.) (Copyright © 1994 Cambridge University Press, reproduced with permission.)

and so these landforms have the longest relaxation times. Sand is more easily moved than gravel and so sandy coastal landforms have shorter relaxation times than gravely ones.

- **Spatial scale** – Relaxation time strongly depends on the volume of sediment being moved, and the spatial scale of a landform is related to the temporal scale of its forcing (see Figure 1.3). Therefore, large coastal landforms such as barriers have longer relaxation times than small landforms such as beach cusps.

The relaxation time of coastal systems varies widely, but is difficult to calculate because it is unlikely that a steady-state equilibrium is ever reached, particularly for large coastal landforms. Ongoing changes in environmental conditions (Figure 1.8) continually perturb the system and interrupt its relaxation.

An example of relaxation over decadal time scales is shown in Figure 1.12a. Moruya beach in southeast Australia suffered major erosion during storms in 1974 and 1975, causing a dramatic decrease in subaerial beach volume from 400 to 200 m^3 per metre beach width (Thom and Hall, 1991). The recovery of the beach to pre-storm conditions took about six years, after which the beach remained relatively stable. Figure 1.12b shows the relaxation time associated with coastal barriers adjusting to relatively stable sea-level conditions at approximately 6000 years before present (BP) (Cowell and Thom, 1994). Following this period, the coastline kept prograding despite the fact that sea level was relatively stable. Figure 1.12 shows that the relaxation time increases with the size of the morphological feature. For Moruya beach, the relaxation time was about six years, while the relaxation time for the coastal barrier was more than 1000 years.

1.4.2 Feedback and self-organisation

Self-organisation is an emergent property of open systems in which feedback leads to internal changes in system dynamics that maintain the system in equilibrium. As a result, self-organisation evolves over time, often in response to cyclic changes in forcing. In coastal systems, the result is an orderly (i.e. 'organised') arrangement of sediments, sedimentary facies and landforms such as coastal dunes and offshore bars.

An example of self-organisation on the event time scale is the formation of **beach cusps**. These are rhythmic shoreline features formed by wave swash action and have steep-gradient, seaward-pointing horns and gentle-gradient, seaward-facing embayments (Figure 1.13). The self-organisation theory of beach cusp formation considers beach cusps to be the result of feedback between morphology and swash flow (Coco and Murray, 2007). In this theory, small topographic depressions on the beachface are amplified by attracting and accelerating water flow, thereby promoting erosion (positive feedback). At the same time, small positive relief features are enhanced by repelling and decelerating water flow, promoting accretion (negative feedback). The combination of positive and negative feedback inhibits erosion and accretion on well-developed cusps and maintains overall equilibrium. The important feature is that the morphological regularity arises from the internal dynamics of the system (Box 1.2). With respect to beach cusps, the dimension of the rhythmic morphology is scaled by the horizontal swash excursion: the more extensive the wave action on the beach, the larger the cusps.

Figure 1.13 Gravel beach cusps (spacing 8 m) in Alum Bay, Isle of Wight, southern England. (From Kuenen, 1948.) (Copyright © 1948 The University of Chicago Press, reproduced with permission.)

Box 1.2 Self-organisation

When clastic sediments (mud, sand and gravel) are subjected to hydrodynamic processes, they tend to become organised over time into distinctive morphological or sedimentary patterns. This order arises from within the system itself, hence the term 'self-organisation' is employed. Self-organising behaviour is a property of many different physical environments (Turcotte, 2007), which suggests that external forcing alone cannot account for self-organising behaviour, and that internal system dynamics are also important.

Computer modelling was used by Coco *et al.* (2000) to investigate the formation and maintenance of beach cusps. In this model, an initial planar beach with small random morphological perturbations is subjected to swash action. Some of these perturbations become enhanced (small holes become big holes; small bumps become big bumps), while others are suppressed. Feedbacks between morphology and swash hydrodynamics eventually result in formation of beach cusps (Figure 1.14). The size and spacing of the cusps is not hard-wired into the model, but emerges as the

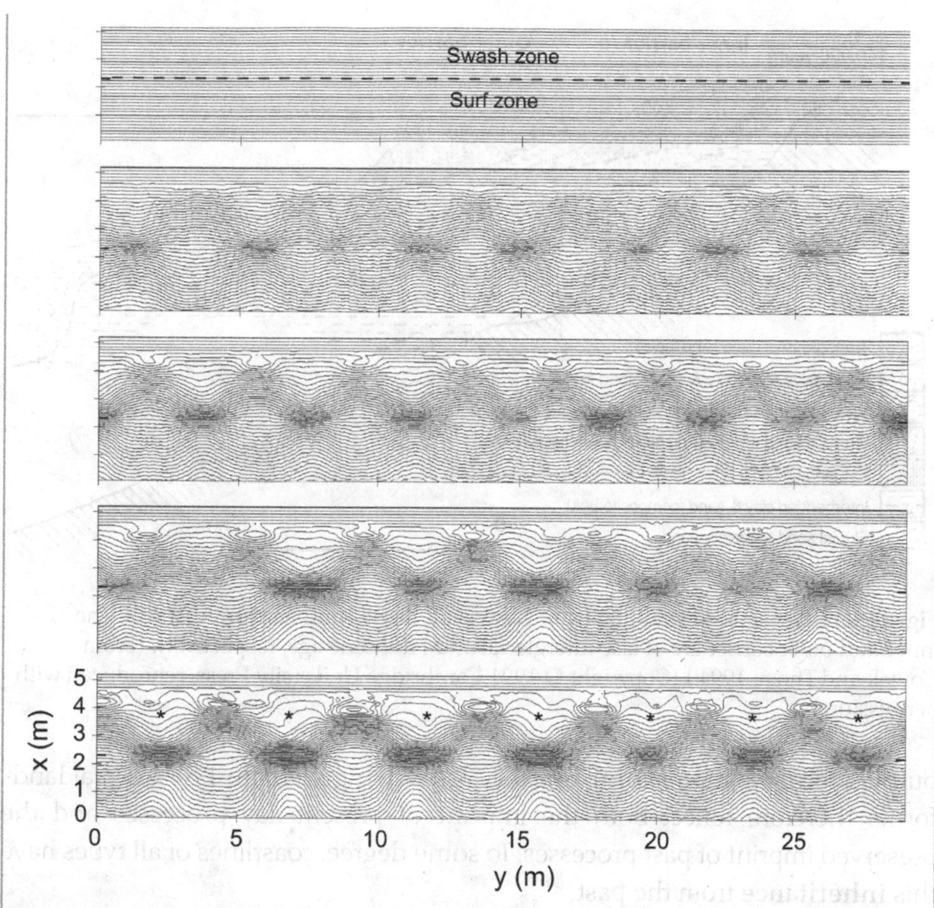

Figure 1.14 Numerical simulation of beach cusp formation. The panels show contour plots of the morphology at the start of the simulation and after 50, 200, 400 and 800 simulated wave cycles. The swash excursion is 1.8 m and the resulting cusp spacing is *c.* 4 m. The small bold ticks in the panels indicate MSL and the asterisks in the bottom panel indicate the location of cusp horns. (Modified from Coco *et al.*, 2000.)

model run progresses. The computer-generated cusps are therefore self-organising features, and it is likely that natural cusps are too (Coco and Murray, 2007). Self-organisation is increasingly used to explain geomorphic systems, and has been applied to other coastal features including gravel barriers, estuaries, coastal dunes and nearshore bars.

1.5 Coastal systems and long-term change

Concepts of relaxation time and self-organisation in coastal systems are contingent on the system maintaining equilibrium over some spatial and temporal scales. In reality, however, coastal evolution is cumulative, because morphological

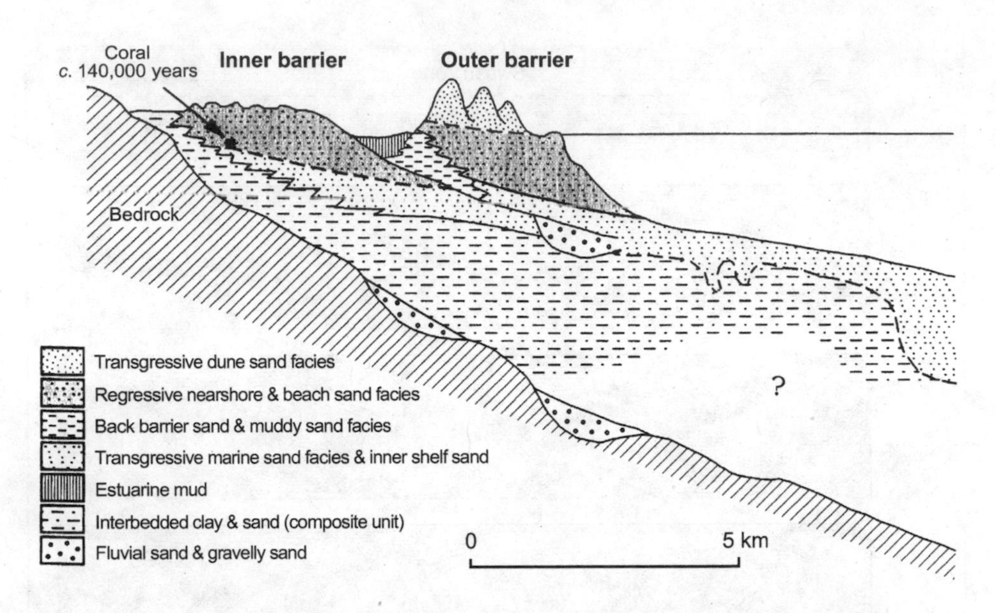

Figure 1.15 Generalised stratigraphy of Newcastle Bight, southeastern Australia. The inner and outer barrier are of last-interglacial and Holocene age, respectively. (From Cowell and Thom, 1994.) (Copyright © 1994 Cambridge University Press, reproduced with permission.)

outputs form the inputs for the next cycle of change (Figure 1.8). Coastal landforms therefore reflect both the imprint of present-day processes and the preserved imprint of past processes. To some degree, coastlines of all types have this **inheritance** from the past.

An example of inheritance from the southeast Australian coast is shown in Figure 1.15. Here, sea-level transgressions took place during the last interglacial (*c.* 122,000–128,000 years BP) and the Holocene. Both episodes caused deposition of transgressive (rising sea level) and regressive (coastal progradation) stratigraphic sequences. The stratigraphy shows that the last-interglacial barrier (the inner barrier) developed on a substrate that was different from that on which the Holocene barrier (outer barrier) was deposited. Cowell and Thom (1994) argue that the Holocene barrier developed under more energetic wave conditions than the older, inner barrier due to a steeper substrate and a more exposed setting, despite the fact that offshore wave conditions may have been similar. In this example, pre-existing conditions of sediment supply and sea-level tendency strongly controlled the response of the coastal system to forcing. This example also highlights the concept of **equifinality**, whereby the two barriers are morphologically similar but were formed at different times and under different conditions.

Coastal sedimentary and morphological systems are an amalgam of elements of different ages and on different scales. As such, coastal landscapes can be described as a **palimpsest**, whereby sediments or landforms from one time period are variously preserved, overprinted or destroyed by later conditions or processes. The net result is a coastal system that has a multi-layered and partial history of environmental forcing and coastal response. The rest of this book explores the basis of these forcing factors and describes their impact on the processes and geomorphology of real coasts.

SUMMARY

- The spatial boundaries of the coastal zone correspond to the limits to which coastal processes have extended during the Quaternary period and include the coastal plain, the shoreface and continental shelf.
- The coastal morphodynamic system consists of processes, sediment transport, morphology and stratigraphy. An essential ingredient to coastal morphodynamics is the feedback between morphology and processes.
- The major environmental conditions affecting coasts are geology, sediment supply/characteristics and external energy sources (wind, waves and tides). Human activity is an emerging forcing factor on most coasts worldwide, and is perhaps the dominant factor on many.
- There is a close coupling between temporal and spatial scales in coastal morphodynamics. Therefore, the modification of large-scale features occurs over long time periods, whereas small-scale features change over short time periods.
- Coastal morphodynamic systems possess a number of fundamental characteristics that make prediction of coastal evolution difficult. These include negative and positive feedback, equilibrium, self-organisation, relaxation time and inheritance.

Reflective questions

These questions are designed to test your comprehension of material covered in this chapter. Answers to these questions can be found on this book's website.

1a. Consider the reasons how and why our definition of the coastal zone and its component parts (Figure 1.1) may change over time. What implications does this have for coastal management?

1b. Why is coastal classification useful?

1c. Figure 1.8 shows the major components of a coastal system. Using this 'box and arrow' method draw a simple diagram showing the major processes contributing to the morphodynamics of a sandy beach.

1d. Describe examples of positive feedback and negative feedback in a coastal system.

Further reading

Cowell, P.J. and **Thom**, B.G., 1994. 'Morphodynamics of coastal evolution'. In: R.W.G. Carter and C.D. Woodroffe (eds) *Coastal Evolution*. Cambridge University Press, Cambridge, 33–86. (An excellent treatment of the principles of coastal morphodynamics.)

Davies, J.L., 1980. *Geographical Variation in Coastal Development* (2nd edn). Longman, New York. (An old text but still provides a good overview of the global distribution of coastal environments and environmental conditions.)

Finkl, C.W., 2004. 'Coastal classification: systematic approaches to consider in the development of a comprehensive scheme'. *Journal of Coastal Research*, 20, 166–213. (This is a detailed examination of different coastal classification schemes, and illustrates the wide range of factors that could be considered in these schemes.)

Wright, L.D. and **Thom**, B.G., 1977. 'Coastal depositional landforms: A morphodynamic approach.' *Progress in Physical Geography*, 1, 412–459. (A benchmark paper outlining the morphodynamic approach to coastal landforms.)

CHAPTER 2

SEA LEVEL

AIMS

This chapter outlines the major controls on and patterns of sea-level change during the Pleistocene and Holocene. The chapter discusses future sea-level rise due to global warming, which is of critical importance in considering how today's coastlines are going to adapt.

2.1 Introduction

The position of the coastline is located at the intersection of the sea's edge with the land. If sea level rises (falls), the coastline is shifted landwards (seawards). This refers to a situation whereby only the sea's height changes. In reality, however, the height of the land also changes through erosion, deposition and tectonics. The net result of variations in sea and land height leads to transgressive and regressive conditions at the coast. Net rising sea level leads to **transgressive conditions** and results in drowning and/or onshore migration of coastal landforms. Net falling sea level leads to **regressive conditions** and results in coastal emergence and/or progradation. Locally and over short time scales (seconds–months), sea level varies with waves, tides, changes in atmospheric pressure and winds. However, when averaged over large spatial and temporal scales, a stable value can be obtained, which is referred to as **mean sea level** (MSL). The elevation of MSL varies considerably from place to place (see Section 2.3).

We can distinguish between two types of sea-level change:

- **Relative sea-level change** can be brought about by a change in the sea level and/or a change in the land level, and operates on a regional/local level.
- **Eustatic sea-level change** (sometimes called absolute sea-level change) refers to a global change in sea level unrelated to local/regional effects and is most commonly caused by changes in ocean water volume and temperature, related in turn to the vigour of the hydrological cycle and climate.

When MSL is measured at a single location over an extended period, only a relative sea-level curve can be derived. Eustatic changes in sea level can only

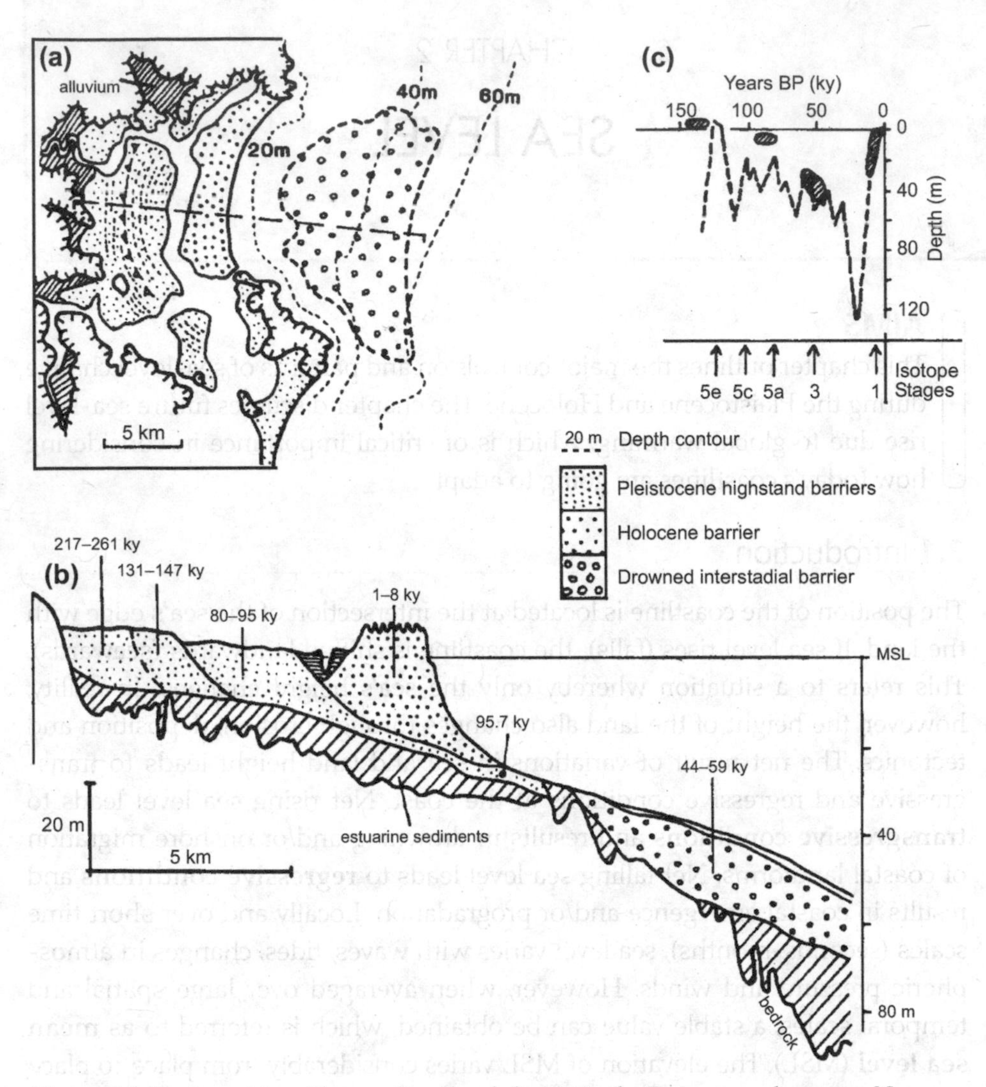

Figure 2.1 Stratigraphic evidence of sea-level change in the Tuncurry embayment, New South Wales, Australia. (a) Distribution of barrier systems onshore and on the inner shelf. (b) Shore-normal cross-section (dashed line in (a)) through the barriers showing their dimensions and ages ('ky' refers to thousand years before present). (c) Group ages of the various barrier systems (shaded) and the estimated positions of sea level when they formed. The numbers indicate the Marine Isotope Stages and the sea-level curve (dashed line) is that of Chappell and Shackleton (1986). (From Roy *et al.*, 1994.) (Copyright © 1994 Cambridge University Press, reproduced with permission.)

be determined from the relative sea-level curve if the vertical movement of the land level is known. If there is a constant relative sea level, this does not imply that both the land level and eustatic sea level are constant – it is more likely that the change in eustatic sea level is balanced by the movement of the land plus other site-specific factors such as sediment compaction and subsidence.

Evidence of past MSL variations, termed **sea-level indicators**, can be found along most coastlines worldwide. Indicators of sea levels higher than present (**sea-level highstands**) are easier to find than those for lower sea levels (**sea-level lowstands**), because evidence of lowstands are generally submerged and may be eroded away or covered by sediment during subsequent transgressions. The most accurate sea-level indicators are marine organisms such as barnacles and corals that live at known tidal levels. Information about sea-level change can be obtained by comparing the elevation of present-day tidal bands (with living organisms) with those of equivalent fossil organisms which now stand at a different level. When preserved within dated sediments, these organisms can help establish a **sea-level curve**, which is a graph tracking the position of MSL over time.

Former sea levels in sedimentary coastal environments are recorded in the coastal geomorphology and stratigraphy. For example, Figure 2.1 shows coastal barriers of different ages. The present shoreline is formed by a relatively young Holocene barrier. Landward of this is an older barrier complex composed of three units that get older in a landward direction. These four periods of barrier formation all took place when sea level was close to present. Evidence of a fifth barrier system can be found at 40 m below present MSL on the inner shelf. By contrast, rocky coasts generally lack depositional features, and indicators of sea-level change often consist of eroded shore platforms, cliffs, notches and benches.

2.2 Causes of sea-level change

Many factors contribute to sea-level change over time and space. On the basis of their spatial extent, we can distinguish between global, regional and local causes of sea-level change.

2.2.1 Global causes: Changes in ocean water volume and thermal expansion

Eustatic sea-level change is caused by a change in the quantity of water in ocean basins. An increase (decrease) in water volume results in sea-level rise (fall). Total global water volume can be considered constant and is expressed by the **global water balance**

$$K = A + O + L + R + S + B + M + U + I \qquad \textbf{(2.1)}$$

where K is total water volume (a constant) and the other variables are indicated in Table 2.1 and expressed as an equivalent water depth. It is apparent that the growth and melting of continental ice sheets during the Pleistocene is of paramount importance for long-term eustatic sea-level changes. During cold glacial periods, sea water is progressively lost from the oceans as it accumulates on the continents as snow and ice. During warm interglacial periods, ice melts and the water is returned to the oceans, resulting in sea-level rise. The effect of ice sheets on sea level is called **glacio-eustasy**. Floating sea ice has no effect on eustatic sea level since the weight of the ice is already supported by the water.

Table 2.1 Estimates of storage volumes and equivalent water depth for components of the world water balance. Note the large uncertainty regarding the volume of water stored as groundwater. (After Pirazzoli, 1996.)

Parameter	Present volume (km^3)	Equivalent water depth
A Atmospheric water	13,000	36 mm
O Ocean and seas	1,370,000,000	3.8 km
L Lakes and reservoirs	125,000	35 cm
R Rivers and channels	1700	5 mm
S Swamps	3600	10 mm
B Biological water	700	2 mm
M Moisture in soils and the unsaturated zone	65,000	18 cm
U Groundwater	4,000,000 to 60,000,000	11 to 166 m
I Frozen water	23,290,000	64.5 m
Antarctica	20,440,000	56.6 m
Greenland	2,635,000	7.3 m
Others	215,000	0.6 m

Even if the quantity of sea water remains constant, sea level may change due to variations in sea water temperature. Sea water density increases with decreasing water temperature. If a volume of water is heated (cooled), it will occupy a larger (smaller) volume, so a decrease (increase) in sea water temperature causes a fall (rise) in sea level. This process is known as **steric** or **thermal expansion** and is a significant contributor to present sea-level rise. For example, an increase of 1 °C over a water depth of 4000 m (approximately the average depth of the Pacific Ocean) produces a rise in sea level of 0.6 m.

Eustatically-high sea levels are characteristic of interglacials when continental ice volume is low and oceans are warm. Beaches, shore platforms and coral reefs formed during previous interglacials are commonly found near present sea level. To calculate the amount of eustatic sea-level change from such features one needs to know the rate of land uplift and the date at which the features formed. Eustatic sea level is calculated as

$$E = P - U \qquad (2.2)$$

where E is the amount of vertical shoreline displacement caused by eustatic change, P is the present elevation of the feature above MSL, and U the amount of tectonic uplift that has occurred. For example, if a shore platform is found at 120 m above MSL, the platform is 100,000 years and uplift rate is 1 mm yr^{-1}, then the platform has been uplifted by 100 m and the eustatic change in sea level is 20 m. In other words, when the shore platform developed, the eustatic sea level was 20 m higher than present.

2.2.2 Regional causes: Isostatic changes

The Earth's crust floats on a denser underlying layer (asthenosphere). This two-layer system is in **isostatic balance** when the total weight of the crust is exactly balanced by its buoyancy. Addition of a load to the crust (e.g. water, ice or sediments) will upset the isostatic balance. To compensate for the increased crustal weight, some of the asthenospheric material will flow away and the land level will eventually fall.

Glacio-isostasy refers to isostatic adjustments of the Earth's crust due to the loading and unloading of ice sheets as they grow and decay respectively. The weight of an ice sheet causes depression of the underlying land surface (Box 2.1). At the margins of the depressed land surface the land rises slightly upward due to the flow of asthenospheric material towards this region and away from the centre of depression. This leads to a raised 'rim' around ice sheet margins, known as a **forebulge**. When the ice melts, the land surface will eventually revert back to its former position, thus the area formerly covered by the ice will rise up and the forebulge will sink down. This process is known as **isostatic readjustment**. Shorelines located in areas formerly covered by ice sheets will therefore experience a postglacial fall in relative sea level, whereas shorelines in the forebulge areas will undergo relative sea-level rise. Patterns of sea-level change in glaciated areas are therefore extremely complex and variable.

Box 2.1 Calculating postglacial rebound

If we ignore the response times of isostatic adjustment, it is relatively easy to determine the amount of postglacial rebound that can be expected following ice unloading. Figure 2.2 shows the depression of the Earth's crust by an ice sheet (Allen, 1997). The **isostatic balance** above the depth of compensation can be performed for a column far from the ice sheet (far-field) and a column through the centre of the ice sheet (near-field), and is

$$\rho_i h_i + \rho_c h_c = \rho_c h_c + \rho_m w_0$$

where ρ_i, ρ_c and ρ_m are the densities of ice, crust and mantle, respectively, and h_i, h_c and w_0 are the thickness of the ice sheet, crust and the maximum deflection of the land surface under the ice sheet, respectively. If the thickness of the crust is considered constant, the isostatic balance can be simplified to

$$\rho_i h_i = \rho_m w_0$$

Assuming an ice density of 800 kg m^{-3} and a mantle density of 3300 kg m^{-3} we find

$$w_0 = \frac{\rho_i}{\rho_m} h_i = 0.24 h_i$$

Thus, depression of the Earth's crust is about a quarter of the ice thickness. The Laurentide ice sheet over the Hudson Bay area of Canada had an average thickness of 5 km during the height of the glaciation, implying that the amount of depression, and consequent rebound during deglaciation, may have exceeded 1 km. This has led to the formation of concentric staircases of shorelines around the Bay as the land has risen up. Uplift of the region today is still taking place at up to 10 mm yr^{-1} (Sella *et al.*, 2007).

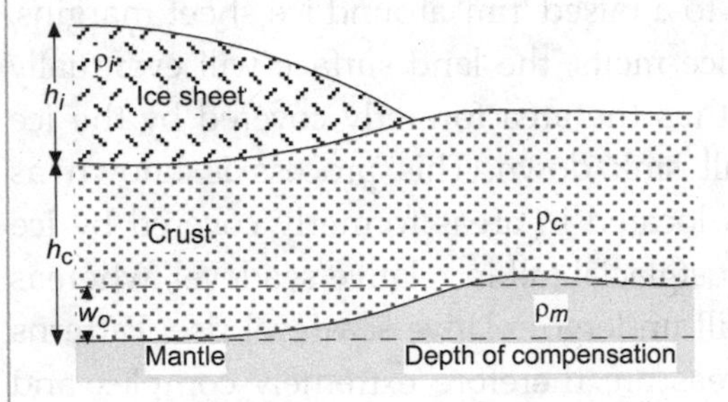

Figure 2.2 Schematic diagram showing isostatic balance. (From Allen, 1997.) (Copyright © 1977 Blackwell Publishers, reproduced with permission.)

There are considerable time lags involved with glacio-isostatic movements due to the rigidity of the Earth's crust. These time lags are so long that, even today, the Earth's crust is still isostatically adjusting to ice sheet melt that took place more than 10,000 years ago. For example, **postglacial uplift** in the Baltic Sea region due to isostatic rebound has amounted to 300 m over the past 10,000 years, and the region is still being uplifted at up to 10 mm yr^{-1} (Figure 2.3). In Britain, too, isostatic adjustments are also still ongoing. Scotland and northern England, which were covered by ice during the last glacial period, are being uplifted at average rates of up to 2 mm yr^{-1}, while southern England, which is located in the forebulge area, is subsiding by up to 2 mm yr^{-1} (Case Study 2.1).

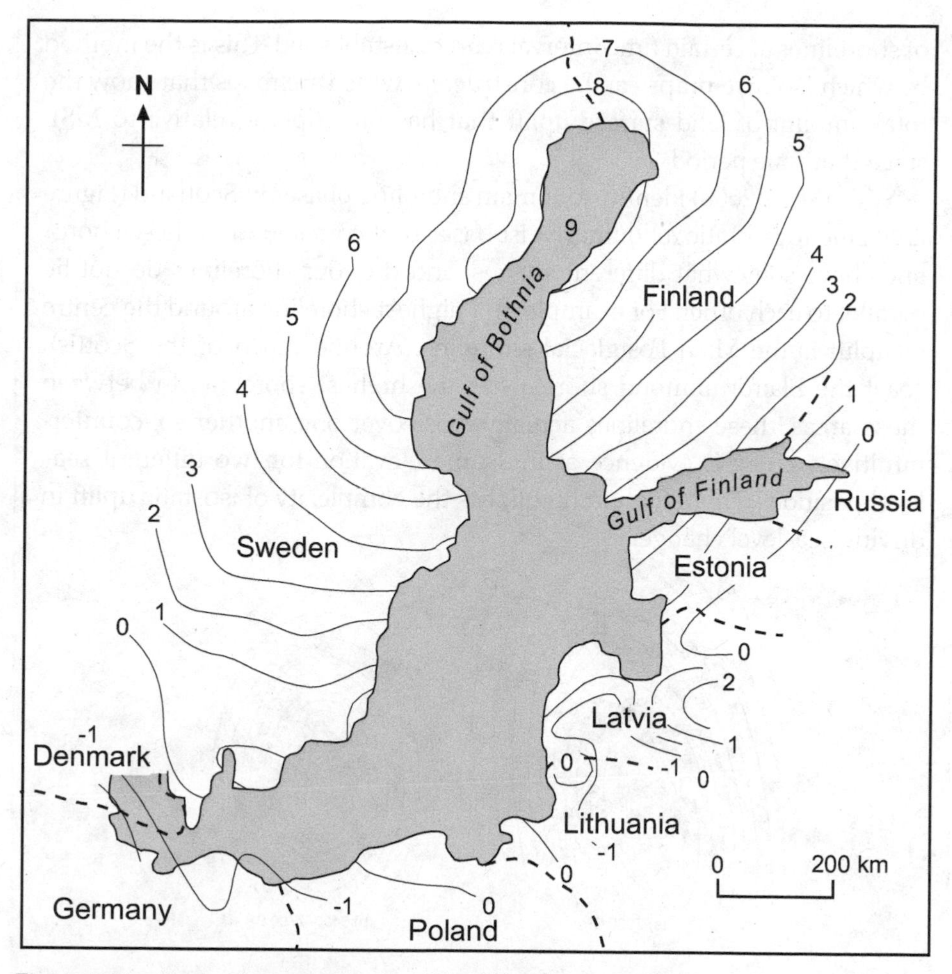

Figure 2.3 Isobases of present-day rates of uplift of land surfaces (mm yr^{-1}) in Sweden and Finland due to postglacial rebound. (Modified from Eronen, 1983.)

Case Study 2.1 Isostatic uplift and sea-level change in Scotland

Scotland was overlain by the thickest extent of ice during the last glaciation in the British Isles and therefore experienced the greatest amount of isostatic depression and subsequent uplift. The area of greatest ice thickness in west-central Scotland has been rising more quickly than peripheral areas where the ice was thinner and the loading not so great. During uplift, the position of the coastline (and therefore the position of MSL) has varied substantially. Evidence for previous MSL position comes from the presence of raised shore platforms and estuarine sediments preserved within bedrock valleys. Where these landforms can be dated, the position of shorelines at certain time intervals can be established. This is the method by which **isobase maps** can be constructed, which are maps that show the total amount of land-surface uplift that has taken place relative to MSL since that time period.

Smith *et al.* (2006) identify four main shoreline phases in Scotland (Figure 2.4). Due to isostatic rebound the isobases that correspond to these shorelines have somewhat different shapes, and the four shorelines do not lie parallel to each other. For example, the highest shoreline around the centre of uplift is the Main Postglacial shoreline. Around much of the Scottish coast the Blairdrummond shoreline is the highest shoreline. In between these areas, these shorelines actually cross over one another so, counter-intuitively, there is evidence at the same elevation for two different sea-level positions. This situation highlights the complexity of isostatic uplift in driving sea-level change.

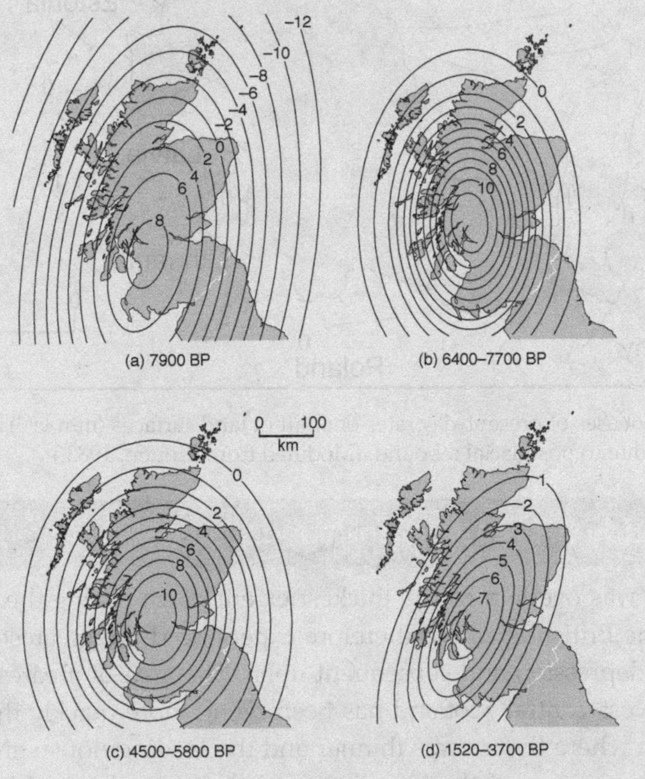

Figure 2.4 Map of Scotland showing isobases (in metres relative to present-day MSL) that correspond to four distinctive shoreline positions, and their likely timings (years BP). (a) The Holocene Storegga tsunami shoreline. (b) The Main Postglacial shoreline. (c) The Blairdrummond shoreline. (d) The Wigtown shoreline. (Modified from Smith *et al.*, 2006.)

During ice sheet decay, meltwater escapes to the ocean, causing sea level to rise, which in turn exerts a load on the flat continental shelves. This results in subsidence of the sea floor due to **hydro-isostasy**. Along coasts with wide continental shelves, the water load across the shelf is not uniform because the deeper part of the shelf near the shelf edge experiences a higher pressure than the shallow part near the coast. As a result, the subsidence of the deeper part of the shelf results in an uplifting of the shallower part and coastal fringe, hence a fall in relative sea level. Most Southern Hemisphere coastlines, where the effects of glacio-isostasy are insignificant or non-existent, have experienced such a fall in relative sea level during the Holocene.

The geophysical effects of glacio- and hydro-isostasy have been investigated using numerical models, and based on these effects Lambeck (1993) distinguishes four zones:

- **Near-field sites** are regions within the limits of former ice sheets.
- **Ice-margin sites** are regions near former ice margins.
- **Intermediate-field sites** are regions just outside former ice margins.
- **Far-field sites** are regions well away from the influence of former ice sheets.

An outcome of global isostatic models is that vertical land movements caused by deglaciation extend more or less all around the globe. This is confirmed by satellites such as GRACE (Gravity Recovery and Climate Experiment) which is able to detect very small-scale crustal responses to glacier unloading (e.g. Argus and Peltier, 2010). Based on these data, computer models can be used to distinguish the relative contributions of glacio-isostasy *versus* thermal expansion to sea-level change.

2.2.3 Local causes: Tectonics and subsidence

Locally, **tectonic activity** can result in changes in land level. Trends of vertical displacement caused by tectonics often appear to be continuous and gradual over the long term, but in detail usually take place in short jumps associated with earthquakes. For example, a large earthquake in Prince William Sound, Alaska, in March 1964 resulted in 250,000 km^2 of the coastal zone being uplifted by up to 11 m. Another local cause of sea-level change is sediment compaction which results in land **subsidence**, especially in deltas and wetlands where sediments can contain up to 50 per cent water. In peat layers, **compaction** may reduce water content to as little as 10 per cent, resulting in a fall in land level. Associated with delta environments is the process of **sediment-isostasy**, whereby the Earth's crust is depressed due to the weight of the sediments on top (similar to glacio-isostasy). The combination of compaction and sediment-isostasy makes deltas very prone to relative sea-level rise. The Mississippi delta, for example, has experienced a drop in land level over the last 10,000 years of *c.* 165 m, equating to a rate of relative sea-level rise of 16.5 mm yr^{-1} (Fairbridge, 1983). Similarly, land subsidence and relative sea-level rise during the Holocene for the Yangtze delta, China, is 1.6–4.4 mm yr^{-1} (Stanley and Chen, 1993). Land subsidence can also result from the extraction of groundwater, oil or gas (see Chapter 12). Land subsidence due to water extraction is 4.6 m in Tokyo and 2.7 m in the Po delta, northeast Italy (Pirazzoli, 1996).

2.3 Pleistocene climate and sea levels

The **Pleistocene**, which started at 2.6 million years BP, comprised at least 17 alternating cold **glacials** and warm **interglacials**. The glacials lasted around 100,000 years and the interglacials around 10,000 years. The major driver of the cyclic pattern of glacials and interglacials is small variations in the trajectory of Earth's orbit around the Sun, which causes small changes in the total amount of the Sun's radiative heat energy received by the Earth (Berger, 1992). There are three main orbital cycles that have different cycle lengths but which, in combination, have given rise to the glacials and interglacials of the Pleistocene. Briefly, these cycles are:

- **Eccentricity** – This refers to the elliptical trajectory of the Earth's orbit around the Sun. Temporal variations in eccentricity have a period of about 100,000 years.
- **Obliquity** – This refers to the inclination of Earth's axial tilt which accounts for the seasonal alternations of summer and winter. In the Northern Hemisphere, summer occurs when the axis of rotation is inclined towards the Sun, while winter occurs when the axis of rotation leans away from the Sun. At present the inclination is nearly 23.5°, but the angle varies between 22.1° and 24.5°. The periodicity of this variation is about 41,000 years.
- **Precession** – This refers to the orientation of Earth's axis of rotation relative to the plane of its orbit. This orientation changes with a periodicity of about 21,000 years, termed precession. At present, the Earth is closest to the Sun (at perihelion) in the Northern Hemisphere winter. However, due to the cycle of precession, perihelion will coincide with the Northern Hemisphere summer in about 11,000 years time.

For more information the interested reader is referred to Berger (1992) and Ruddiman (2006).

The last glacial period started about 70,000 years BP and finished around 11,500 years BP. During glacial periods ice sheets grew and sea level fell, while during the interglacials ice sheets melted and sea level rose. Interspersed within the glacials and interglacials were colder and warmer intervals of shorter duration (*c.* 1500 years), known as **stadials** and **interstadials**, respectively. These shorter warming/cooling episodes also caused eustatic sea-level fluctuations. Along coastlines that have undergone steady tectonic uplift over the Pleistocene, such as Barbados and New Guinea, evidence of interglacial sea levels can be observed in the form of raised shore platforms or raised coral reefs. More commonly, however, any traces of the lower sea levels have been obliterated by subsequent rises of sea level. This means that the most detailed sedimentary and morphological evidence only exists for the most recent glacial cycles.

2.3.1 Evidence from ice cores for Pleistocene climate change

The highest resolution evidence for Pleistocene climate change comes from **ice cores** recovered by drilling through the ice sheets of Greenland and Antarctica. Here, physical and chemical properties of annually-layered snow and ice can be used to reconstruct past climate, and can be compared with sea-level curves derived from deep-sea cores or with biological evidence on land. Key climatic data from ice cores include the ratios of hydrogen and oxygen isotopes in the ice, which provide an index of former temperatures, and the chemistry of air that is contained within sealed bubbles inside the ice, which can show the chemical make-up of past atmospheres, in particular the concentration of greenhouse gases such as carbon dioxide (CO_2). Important ice cores are from the Vostok and Dome C drilling sites in East Antarctica, and the GRIP

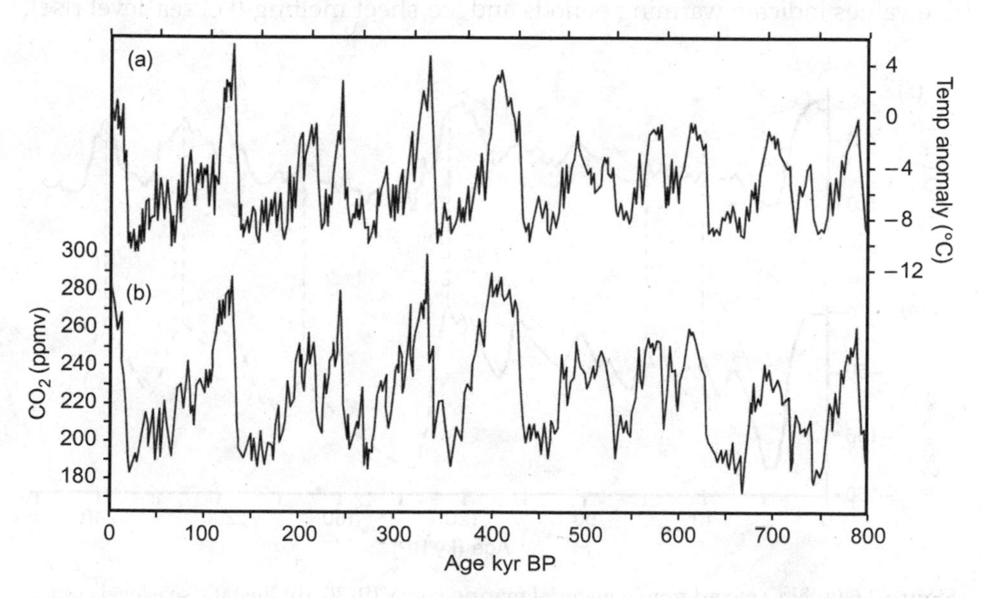

Figure 2.5 Variations in (a) temperature (b) CO_2 during the past 800,000 years from the EPICA ice core from Antarctica. Note the strong positive correlation between CO_2 concentration and temperature. (Modified from Lüthi *et al.*, 2008.)

(Greenland Ice-core Project), North GRIP and GISP2 (Greenland Ice-Sheet Project) sites in central Greenland.

The longest Pleistocene ice core record yet recovered comes from the EPICA core (European Project for Ice-Coring in Antarctica) at Dome C in Antarctica. The basal ice recovered at this site is 890,000 years old and shows the last eight glacial–interglacial cycles (EPICA, 2004). Each cycle has a similar succession of changes (Figure 2.5). The glacial periods show progressive but intermittent cooling followed by very rapid warming at the start of the interglacials. The amplitude of the temperature change is *c.* 12 °C and the coolest part of each glacial occurs just before the onset of the next interglacial. The temperature is strongly coupled to atmospheric CO_2 concentration. High temperatures during interglacials are associated with large CO_2 concentrations (270–280 parts per million, ppm) and low temperatures during the glacials with low CO_2 concentrations (180–200 ppm).

2.3.2 Evidence from deep-sea cores for Pleistocene sea-level change

Evidence for changes in sea water temperature and eustatic sea level during the Pleistocene can be derived from **deep-sea cores** by analysing the **oxygen isotope ratio** $\delta^{18}O$ (Box 2.2). This ratio can be used as a proxy for the volume of water stored in ice sheets and therefore global sea level. Figure 2.6a shows the $\delta^{18}O$ record for the last 250,000 years. More negative values of $\delta^{18}O$ represent cooling periods and ice sheet expansion (i.e. sea-level fall), whereas more positive values indicate warming periods and ice sheet melting (i.e. sea-level rise).

Figure 2.6 (a) $\delta^{18}O$ record from equatorial marine core V19-30. (b) Eustatic sea-level curve from the Huon Peninsula, New Guinea, derived from a series of raised coral reefs. The time-axis of the sea-level curve was recalculated using the $\delta^{18}O$ record. The last interglacial (5e) is marked. (Modified from Chappell and Shackleton, 1986.)

In order to distinguish between the different Pleistocene cooling/warming cycles they are numbered back in time from present, with odd numbers representing sea-level highstands and warm periods, and even numbers indicating sea-level lowstands and cold periods. This numbering system is referred to as **Marine Isotope Stages (MIS)**. The present Holocene interglacial is MIS 1 whereas the last interglacial of similar warmth (*c.* 122,000–128,000 years BP) was substage (e) of MIS 5, denoted 5e in Figure 2.6.

Box 2.2 Oxygen isotope ratios

Analysis of oxygen isotope ratios makes use of the relative proportion of the oxygen isotopes ^{18}O and ^{16}O in the skeletal material of calcareous organisms. The ratio $^{18}O/^{16}O$ in these micro-organisms is controlled by temperature-dependent fractionation of the isotopes from sea water into the skeleton of the micro-organism, and the isotopic composition of the sea water. The latter is dependent on the amount of water stored in ice sheets. The oxygen isotope ^{16}O is lighter than ^{18}O and evaporation preferentially removes ^{16}O. Growing ice sheets imply a net transfer of water from the ocean to the ice sheets, resulting in an enrichment of sea water with the isotopically-heavier isotope ^{18}O. Melting of the ice sheets on the other hand causes an enrichment of sea water with the isotopically-lighter isotope ^{16}O.

Oxygen-isotopic compositions are measured relative to a global standard known as the **Standard Mean Ocean Water** (SMOW). The oxygen isotopic composition of a sample can then be expressed as per mille differences relative to SMOW

$$\delta^{18}O = \left[\frac{(^{18}O/^{16}O)_{sample} - (^{18}O/^{16}O)_{SMOW}}{(^{18}O/^{16}O)_{SMOW}} \right] * 1000$$

Positive values of $\delta^{18}O$ indicate enrichment of the sample in the heavier isotope, whereas negative values indicate depletion. Changes in surface water temperature reflect to a large extent changes in air temperature. Therefore, $\delta^{18}O$ of planktonic (surface-living) micro-organisms such as foraminifera can serve as a proxy for surface water or air temperature. In the deep ocean, however, temperatures are likely to have fluctuated little during glacial–interglacial cycles, so $\delta^{18}O$ of deep-ocean (benthic) foraminifera is likely to represent changes in $\delta^{18}O$ of sea water due to changes in ice volume. An increase in $\delta^{18}O$ indicates an increase in global ice volume and hence a fall in sea level, whereas a decrease in $\delta^{18}O$ represents a decrease in global ice volume and hence a rise in sea level.

As a rule of thumb, a difference of 0.1 in $\delta^{18}O$ is equivalent to 10 m of sea-level change. From Figure 2.6a we can infer that sea level fluctuated during glacial–interglacial cycles over more than 100 m, which is borne out by computer models and field evidence. However, we know that deep-sea temperatures have changed as a result of changes in deep-ocean circulation, so it is more desirable to derive a palaeo sea-level curve directly from stratigraphic evidence, rather than the $\delta^{18}O$ record alone. Chappell and Shackleton (1986) determined a **eustatic sea-level curve** over the last 250,000 years from a series of raised coral terraces on the Huon Peninsula, New Guinea (Figure 2.6b). Their sea-level curve indicates that sea level during the last interglacial was very similar to present, but during glacial cooling, sea level fluctuated by at least 20 m. At the period of maximum ice sheet extent at 19,000 years BP, sea level was around 135 m lower than present (Yokoyama *et al.*, 2000). However, sea-level rise to the onset of the Holocene did not taken place at a constant rate but shows a step-like pattern with short episodes of very rapid rise separated by longer periods of much slower rise. For example, radiocarbon-dated coral reefs in the Caribbean were drowned very suddenly by periods of rapid sea-level rise at 14,200 years, 11,500 years and 7600 years BP when sea-level rose by over 45 mm yr^{-1} for several centuries and then slowed down dramatically (Blanchon and Shaw, 1995). This unsteady nature of sea-level rise has been linked to periods of drainage from large lakes in front of (and also underneath) the decaying Laurentide ice sheet in North America. A lake drainage event *c.* 8200 years BP has been linked to a jump in European sea level around the same time of 2.11±0.89 m (Hijma and Cohen, 2010). Such changes in sea level have the power to rapidly drown low-lying shelves and islands, change a coastline's palaeo-geography and significantly alter patterns of nearshore water and sediment circulation, including tide regime.

2.4 Holocene sea-level transgression

2.4.1 Eustatic sea level

The present **Holocene** interglacial started at 11,500 years BP. Holocene sea-level rise, known as the **Holocene transgression**, reflects the melting histories of the different ice sheets, and so the rate of sea-level rise has been very variable, slowing to the present-day. A typical global eustatic sea-level curve for the late Pleistocene–Holocene transgression is shown in Figure 2.7. Rapid sea-level rise of about 0.5 cm yr^{-1} took place at the end of the Pleistocene (18,000–11,500 years BP), speeding up to about 2 cm yr^{-1} at the start of the Holocene (11,500–7000 years BP), then slowing over the mid-Holocene (7000–5000 years BP). In some midlatitude and subtropical areas, sea level peaked in the mid-Holocene

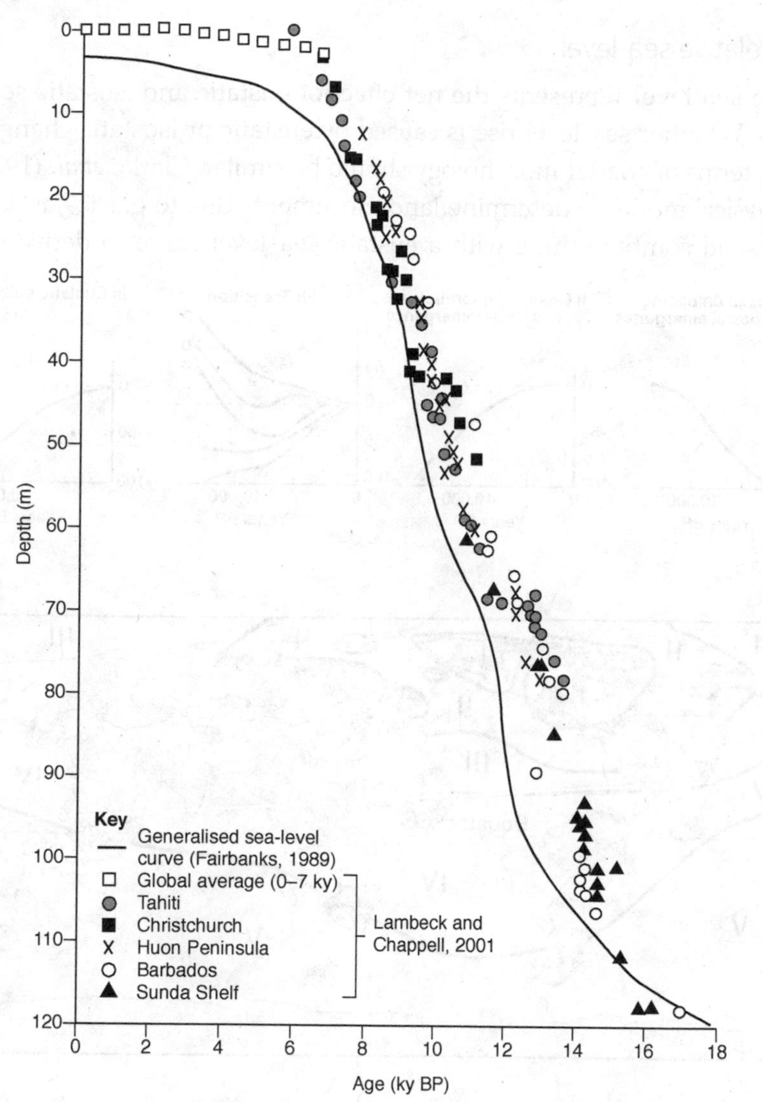

Figure 2.7 Late Pleistocene and Holocene sea-level curve according to Fairbanks (1989), with individual data points for different locations marked after Lambeck and Chappell (2001). Note the wide scatter of these data points with respect to Fairbanks' generalised curve.

at up to 1–2 m above present, corresponding to both warm temperatures and aridity in many areas at that time. The precise shape of the sea-level curve varies locally due to glacio-isostatic as well as eustatic factors (Gehrels, 2010), and as a result the definition of eustasy as a world-wide, simultaneous, uniform change in sea level has been questioned. For example, Mörner (1987) has argued that eustatic curves should only be defined for regions, not for the whole Earth. Milne and Mitrovica (2008) show, using geophysical models, that the concept of eustasy is only valid for tectonically-stable areas far away from former ice sheets.

2.4.2 Relative sea level

Relative sea level represents the net effect of eustatic and isostatic sea-level changes. Whether sea-level rise is caused by eustatic or isostatic changes, the effect in terms of coastal morphology should be similar. Clarke *et al.* (1978) use a geophysical model to determine land movements due to glacio- and hydro-isostasy, and combine these with a eustatic sea-level curve to derive relative

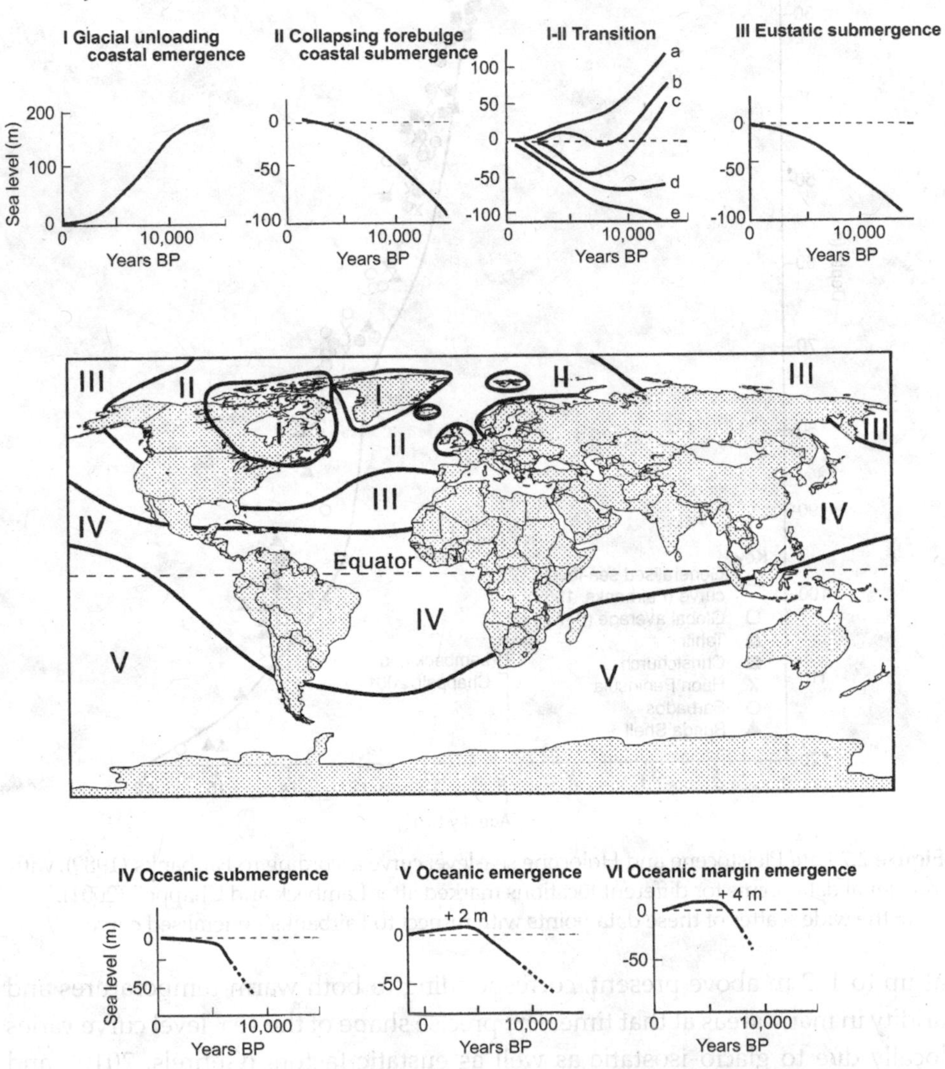

Figure 2.8 Global variation in relative sea-level rise derived using an isostatic model under the assumption that no eustatic change has occurred since 5000 years BP and that the Earth's mantle has a constant viscosity. Zone VI includes the landward margins of continental shelves in zones III, IV and V. Emerged beaches are predicted for Zones I, III, V and VI. (From Komar, 1998; modified from Clarke *et al.*, 1978.) (Copyright © 1978 Academic Press, reproduced with permission.)

sea-level changes during the Holocene (Figure 2.8). The results indicate different types of sea-level curves that are similar to the four different regions identified by Lambeck (1993):

- At near-field sites (Zone I in Figure 2.8) sea-level changes are dominated by ice-load effects, meaning that late-glacial and postglacial relative sea level is falling because of the rising land. This characterises many areas of Atlantic Canada and northern Europe (e.g. Figure 2.3).
- The intermediate-field sites (Zone II in Figure 2.8) correspond to the forebulge region around large Northern Hemisphere ice sheets. The forebulge both decreases in size and migrates back towards the ice sheet centre following ice retreat. In areas affected by a forebulge, relative sea level rises into the Holocene although at gradually decreasing rates. Forebulge areas have the most complex late Pleistocene sea-level records.
- At ice-margin sites (Transition Zone I–II in Figure 2.8) the relative sea-level curves vary greatly depending on location, and range from a progressive fall (curve I–IIa) to a progressive rise (curve I–IIe). This is typical of many sea-level curves around Britain and Ireland.
- In far-field sites (Zones III–VI in Figure 2.8) glacio-eustatic changes in sea level are considerably greater than glacio- and hydro-isostatic effects. This is the reason why eustatic sea-level curves use data from far-field locations like Barbados and New Guinea. Here, relative sea-level rise predominates during deglaciation, often followed by a slight relative sea-level fall of hydro-isostatic origin during the late Holocene (e.g., Figure 2.7).

Holocene sea-level rise has played a significant role in reworking sediment from the continental shelf and building barrier islands, beaches and sand dunes, and infilling estuaries. Many of the landforms found along glaciated coasts in Europe, New Zealand and North America first started to accumulate around the mid-Holocene when sea levels stabilised. Coincidently, this is also the period of human development when prehistoric peoples were occupying the coastal zone and exploiting its food and resources, and developing coastal and river-mouth settlements (see Chapter 12). This highlights the close relationship between Holocene sea-level rise, coastal change and human activity.

2.5 Present sea-level rise

The geomorphology and sedimentology of the present coastline are primarily the result of the Holocene sea-level history. Of most interest to coastal communities, however, are present and future patterns of sea-level change. Over the last 100 years a global network of tide gauges has developed, allowing historic sea-level changes to be plotted. At present, around 2000 tide gauges are

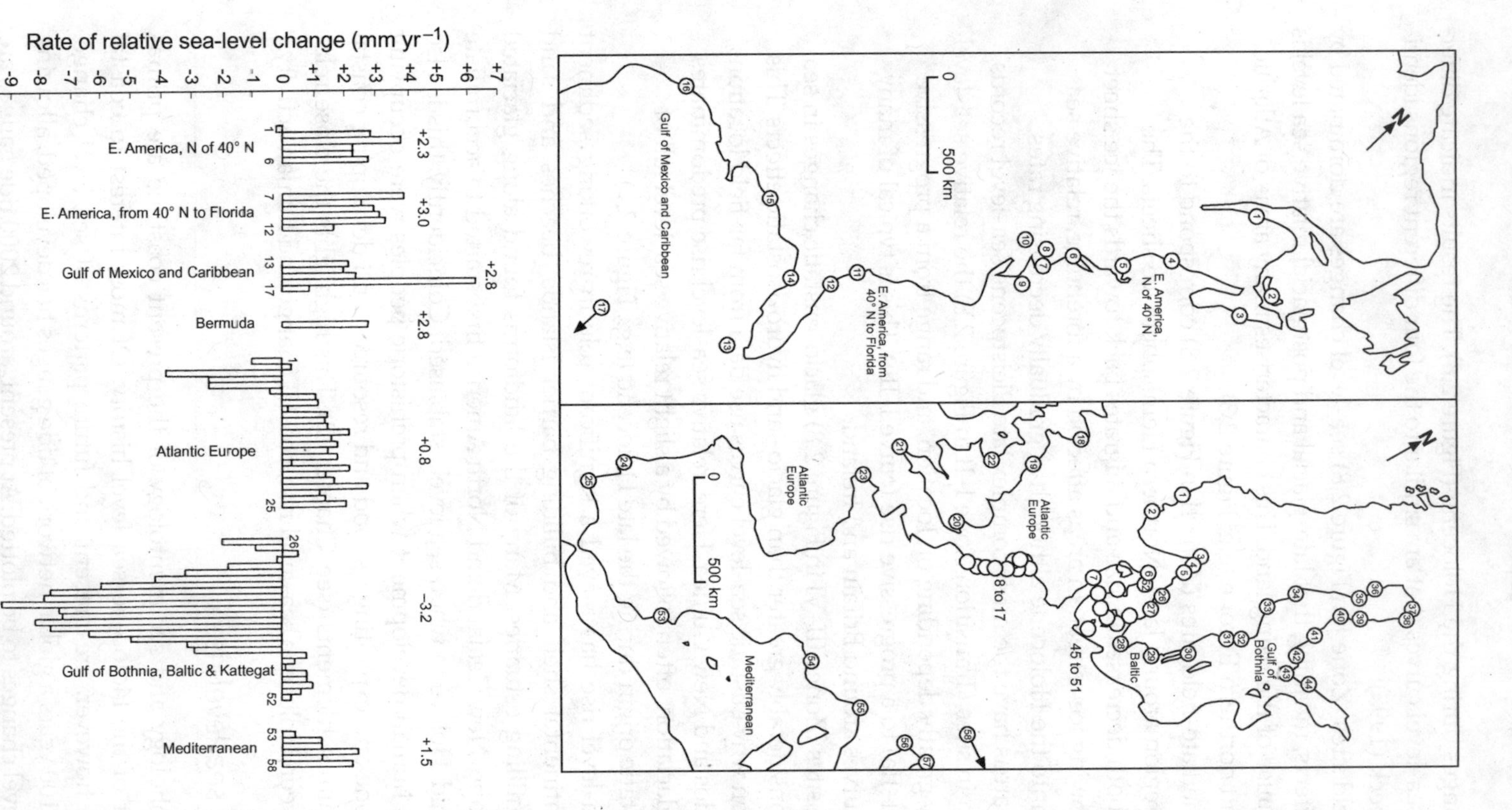

Figure 2.9 Locations of tide gauge stations with records starting earlier than 1925 and the associated trends of relative sea-level change. The numbers above the sea-level change graph are average rates of change in each region (+ = relative sea level rise; – = relative sea level fall). The numbers near the zero line in the sea-level change graph correspond to the tide gauge numbers in the location graph. (Modified from Pirazzoli, 1989.)

operative worldwide, although the longest records are almost all located in the Northern Hemisphere. Monthly and yearly means from most tide-recording stations are stored in the data base of the Permanent Service for Mean Sea Level (PSMSL) at Birkenhead, UK, and are publicly available from their web site (www.psmsl.org).

Tide gauges record changes in relative sea level at the coast, and a central problem in identifying trends in eustatic sea level from tide gauge data is in removing the isostatic trend from the data. For example, Figure 2.9 shows local trends in sea level derived from a large number of long-term tidal records. Most stations exhibit a rise in sea level across these records, with rates of 1–3 mm yr^{-1}, but the tide record in the Baltic Sea region indicates a fall in sea level of up to 9 mm yr^{-1} caused by isostatic uplift.

Due to uncertainties in determining the isostatic effects, it is difficult to obtain a reliable figure for eustatic sea level rise. Nevertheless, isostatic land-level changes can be accounted for by using geological data directly from sites adjacent to tide gauges and subtracting trends in land level from the relative sea-level change. Figure 2.10 shows the results of two methods of estimating the eustatic component of sea-level rise. The first method takes the mean of 130 station trends (corrected for changes in land level), resulting in a rate of sea-level rise over the last 100 years of 1.2 mm yr^{-1}. In the second method, the corrected data are averaged annually into a composite global mean sea-level curve and the slope of the curve is estimated to be 1.0 mm yr^{-1}.

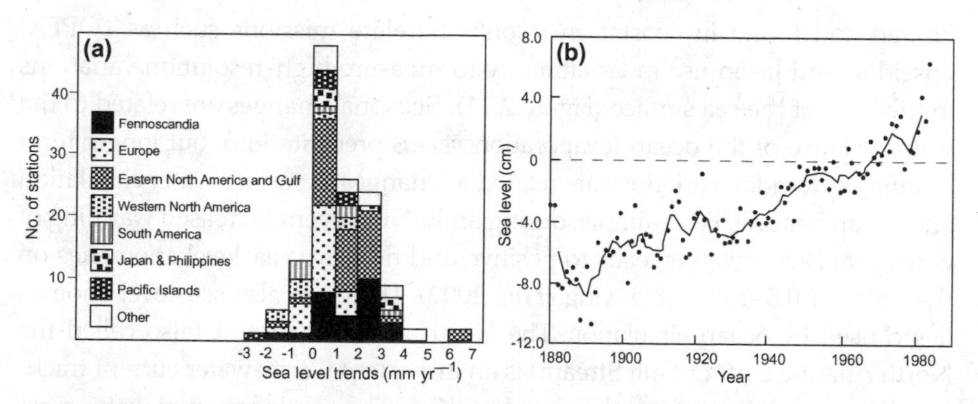

Figure 2.10 (a) Histogram of the number of tide gauge stations versus sea-level trends. (Modified from Gornitz, 1993.) (b) Composite eustatic mean sea-level curve over the period 1880–1980 (solid dots) with five-year running mean (solid line). The period 1951–70 has been used as the reference period. (Modified from Gornitz, 1995.)

More recently, satellite data have been used to estimate sea-level change. The main advantages of satellites over tide gauges are that satellites can have global coverage and include the open ocean, whereas the number of tide gauges is

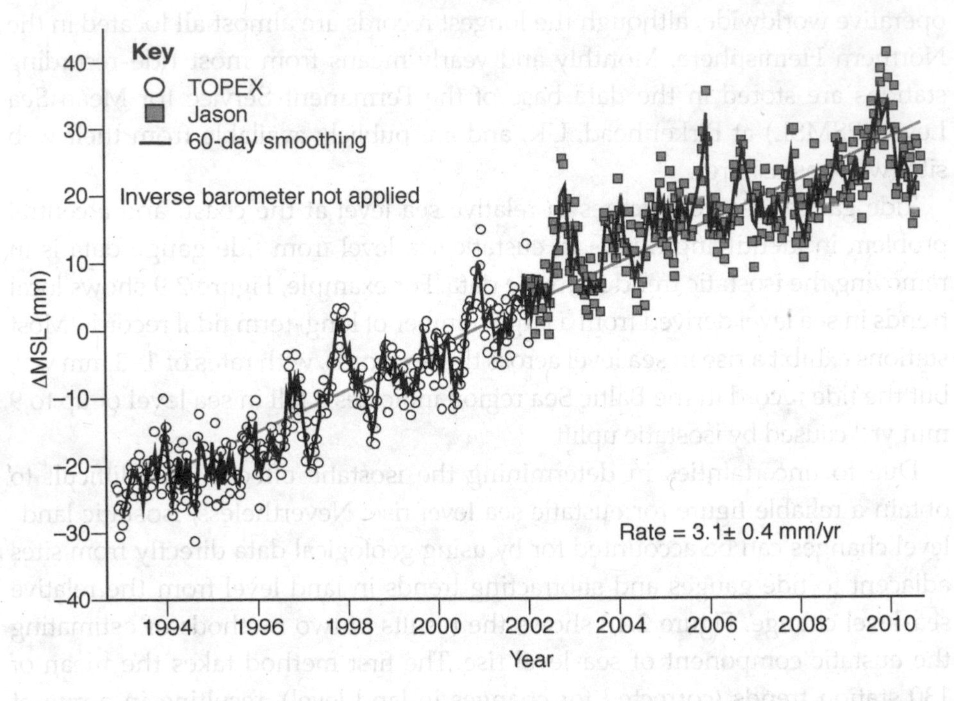

Figure 2.11 Variations in sea level in the period 1993–2010 from TOPEX and Jason satellite data with the seasonal signal removed. Data are available from the TOPEX website at http://sealevel.colorado.edu/results.php (Copyright © 2010 University of Colorado, reproduced with permission.)

limited and found in coastal areas only. Satellite missions such as TOPEX/Poseidon and Jason use radar altimetry to measure high-resolution variations in the level of the sea surface (Figure 2.11). Seasonal changes are related to the water balance of the ocean (evaporation *versus* precipitation), but longer term (annual to decadal) variations are related to changes in surface water circulation and ocean heating. For example, during an El Niño event increased warming of equatorial Pacific waters leads to positive and negative sea-level anomalies on the order of 0.5–1.0 m (Schwing *et al.*, 2002). There are also sea-level anomalies caused by ocean circulation. The North Atlantic Current (also called the North Atlantic Drift or Gulf Stream) is an important warm-water current tracking across the Atlantic from the Caribbean Sea. The combination of water mass transport and warmer sea surface temperatures mean that sea level within the current rise about 15 cm higher than sea level outside it (Fu *et al.*, 1987).

2.6 Present and future sea-level changes

The coastal zone is one of the most densely populated regions of the Earth and the ability to predict future change in sea level is of importance to the global community. Not surprisingly, much research effort and funding is directed towards making such predictions. A simple extrapolation of existing trends in sea-level rise (1–2 mm yr^{-1}) into the future is inadequate, not least because recent records show a non-linear trend in sea-level rise. This approach also ignores time lag effects in the atmosphere and ocean to increased greenhouse gas concentrations, and does not consider future greenhouse gas emission rates.

Tide gauge data show the upward trajectory of sea-level change over recent decades. The Fourth Assessment Report (published in 2007) of the Intergovernmental Panel on Climate Change (IPCC) concludes that the average rate of eustatic sea-level rise during the twentieth century as a whole was 1.7 mm yr^{-1} but that this rate increased dramatically from the 1990s, with sea-level rise in the period 1993–2003 amounting to 3.1±0.7 mm yr^{-1} (Bindoff *et al.*, 2007). Comparison of the present rate of sea-level rise with the geological rate over the last 2000 years (0.1–0.2 mm yr^{-1}) implies a relatively recent acceleration in the rate of sea-level rise. Of this rise, approximately half is by thermal expansion of the ocean due to higher water temperatures (both on the surface and at

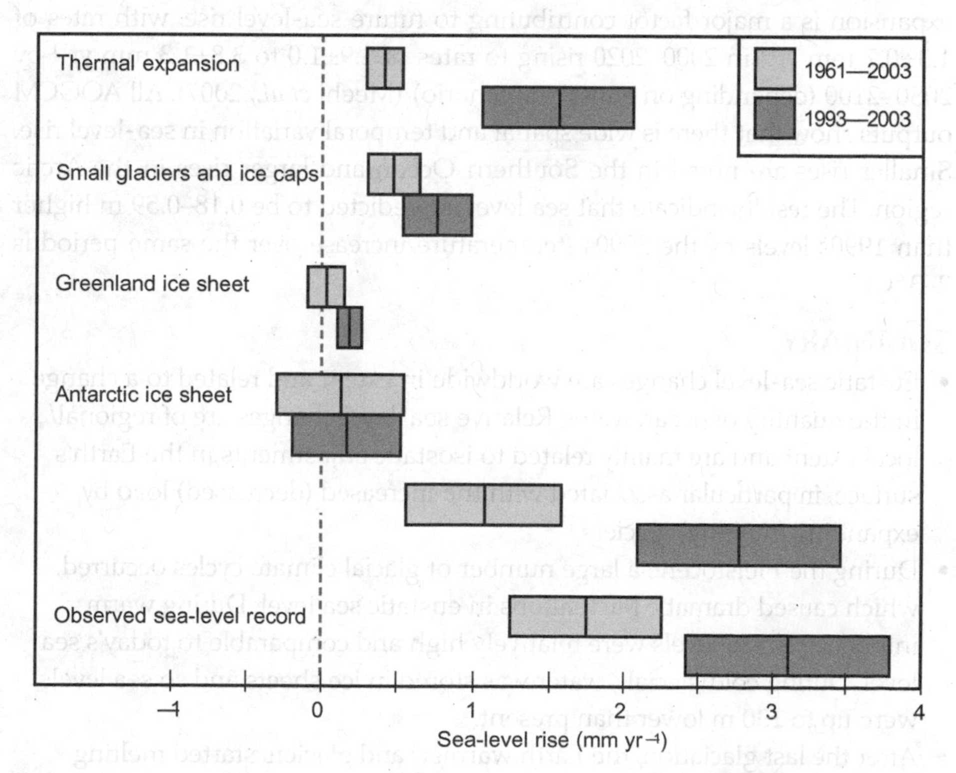

Figure 2.12 Estimates of the contribution to total sea-level rise by various factors for total sea level in the years 1961–2003 (light grey boxes) and 1993–2003 (dark grey boxes). The vertical line is the mean estimated value, with the width of the box representing the degree of error. (Redrawn and modified from Table 5.3 in Bindoff *et al.*, 2007.)

depth). This reflects the increase in global surface temperature of 0.74 °C over the century 1906–2005. Approximately one quarter of the total sea-level rise is through enhanced melting of small glaciers and ice caps (Figure 2.12). The relatively low (or even negative) contributions of the Greenland and Antarctic ice sheets may reflect their larger size which makes them less vulnerable to present climate change, and a possible increase total ice volume due to increased precipitation. However, more recent analysis by Rignot *et al.* (2011) shows that ice mass loss from Greenland and Antarctica is increasing and will be the major contributor to sea-level rise over the next decades.

The importance of obtaining good estimates of **future sea-level rise** has been one of the main aims of the IPCC. To obtain reliable predictions of future sea-level rise, coupled **atmosphere-ocean general circulation models** (AOGCMs) have been run for a range of CO_2 emission scenarios. These models consider interactions between the ocean, atmosphere and land surface with respect to global energy balance, biosphere and hydrological cycle, including future patterns of temperature and precipitation. Results show that thermal expansion is a major factor contributing to future sea-level rise, with rates of 1.3±0.7 mm yr^{-1} in 2000–2020 rising to rates of 1.9±1.0 to 3.8±1.3 mm yr^{-1} by 2080–2100 (depending on emission scenario) (Meehl *et al.*, 2007). All AOGCM outputs show that there is wide spatial and temporal variation in sea-level rise. Smaller rises are noted in the Southern Ocean and larger rises in the Arctic region. The results indicate that sea level is predicted to be 0.18–0.59 m higher than 1990s levels by the 2090s (temperature increase over the same period is 2–3 °C).

SUMMARY

- Eustatic sea-level changes are worldwide in extent and related to a change in the quantity of ocean water. Relative sea-level changes are of regional/local extent and are mainly related to isostatic adjustments in the Earth's surface, in particular associated with the increased (decreased) load by expanding (melting) glaciers.
- During the Pleistocene a large number of glacial climate cycles occurred, which caused dramatic fluctuations in eustatic sea level. During warm interglacials, sea levels were relatively high and comparable to today's sea level. During cold glacials, water was stored in ice sheets and so sea levels were up to 130 m lower than present.
- After the last glaciation, the Earth warmed and glaciers started melting leading to eustatic sea-level rise. This was initially rapid (2 cm yr^{-1} from 11,000–7000 years BP), but slowed down from 7000–5000 years BP, reaching present sea level at about 6000 years BP.

- Eustatic sea level is presently rising by 2–3 mm yr^{-1} as a result of global warming. This is expected to accelerate throughout the present century, and the Intergovernmental Panel on Climate Change predicts that by the 2090s sea level will be 0.18–0.59 m higher than it was in the 1990s.

Reflective questions

These questions are designed to test your comprehension of material covered in this chapter. Suggested answers to these questions can be found on this book's website.

2a. Describe the major controls on sea-level change in (1) arctic and (2) equatorial environments.

2b. Outline some of the problems in distinguishing between eustatic and isostatic controls on sea level.

2c. What can coastal responses to past sea-level change tell us about likely responses to future sea-level change?

2d. Why might there be uncertainty in IPCC estimates of future contributions to sea-level rise?

Further Reading

Bindoff, N.L., **Willebrand**, J., **Artale**, V., **Cazenave**, A., **Gregory**, J., **Gulev**, S., **Hanawa**, K., **Le Quéré**, C., **Levitus**, S., **Nojiri**, Y., **Shum**, C.K., **Talley**, L.D. and **Unnikrishnan**, A., 2007. 'Observations: Oceanic Climate Change and Sea Level'. In: S. Solomon, D. Qin, M. Manning, Z Chen, M. Marquis, K.B. Averyt, M. Tignor and H.L. Miller (eds) *Climate Change 2007: The Physical Science Basis. Contribution of Working Group I to the Fourth Assessment Report of the Intergovernmental Panel on Climate Change*. Cambridge University Press, Cambridge, 386–432. (This is the IPCC chapter that relates specifically to sea-level change, available from www.ipcc.ch/.)

Gehrels, R., 2010.'Sea-level changes since the Last Glacial Maximum: an appraisal of the IPCC Fourth Assessment Report'. *Journal of Quaternary Science*, 25, 26–38. (This is a useful commentary alongside the IPCC chapter, above.)

Chappell, J. and **Shackleton**, N.J., 1986. 'Oxygen isotopes and sea level'. *Nature*, 324, 137–140. (This is a classic paper describing glacial–interglacial sea-level cycles.)

Pirazzoli, P.A., 1996. *Sea-level changes: The last 20,000 years*. Wiley, Chichester. (Comprehensive review of Pleistocene and Holocene sea-level changes.)

CHAPTER 3

TIDES

AIMS

This chapter describes how tides are generated by forces that arise through interactions of the Sun, Moon and Earth. The physical laws describing these forces are discussed with respect to the Equilibrium and Dynamic theory of tides. Tidal dynamics are significant for landform and sediment patterns.

3.1 Introduction

The tidal rise and fall of the ocean surface due to the gravitational attraction between the Earth, Moon and Sun is barely noticeable in the open ocean, but tidal processes are most dominant on shallow continental shelves, near coasts and within estuaries. For example, Porter-Smith *et al.* (2004) found that there is a good correspondence between shelf sediment type and areas of wave- or tide-dominance around the coast of Australia. Tidal currents are the dominant force in transporting sediment across 41 per cent of the total shelf area, whereas waves are most dominant over 31 per cent of the shelf. On 27 per cent of the shelf neither tides nor waves are strong enough to transport sediment. In tide-dominant areas shelf sediments are much finer and have a higher mud content than wave-dominated areas. As a result, tidal processes are associated with coastal landforms with a high fine-sediment content including mud flats, salt marshes, mangroves, back-barrier lagoons and estuaries. In order to consider the role of tidal processes in these different environments, it is important to appreciate the fundamentals of tide-generation.

Our understanding of tides is based largely on the works of the mathematicians Isaac Newton (1643–1727) and Pierre-Simon Laplace (1749–1827). In his *Principia Mathematica,* Newton derived the fundamental astronomical forces that produce forced waves on a uniform, infinitely deep ocean. In his *Mecanique celeste,* Laplace derived the fundamental hydrodynamic equations that govern the behaviour of forced long waves on a rotating Earth with oceans of finite depth. William Thomson, Lord Kelvin (1824–1907), then demonstrated that Laplace's equations could describe tides in natural ocean basins surrounded by continental margins. Our discussion of tides broadly follows this historical development of the topic. We first describe the tide-generating force originally derived by Newton upon which the Equilibrium Theory of tides is based. We then describe the Dynamic Theory of tides which emerged from the work of Laplace and Lord Kelvin. Tidal processes in coastal bays and estuary channels are discussed in Chapter 7. Coughenour *et al.* (2009) and Cherniawsky *et al.* (2010) give up-to-date summaries of the forcing factors of tides and their changes over time.

3.2 The astronomical tide-generating force

The **gravitational attraction** between the Earth and other bodies in the solar system produces what is termed the **astronomical tide**. Gravitational attraction can be thought of as a force F_g, which is written as

$$F_g = G\frac{m_1 m_2}{R^2} \qquad (3.1)$$

where G is the universal gravitational constant (6.6×10^{-11} N m^2 kg^{-2}), m_1 and m_2 are the respective masses of the two bodies we are interested in, and R is the distance between the centres of mass of each body. Equation 3.1 shows that the greater the mass of the two bodies, the larger the gravitational attraction between them. It also shows that the smaller the distance between the two bodies, the larger the gravitational attraction between them. In principle all planetary bodies can influence the behaviour of tides on Earth, but in practical terms their attractive effect is negligible, due either to their small size or large distance from Earth. The astronomical tide-generating force on Earth is dominated by the gravitational attraction of the Moon and the Sun, because of the Moon's close proximity and the Sun's large mass (Table 3.1).

Table 3.1 Properties of the Earth, Moon and Sun relevant to the derivation of the tide-generating force.

	Mass (tonnes)	Radius (km)	Average distance from Earth (km)
Earth	5.97×10^{21}	6378	0
Moon	7.35×10^{19}	1738	384,400
Sun	1.99×10^{27}	696,000	149,600,000

We first consider the Moon's role in generating tides. It is a common misconception that the Moon simply revolves around the Earth, but in fact, they work as a system that rotates anticlockwise around their combined centre of mass, known as the **barycentre**. Due to the large size of the Earth in comparison to the Moon (Table 3.1), the location of the barycentre is approximately 4700 km from the Earth's centre (Figure 3.1). The time taken for one rotation of the

Earth–Moon system around its barycentre is 27.32 days – one **sidereal month**. The orbital path that the Moon traces during one rotation has a radius equal to the distance between its own centre and the barycentre, on average 379,700 km. This path is considerably larger than Earth's circumference, which is why it appears (incorrectly) that the Moon is revolving around the Earth's centre of mass. During one rotation of the Earth–Moon system, the Earth will also trace an orbital path around the system's barycentre. The Earth's path has a radius also equal to the distance between its centre and the system's barycentre, so as the Earth's path is quite small, we are largely unaware of this motion. As the Earth–Moon system revolves around the Sun, the system's barycentre follows a smooth orbital path, whereas the individual paths of the Earth and Moon waver.

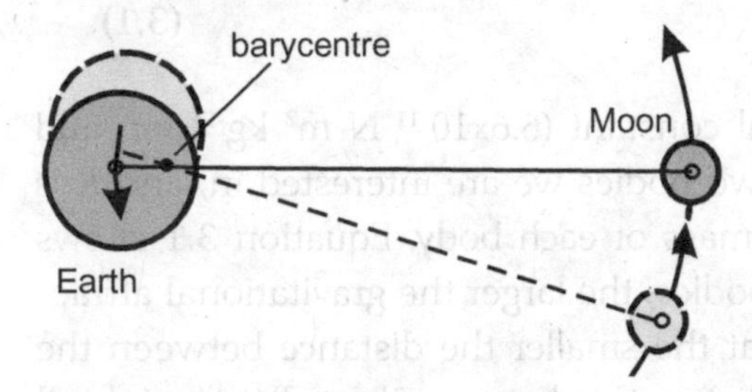

Figure 3.1 The Earth–Moon system rotates on its common centre-of-mass, known as the barycentre. The barycentre (solid dot) is located approximately 4700 km out from the Earth's own centre of mass (open dot). View is looking down upon the Earth's north pole, hence the sense of rotation is anticlockwise.

Newton's Laws of Motion state that a force is the product of a body's mass and acceleration. Acceleration is therefore the ratio of force to mass, and takes place in the direction of the force. As the Earth and Moon are orbiting their barycentre, the acceleration of every particle on the Earth (and Moon) experiences a gravitational force in order to maintain its orbital motion. This gravitational force is termed the **centripetal force** F_c. It follows from Equation 3.1 that the centripetal force for the Earth-Moon system is

$$F_c = G\frac{m_E m_M}{R^2} \tag{3.2}$$

where m_E and m_M are the masses of the Earth and Moon, respectively. Since every particle on Earth follows the same orbit and therefore experiences the same acceleration, the centripetal force must be acting with the same magnitude and direction on every particle. The direction in which the centripetal force acts is always parallel to the plane of rotation of the Earth–Moon system (Figure 3.2a).

The local **gravitational force**, however, does depend on location. Following Equation 3.1, the local gravitational force F_{lg} is

$$F_{lg} = G\frac{m_E m_M}{(R \pm r)^2} \tag{3.3}$$

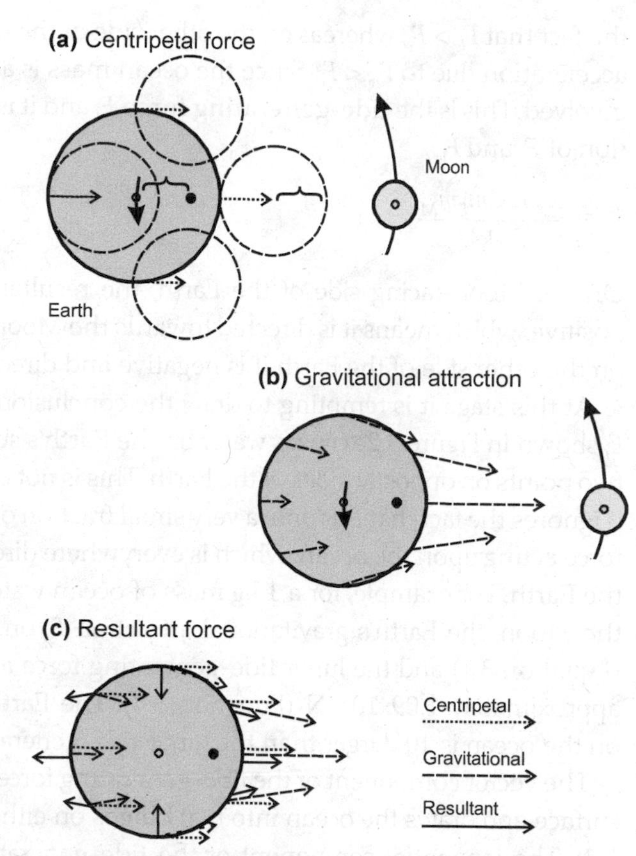

Figure 3.2 The view in all three panels is looking down upon the Earth's north pole. (a) The dashed circles trace the paths of positions on the Earth's surface as the Earth–Moon system completes one full rotation. The radius of the circles is equal to the distance between the Earth's own centre-of-mass and the barycentre (solid dot). The magnitude and direction of the centripetal force is indicated by the length and direction of the dotted arrows. (b) The magnitude and direction of the gravitational attraction is indicated by the length and direction of the dashed arrows. (c) The resultant tide-generating force (vector addition of the centripetal and local gravitational forces) is indicated by the length and direction of the solid arrows.

where r is the distance between the Earth's centre and the point of interest on the Earth's surface ($-r$ for points on the Moon-facing side of the Earth and $+r$ for points on the opposite side). Locations on the Moon-facing side will experience a local gravitational force that is larger than that experienced at locations on the opposite side, due to their closer proximity to the Moon (Equation 3.3). The direction in which the local gravitational force acts is towards the centre of the Moon's mass, which in most cases is at an angle to the centripetal force (Figure 3.2b).

In order for the rotation of the Earth–Moon system to remain stable, the average centripetal force per unit of mass must equal the average gravitational force. If these forces were not equal then the Earth and Moon would either accelerate towards each other or away from each other. Equations 3.2 and 3.3 show that the centripetal force per unit mass is the same everywhere on or inside the Earth, but the gravitational force per unit mass varies locally. If F_c and F_{lg} were equal everywhere then there would be no tide-generating force, so local differences of F_c and F_{lg} are ultimately responsible for the tides. The ocean on the Moon-side of the Earth experiences a small acceleration due to the fact that $F_{lg} > F_c$, whereas on the other side of the Earth it experiences a small acceleration due to $F_{lg} < F_c$. Since the ocean mass is accelerated a force must be involved. This is the **tide-generating force** F_t and it is the resultant vector addition of F_c and F_{lg}

$$F_t = \frac{(\pm r)2Gm_E m_M}{R^3} \tag{3.4}$$

On the Moon-facing side of the Earth, the resultant tide-generating force is positive, which means it is directed towards the Moon (Figure 3.2c). Conversely, on the other side of the Earth it is negative and directed away from the Moon.

At this stage it is tempting to draw the conclusion that the local variation in F_t shown in Figure 3.2c causes water on the Earth's surface to be drawn towards two points on opposite sides of the Earth. This is not completely correct, because it ignores the fact that F_t is only a very small fraction of the Earth's own attractive force acting upon the ocean, which is everywhere directed towards the centre of the Earth. For example, for a 1 kg mass of ocean water located directly beneath the Moon, the Earth's gravitational attraction F_g on this water mass is 9.68 N (Equation 3.1) and the lunar tide-generating force acting on the water mass is approximately 1.09×10^{-6} N (Equation 3.4). The Earth's gravitational attraction on the ocean is 10^7 larger than the lunar tide-generating force.

The vector component of the tide-generating force is tangential to the Earth's surface and draws the ocean into two bulges on either side of the Earth (Figure 3.3). The tangential component of the tide-generating force is known as the **tractive force** F_T

$$F_T = \frac{3m_M r^3}{2m_E R^3} g \sin 2\theta \tag{3.5}$$

where θ is the angle between the point of interest on the Earth's surface and the line joining the centres of the Earth and Moon (Figure 3.3). The tractive force is smallest for θ approaching 0° and 90°, which is where the Earth's gravitational attraction directly opposes the lunar tide-generating force, and largest for θ = 45° where it does not.

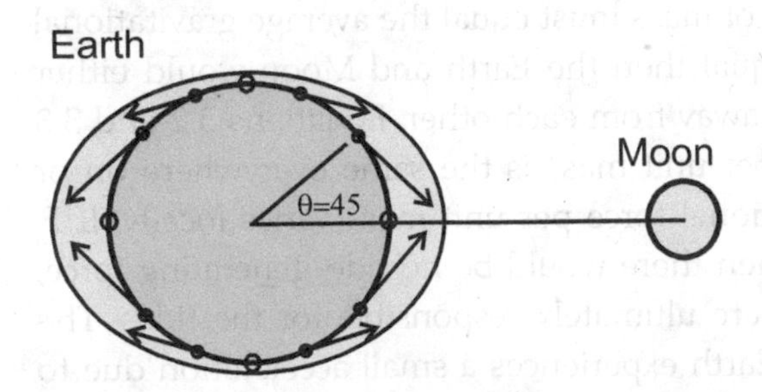

Figure 3.3 The magnitude and direction of the tractive force causing tidal bulges on opposite sides of the Earth that are aligned with the Moon. The tractive force is maximum at an angle of θ = 45° and decreases towards zero as θ approaches both 0° and 90°.

To summarise up to this point, in order for the Earth–Moon system to be in stable rotation around its barycentre, the average centripetal force per unit mass in the system must equal the average gravitational attractive force. Because the distance to the Moon varies with location on the Earth's surface there are local differences between these forces. The vector sum of the forces acting on the ocean is the tide-generating force. The small tangential component of the tide-generating force (i.e. the tractive force), which is largely unopposed by the gravitational attraction of the Earth, causes water to form two **tidal bulges** on opposite sides of the Earth and aligned with the Moon. The sloping water surface of these bulges is maintained through a balance between the tractive force and the horizontal **pressure-gradient force** (Box 3.1).

Box 3.1 Derivation of the pressure-gradient force

The hydrostatic pressure P at any given depth is the weight of the overlying water acting on a unit area and is

$$P = \rho g h$$

where ρ is water density, g is gravitational acceleration and h is the height of water above the location of interest. A sloping water surface is effectively the same as a pressure gradient. Consider the sloping water surface in Figure 3.4 and assume a uniform water density. The hydrostatic pressure at Site A and Site B is, respectively

$$P_A = \rho g h \quad \text{and} \quad P_A = \rho g(h + \Delta h)$$

The pressure gradient between the two sites is

$$\frac{\Delta P}{\Delta x} = \frac{P_B - P_A}{\Delta x} = \rho g \frac{\Delta h}{\Delta x} = \rho g \tan\theta$$

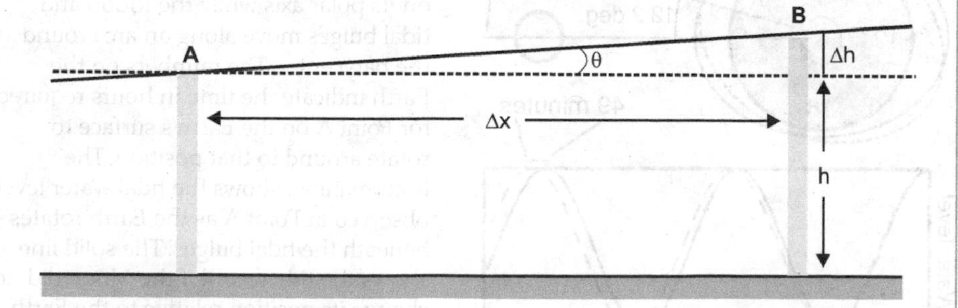

Figure 3.4 Schematic diagram of a sloping water surface, which defines the terms used to derive the pressure gradient force.

where θ is the water surface slope. The pressure gradient force (ρgtanθ) opposes the tractive force (Equation 3.5) and thus maintains the sloping water surface that represents the tidal bulge. The pressure gradient also drives coastal tidal currents, as water is forced to flow horizontally from areas of high to low pressure (Section 7.3).

3.3 The Equilibrium Theory of tides

The three principal assumptions of the **Equilibrium Theory of tides** are: (1) the Earth has no continental land masses but is covered by an ocean of uniform depth; (2) there is no inertia in the system and oceans respond immediately to the tide-generating force; and (3) the Coriolis and friction effects can be neglected. The basis of the Equilibrium Theory is the tidal bulges shown in Figure 3.3. If there is no system inertia, these bulges will follow the Moon around the Earth.

3.3.1 Earth rotation and the semi-diurnal tide

In addition to the Earth and Moon rotating around their common barycentre, the Earth also rotates in an anticlockwise direction on its polar axis. The time it takes to spin through one rotation is one **solar day** (24 hours). Consider an observer at Point A on the Earth's surface (Figure 3.5). The observer would experience two high tides separated by two low tides as the Earth spins through one rotation beneath the two tidal bulges. While the Earth is rotating on its polar axis the Moon is also moving in the same direction around the barycentre. During one rotation of the Earth on its axis the Moon has advanced along its path by about 12.2°, shifting the location of the tidal bulges by the same

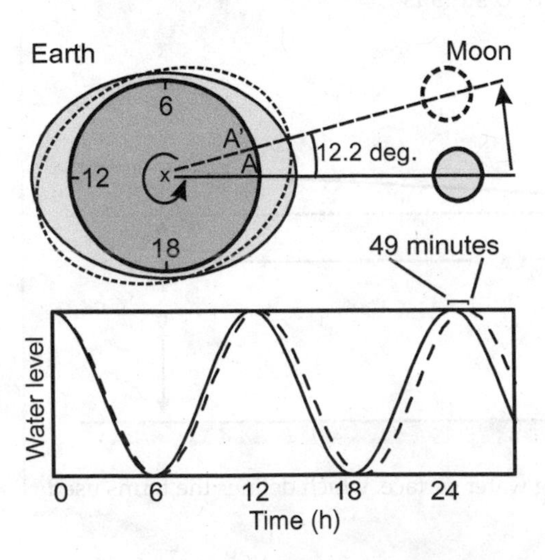

Figure 3.5 The view in the top panel is looking down upon the Earth's north pole. The Earth completes one rotation on its polar axis while the Moon and tidal bulges move along an arc around the barycentre. The numbers on the Earth indicate the time in hours required for Point A on the Earth's surface to rotate around to that position. The bottom panel shows the tidal water level observed at Point A as the Earth rotates beneath the tidal bulges. The solid line shows the tide record if the Moon did not change its position relative to the Earth, and the dashed line shows the true tide record, with a daily offset of 49 minutes.

amount. It therefore takes an additional 49 minutes longer than a solar day for Point A to reach the first bulge again (now located at A′). That is why the times of high and low water are approximately an hour later than on the previous day.

3.3.2 Lunar declination and the diurnal inequality (tropical and equatorial tides)

So far we have assumed that the Moon is aligned with the Earth's equator. In fact the orbital plane of the Earth–Moon system is tilted up to 5° from the Earth's equatorial plane. This tilt is known as the **lunar declination**. As the moon revolves around the barycentre on the tilted orbital plane, its position above the Earth varies between latitudes 28.5° north and south of the equator (Figure 3.6). When the Moon is above the equator the tidal bulges are aligned with the equator, and when the Moon is north or south of the equator the tidal bulges are tilted relative to the equatorial plane.

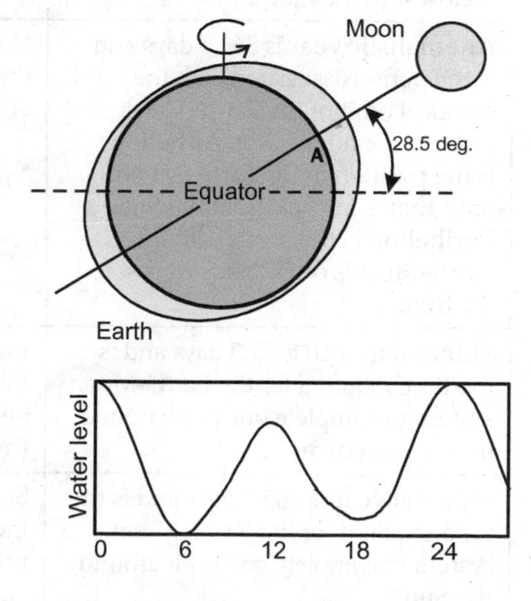

Figure 3.6 The view in the top panel is looking side-on to the Earth. The angle of declination between the orbital plane of the Earth–Moon system (solid line connected to the Moon) and the equator (dashed line) is 28.5°. The bottom panel shows the tide record that would be observed as Point A rotates beneath the tidal bulges when they are at maximum declination. Note the high-high tide at 0 and 24 h and low-high tide at 12 h.

As the Earth spins on its polar axis, an observer at Point A in Figure 3.6 would experience a diurnal (daily) variation in the magnitude of high and low tide. Initially Observer A experiences a high-high tide, but when the observer reaches the tidal bulge facing away from the Moon a low-high tide occurs. When the observer returns to the starting point approximately one day later, a high-high tide occurs again. Maximum diurnal variation in tidal range within a monthly lunar cycle occurs when the combined declinations position the Moon somewhere between the Tropics of Capricorn (Southern Hemisphere) and Cancer (Northern Hemisphere). For this reason, tides that display a diurnal inequality are referred to as **tropical tides**. There is minimal diurnal variation in the tide during those times of the monthly lunar cycle when the Moon is positioned over the Equator, so tides that display least diurnal inequality are referred to as **equatorial tides**. We therefore observe two tropical and equatorial tides in a sidereal month (sec Table 3.2 for terminology).

Table 3.2 Definition of astronomical terms and their relationship to the ocean tide.

Terms	Related tide behaviour
Anomalistic month is 27.6 days and is the period associated with the elliptical orbit of the Moon. **Apogee** is the position in the Moon's orbit that is furthest from the Earth. **Perigee** is the position in the Moon's orbit that is closest to the Earth.	Variation in the Moon's distance to the Earth causes the tides to be larger at perigee and smaller at apogee.
Anomalistic year is 366.5 days and is the period associated with the elliptical orbit of the Earth–Moon system around the Sun. **Aphelion** is the position in the Earth–Moon's orbit that is furthest from the Sun. **Perihelion** is the position in the Earth–Moon's orbit that is closest to the Sun.	Variation in the Sun's distance from the Earth causes the tides to be larger at perihelion and smaller at aphelion.
Sidereal month is 27.3 days and is the time required for the Earth–Moon system to complete one orbit around their **barycentre**.	Lunar declination causes the two tides each day to alternate between being equal (equatorial) and unequal (tropical) twice over the month.
Sidereal year is 365.3 days and is the time required for the Earth–Moon system to complete one orbit around the Sun.	Solar declination causes diurnal inequality in the tides to be most pronounced in June and December.
Synodic month is 29.5 days and is the time between successive conjunctions of the Earth, Moon and Sun.	Relative positions of the Moon and Sun cause larger (smaller) than average spring (neap) tides to occur twice a month.

3.3.3 Solar–lunar interaction and the spring–neap tide cycle

The same line of reasoning to explain the tide-generating force of the Moon is equally applicable to the tide-generating force of the Sun. If we make appropriate substitutions from Table 3.1 into Equation 3.4 we find that the

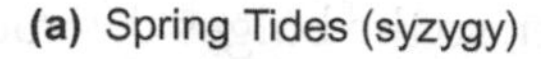

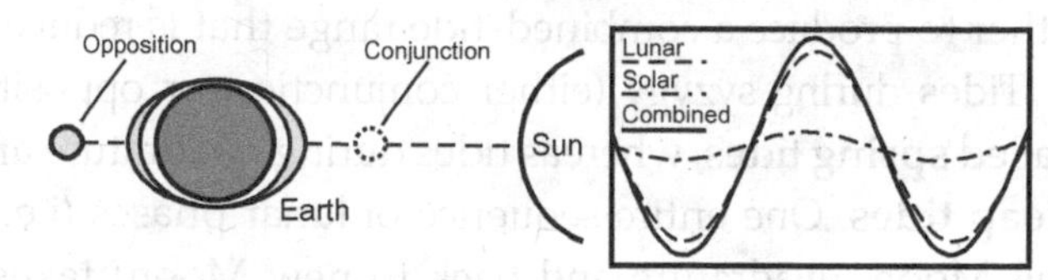

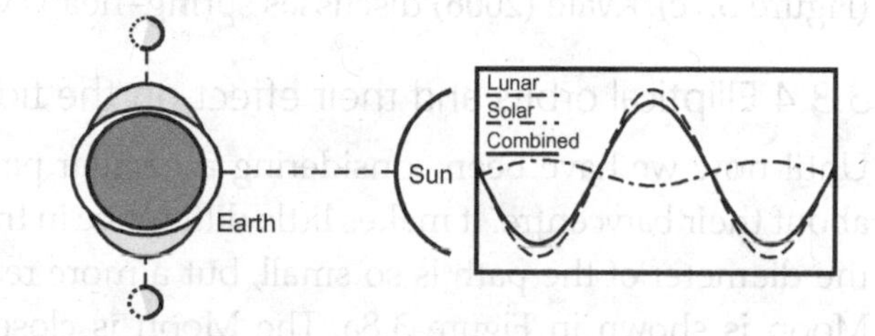

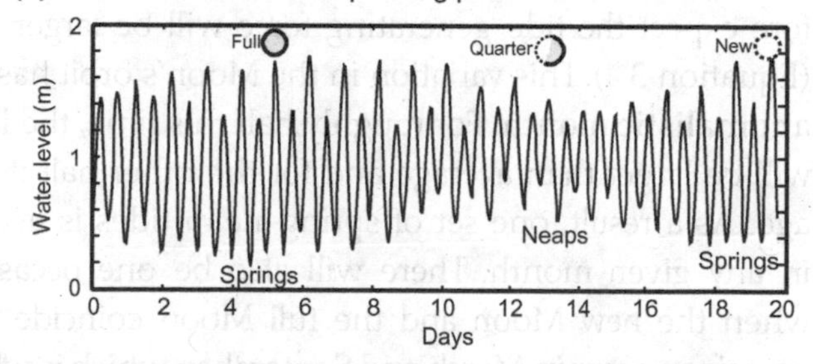

Figure 3.7 The view in the top two panels is looking down upon the Earth's north pole. (a) When the Moon is in conjunction (new Moon phase) or opposition (full Moon phase) with the Sun the lunar (grey) and solar (white) tidal bulges are aligned. The panel on the right shows the resultant tide record. (b) When the Moon is in quadrature (quarter Moon phase) the lunar and solar tidal bulges are at 90° to one another. (c) A tide record covering nearly one and a half spring–neap tide cycles, demonstrating the relationship with the Moon's phases.

tide-generating force of the Sun is about 46 per cent that of the Moon due to its greater distance from Earth.

Figure 3.7 shows the relative magnitudes of the lunar and solar tidal bulges and their positions for different phases of the Moon. When the Moon is new it is located in a line between the Earth and Sun, and is barely visible from Earth because it is not being illuminated directly by the Sun (Figure 3.7a). When the Moon is full it is located on the opposite side of the Earth from the Sun and is illuminated directly by the Sun. When the Earth, Moon and Sun are all aligned, during either a full or new moon, the three bodies are said to be in **syzygy**. When the Moon is at a right-angle to the Earth with respect to the Sun, the bodies are said to be in **quadrature** (Figure 3.7b).

When the Sun and Moon are in syzygy their respective tidal bulges are aligned. This results in a combined tidal bulge that is the sum of the individual contributors because their crests and troughs are aligned and constructively interfere with each other (Figure 3.7a). The result is an amplified combined-tide range, in which the high tide is higher and the low tide is lower than the lunar tide alone. When the Sun and Moon are in quadrature their respective bulges are at right-angles to each other, so the bulges destructively interfere with each other to produce a combined-tide range that is reduced (Figure 3.7b).

Tides during syzygy (either conjunction or opposition) are largest and are called **spring tides**, whereas tides during quadrature are smallest and are called **neap tides**. One entire sequence of lunar phases (i.e. new Moon, quadrature, full Moon, quadrature and back to new Moon) takes 29.5 days – a **synodic month**. A set of spring–neap tides therefore occurs approximately every 15 days (Figure 3.7c). Kvale (2006) discusses spring–neap cycles in some detail.

3.3.4 Elliptical orbits and their effect on the tide

Until now we have been considering a circular path for the Earth and Moon about their barycentre. It makes little difference in the case of the Earth, because the diameter of the path is so small, but a more realistic elliptical path for the Moon is shown in Figure 3.8a. The Moon is closest to the Earth at **perigee** (357,000 km) and farthest from the Earth at **apogee** (407,000 km). We therefore expect the tide-generating force will be larger at perigee than at apogee (Equation 3.4). This variation in the Moon's orbit has a period of 27.6 days – an **anomalistic month**. For roughly half this time, the lunar tide-generating force will be larger than average and for the other half it will be smaller than average. As a result, one set of spring–neap tides is usually larger than the other in any given month. There will also be one occasion each during the year when the new Moon and the full Moon coincide with lunar perigee. These occasions occur in March and September, which is when maximum spring tides occur.

The orbit of the Earth–Moon system around the Sun is also elliptical (Figure 3.8b). The Earth is closest to the Sun at **perihelion** (148,500,000 km) and farthest away at **aphelion** (152,200,000 km). This variation in the orbit of the Earth–Moon system has a period of 366.5 days – an **anomalistic year**. For roughly half this time the solar tide-generating force will be larger than average

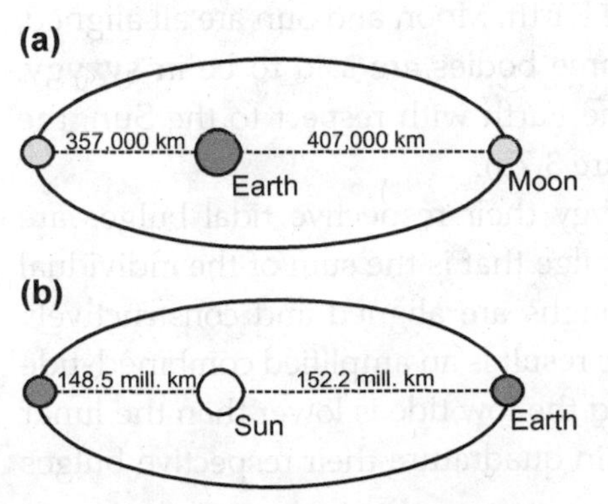

Figure 3.8 The elliptical orbit of (a) the Moon around the Earth–Moon barycentre and (b) the Earth around the Sun. Maximum and minimum distances from the Earth are also shown.

and for the other half it will be smaller than average. Thus the tides are marginally larger in the six months of the year centred on January (at perihelion), and smaller in the six months centred on July (at aphelion).

3.3.5 Solar declination and the seasonal inequality of the tide

The plane of the Earth–Moon system's orbit around the Sun is tilted at an angle of 23.5° to the Earth's equatorial plane. The effect of this **solar declination** is similar to that described for the lunar declination (Figure 3.5). The Sun's position over the Earth varies between the Tropics of Cancer and Capricorn and back again in the course of 365.25 days – a **sidereal year**. The Sun is located over the tropics during the solstices (21 June and 21 December) and over the equator during the equinoxes (21 March and 21 September). During the solstices, the solar tidal bulge will add a small amount to the diurnal inequalities in tide range produced by the lunar declination, whereas during the equinoxes it will not. There is also a slow change (or **precession**) of the lunar declination with respect to the solar declination, which produces an 18.6-year periodicity in the tides (Cherniawsky *et al.*, 2010).

3.3.6 Shortcomings of the Equilibrium Theory

While the Equilibrium Theory can explain most features of the tide there are four significant shortcomings: (1) the predicted tide range is typically smaller than the observed range; (2) the predicted tidal range is not constant but varies with location; (3) the timing of high water is generally several hours before or after the time of transit of the Sun and Moon; and (4) the timing of spring and neap tides does not always coincide with syzygy or quadrature, but is typically a day or more different. The first two shortcomings suggest that local characteristics are significant, and the second two suggest that the assumptions of zero inertia and zero friction are too restrictive. For these reasons the Equilibrium Theory cannot be used to precisely predict the tide at a given location. We now consider in more detail the Earth-based factors that influence tide behaviour.

3.4 The Dynamic Theory of tides

The main premise of the **Dynamic Theory of tides** is that the two tidal bulges discussed in Section 3.3 actually behave as waves. Since the bulges are on opposite sides of the Earth, the wavelength is equal to half the Earth's circumference, some 20,000 km. Because the wavelength is very large compared with the water depth, the tidal bulges behave as **shallow water waves** or **long waves** (see Section 4.4). Another example of long waves are tsunami waves, sometimes incorrectly referred to as tidal waves (Case Study 3.1). In the Dynamic Theory there is a long wave associated with each contributor to the tide-generating force (principally the Moon and Sun). Since the Earth is spinning in an anti-clockwise direction, the relative movement of the long waves is from east to west. The tide-generating force is always present, thus the long wave is always being driven by it. We refer to such waves as **forced waves**.

Case Study 3.1 Tsunami! When a tidal wave is not a tidal wave

It is common for waves produced by a **tsunami** (plural tsunami) to be referred to as a tidal wave, but this is incorrect because they are generated in different ways. Tide waves are generated by astronomical forces involving the Earth, Moon and Sun. Tsunami are generated by external disturbance of the sea bed or sea surface, the energy from which travels outwards through the water column and away from the disturbance site. There are several mechanisms that can trigger a tsunami (see Bryant, 2001, for a review). The 1883 Krakatau (Java) submarine volcanic eruption produced a tsunami with waves over 35 m high that killed more than 36,000 people. Tsunami waves from this event were registered on tide gauges in South Africa (+30 cm) and the English Channel (+5 cm). The submarine Storegga land-slides off the coast of Norway *c.* 7900 years BP caused tsunami waves that ran up to 25 m above sea level on the coast of Scotland. Tsunami take place most commonly as a result of sub-sea earthquakes. The December 2004 Banda Aceh earthquake in the Indian Ocean produced tsunami waves up to 25 m high, killing over 283,000 people across the region.

In the open ocean, tsunami waves typically have a wavelength of a few hundred kilometres and a height of less than a few metres. This results in very small water surface slopes, so the waves can travel across the ocean basin at speeds of around 800 km hr^{-1} and hardly be noticed. When they enter shallower water, however, they begin to shoal, which means they slow down, become shorter in wavelength and larger in height (see Section 4.5.2 for a discussion of shoaling). By the time tsunami waves reach the coast-line they can be several tens of metres high and cause enormous property damage and loss of life. Tsunami monitoring networks across the world's oceans will act as an early-warning system to alert coastal communities to potential tsunami hazards.

Because the tide is a long wave forced largely by the Moon we might expect the crest of the tide to always be located beneath the Moon, just like the tidal bulges in the Equilibrium Theory. For this to occur, the crest of the long wave must be

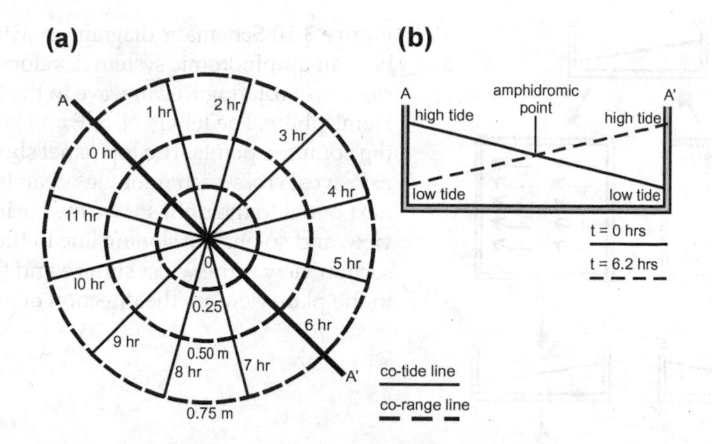

Figure 3.9 (a) Plan view of an idealised amphidromic system in the Southern Hemisphere. Clockwise rotation is indicated by co-tide lines, which mark arrival times of the wave crest (high tide) as it rotates around the amphidrome. The co-range lines show that minimum and maximum tide range occurs at the centre and outer edge of the system, respectively. Note that the circulation is in an anticlockwise direction in the Northern Hemisphere. (b) Cross-section through the amphidrome along the line A–A′, and showing the system to be confined by solid boundaries representing the edge of the ocean basin. When the wave crest (high tide) is on one side of the basin the wave trough (low tide) is on the opposite side of the basin. (Modified from Pinet, 2000.)

capable of travelling at the same linear velocity that the Earth's surface is spinning relative to the Moon's position. The linear velocity of the Earth's surface is at a maximum of 449 m s^{-1} at the equator and diminishes to zero at the poles. The velocity of a long wave is equal to $\sqrt{gh}$ (Equation 4.9), where g is gravitational acceleration (9.81 m s^{-2}) and h is water depth. For an average ocean water depth of 4000 m, the long wave velocity is approximately 198 m s^{-1}. This means that the forced long wave of the tide can only remain in equilibrium with the lunar forcing at latitudes of 65° or higher because at lower latitudes the long wave cannot travel fast enough to keep up with the lunar forcing. The limited ocean depth and the fact that the global ocean is broken into deep basins separated by shallow shelves and land masses, mean that the forced long waves are effectively broken up into smaller systems, which are called **amphidromes**.

An idealised representation of how amphidromes operate is shown in Figure 3.9. The forced long wave rotates around the central point (the node) of the amphidrome, achieving one circuit in a time period that corresponds with the astronomical forcing. The sense of rotation is clockwise in the Southern Hemisphere and anticlockwise in the Northern Hemisphere. The sense of rotation can be determined from the **co-tide lines**, which show the location of the wave crest at successive hours. When the wave crest (high tide) is at A, the wave trough (low tide) is on the directly opposite side of the amphidrome at A′. The

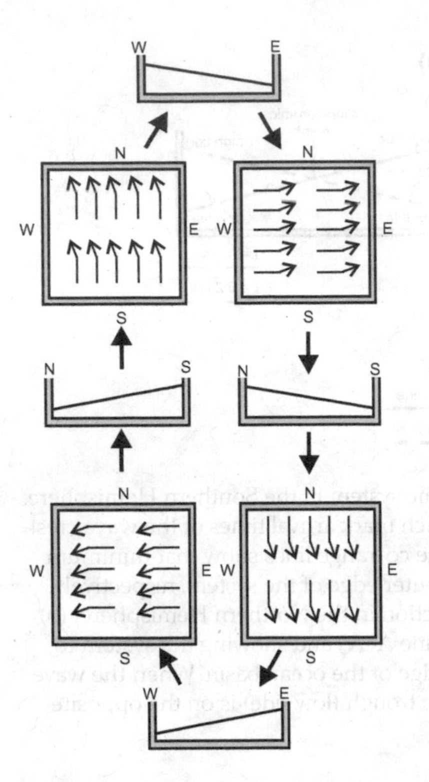

Figure 3.10 Schematic diagram showing how an amphidromic system develops a clockwise rotating Kelvin wave in the Southern Hemisphere. The letters N, S, E and W denote the compass points. The top panel shows an east–west cross-section of the ocean basin, the next panel to the right shows the basin in plan view, and so on. The sloping line in the cross-section view is the water surface and the arrows in the plan-view are the direction of water flow.

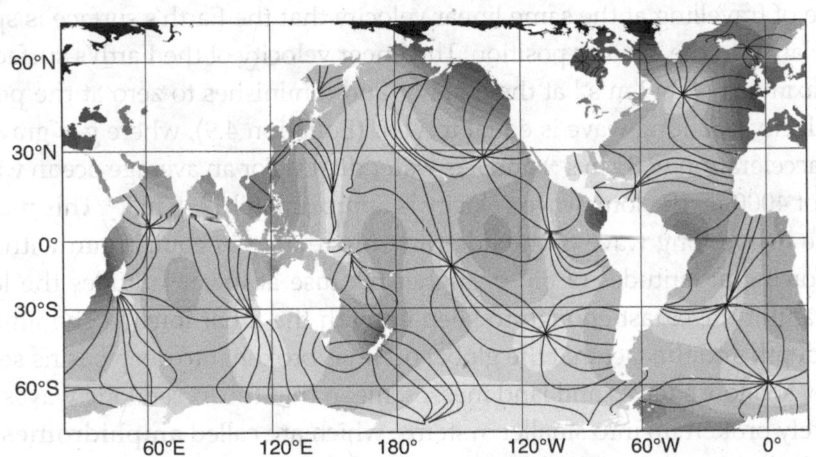

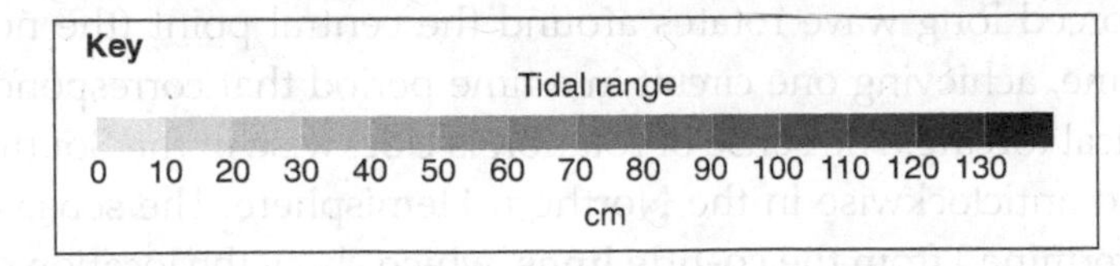

Figure 3.11 Amphidromic systems in the world's oceans identified by co-tidal lines. Note that tidal range is greatest around ocean basin margins. (From Coughenour *et al.*, 2009.) (Copyright © 2009 Elsevier Science, reproduced with permission.)

minimum tide range is zero at the centre of the amphidrome and reaches a maximum at the outer edge as indicated by **co-range lines**.

Amphidromic systems in a fully enclosed ocean basin arise from the forced tidal long wave which travels from east to west, causing sea level to be elevated against the western margin of the basin (Figure 3.10). The resultant sloping surface produces a pressure gradient force that causes the water to flow eastward (see Box 3.1). For an ocean basin in the Southern Hemisphere (Figure 3.10), eastward flow is deflected to the left (north) by the Coriolis force, which causes water to be elevated against the northern margin of the basin. Subsequently, the pressure gradient force drives water southwards from which it is deflected to the east, and so on. In this way a wave crest rotates around the ocean basin in a clockwise direction in the Southern Hemisphere and anticlockwise direction in the Northern Hemisphere. This wave is called a **Kelvin wave** named after its discoverer, Lord Kelvin.

The major amphidromic systems of the Earth's oceans and their circulation patterns are shown in Figure 3.11. Many of the systems in large ocean basins are similar to the idealised system shown in Figure 3.9a, but are more complicated in smaller, shallower basins that have an open connection to the larger basins. In reality the ocean tides are a complex interaction between co-oscillating forced Kelvin waves (amphidromic systems) in large basins, with reflected forced- and free-waves that can be amplified by complex basin topography (Figure 3.12).

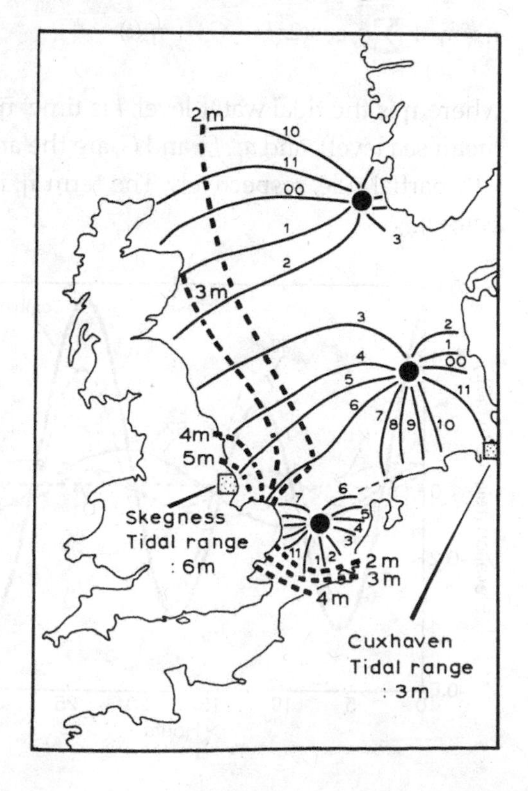

Figure 3.12 Map showing amphidromic systems in the North Sea. Tides along the North Sea coast are a result of the interactions of three Kelvin waves with different heights (tide ranges). At a particular coastal location the observed tide range depends on which is the influential Kelvin wave and how far it is away from the amphidromic point. The numbers refer to the hour of the high tide. (From Pethick, 1984.) (Copyright © 1984 Edward Arnold, reproduced with permission.)

3.5 Predicting tides

The periodicities of the tide-generating forces are obtainable from the Equilibrium Theory and they can be used in solving Laplace's equations (Dynamic Theory) to forecast tides for locations well away from basin margins. However, the accuracy of this approach diminishes markedly near the coast where tidal energy is reduced due to friction with the sea bed. Since we are concerned with the coastal zone, we need to adopt an empirical approach based on past tidal measurements in order to predict future tides.

3.5.1 Harmonic analysis

The fact that real tides are highly periodic means that they can be modelled as the summation of several **partial tides**. This is illustrated graphically in Figure 3.13 where five partial tides are shown together with the combined tide, which is the numerical sum of the partial tides. Unlike the real tide, partial tides have a single amplitude and frequency that can be linked to physical processes. For example, the fundamental semi-diurnal tide produced by the Earth–Moon system rotating on its barycentre has a frequency of 0.081 hour^{-1}. The partial tides are **harmonics** because they can be expressed in terms of a cosine (wave-like) function that includes time. A measured tide record can be expressed mathematically as

$$\eta = \bar{\eta} + \sum_{i=1}^{n} a_i \cos(2\pi f_i t - G_i) + \eta_r(t) \qquad (3.6)$$

where η is the tidal water level, t is time, $\bar{\eta}$ is the time-averaged water level (or mean sea level), and a_i, f_i, and G_i are the amplitude, frequency and phase of the i-th partial tide, respectively. The term η_r is the residual water level (discussed below).

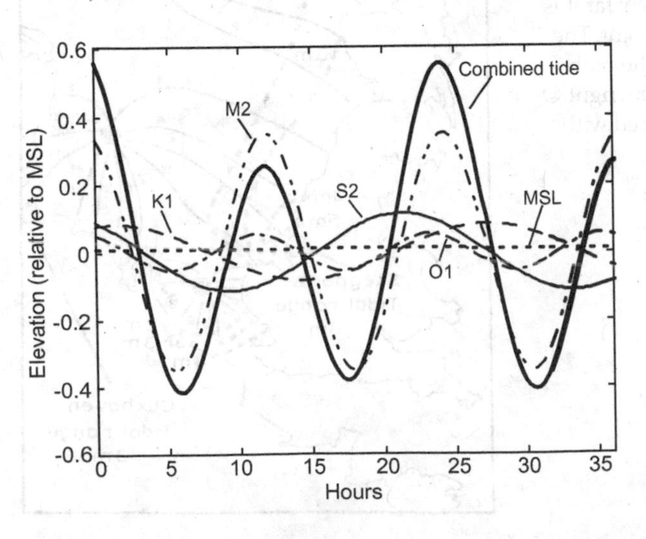

Figure 3.13 Graph showing five partial tides with differing amplitudes, frequencies and phases that, when summed, produce the combined tide (bold line). The partial tides are listed in Table 3.3.

In order to predict future tides at a given location using Equation 3.6 it is necessary to know the amplitudes, frequencies and phases of the partial tides. The frequencies of the partial tides are fixed, and can be obtained from the Equilibrium and Dynamic theories. However, the amplitudes and phases vary with location and must be obtained from field measurements. Tidal prediction by harmonic analysis typically involves a least-squares fitting of the principal partial tides to a tide record for each locality. The amplitudes, frequencies and phases are then substituted into Equation 3.6 to forecast the tide into the future. Tidal harmonics of the past can also be reconstructed from sediments, termed **tidalites**, which are deposited under certain tidal states and preserved in the geological record (Box 3.2).

Box 3.2 Tidalites and tidal harmonics

Sediments deposited under the influence of tidal processes (tidalites) are often rhythmically-bedded with alternating coarse (sand) and fine (clay/silt) layers. A single coarse layer and the fine layer that conformably overlies it is called a tidal **couplet**. The coarse layer is deposited by water moving on either the incoming (flood) or outgoing (ebb) tide and the fine upper layer is deposited at slack water. Variations in couplet thickness reflect both variations in sediment supply and, where tidalites accumulate and are preserved over very long time scales, variability in the orbital harmonics that drive the

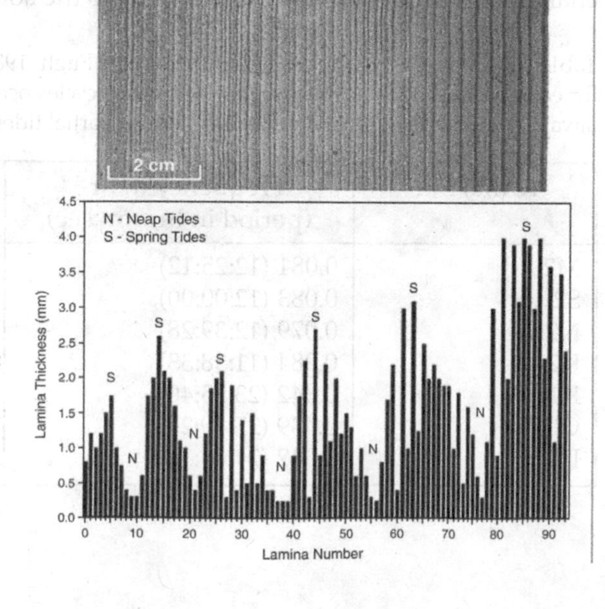

Figure 3.14 Photo of foreset laminae from the Hindostan Whetstone Beds (Pennsylvanian) of Indiana, USA, showing neap–spring cyclicity. The graph shows measurements of individual sandy laminae thicknesses that pick out these cycles. (From Coughenour *et al.*, 2009; after Kvale *et al.*, 1999.)

tides. These harmonics can be determined by measuring the thickness of the couplets and performing a Fourier transform analysis which takes the sequence of numbers in a dataset and decomposes them into cycles (or harmonics) of different lengths (or wavelength). When plotted graphically, the higher the spectral density peak the more important is that particular harmonic's cycle length in explaining the data (Figure 3.14). These spectral peaks can be matched with tidal cycles of known length.

Frouin *et al.* (2006) used Fourier analysis to examine tidal rhythmites within the Seine estuary, northern France, and found a close correspondence to the anomalistic monthly cycle and the fortnightly spring–neap cycle. Kvale *et al.* (1995) argue that at least 95 per cent of tidal events need to be preserved in tidalites in order to deduce changes in orbital period, but identifying the major tidal harmonics can be achieved with just a few hundred data points.

There are seven fundamental partial tides (Table 3.3) that will in most cases forecast the tide to within 10 per cent of its true value for the coming month. The more complicated the shelf morphology, and the further into the future one wishes to forecast, the larger the number of partial tides required. Prediction of future tides at coastal ports involves analysing one month of the historic tidal record. In complicated situations such as in estuaries, up to 60 partial tides are used, which requires a record length of one year. The most accurate tidal predictions are based on partial tides calculated from historical records spanning 19 years, and which include the 18.6-year tidal periodicity produced by the change of lunar declination with respect to the solar declination.

Table 3.3 List of the principal partial tides. (After Pugh, 1987.) The number in the symbol for each partial tide indicates roughly how many cycles occur each day. Diurnal partial tides always contain the number 1 and semi-diurnal partial tides contain the number 2.

Partial tide	Frequency in hr^{-1} (period in hr:min:sec)	Description
M2	0.081 (12:25:12)	principal lunar
S2	0.083 (12:00:00)	principal solar
N2	0.079 (12:39:28)	elliptical lunar
K2	0.084 (11:58:33)	lunar/solar declinations
K1	0.042 (23:55:40)	principal lunar/solar
O1	0.039 (25:49:26)	principal lunar
P1	0.042 (24:04:32)	principal solar

3.5.2 The residual tide: Radiational tides and storm surge

Often, when predicting tides using harmonic analysis, there will be a small but significant difference between the predicted and observed tidal records. This difference is referred to as the **residual tide** and is represented by the last term on the right-hand side of Equation 3.6. The residual tide can display both periodic and random fluctuations in water level. Sometimes periodic fluctuations in the residual tide are due to an insufficient number of partial tides employed in the prediction, but it can be also due to factors unrelated to tidal forces. One such factor is **radiational forcing**, where weather cycles cause atmospheric pressure to fluctuate, thus loading and unloading the ocean surface. As a general rule-of-thumb sea level will rise (fall) 1 cm for every 1 millibar fall (rise) in air pressure, and is termed the **inverse barometer effect**. Diurnal and annual periodicities in the residual tide are often due to radiational forcing.

Non-periodic fluctuations in the residual tide are typically associated with the effect of wind on sea level. If the wind vector is directed shoreward it can pound water against the coast, causing an increase in the residual tide. The magnitude of the wind effect is related to fetch length, wind strength and duration, and is strongest during storms, hence the term **storm surge** (Figure 3.15). The strongest winds are often associated with low-pressure systems, thus storm surges combine with the inverse barometer effect to produce a residual tide that can lead to widespread coastal inundation (Case Study 3.2).

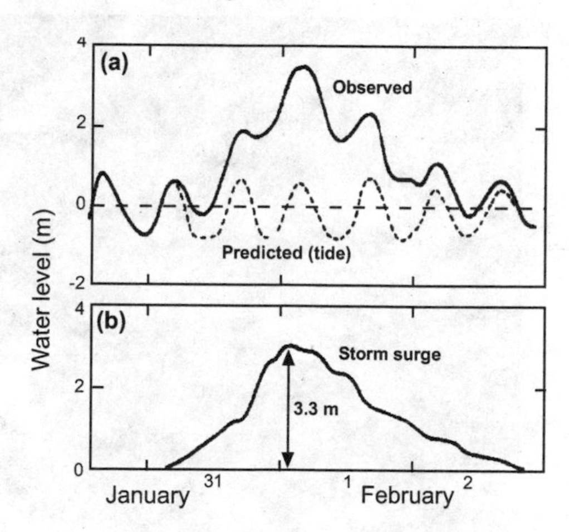

Figure 3.15 (a) Graph showing the predicted (dashed line) and observed (solid line) water level at the Hook of Holland in January–February 1953. (b) Graph showing the residual tide, which is attributed to a storm surge of 3.3 m. (Modified from Wemelsfelder, 1953.)

Case Study 3.2 Deadly storm surges

Hurricanes (also known as typhoons or tropical cyclones) are low-pressure systems that develop in low latitudes above areas of very warm sea surface temperatures (> 26.5 °C). As tropical storms grow in intensity, air pressure above the sea surface decreases leading to sea-level rise through the inverse barometer effect. In addition, wind speed increases. As a result, hurricanes are characterised by strong onshore winds and elevated sea levels during the associated surge. Storm surges are also amplified where they coincide with high tide, as they did during the January 1953 event that flooded large areas of the southern North Sea including eastern England and the Netherlands, killing around 2400 people (Figure 3.15).

Similar conditions were also present during Hurricane Katrina (August 2005) which made landfall on the coast of Louisiana (Gulf of Mexico). Here, elevated sea levels caused by very strong winds (reaching 280 km hr^{-1}) and deep low pressure (902 mb) helped overtop the levees protecting the city of New Orleans, leading to catastrophic flooding. The magnitude and impacts of the storm surge were very variable, depending on location with respect to the hurricane eye, but varied between 3 and 8 m above normal levels. As well as coastal flooding, barrier islands were breached, beaches eroded, and sediments carried inland (Figure 3.16). The economic cost of the disaster was estimated at $81.2 bn (2005) and over 1800 people were killed.

Figure 3.16 Photograph showing the impact of Hurricane Katrina on Bay St Louis, Mississippi, USA. (Copyright © National Oceanic and Atmospheric Administration, USA.)

3.6 Classification of tides

A simple scheme was introduced in Chapter 1 that classified tidal environments according to the spring tide range (Figure 1.5). We can now introduce a more complex classification based on tidal period. Examples of tidal records from four coastal ports are shown in Figure 3.17. The two end-members are **semi-diurnal** and **diurnal tides**. Between these end-members lie **mixed tides**, which display both diurnal and semi-diurnal periodicities. Tidal records can be classified quantitatively using the **tidal form factor** F

$$F = \frac{a_{K1} + a_{O1}}{a_{M2} + a_{S2}} \tag{3.7}$$

where a is the amplitude of the partial tides identified by the subscripts (Defant, 1958). The tidal form factor is effectively the ratio of the amplitudes of the major diurnal and semi-diurnal partial tides. The ranges of F associated with each tidal type are:

- $F = 0.00\text{-}0.25$ Semi-diurnal tide
- $F = 0.25\text{-}1.50$ Mixed, dominantly semi-diurnal tide
- $F = 1.50\text{-}3.00$ Mixed, dominantly diurnal tide
- $F > 3.00$ Diurnal tide

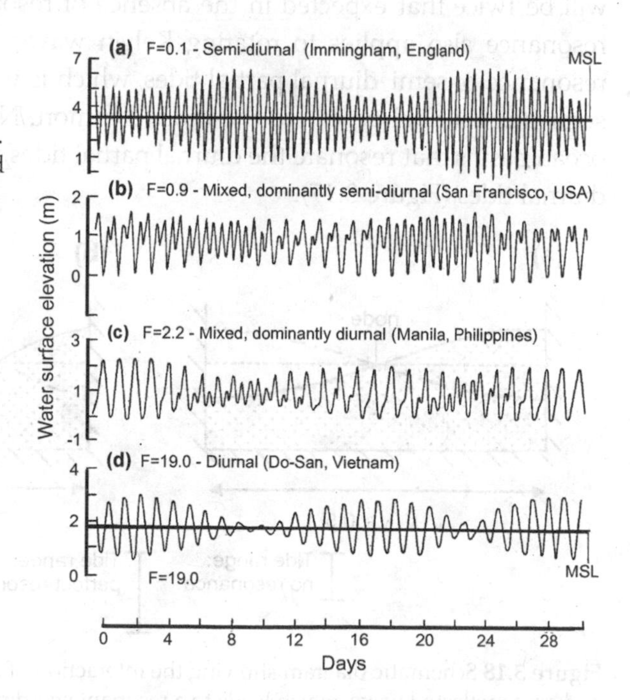

Figure 3.17 Monthly tidal records from four coastal ports showing examples of semi-diurnal, mixed and diurnal tides. The tidal form factor F is defined in Equation 3.7. (Modified from Defant, 1958.)

Worked Example 1

The type of tide found at any one location depends on the local predominance of semi-diurnal versus diurnal partial tides, which arises as a result of how tides of given wavelength respond to ocean basins of variable length. There is a maximum amplification of the tide for basin lengths that are integer multiples of a quarter of the tidal wavelength. It turns out that the major ocean basins have a dimension that preferentially amplifies the semi-diurnal partial tides, which is why it is globally the most common tide. This amplification is termed **tidal resonance**.

The mechanism by which tidal resonance takes place is shown in Figure 3.18 where the basin length λ is half the tidal wavelength *L*, in other words, where one half of one wavelength fits into the basin. The partial tide is reflected when it encounters the basin margin. The incident (solid curve) and reflected (dash-dot curve) waves interact to produce a standing wave. In Figure 3.18a the single node of the standing wave is at the centre of the basin and two antinodes (the highest and lowest points of the wave) are at the basin margins. Note that the tide range is zero at the node, but at the antinodes the tide range is twice the value it would be if there was no resonance. In Figure 3.18b we have a similar situation, but here the wavelength and basin length are equal so there will be two nodes and three antinodes. Again the tide range at the basin margins will be twice that expected in the absence of resonance. The concept of tidal resonance also applies to rotating Kelvin waves. Most ocean basins favour resonance of semi-diurnal partial tides, which is why semi-diurnal and mixed semi-diurnal tidal types are the most common. Nevertheless there are some ocean basins that resonate the diurnal partial tides, producing diurnal or mixed diurnal tides (Figure 3.17).

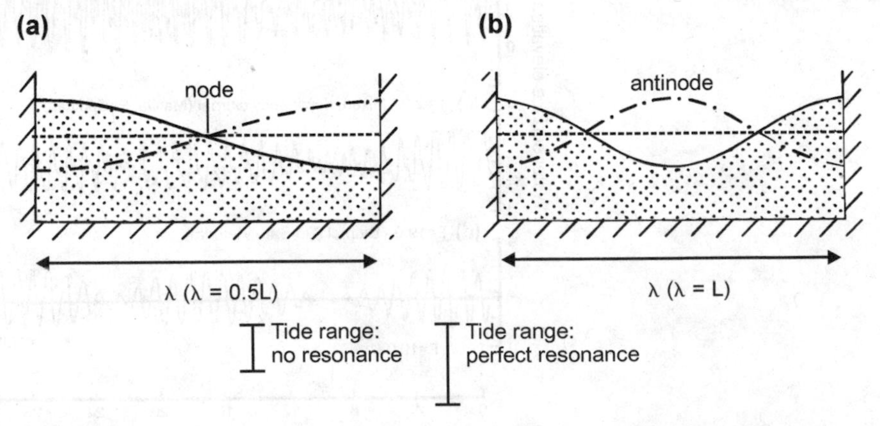

Figure 3.18 Schematic diagram showing the interaction of a forced wave (the partial tide) and a free reflected wave, which leads to a resonant standing wave. Two different basin lengths are shown: (a) basin length = half a tidal wavelength; and (b) basin length = one tidal wavelength. (Modified from Pond and Pickard, 1983.)

SUMMARY

- The magnitude of the tide-generating force exerted by the Moon and Sun is directly related to their mass and inversely related to their distance from the Earth. The tide-generating force of the Sun is 46 per cent that of the Moon.
- The Equilibrium Theory of tides is based on the premise that the tangential component of the tide-generating force (i.e. the tractive force) causes the water to form two tidal bulges on opposite sides of the Earth and aligned with the Moon and Sun. The Equilibrium Tide is the resulting rise and fall of the water level as the Earth spins on its polar axis beneath the bulges. The Equilibrium Theory does not account for lags between the timing of high tide and the transits of the Sun and Moon, or for variation in observed tidal range.
- The Dynamic Theory of tides is based on the premise that the tides are Kelvin waves rotating around amphidromic systems whose dimensions correspond to the size of ocean basins. The Dynamic Tide is the resulting rise and fall of the water as the crest and trough of the Kelvin waves pass by. The tidal period is determined by the size of the ocean basin, which favours the resonance of diurnal or semi-diurnal partial tides. The Dynamic Theory cannot deal with the effect of complex coastal topography.
- In order to predict tides in coastal waters an empirical approach is required based on harmonic analysis, which uses past tidal records to determine the amplitude and phase of partial tides obtained from the Equilibrium Theory. Once the amplitude, frequency and phase of the partial tides have been determined, the modelled partial tides are summed to forecast the total tide into the future.

Reflective questions

These questions are designed to test your comprehension of material covered in this chapter. Suggested answers to these questions can be found on this book's website.

3a. In your own words, describe the significance of the barycentre for tidal generation.

3b. Explain the reasons why spring and neap tides form and how they change throughout the year.

3c. How can our knowledge of amphidromic points be used to predict tidal behaviour around an ocean basin?

3d. Consider the ways in which tidal range can have an influence on coastal sediments and geomorphology.

Further reading

Cartwright, D.E., 1999. *Tides: A Scientific History*. Cambridge University Press, Cambridge. (A fascinating account of the historical developments leading to modern tidal theory and observation.)

Komar, P.D., 1998. *Beach Processes and Sedimentation*. (2nd edn.) Prentice Hall. (Chapter 4 of this text presents a very readable account of tides that goes into more depth than the level presented here.)

Open University, 2000. *Waves, Tides and Shallow-Water Processes*. Butterworth-Heineman, Oxford. (Chapter 2 of this text gives a very readable account of tides.)

Pugh, D.T., 1987. *Tides, Surges and Mean Sea Level*. John Wiley and Sons, London. (Simply the best book around on tidal processes, but certainly not an introductory text. This is available free online at http://eprints.soton.ac.uk/19157/1/sea-level.pdf)

CHAPTER 4

WAVES

AIMS

This chapter presents the major processes associated with the formation, development and breaking of waves, and are described with respect to their geometric properties. Wave development in deep and intermediate water is first examined, and then the changes in wave geometry and their eventual breaking in shallow water. Wave processes in deep, intermediate and shallow water provide the context for understanding the contribution of waves to coastal morphological change.

4.1 Introduction

Along most coastlines, waves represent the dominant source of nearshore energy. For this reason, the global distribution of wave environments (Figure 1.5) is useful in classifying coastal environments because it broadly identifies the amount of wave energy available for coastal sediment transport. Part of the incoming wave energy is reflected at the shoreline and propagated back to the open sea. Most of the incoming energy, however, is transformed through its interaction with the coastline to generate nearshore currents and sediment transport, and is

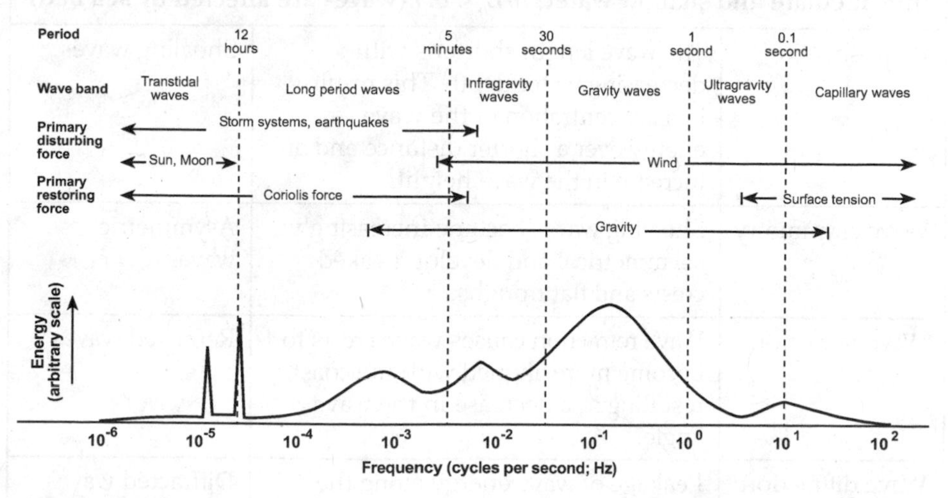

Figure 4.1 Schematic representation of the energy contained in the surface waves of the oceans. (Modified from Kinsman, 1984.)

ultimately the driving force behind morphological change. Knowledge of wave dynamics is therefore fundamental to understanding coastal morphology.

Ocean waves can be classified by wave period *T*, or by wave frequency *f*, which is the reciprocal of the period ($f = 1/T$). Alternatively, we can classify waves by the disturbing force that generates them, or by the restoring force that dampens the wave motion. Figure 4.1 shows a schematic wave spectrum, which plots wave energy as a function of frequency, and indicating different types of ocean waves. Those that are addressed in this chapter are gravity waves, which are the most energetic of the wave types and are part of the wind-forcing wave spectrum. **Gravity waves** have periods of 1–30 seconds and frequencies of 0.033–1 cycles per second (or Hertz [Hz]). They are generated by wind and their main restoring force is gravity. Most of the wave terms described in this chapter are shown in Table 4.1.

Table 4.1 Overview of wave processes and their terminology (h = water depth, L_o = deep water wave length, H_o = deep water wave height).

Process	Description of process	Types of waves
Deep water: $h/L_o > 0.5$ (waves are unaffected by sea bed)		
Wave generation	Waves are generated by wind. Wave height and period increase with increase in wind speed and duration.	Sea
Wave dispersion	In deep water long-period waves travel faster than short-period waves. This results in a narrowing of the wave spectrum.	Swell
Intermediate and shallow water: $h/L_o < 0.5$ (waves are affected by sea bed)		
Wave shoaling	The wave length shortens with decreasing water depth. This results in a concentration of the wave energy over a shorter distance and an increase in the wave height.	Shoaling waves
Wave asymmetry	Shoaling waves become increasingly asymmetrical and develop peaked crests and flat troughs.	Asymmetric waves
Wave refraction	Wave refraction causes wave crests to become more aligned with the coast, resulting in a decrease in the wave angle.	Refracted waves
Wave diffraction	Leakage of wave energy along the wave crests into shadow areas.	Diffracted waves

Table 4.1 continued

Surf zone: $h < H_o$ (regular waves) or $h < 2H_o$ (irregular waves)		
Wave breaking	When the horizontal velocities of the water particles in the wave crest exceed the wave velocity, the water particles leave the wave form and the wave breaks.	Spilling, plunging, surging breakers
Energy dissipation	As breaking/broken waves propagate through the surf zone, they progressively decrease in height due to wave energy dissipation.	Surf zone bores
Wave reflection	On steep beaches a significant part of the incoming wave energy is not dissipated in the surf zone, but is reflected back to sea.	Standing waves
Infragravity wave energy	A significant part of the wave energy in the surf zone is at very low (i.e. infragravity) frequencies. Infragravity waves are particularly energetic during storms.	Infragravity waves, long waves, edge waves
Swash zone: $h = 0$		
Runup	The maximum water level attained on a beach is higher than the still water level. This vertical displacement of the water level is known as runup and consists of a steady (wave set-up) and a fluctuating component (swash).	Wave runup, swash, wave set-up

4.2 Characteristics and analysis of natural waves

It is first important to distinguish between regular (or monochromatic) and irregular (or random) waves. This is because theoretical discussions on wave processes most commonly assume regular wave motion, whereas natural waves are characteristically irregular. The motion of **regular waves** is periodic, so their motion is repetitive over time and space, as is shown in Figure 4.2. Regular waves can be described in terms of a single representative wave height H, wave length L and wave period T. The **wave height** is the difference in elevation between the wave crest and wave trough, the **wave length** is the distance between successive crests (or troughs), and the **wave period** is the time it takes for the wave to travel a distance equal to its wave length. An important characteristic of natural waves is that they are highly irregular and that a range of wave heights and periods are present (Figure 4.3a).

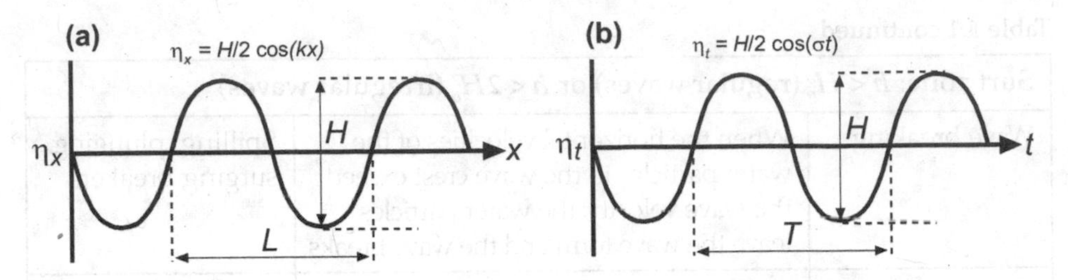

Figure 4.2 Schematic diagram showing a regular wave train. (a) The spatial variation in water level η_x is measured at a single moment in time along the direction of wave travel. From such data the wave length L can be derived. (b) The temporal variation in water level η_t is measured at a single location in space over a representative time period. Such data can determine the wave period T. The wave height H can be derived from both types of data.

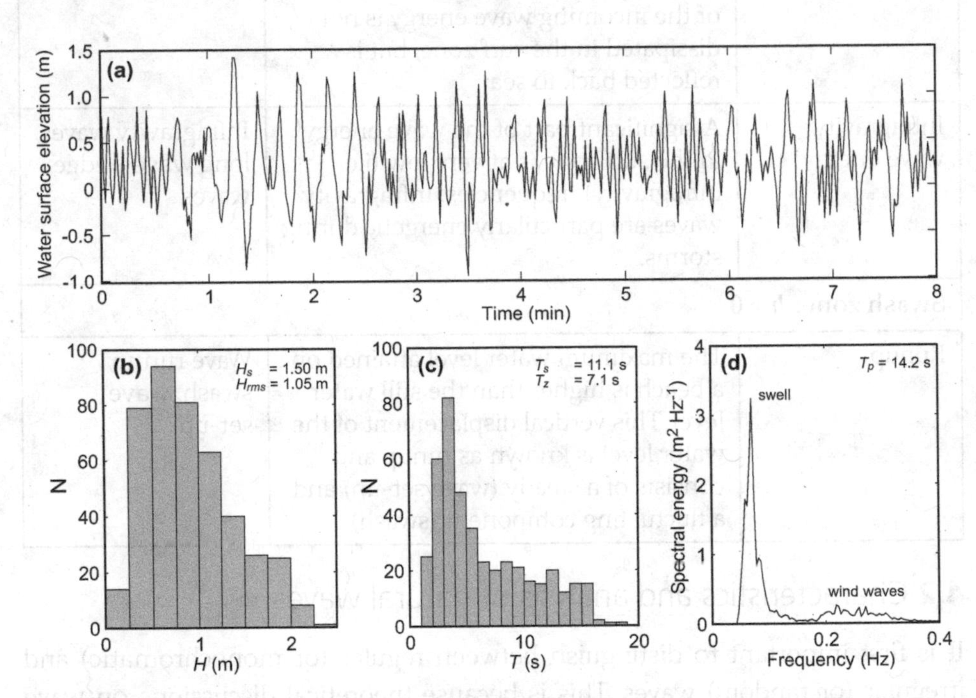

Figure 4.3 Analysis of wave data collected at 48 m water depth off the coast of Perth, Western Australia: (a) time series of 8 minutes of water surface elevation data; (b) frequency distribution of the wave heights in the data record; (c) frequency distribution of the wave periods in the data record; and (d) wave spectrum. The wave data represent a mixture of short-period wind waves and long-period swell waves, accounting for the very irregular nature of the time series, the wide distribution of wave periods and the bimodal wave spectrum.

To quantitatively describe the wave conditions of **irregular waves**, statistical techniques are required. **Wave-by-wave analysis** is a common approach which identifies the individual waves in the wave record and determines representative wave parameters from the sub-set of wave heights and periods (Figure 4.3b, c). Wave parameters commonly used are:

- H_s (or $H_{1/3}$) – The **significant wave height** is the average wave height of one-third of the highest waves in the record. Significant wave height approximately corresponds to visual estimates of wave heights, and has been used by coastal engineers for practical design purposes.
- H_{rms} – The **root-mean-square wave height** is obtained by taking the square root of the mean squared wave height, using all the waves in the wave record. As a rule of thumb, $H_s = 1.41H_{rms}$.
- T_z – The **mean wave period** is the mean period of all waves in the wave record. It is also referred to as the zero-crossing wave period.
- T_s (or $T_{1/3}$) – The **significant wave period** is the mean wave period of one-third of the highest waves in the record.

An alternative method to describe the properties of irregular waves is through **spectral analysis**. This method (described in Box 3.2) can identify the dominant wave frequencies (or periods) in a wave record. Spectral analysis produces

Case Study 4.1 Generating extreme waves

Extremely high waves are generated by strong winds that blow over long time periods and across the widest oceans. The largest wave ever reliably reported had an estimated height of 34 m and was encountered on 7 February 1933 by the USS *Ramapo* travelling from Manila to San Diego (Bascom, 1980). However, 45 per cent of ocean waves are less than 1.2 m high, 80 per cent are less than 4 m high, and only 10 per cent are greater

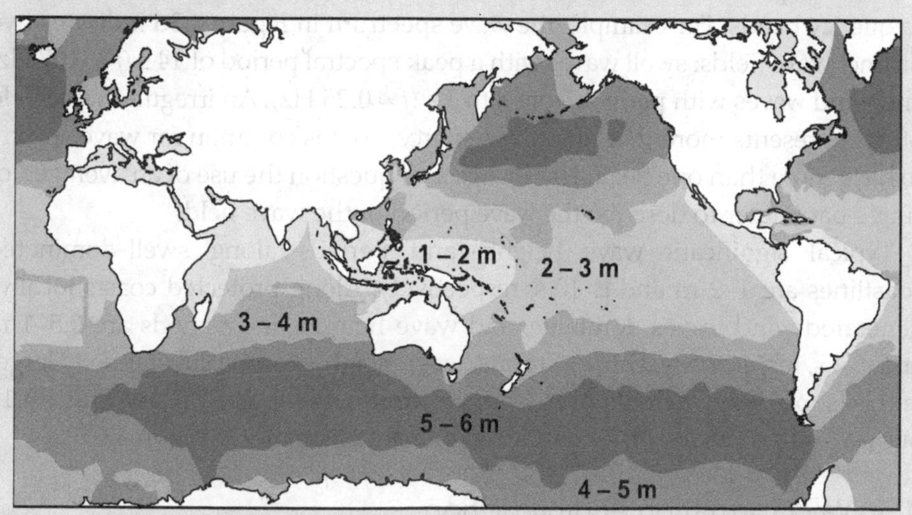

Figure 4.4 Global values of significant wave height which will be exceeded 10 per cent of the time. (From Short, 1999; modified from Young and Holland, 1996.) (Copyright © 1999 John Wiley & Sons, reproduced with permission.)

than 7 m (Kinsman, 1984). Figure 4.4 shows a map of the global distribution of significant wave height. The most energetic wave regions are the Indian Ocean, South Pacific and North Atlantic, where a wave height of 5 m is exceeded 10 per cent of the time.

Apart from storm forcing, extreme waves may also occur due to constructive interference involving different wave fields. For example, wave refraction around an island results in two wave trains wrapping around the island. There will be a region in the lee of the island where the two wave trains meet and where their addition results in relatively rough conditions with large waves. Large singular waves (defined as more than twice that of H_s), sometimes referred to as 'king waves', 'freak waves' or 'rogue waves' also occur and these waves are often held responsible for sweeping sea fishermen off rock ledges. These extreme waves develop when two (or more) large waves from different wave fields coincide, resulting in a new wave with a much increased height. The occurrence of such waves is virtually unpredictable (Chalikov, 2009). Very large waves, including those related to tsunami, are significant for the transport of very large boulders (see Chapter 10).

a **wave spectrum** (Figure 4.3d) which shows the proportion of wave energy received at different frequencies. The **peak spectral period** T_p refers to the level at which maximum wave energy is achieved across the wave spectrum. The wave spectrum also shows the partitioning of wave energy over distinct frequency bands. For example, the wave spectrum in Figure 4.3d indicates two distinct wave fields: swell waves with a peak spectral period of 14 s ($f \approx 0.07$ Hz) and wind waves with periods from 3 to 5 s ($f \approx 0.25$ Hz). An irregular wave field often represents more than one wave source, so it is common for wave spectra to have more than one peak. This brings into question the use of an averaged or single parameter to describe the wave period of the wave field.

Typical significant wave heights and periods along swell-dominated coastlines are 1–2 m and *c.* 10 s, respectively. Along protected coasts locally-generated wind waves dominate, and wave heights and periods are 0.5–1 m and *c.* 4 s, respectively. During storms, wave heights can be substantially larger and offshore wave heights over 10 m are routinely reported (Case Study 4.1), but waves that break at the coast are significantly smaller (Section 4.6.5).

4.3 Wave formation and prediction

Waves are generated by the transfer of wind energy on to a water surface, and the growth of wind waves is described by the combined Miles-Phillips mechanism. This involves an initial slow growth phase that accounts for wave formation on a calm water surface, and then an ongoing rapid growth of the waves due the interactive coupling between wind and waves. In a developing wave field, wave height and wave period steadily increase (i.e. experience positive feedback).

Wave growth is not infinite, however, and in the open ocean is limited by the ratio between wave height and wave length. This ratio is the **wave steepness** H/L and increases progressively during wave formation. When waves reach their limiting steepness ($H/L \approx 1/7$ in deep water) they start to break in the form of **white caps**. An equilibrium can eventually be achieved when the energy losses by wave breaking are balanced by the addition of new energy transferred from the wind to the waves. Such an equilibrium wave field is referred to as a **fully arisen sea**. Its development may take several days or hundreds of kilometres of wave travel under strong wind conditions, but may develop relatively quickly (less than 10 hours) under mild winds.

The characteristics of locally-generated waves can be predicted based on existing wind conditions (it is less straightforward to predict wave characteristics

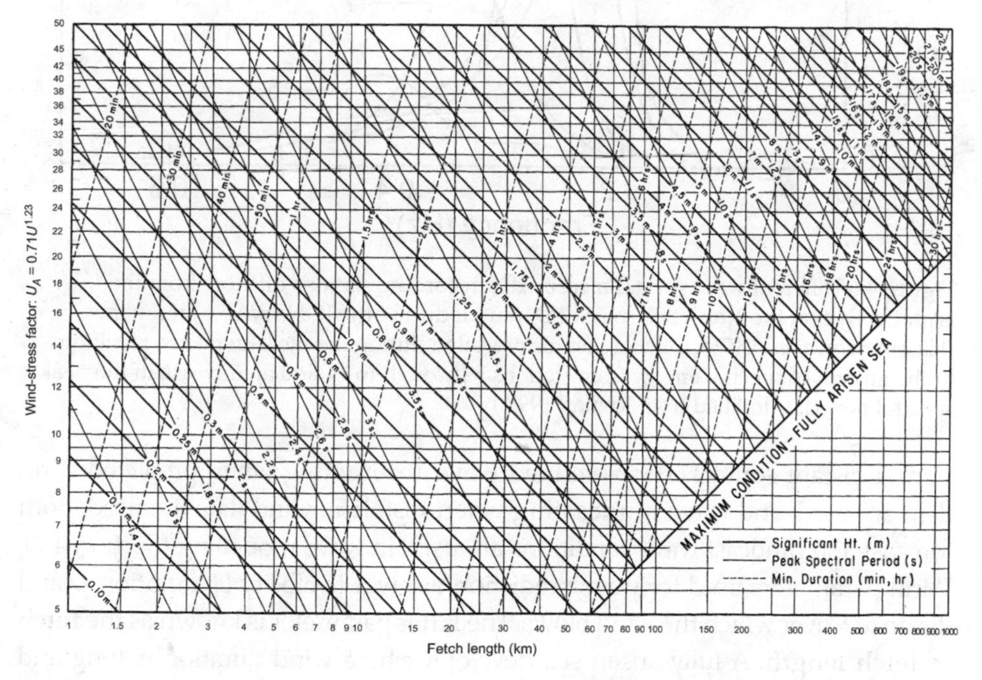

Figure 4.5 Nomogram to predict significant wave height H_s and peak spectral period T_p from wind speed U, wind duration D and fetch F. Note that before using the diagram, the measured or estimated wind speed U has to be converted to a wind-stress factor U_A given by $U_A = 0.71U^{1.23}$. (From CERC, 1984.) (Reproduced with permission from US Army Corps of Engineers.)

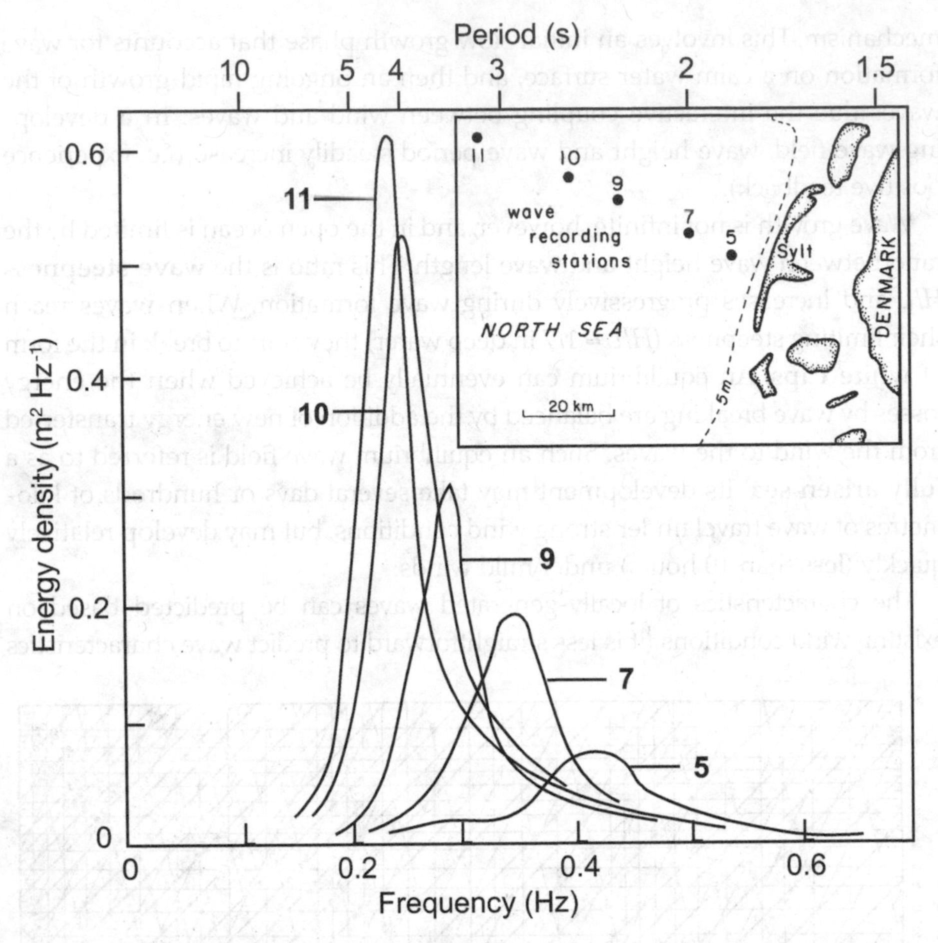

Figure 4.6 Different wave spectra measured at recording stations offshore from the island of Sylt on the North Sea coast of Denmark during the JONSWAP experiments (Hasselmann *et al.*, 1976). Offshore winds prevailed during the measurements, resulting in the growth of spectral energy from stations 5 through to 11 and an increase in the peak spectral period. (Modified from Komar, 1998.)

from a distant source). This is termed **wave forecasting** when predicted wind data are used, and **wave hindcasting** when historical wind data are used. Both forecast and hindcast wave conditions are used in wave modelling (Section 4.7). Wind parameters used for wave prediction are wind velocity *U*, duration *D* and distance *F* over which the wind blows. The latter parameter is known as the **fetch** or **fetch length**. A fully arisen sea develops where wind duration is long and fetch is large. If the fetch is too short, as in the case of a small lake or a sheltered coastal environment, the wave conditions are said to be **fetch-limited**.

The simplest method for wave prediction is the significant wave approach, abbreviated to the *S-B-M* method after the names of its developers (Sverdrup and Munk, 1946; Bretschneider, 1952). Using the *S-B-M* method, the significant wave height H_s and peak spectral period T_p of a wave field can be predicted from *U*, *D* and *F* using a nomogram (Figure 4.5). For example, for a wind speed of 10 m s^{-1}, unlimited fetch and wind duration of 10 hours, the resulting significant wave height and peak spectral period are 2.2 m and 7.3 s, respectively. For the same wind speed, unlimited wind duration and a fetch length of 20 km, the resulting significant wave height and peak spectral period are 0.87 m and 3.9 s, respectively.

The wave spectrum provides a more comprehensive description of an entire wave field than a single measure of wave height and period alone. Therefore, a more sophisticated method to predict wave characteristics from wind conditions is the wave spectrum approach, which characterises the growth of a wave field according to the development of its wave spectrum over time or distance. One of the most commonly used spectral formulations is the **JONSWAP spectrum** based on measurements conducted in the North Sea (Hasselmann *et al.*, 1976). Wave spectra of a developing wave field indicate a progressive increase in wave energy level and period (Figure 4.6). In addition, the wave field becomes more narrow-banded, meaning that the range of wave frequencies (periods) present in the wave field decreases. As a result, the waves become more organised and regular the longer the wind blows.

4.4 Linear wave theory

Wave theories are mathematical formulations that predict the change in wave properties, such as water particle velocity, wave height and wave energy, with water depth. They allow the derivation of nearshore wave characteristics from offshore wave conditions, and are a valuable tool for coastal engineering applications. The simplest and most widely used theory is **linear wave theory**, also known as **Airy wave theory**, named after the mathematician George Biddell Airy (1801–92). The linear wave theory adequately describes important wave processes such as shoaling and refraction. Since shoaling and refraction are significant for nearshore processes, linear wave theory is discussed in detail here.

4.4.1 Dispersion equation

According to linear wave theory, the fluctuation of water surface elevation with time $\eta(x,t)$ is described by

$$\eta(x,t) = \frac{H}{2}\cos(kx - \sigma t) \qquad \textbf{(4.1)}$$

where x is the coordinate axis in the direction of wave propagation, t is time, H is wave height, $k = 2\pi/L$ is wave number (L is the wave length) and $\sigma = 2\pi/T$ is the wave radian or angular frequency (where T is the wave period). Linear wave theory considers natural waves to be simple sinusoids such as those shown in Figure 4.2.

An important relationship derived from linear wave theory is the **dispersion equation**, which expresses the relation between wave length and wave period

$$\sigma^2 = gk \tanh(kh) \qquad \textbf{(4.2)}$$

which can be rewritten as

$$L = \frac{g}{2\pi} T^2 \tanh\left(\frac{2\pi h}{L}\right) \qquad \textbf{(4.3)}$$

where h is water depth and g is gravitational acceleration. The tanh term is referred to as the hyperbolic tangent. Over a single wave period T, a wave travels a distance equal to its wavelength L. The speed at which the wave travels, the **wave phase velocity** C, is therefore given by the ratio L/T according to

$$C = \frac{L}{T} = \frac{g}{2\pi} T \tanh\left(\frac{2\pi h}{L}\right) \qquad \textbf{(4.4)}$$

The wave phase velocity is also referred to as the **wave celerity**. In all except very shallow water, wave celerity and wave phase velocity refer to the rate of energy transfer through movement of the wave form. It does not refer to the net movement of water particles nor of the wave form itself (see Section 4.4.4).

The dispersion equation (Equation 4.3) cannot be solved directly because it contains L on either side of the equation and therefore has to be solved using an iterative process. However, Fenton and McKee (1990) derived a function which is accurate to 1.7 per cent and is given by

$$L = L_o \left(\tanh\left[\left(\frac{4\pi^2 h}{gT^2} \right)^{3/4} \right] \right)^{2/3} \qquad \textbf{(4.5)}$$

where L_o refers to the deep water wave length (Equation 4.6). More accurate equations for approximating the wave length are available, but these tend to be more complicated.

4.4.2 Deep and shallow water approximations

Computers make calculations of the general dispersion relationship (Equations 4.2 and 4.3) relatively straightforward, especially if Equation 4.5 is used. In the past, however, simpler approximations to the dispersion relationship were desirable and these were made possible due to the characteristics of the tanh function shown in Figure 4.7. If $kh = 2\pi h/L$ becomes large (in relatively deep water), $\tanh(kh) \approx 1$, and Equations 4.3 and 4.4 reduce to

$$L_o = \frac{gT^2}{2\pi} \tag{4.6}$$

and

$$C_o = \frac{gT}{2\pi} \tag{4.7}$$

Equations 4.6 and 4.7 are known as the **deep water approximations** and are valid for $kh > \pi$ (or $h/L_o > 0.5$, *i.e.*, where water depth is greater than twice the wave length). The subscript 'o' denotes deep water conditions. However, if $kh = 2\pi h/L$ approaches zero (in relatively shallow water), $\tanh(kh) \approx kh$ and Equations 4.3 and 4.4 reduce to

$$L_s = T\sqrt{gh} \tag{4.8}$$

and

$$C_s = \sqrt{gh} \tag{4.9}$$

Equations 4.8 and 4.9 are known as **shallow water approximations**, shown by the subscript 's', and are valid for $kh < 0.1\pi$ (or $h/L_o < 0.05$, *i.e.*, where water depth is less than 1/20th of the wave length). Intermediate depth conditions prevail for $0.05 < h/L_o < 0.5$ and here the general wave equations (Equations 4.3 and 4.4), or an approximation such as Equation 4.5, must be used.

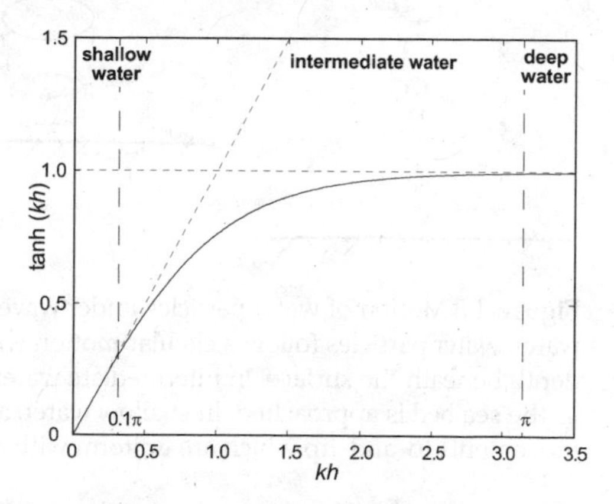

Figure 4.7 Properties of the tanh function. *kh* is defined in the text.

Wave conditions in deep, intermediate and shallow water can be predicted based on these equations. For example, consider a wave with a period of 10 seconds. In deep water, the wave length and velocity are only dependent on the wave period, and application of Equations 4.7 and 4.8 gives $L_o = 156$ m and

C_o = 15.6 m s^{-1}. Application of Equation 4.6 for an intermediate water depth of 10 m yields L = 91 m and C = 9.1 m s^{-1}. In shallow water, only the water depth controls the wave length, and for a water depth of 1 m and a wave period of 10 s Equations 4.8 and 4.9 yield L_s = 31 m and C_s = 3.1 m s^{-1}. These examples clearly show that waves get closer together and slow down as they propagate from deep to shallow water.

4.4.3 Wave orbital motion

The distinction between deep, intermediate and shallow water is not simply an artefact of the characteristics of linear wave theory (and in turn the tanh function), but has a physical explanation that becomes apparent when the **water particle motion** is considered. As waves propagate across the sea surface, the water particles beneath the wave undergo an almost closed circular path. The water particles move forward under the wave crest and seaward under the wave trough. Because the water particle velocities decrease with depth, the forward velocity at the top of the orbit is slightly greater than at the bottom. Consequently, there is a net drift of water in the direction of wave travel, known as **Stokes drift**. For our purpose, we can ignore Stokes drift and focus on how the characteristics of water particle motion vary with depth according to linear wave theory (Figure 4.8).

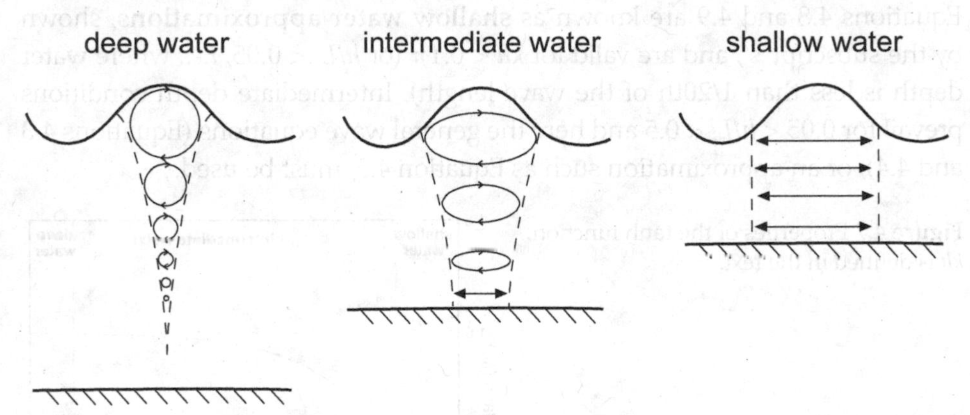

Figure 4.8 Motion of water particles under waves according to linear wave theory. In deep water, water particles follow a circular motion with the radius of the orbits decreasing with depth beneath the surface. In intermediate water, the orbits are elliptical and become flatter as the sea bed is approached. In shallow water, all water motion consists of horizontal movements to-and-fro which are uniform with depth.

In deep water the water particles under the wave move in a circular path with the diameter of the circles decreasing with increasing depth beneath the surface according to

$$d = He^{-kz} = He^{-2\pi z/L} \quad \textbf{(4.10)}$$

where d is diameter of the circular orbit and z is depth beneath the water surface. The maximum velocity of the rotating water particle u_m also decreases with increasing depth beneath the surface and is given by

$$u_m = \frac{\pi H}{T} e^{-kz} = \frac{\pi H}{T} e^{-2\pi z/L} \tag{4.11}$$

According to Equations 4.10 and 4.11, at depths greater than half the wave length ($h/L > 0.5$; the deep water region), the orbital diameter and velocities are less than 5 per cent of those at the sea surface. Therefore, it can be inferred that the wave motion in deep water is hardly experienced near the sea bed, and that deep water waves are not significantly affected by the presence of the sea bed.

In contrast, the wave motion extends to the sea bed at intermediate water depths and the surface waves 'feel' the presence of the bed. As a result, the water particles follow an elliptical path with the ellipses becoming flatter and smaller as the sea bed is approached (Figure 4.8). At the bed, the water particles merely undergo a horizontal to-and-fro motion. The excursion of this motion at the bed d_0 is given by

$$d_0 = \frac{H}{\sinh (kh)} \tag{4.12}$$

and the associated maximum flow velocity u_0 is

$$u_0 = \frac{\pi}{T} d_0 = \frac{\pi H}{T \sinh (kh)} \tag{4.13}$$

where the sinh term is referred to as the hyperbolic sine.

In shallow water all water motion consists of horizontal movements to-and-fro which are uniform with depth. Because in shallow water $\sinh(kh) = kh$, Equations 4.12 and 4.13 reduce to

$$d_0 = \frac{H}{kh} = \frac{HT}{2\pi} \sqrt{\frac{g}{h}} \tag{4.14}$$

and

$$u_0 = \frac{\pi}{T} d_0 = \frac{H}{2} \sqrt{\frac{g}{h}} \tag{4.15}$$

As kh always has values of 0.6 or less in shallow water, the excursion of the water motion will always be greater than the wave height.

When the waves' orbital motions are able to disturb the sea bed, sediment movement then becomes possible. The to-and-fro motion can lead to sea bed disturbance, increased sediment mobility and size sorting. These processes are accentuated as water shallows. Where the sea bed is sloping, which is the situation as waves approach the coast, sediment can be brought into suspension within the water column, accumulate in depressions on the sea bed, and/or undergo net transport towards the coast in the direction of wave energy transfer. Wave-influenced bedforms such as ripples may develop (see Section 5.6).

4.4.4 Wave energy and wave energy flux

We have already seen that there is only limited net movement of water associated with wave motion because the water particles follow almost-closed orbits in deep and intermediate water, and undergo to-and-fro motion in shallow water. However, the propagation of the wave form itself constitutes a transfer of energy over the sea surface. Waves have potential energy associated with the deformation of the water surface from a level plane (i.e. trough/crest of waves), and kinetic energy due to the orbital motion of the water particles. Applying linear wave theory, the two forms of energy are equal and the total **wave energy** E is given by

$$E = \frac{1}{8}\rho g H^2 \qquad \textbf{(4.16)}$$

where ρ is water density. The wave energy is expressed as the amount of energy per unit area (N m^{-2}) and is more correctly referred to as the wave energy density. Wave energy depends on the square of the wave height, so a doubling of wave height will result in a fourfold increase in wave energy. Because E is the amount of energy per unit area, the total energy associated with a long-period wave is greater than that of a short-period wave because the long-period wave has a larger wave length.

The rate at which wave energy is carried along by the moving waves, known as the **wave energy flux** P, is given by

$$P = EC_g \qquad \textbf{(4.17)}$$

where C_g is the wave energy speed and is known as the **wave group velocity**, because it represents the speed at which wave groups travel (Box 4.1). Wave groups do not travel at the same speed as individual waves and the wave group velocity C_g is related to the speed of individual waves C according to

$$C_g = Cn \qquad \textbf{(4.18)}$$

where n is given by

$$n = \frac{1}{2}\left[1 + \frac{2kh}{\sinh(2kh)}\right] \qquad \textbf{(4.19)}$$

The parameter n increases from 0.5 to 1 from deep to shallow water, in other words, deep water waves travel at twice the speed of the wave groups ($n = 0.5$), whereas shallow water waves propagate at the same speed as the wave groups ($n = 1$). Inserting Equation 4.18 into Equation 4.17 results in

$$P = ECn \tag{4.20}$$

which is more commonly used than Equation 4.17.

Box 4.1 Formation of wave groups

If two sets of waves (or wave trains) with slightly different wave lengths (and frequencies) are present at the same time, they will interfere and produce a single set of resultant waves (Figure 4.9). Where the wave trains are in phase (i.e. crests and troughs of both wave trains coincide), the wave amplitudes are added together, and the resultant wave has an amplitude of the sum of the two original waves. Where the wave trains are out of phase (i.e. crests of one wave train coincide with the troughs of the other, and *vice versa*), the amplitudes cancel out and the water surface displacement is minimal. The two wave trains thus interact, each losing its individual identity, and combine to form a series of **wave groups**, separated by regions almost free of waves.

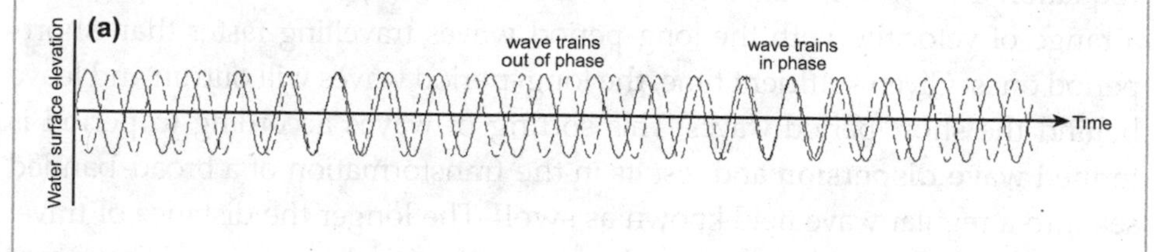

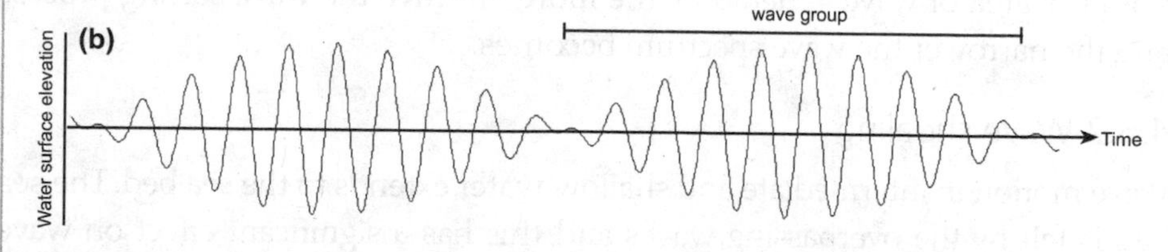

Figure 4.9 The merging of two wave trains, of different wave lengths but the same amplitude, to form wave groups. (Modified from Open University, 1994.)

The velocity of the resultant wave group can be derived from the properties of the two sets of waves that generated the group. Using the deep water wave equations and some manipulation, the wave group velocity C_g can be expressed in terms of the respective speeds C_1 and C_2 of the wave trains according to

$$C_g = \frac{C_1 C_2}{C_1 + C_2}$$

If C_1 is nearly equal to C_2 then the simple result $C_g = C^2/2C = C/2$ is obtained, where C is the average speed of the wave trains. In other words, the wave group travels at half the speed of the individual waves.

4.5 Wave processes outside the surf zone

A number of wave processes take place outside the surf zone, including wave dispersion, wave shoaling, development of wave asymmetry, wave refraction and wave diffraction. These processes modify the characteristics of offshore waves during their travel across the open ocean, and are discussed below.

4.5.1 Wave dispersion

A developing wave field has a broad-banded wave spectrum, indicating that a wide range of wave periods is present in the wave field. Such a wave field is referred to as **sea**. In deep water, wave velocity increases with wave period (Equation 4.7). In a broad-banded wave field, therefore, waves propagate at a range of velocities with the long-period waves travelling faster than short-period ones. Given sufficient time, the long-period waves will outrun and leave behind the short-period waves. This sorting of waves according to period is termed **wave dispersion** and results in the transformation of a broad-banded sea into a regular wave field known as **swell**. The longer the distance of travel from the area of wave generation, the more effective the wave sorting process and the narrower the wave spectrum becomes.

4.5.2 Wave shoaling

Wave motion in intermediate and shallow water extends to the sea bed. The sea bed is felt by the overpassing waves and this has a significant effect on wave motion. Some of the variations in wave properties that take place as waves move from deep to shallow water are summarised in Figure 4.10. It is notable that wave period is the only property that remains constant as waves move to shallow water.

The changes in wave length, wave group velocity, wave velocity and n that occur when waves propagate from deep to shallow water have a profound effect on wave height. The variation in height of shoreward-propagating waves

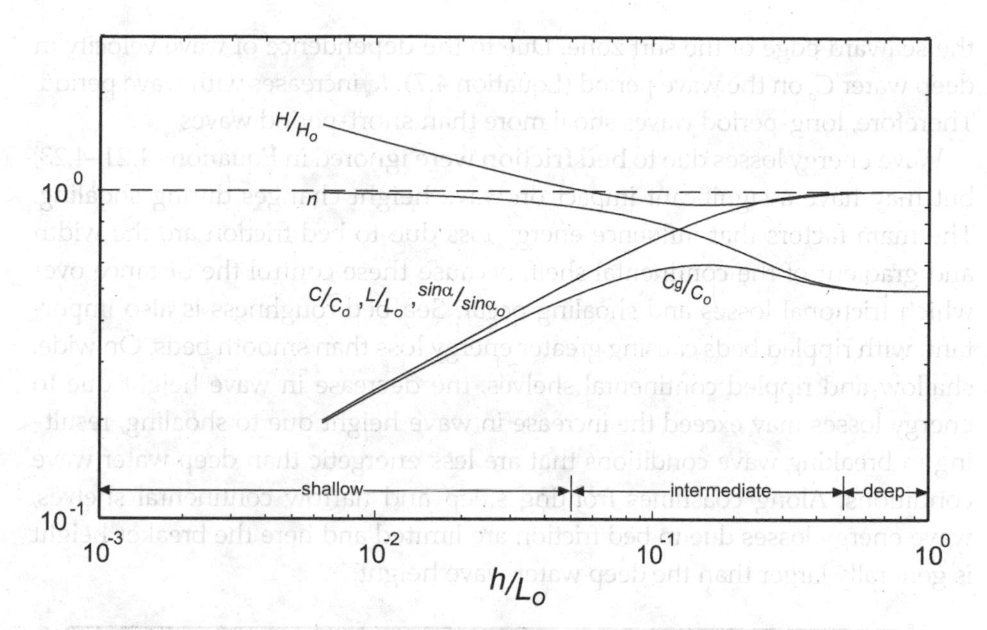

Figure 4.10 Shoaling transformations for linear waves as a function of the ratio of the water depth h and the deep-water wave length L_o.

can be calculated from consideration of the wave energy flux P. We assume that energy losses due to bed friction can be ignored, therefore the wave energy flux $P = ECn$ remains constant during wave propagation. This can be expressed as

$$P = (ECn)_1 = (ECn)_2 = \text{constant} \tag{4.21}$$

where the subscripts '1' and '2' indicate two different locations along the path of wave travel ($h_1 > h_2$). Substituting the wave energy $E = 1/8\rho g H^2$ in Equation 4.21 and re-arranging the result yields

$$H_2 = \left(\frac{C_1 n_1}{C_2 n_2}\right)^{1/2} H_1 \tag{4.22}$$

The ratio of the local wave height H (at depth h_2) to the deep water wave height H_o (at depth h_1) can now be derived from Equation 4.22

$$\frac{H}{H_o} = \left(\frac{1}{2n}\frac{C_o}{C}\right)^{1/2} = K_s \tag{4.23}$$

where K_s is referred to as the **wave shoaling coefficient**.

The ratio H/H_o according to Equation 4.23 is plotted in Figure 4.10 and illustrates that wave height initially decreases while entering intermediate water depths, followed by a rapid increase. This increase in wave height is known as **wave shoaling** and is particularly pronounced just before waves break at the seaward edge of the surf zone. Due to the dependence of wave velocity in deep water C_o on the wave period (Equation 4.7), K_s increases with wave period. Therefore, long-period waves shoal more than short-period waves.

Wave energy losses due to **bed friction** were ignored in Equations 4.21–4.23, but may have a significant impact on wave height changes during shoaling. The main factors that influence energy loss due to bed friction are the width and gradient of the continental shelf, because these control the distance over which frictional losses and shoaling occur. Sea bed roughness is also important, with rippled beds causing greater energy loss than smooth beds. On wide, shallow and rippled continental shelves, the decrease in wave height due to energy losses may exceed the increase in wave height due to shoaling, resulting in breaking wave conditions that are less energetic than deep water wave conditions. Along coastlines fronting steep and narrow continental shelves, wave energy losses due to bed friction are limited and here the breaker height is generally larger than the deep water wave height.

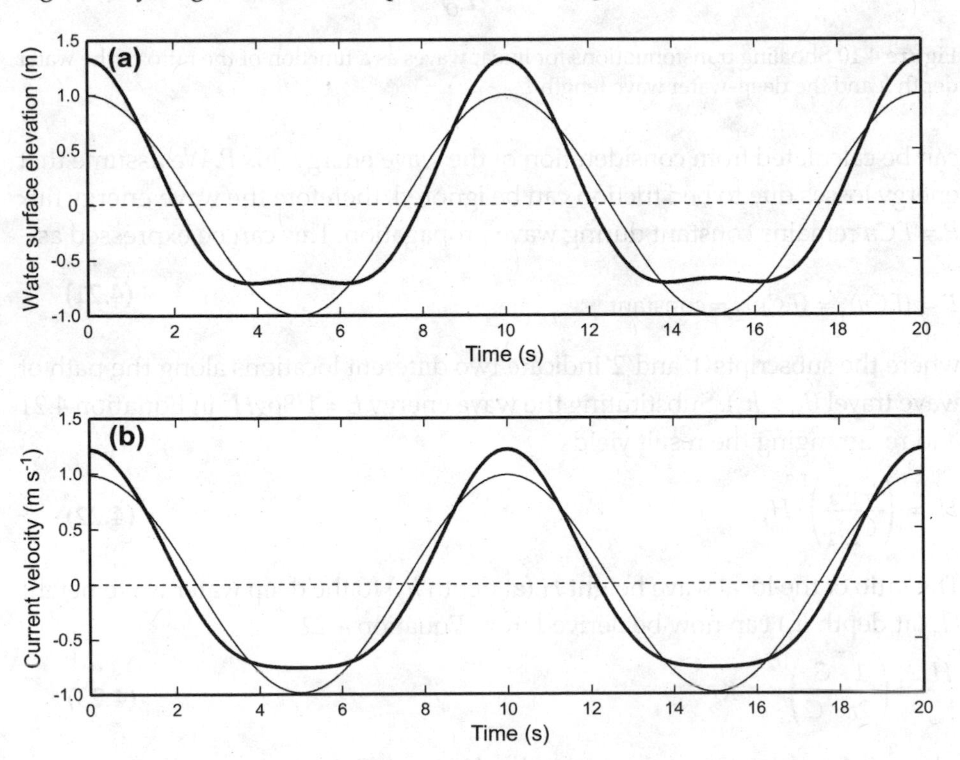

Figure 4.11 Comparison of the theoretical wave motion according to linear wave theory (thin line) and second-order Stokes wave theory (thick line): (a) wave form and (b) orbital velocity at the sea bed where positive (negative) values represent forward (backwards) motion. The wave motion is calculated using H= 2 m, T = 10 s and h = 8 m. The linear wave describes a sinusoidal variation in the water level and flow velocities, whereas the Stokes wave is highly asymmetric and is characterised by a peaked crest and a flat trough.

4.5.3 Development of wave asymmetry

In deep water, waves have a sinusoidal shape and water particle velocities associated with the wave motion are symmetrical, meaning that the onshore velocities are of equal strength and duration as the offshore velocities. However, as the waves enter intermediate water depths, they become increasingly asymmetric and develop peaked crests and flat troughs. The associated flow velocities also become asymmetric with the onshore stroke of the wave being stronger, but of shorter duration, than the offshore stroke. Linear wave theory can no longer be used to accurately describe the dynamics of such waves and so second-order wave theories are necessary such as the Stokes and Cnoidal wave theories (Komar, 1998).

Figure 4.11 shows the wave profile and associated water velocities at the sea bed for a wave in intermediate water depth according to linear wave theory and the more advanced second-order **Stokes wave theory**. The velocity field under the linear wave is symmetrical, whereas the maximum onshore water velocity under the Stokes wave is twice that of the maximum offshore water velocity. This wave asymmetry is important because it promotes the onshore transport of sediment particles (see Chapter 8).

4.5.4 Wave refraction

When a wave approaches the coast with its crest at an angle to the bottom contours, the water depth will vary along the wave crest. If the wave is in intermediate or shallow water, the wave velocity C will also vary along the wave crest with the deeper water part of the wave propagating at a faster rate than the shallower water part of the wave (Equations 4.4 and 4.9). This results in rotation of the wave crest with respect to the bottom contours (Figure 4.12). This process of **wave refraction** is of great relevance to nearshore currents, sediment transport and coastal morphology. Wave refraction in different coastal settings causes the waves to become aligned more parallel to the shoreline as they propagate into shallow water (Figure 4.13).

Figure 4.12 Photo showing wave refraction as waves enter into an embayment. (Photo: J. Knight.)

Figure 4.13 (a) Aerial photograph showing the refraction pattern of waves approaching a straight beach under a small angle to the shoreline ($\alpha \approx 20°$). The waves clearly bend so that they become more parallel to the shoreline as they enter shallow water. (From Komar, 1998.) (b) Oblique photograph illustrating the wave refraction patterns on approach to a gravel barrier under a large angle with the shoreline ($\alpha \approx 45°$). The bending of the waves is not complete and as a result the angle between the shoreline and the wave crests remains relatively large. (From Orford *et al.*, 1996; photo J.D. Orford.) Copyright © 1996 Coastal Education & Research Foundation , reproduced with permission.)

The change in wave direction by wave refraction is related to a change in wave velocity C and can be described in a similar way to the bending of light rays according to Snell's law

$$\frac{\sin\alpha_1}{C_1} = \frac{\sin\alpha_2}{C_2} = \text{constant} \tag{4.24}$$

where α refers to the angle between the wave crest and the bottom contours and the subscripts '1' and '2' indicate two different locations along the path of wave travel ($h_1 > h_2$) (Figure 4.14). For a straight coast with parallel bottom contours, the angle at a given depth (at depth h_2) is related to the angle of wave approach in deep water α_o (at depth h_1) according to

$$\sin\alpha = \frac{C}{C_o}\sin\alpha_o \tag{4.25}$$

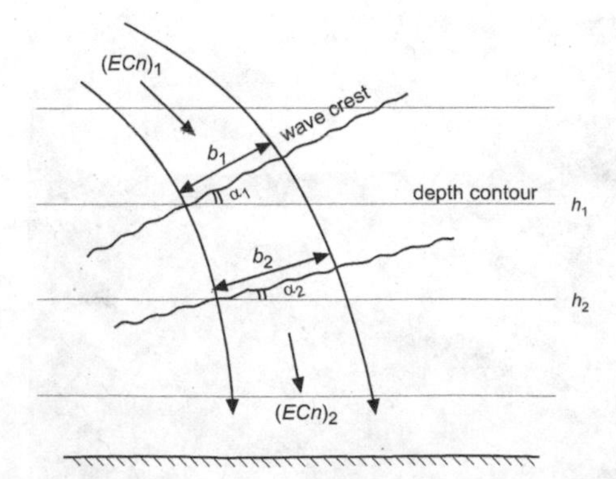

Figure 4.14 Schematic diagram showing wave refraction.

Equation 4.25 shows that as the wave velocity decreases in shallow water, the angle between the wave crest and the bottom contour also decreases.

The refractive bending of the wave rays also causes them to spread out, i.e. the distance between rays increases as the waves are being refracted (Figure 4.14). If b is the spacing between wave rays, then the energy flux between the wave rays at two different depths should be constant so that

$$P = (ECnb)_1 = (ECnb)_2 = \text{constant} \quad (4.26$$

Inserting $E = 1/8\rho g H^2$ in Equation 4.26 then yields

$$H_2 = \left(\frac{n_1 C_1}{n_2 C_2}\right)^{1/2} \left(\frac{b_1}{b_2}\right)^{1/2} H_1 \quad (4.27)$$

where

$$\frac{b_1}{b_2} = \frac{\cos\alpha_1}{\cos\alpha_2} \quad (4.28)$$

for a straight coast with parallel contours. The ratio of the local wave height H (at depth h_2) to the deep water wave height H_o (at depth h_1) can be derived from Equation 4.27

$$\frac{H}{H_o} = \left(\frac{1}{2n}\frac{C_o}{C}\right)^{1/2} \left(\frac{b_o}{b}\right)^{1/2} = K_s K_r \quad (4.29)$$

where K_s is the shoaling coefficient discussed earlier (Equation 4.23) and K_r is referred to as the **wave refraction coefficient**. The wave refraction process is not affected by wave height, but does depend on wave period. Long-period waves feel the bottom earlier than short-period waves. They therefore refract more than short-period waves and become more aligned parallel to the coast.

Irregular bottom topography can cause waves to be refracted in complex ways and produce significant variations in wave height and energy along the coast. Spreading of the wave rays, referred to as **wave divergence**, occurs when waves propagate over a localised area of deeper water such as a depression in the sea floor. Wave divergence is characterised by an increase in the spacing of wave rays ($K_r < 1$) and this causes a reduction in wave energy and wave height. Focusing of the wave rays by **wave convergence** occurs when waves travel over a localised area of shallower water such as a shoal on the sea floor. The resulting decrease in spacing of the wave rays ($K_r > 1$) causes an increase in wave energy and wave height. An example of wave refraction resulting in wave divergence and convergence is given in Figure 4.15.

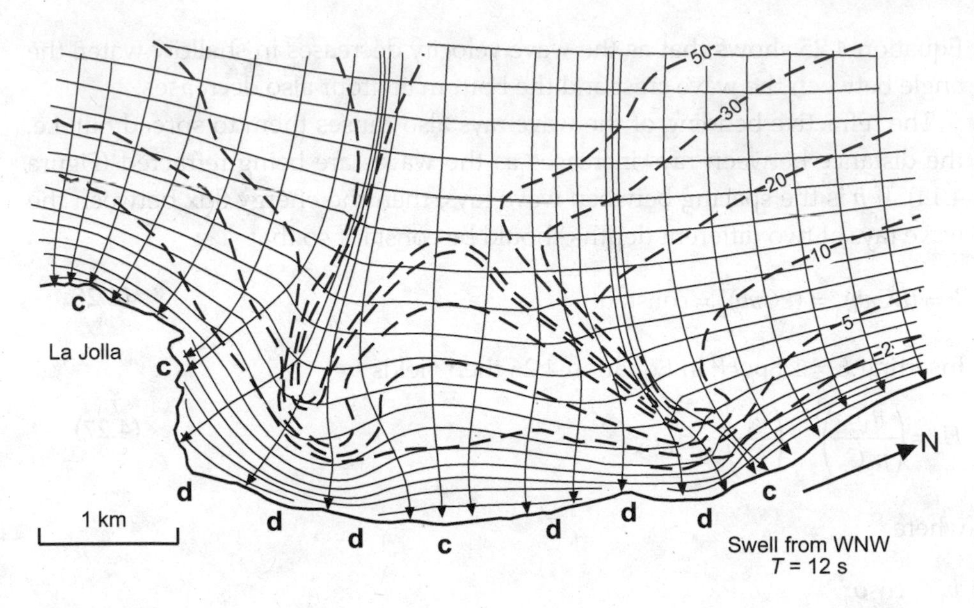

Figure 4.15 Wave refraction over submarine canyons and along the headland of La Jolla, California. The letters 'd' and 'c' refer to wave divergence and convergence, respectively. Depth values are in fathoms. (Modified from Munk and Traylor, 1947.)

4.5.5 Wave diffraction

Wave diffraction is the process of wave energy transfer along the wave crest rather than in the direction of wave propagation, and occurs irrespective of water depth. Wave diffraction occurs when an otherwise regular train of waves encounters a feature such as an island, breakwater or offshore reef. A wave shadow zone is created behind the obstacle, but diffraction causes wave energy to spread into this zone. Wave diffraction also enables wave energy to enter into confined bays and harbours. Wave diffraction is fundamentally different to wave refraction, but both mechanisms often operate together. Figure 4.16 shows the combined refraction/diffraction pattern of waves entering an irregular bay. The wave crests in the bay are bent due to refraction and show that the fastest wave celerities and thus greatest water depths are in the centre of the bay. The margins of the bay are in the wave shadow zone, but wave diffraction results in some wave energy penetrating into these sheltered regions.

4.6 Wave processes in the surf and swash zone

4.6.1 Wave breaking

At some point during wave shoaling, the water depth becomes too shallow for a stable wave form to exist, and the wave will break. **Wave breaking** occurs

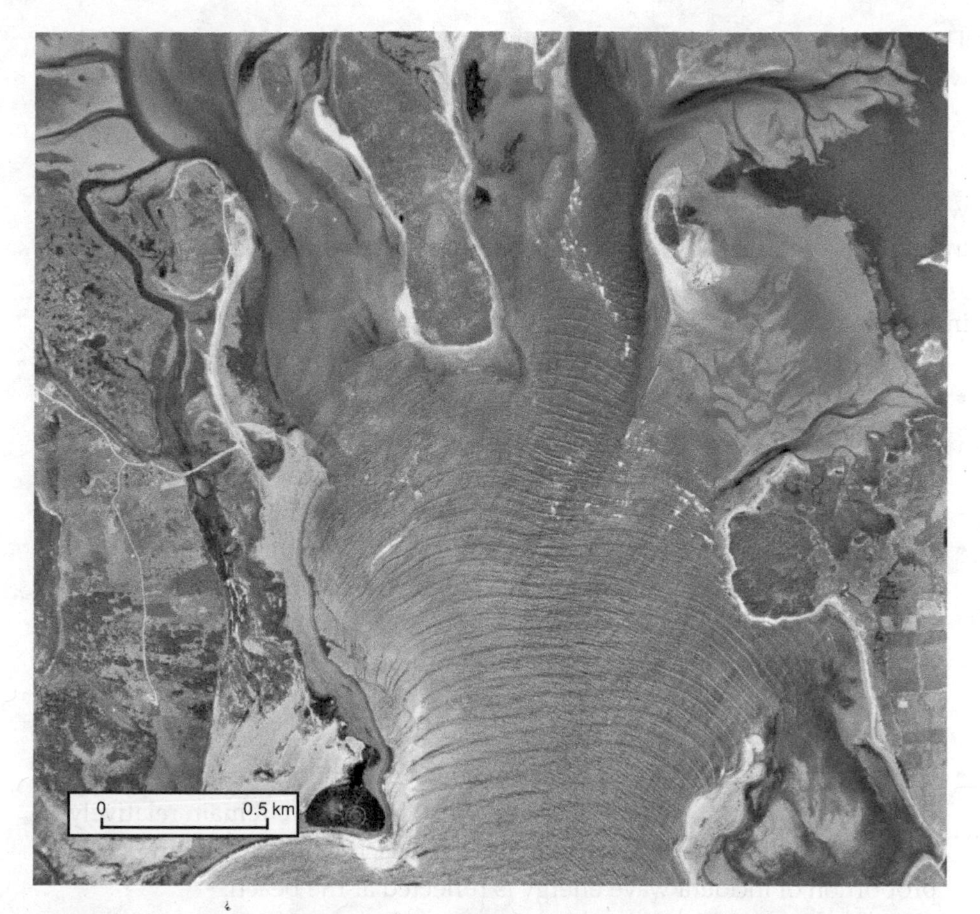

Figure 4.16 Aerial photograph showing the combined refraction/diffraction patterns in an irregular bay. (From Orford *et al.*, 1996.) (Copyright © 1954 Her Majesty the Queen in Right of Canada (airphoto A14288-148). Reproduced with permission from Natural Resources Canada.)

when the horizontal velocities of the water particles in the wave crest exceed the velocity of the wave form. Consequently, the water particles leave the wave form which therefore disintegrates into bubbles and foam. Wave breaking is an important process allowing energy to be released, and can result in the generation of nearshore currents and sediment erosion and transport. The reduction of wave energy due to wave breaking is referred to as **wave energy dissipation**.

Worked Example 2

The water depth at which waves break is related to the height of the breaking wave according to

$$H_b = \gamma h_b \tag{4.30}$$

where H_b is breaking wave height, h_b is mean water depth at the point of breaking and γ is the **breaker index** or **breaking criterion**, which typically has a value of 0.78. When the height-to-depth ratio of the wave exceeds the breaker index, breaking will occur.

Three main breaker types are commonly recognised (Figure 4.17):

- **Spilling breakers** are associated with gentle beach gradients and steep incident waves (large wave height relative to wave length). They are characterised by a gradual peaking of the wave until the crest becomes unstable, resulting in a gentle forward spilling of the crest.
- **Plunging breakers** tend to occur on steeper beaches than spilling breakers, with waves of intermediate steepness. The shoreward face of plunging waves becomes vertical, curled over, and plunges forward and downward as an intact mass of water. Plunging breakers are the most desirable waves for surfers, because they offer the fastest rides and produce 'tubes'.
- **Surging breakers** are found on steep beaches with low steepness waves. In surging breakers, the front face and crest of the wave remain relatively smooth and the wave slides directly up the beach without breaking. A large proportion of incident wave energy is reflected at the beach.

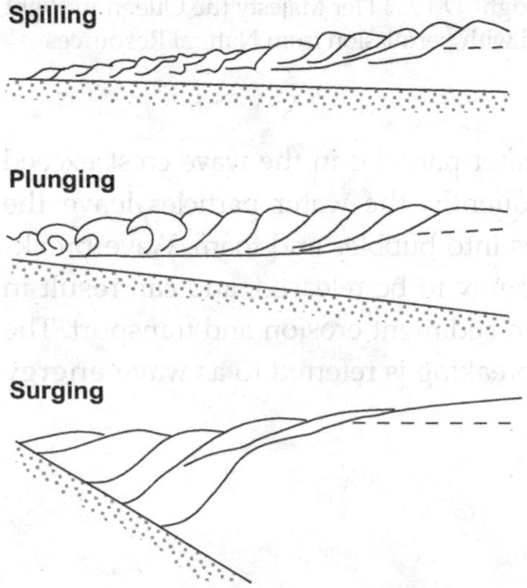

Figure 4.17 Illustration of spilling, plunging and surging breakers. (Modified from Galvin, 1968.)

Dimensionless parameters can be used to predict the type of breaker. The most widely-used formulation is the **Iribarren Number** ξ (Battjes, 1974)

$$\xi = \frac{\tan \beta}{\sqrt{H_b/L_o}} \quad (4.31)$$

where tanβ is the gradient of the beach and the subscripts 'b' and 'o' indicate breaker and deep water conditions, respectively. Small values for ξ (< 0.4) are attained when the beach has a gentle gradient and the incident wave field has a large wave height and short wave length (or short wave period). Such conditions promote the formation of spilling breakers. Large values, where ξ > 1, are found when the beach is steep and the incident wave field has a small wave height and long wave length (or long wave period) and favour the formation of surging breakers. Plunging breakers prevail when ξ = 0.4–1.

4.6.2 Wave reflection

Under certain conditions, waves do not break in shallow water but instead reflect at the shoreline. The interaction of reflected and incoming waves results in the formation of a standing wave. **Wave reflection** occurs when ξ > 1 and is associated with surging breakers. Swell and wind waves only reflect off very steep beaches or vertical shorelines formed by sea walls and cliffs (Figure 4.18), but infragravity waves (Section 4.6.4) may even reflect off shallow, sandy beaches (Box 4.2). Wave reflection is an important process and standing wave motion is significant in the formation of nearshore bars (Section 8.3.3).

Figure 4.18 Photo showing wave reflection from a sea wall, and interaction of the reflected wave with incoming waves. (Photo: G. Masselink.)

Box 4.2 Wave reflection of infragravity waves

Unlike wave reflection from a hard or vertical shoreline that propagates the reflected wave back to sea (Figure 4.18), reflection from low-gradient beaches is more difficult to identify in the field. However, if nearshore bars are present, reflected waves can sometimes be seen breaking on the landward face of the bar.

In the surf zone, wave reflection can produce **standing waves**, characterised by nodes and antinodes. Figure 4.19 compares the movement of **progressive waves** (i.e. waves that propagate) and standing waves. It is customary to indicate the different phases of the wave motion on a scale from 0 to 360° (or 0 to 2π radians). Thus, if half the wave cycle is completed, the phase of the wave is 180° (or π). In a progressive wave, the wave shape remains constant and travels a distance equal to one wave length over one wave period (Figure 4.19a). In contrast, the wave shape associated with the standing wave does not propagate and the water level simply moves up and down over one complete wave period (Figure 4.19b). The up-and-down movement of the water surface of the standing wave is maximum at the antinodes and stationary at the nodes.

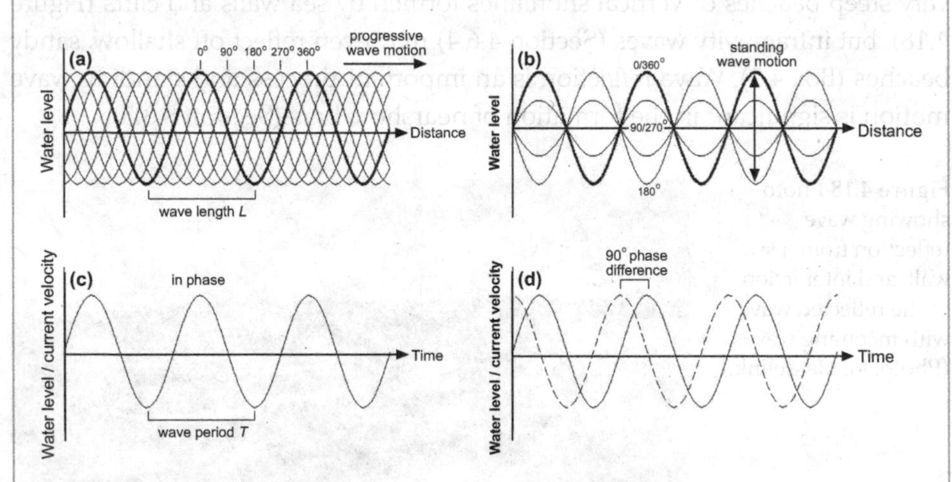

Figure 4.19 Progressive and standing wave motion. See text for explanation.

Wave reflection also has implications for the generation of cross-shore currents. For a progressive wave, only one wave signal is apparent because the two time series are in phase, i.e. the crest (trough) of the progressive wave coincides with the maximum onshore (offshore) directed velocity (Figure 4.19c). For a standing wave, cross-shore currents are more complicated. At nodes, the cross-shore current velocity is maximum but the water level does not change. At antinodes, the water level change is at its maximum but the cross-shore current is zero. When water level and cross-shore current are measured between a node and antinode, they therefore display a 90° ($\pi/2$) phase difference (Figure 4.19d). This phase difference implies that the strongest cross-shore current occurs when the water level crosses mean sea level.

These controls on wave reflection can be explored by calculation of the Iribarren Number ξ (Equation 4.31). Wave reflection is dominant when $\xi > 1$ and is promoted by steep beach gradients, long wave periods and small wave heights. Therefore the potential for wave reflection and formation of standing waves is greatest for longer period waves, in particular infragravity waves. The infragravity waves do not generally break on natural beaches. For example, for an infragravity wave height of 0.5 m and a wave period of 40 s, the beach slope required to cause the wave to break rather than reflect can be obtained by re-arranging Equation 4.31 and inserting the appropriate numbers

$$\tan \beta = \xi \sqrt{H_b/L_o} = \sqrt{0.5/2496} = 0.014$$

Thus, the wave under question will only break if the beach gradient is less than 0.014, which is not typical of natural beaches.

4.6.3 Broken waves in the surf zone

A natural wave field has a range of wave heights and periods so therefore not all the waves arriving at the shore will break at the same location. The largest waves will break farthest offshore, while the smaller waves approach closer to the shore before they break. This means that, at any position within the surf zone, some waves are breaking, while other waves are still undergoing their transformations leading to initial breaking. The proportion of breaking waves increases in a shoreward direction across the surf zone.

Field measurements show that in the inner part of the surf zone, where most waves are broken, the wave height is a direct function of local water depth (Thornton and Guza, 1982) and can be expressed by

$$H_s = \langle \gamma \rangle h \quad \textbf{(4.32)}$$

The coefficient $\langle \gamma \rangle$, sometimes referred to as the relative wave height, ranges from 0.4 to 0.6 and increases with beach slope (Sallenger and Holman, 1985). A surf zone in which the wave height is limited by the local water depth, as prescribed by Equation 4.32, is referred as a **saturated surf zone**.

4.6.4 Infragravity wave motion

Wave and current spectra derived from hydrodynamic measurements in the surf zone are usually characterised by the presence of energy at frequencies lower than those of the incident swell and wind waves (Figure 4.20). This motion is referred to as **infragravity wave motion** and is generally characterised by periods ranging from 20 seconds to several minutes. Certain types of infragravity waves, termed **edge waves**, may be responsible for the formation of rip channels and multiple bars.

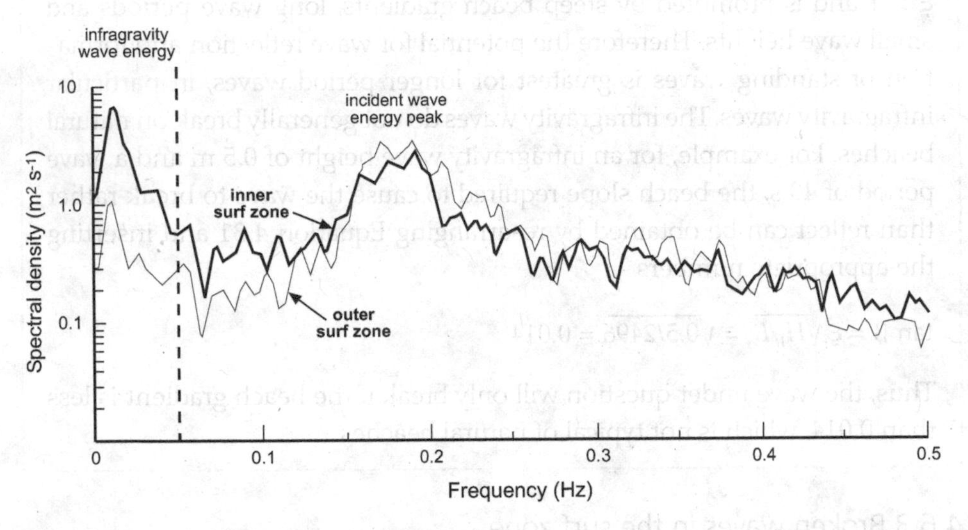

Figure 4.20 Examples of cross-shore velocity spectra from the surf zone demonstrating the presence of large amounts of infragravity energy at frequencies below 0.05 Hz (periods larger than 20 s). The two spectra were recorded simultaneously and indicate that infragravity waves in the inner surf zone are more energetic than in the outer surf zone. (From Aagaard and Masselink, 1999.) (Copyright © 1999 John Wiley & Sons, reproduced with permission.)

Infragravity waves are also important because they are forced directly or indirectly by the incoming waves and so their energy is proportional to the energy of these incident waves. Typically, the height of infragravity waves in the surf zone is 20–60 per cent of offshore wave height (Guza and Thornton, 1985). The amount of infragravity wave energy increases significantly across the surf zone in the onshore direction. This increase is mainly ascribed to shoaling of the infragravity wave as it enters shallow water. At the same time, the incident wave energy level decreases in the onshore direction across the surf zone due to wave breaking.

The dependence of infragravity wave height on the incident wave height, and the opposing cross-shore trends in infragravity and incident wave energy levels, have major ramifications for the behaviour of beaches during storms. During storms, the surf zone will become wider, but conditions at the shoreline will not become more energetic due to incident wave energy dissipation (Figure 4.21). In other words, the larger incident storm waves would simply break farther offshore, dissipating their energy through the wider surf zone. Infragravity energy, on the other hand, is not dissipated in the surf zone because the long wave length of the infragravity wave inhibits wave breaking (Box 4.2). Because the infragravity energy level is proportional to offshore wave height, infragravity energy may dominate the water motion in the inner surf zone during storms. Furthermore, infragravity waves periodically raise the mean water level at the beach, enabling the incident waves to reach the back beach and the toe of the dunes (Figure 4.21).

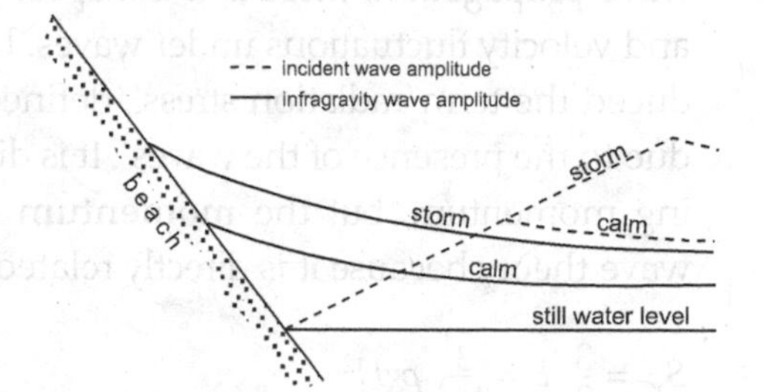

Figure 4.21 Schematic diagram of variations in incident and infragravity wave energy across the nearshore zone during storm and calm weather conditions. (Modified from Holman, 1983.)

4.6.5 Wave set-up, swash and runup

When waves break on the beach, they produce **wave set-up**, which is a rise in the mean water level above the still water level elevation of the sea (Figure 4.22). Wave set-up can be considered as a piling up of water against the shoreline and is caused by the breaking waves driving water shoreward. However,

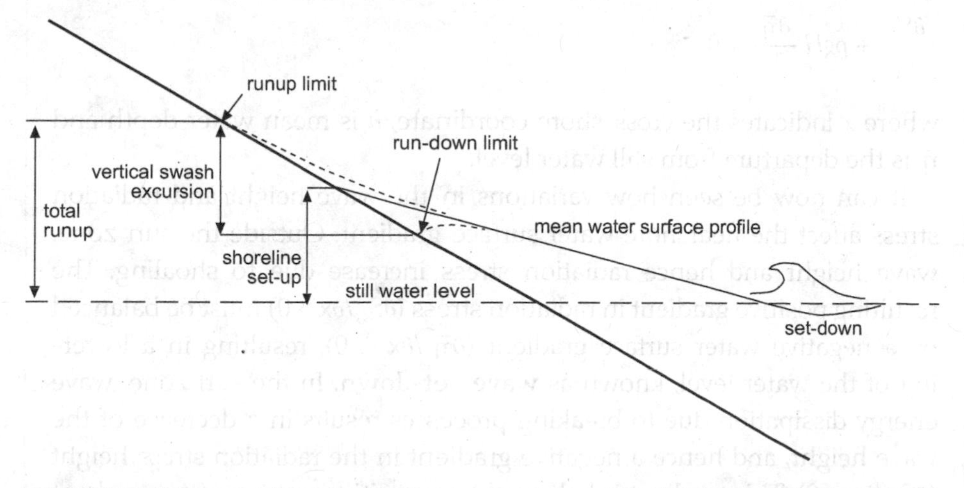

Figure 4.22 Schematic diagram showing wave runup, swash excursion, wave set-up and set-down. (Modified from Komar, 1998.)

this simple explanation does not take into account the forces required to drive wave set-up (Box 4.3). The wave set-up at the shoreline $\bar{\eta}_s$ can be defined as the intersection of the sloping mean water surface (set-up gradient) with the beach. As a rule of thumb, the shoreline set-up is 20 per cent of the offshore significant wave height.

Box 4.3 Momentum flux, radiation stress and wave set-up

Wave set-up is the super-elevation of the nearshore water level due to the presence of waves, and was theoretically explained by Longuet-Higgins and Stewart (1962, 1964) using the **radiation stress** concept. Associated with wave propagation, there is a transport of momentum due to the pressure and velocity fluctuations under waves. Longuet-Higgins and Stewart introduced the term 'radiation stress', defined as the 'excess flow of momentum due to the presence of the waves'. It is difficult to perceive waves as possessing momentum, but the **momentum flux** can be computed from linear wave theory because it is directly related to the wave energy density E

$$S_{xx} = \frac{3}{2}\ E = \frac{3}{16}\ \rho g H^2$$

where S_{xx} represents the cross-shore component of the radiation stress for waves approaching the beach with their crests parallel to the shoreline. Newton's second law of motion states that changes in momentum flux must be balanced by an opposing force. When this law is applied to the surf zone we obtain the following result

$$\frac{\partial S_{xx}}{\partial_x} + \rho g H \frac{\partial \bar{\eta}}{\partial_x} = 0$$

where x indicates the cross-shore coordinate, h is mean water depth and $\bar{\eta}$ is the departure from still water level.

It can now be seen how variations in the wave height and radiation stress affect the nearshore water surface gradient. Outside the surf zone, wave height and hence radiation stress increase due to shoaling. The resulting positive gradient in radiation stress ($\partial S_{xx}/\partial x > 0$) must be balanced by a negative water surface gradient ($\partial \bar{\eta}/\partial x < 0$), resulting in a lowering of the water level, known as **wave set-down**. In the surf zone, wave energy dissipation due to breaking processes results in a decrease of the wave height, and hence a negative gradient in the radiation stress height ($\partial S_{xx}/\partial x < 0$). This gradient is balanced by a positive water surface gradient

($\partial\bar{\eta}/\partial x < 0$), resulting in an elevated water level in the surf zone referred to as **wave set-up**. Figure 4.23 shows laboratory results that confirm the radiation stress theory.

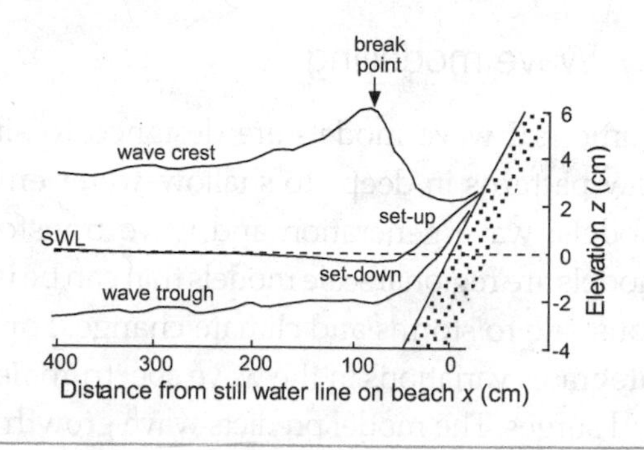

Figure 4.23 Wave set-up and set-down measured in the laboratory, with a wave height of 6.45 cm, wave period of 1.14 s and beach gradient of 0.082. SWL refers to still water level. (Modified from Bowen *et al.*, 1968.)

According to the saturated surf zone concept, the wave height at the shoreline should be reduced to zero due to wave breaking (Equation 4.32). However, in reality there will always be a residual wave at the shoreline and this wave will propagate onto the 'dry' beach in the form of swash. **Swash** motion consists of an onshore phase with decelerating flow velocities (uprush or swash) and an offshore phase characterised by accelerating flow velocities (downrush or **backwash**). Generally, swash motion is asymmetrical with the magnitude of uprush velocities exceeding that of backwash. It is the onshore asymmetry in the swash motion that is mainly responsible for maintaining the beach gradient (see Section 8.3).

Collectively, wave set-up and swash produce a landward and upward displacement of the shoreline allowing waves to act at higher levels on the beach. The vertical displacement can be measured relative to the still water level and is referred to as **wave runup** (Figure 4.22). Wave runup can be substantial under energetic wave conditions and significantly increases the potential for shoreline erosion by storms, as well as forming features such as storm ridges. In a study by Stockdon *et al.* (2006), analysis of runup data obtained from video cameras deployed on a range of beaches resulted in the following equation

$$R = 1.1\left(0.35 \tan\beta (H_o L_o)^{1/2} + \frac{\left[H_o L_o (0.563(\tan\beta)^2 + 0.004)\right]^{1/2}}{2}\right) \quad \textbf{(4.33)}$$

where R is the wave runup height exceeded by 2 per cent of the runup events, H_o and L_o are the (significant) wave height and wave length in deep water, and $\tan\beta$ is the beachface gradient. The runup is made up of shoreline set-up (first component in the large brackets) and the vertical swash excursion (second component in the large brackets), and these can be calculated separately. For reasonably energetic conditions ($H_o = 1.5$ m and $T = 8$ s) and a moderately steep beachface gradient ($\tan\beta = 0.06$), Equation 4.33 yields a runup height of 0.81 m.

4.7 Wave modelling

Numerical wave models are designed to simulate two- or three-dimensional flow patterns in deep- to shallow-water environments. There are two types of models: wave generation and wave transformation models. Wave generation models are regional scale models that can be used to forecast regional wave conditions due to storms and climate change. For example, the WaveWatch III model integrates variations in the wave spectrum field in response to water depth, tides and surges. The model predicts wave growth and decay in response to variations in wind patterns, bottom friction, and wave whitecapping and breaking. Hanson *et al.* (2009) evaluate the output of wave height, period and direction, predicted from the WaveWatch III model, against monthly wave data from buoys across the Pacific Ocean. They find that generally the model produces good agreement with the buoy data, with most error arising from atmospheric effects on swell wave generation. The WaveWatch III model is available at http://polar.ncep.noaa.gov/waves/wavewatch/wavewatch.shtml. Other models are also available.

Wave transformation models are more commonly used for smaller scale studies in shallower water where detailed wave information is required for engineering structures or for understanding of coastline change. For example, the SWAN model (Simulating WAves Nearshore) examines wave propagation and changes in wave properties due to refraction, diffraction, shoaling and interaction with currents. Khalifa *et al.* (2009) use offshore wave data in the Mediterranean Sea as input into the SWAN model, in order to predict nearshore wave climate. They show that modelled nearshore wave conditions match closely to those observed, and that the wave conditions drive patterns of longshore drift along the Egyptian coast. The SWAN model is available at http://www.swan.tudelft.nl/. Other models are also available.

Such numerical models are useful because they can predict patterns of whitecapping, storm surge runup or to hindcast wave conditions from wind datasets. Problems with models include identifying the correct boundary conditions, energy dissipation within the model, and the role of swell waves generated outside of the model's domain. The relationship between wave height and wave power (Equation 4.17) suggests that even small variations in predicted wave height, produced from these models, can have significant implications for wave power. Significant wave height and direction of wave propagation are also important for coastal sediment transport.

SUMMARY

- Waves are generated by wind, and wave properties such as height and period can be predicted using wind speed, wind duration and fetch length. In a developing wave field, wave height and wave period increase to an equilibrium state referred to as a fully arisen sea. The wave field can be described using spectral analysis. Linear wave theory is the most widely used theory to describe wave behaviour. A fundamental relationship derived from linear wave theory is the dispersion equation, which expresses the relation between wave length and wave period.
- In deep water, wave motion does not extend to the sea bed and the wave velocity is only dependent on wave period. Deep water wave velocity increases with wave period and gives rise to wave dispersion, whereby the longer-period waves in a broad-banded wave field move more quickly than shorter-period waves.
- As waves propagate into shallower water, the waves start to 'feel' the sea bed during shoaling, which slows down the waves, decreasing the wave length and increasing the wave height. During shoaling, wave energy may be lost due to bed friction. In intermediate and shallow water, wave velocity decreases, resulting in wave refraction whereby the wave crest rotates with respect to bottom contours and becomes more aligned with the coastline.
- Waves will break in water that has a depth similar to the wave height. In the surf zone, where most waves are breaking, wave energy is either dissipated by wave breaking or reflected at the shoreline. Infragravity wave motion, wave set-up and swash produce a landward and upward displacement of the shoreline allowing waves to act at higher levels on the beach. This displacement is particularly significant during storms and may promote beach erosion.

Reflective questions

These questions are designed to test your comprehension of material covered in this chapter. Suggested answers to these questions can be found on this book's website.

4a. Describe the relationships between the different geometrical properties of waves.

4b. Outline how and why waves change when they move into the surf zone.

4c. What is the role of coastal topography and bathymetry in the process of wave refraction?

4d. Consider how swash and backwash processes may differ on (1) steep and shallow, and (2) sandy and rocky coastlines.

Further reading

Komar, P.D., 1998. *Beach Processes and Sedimentation* (2nd edn). Prentice Hall, New Jersey. (Chapters 5 and 6 of this text provide an excellent and comprehensive treatment of wave processes.)

Open University, 2000. *Waves, Tides and Shallow-Water Processes*. Pergamon Press, Oxford. (Chapter 1 of this text gives a very readable account of the dynamics of ocean waves.)

Pond, S. and **Pickard**, G.L., 1995. *Introductory Dynamical Oceanography* (2nd revised edn), Butterworth-Heinemann Ltd. (A thorough presentation pitched at students with some physics background.)

Thurman, H.V. and **Burton**, E.A., 2001. *Introductory Oceanography* (9th edn), Prentice Hall. (A useful introductory text including a chapter on waves.)

CHAPTER 5:

SEDIMENTS, BOUNDARY LAYERS AND TRANSPORT

AIMS

This chapter discusses the physical properties of sediments that are found in coastal environments. Sediment properties affect the modes of sediment transport and deposition that take place along coasts. Fluid properties that drive sediment transport are described with respect to shear stress and the development of turbulent flow. The development and geometry of the most common bedform types (ripples and dunes) are directly related to fluid dynamics.

5.1 Introduction

The term sediment refers to both organic and inorganic loose material that can be moved by physical agents including wind, waves, currents and under gravity. The sediments found in coastal environments can be either imported from outside the region (allochthonous) or locally produced (autochthonous). **Allochthonous sediments** are generally derived from the breakdown and transport of rocks into smaller particles, and along coasts commonly include minerals such as quartz, and clay minerals such as illite and montmorillonite. **Autochthonous sediments** include materials derived from the mechanical breakdown of rocky shorelines, but more commonly consist of broken-up shells of coastal organisms and/or the chemical precipitates of dissolved minerals within coastal waters and include biogenic carbonate and silica. Globally, allochthonous sediments account for about 92 per cent of sediment in the modern coastal zone. Processes by which this sediment is delivered to the coast are, in decreasing order of importance, river, glacial and wind transport and volcanic eruptions.

Coastal landforms result from patterns of erosion and deposition that take place within larger coastal sediment systems. The wide range of spatial and temporal scales evident in coastal geomorphology (Figure 1.3) indicates that in order to fully understand large-scale morphodynamic processes it is also necessary to understand smaller-scale processes. The sequence of processes that control localised sediment movement are: (1) erosional **entrainment** of sediment into the flow via fluid-induced stresses and forces acting on the bed; (2) **transport** of sediment via momentum transfer from the fluid to the sediment; and (3) **settling** or **deposition** of sediment back on the bed via gravity (Figure 5.1). Depending on the speed at which sediment settles to the bed relative to the speed at which flow conditions change, this sequence can re-initialise in one of two ways, by re-entrainment of sediment that is at rest on the bed, or by remobilisation of sediment that has not yet completed its return to the bed. The detailed mechanics of this cycle vary, depending on the intrinsic properties of the sediment and the fluid.

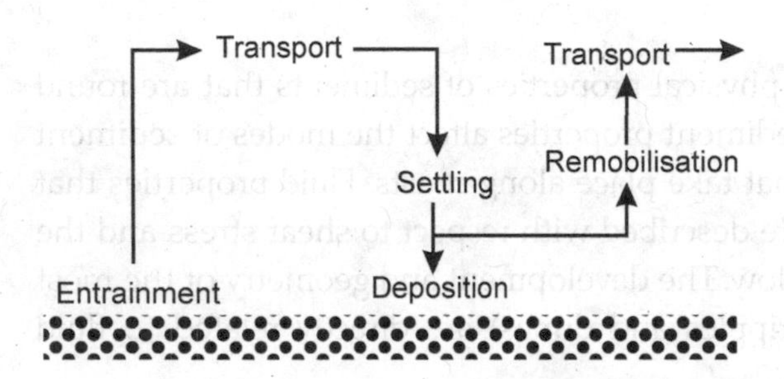

Figure 5.1 Schematic representation of the fundamental processes that together constitute sediment dynamics.

Both air and water are important agents for coastal sediment transport. While many concepts presented in this chapter apply to both fluid types, they differ markedly in their density and viscosity, and hence their ability to move sediment. This chapter focuses on hydrodynamics and sediment transport by water. Aerodynamics and sediment transport by wind are described in Chapter 9. Sediment erosion, transport and deposition processes by water or wind are the mechanisms by which coastal landforms are developed and destroyed. Therefore sediment–fluid interactions are critical to understanding how coastal landforms respond to forcing by tides, waves and climate.

5.2 Sediment properties

5.2.1 Grain size

Grain size is the sediment property most widely measured by coastal geomorphologists since it is important in a wide range of coastal processes. The simplest measurements of a grain's size are the lengths of the longest, intermediate and shortest axes which are termed the *a*, *b* and *c* axes, respectively. By convention the *b*-axis and *c*-axis are measured at right-angles to the *a*-axis. The axial dimensions of large grains can be measured directly with callipers, but smaller grains (of sand size or below) are measured indirectly, usually by sieving or laser granulometry. Sieving effectively measures only the *b*-axis length. In laser granulometric techniques, the sediment sample is mixed with a circulating water source so that the grains are in motion as they move past the laser device. As a result, this technique has equal likelihood of recording any axis of the sample, so the averaged b-axis is a good approximation. Goudie (1990) describes various methods for grain-size analysis.

Grains are often classified by their b-axis length using the **Udden-Wentworth Scheme** (Table 5.1). Individual pebbles, cobbles and boulders are often termed **clasts**. Note that the class boundaries in Table 5.1 are presented both on a millimetre-scale and a **phi-scale** (ϕ-scale), which is discussed below. The conversion between grain diameter D on the ϕ-scale to diameter on the millimetre-scale is

$$D = 2^{-\phi} \tag{5.1}$$

and *vice versa*

$$\phi = -\log_2 D \tag{5.2}$$

It should be evident from both Table 5.1 and these two equations that a change from one phi class to the next involves a doubling of the axis length.

While it is common to use a single number (often the arithmetic mean) to represent the grain size of a sediment sample, this number is typically obtained from a statistical analysis of the b-axis length of all or a subsample of the indi-

Table 5.1 The Udden-Wentworth Scheme of grain size classification.

mm	ϕ	Class terms	
		Boulders	
256	−8		
128	−7	Cobbles	
64	−6		
32	−5		
16	−4	Pebbles	
8	−3		
4	−2	Granules	
2	−1		
1	0	Sand	very coarse
0.5	1		coarse
0.25	2		medium
0.125	3		fine
			very fine
0.062	4		
0.031	5	Silt	coarse
0.016	6		medium
0.008	7		fine
			very fine
0.004	8		
		Clay	

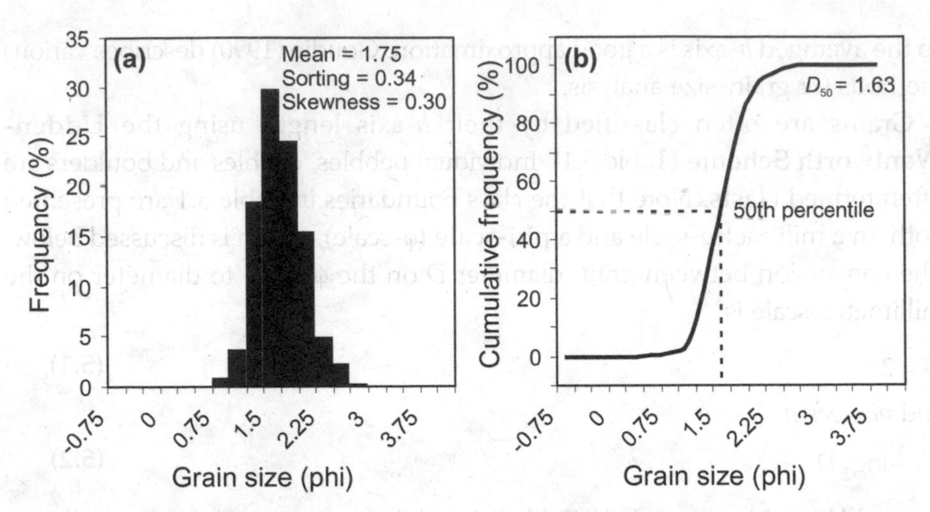

Figure 5.2 Two methods of presenting grain size data: (a) frequency histogram and (b) cumulative-frequency curve. The example shown is for a sample of medium sand that is very well sorted and fine-skewed (*cf.* Tables 5.1 and 5.2).

vidual grains in that sample. The procedure typically begins with the grain size measurements from a sediment sample being presented as a **frequency histogram** and **cumulative-frequency curve** (Figure 5.2). Typically the histogram approximates a log-normal distribution, so when it is plotted on a logarithmic scale such as the ϕ-scale, the histogram then appears normally distributed (Figure 5.2a). It is for this reason that a log transformation of grain size measurements (performed either graphically by plotting on the ϕ-scale or by applying Equation 5.2) is performed prior to calculating grain size statistics. It is important to note that large positive ϕ-values indicate finer grain sizes and large negative ϕ-values indicate coarser grain sizes.

Before the widespread availability of computers, grain size statistics were calculated by graphical means. The most widely used formulae are (Folk and Ward, 1957)

$$Median = \phi_{50} \quad (5.3)$$

$$Mean = \frac{\phi_{16} + \phi_{50} + \phi_{84}}{3} \quad (5.4)$$

$$Sorting = \frac{\phi_{84} - \phi_{16}}{4} + \frac{\phi_{95} - \phi_{5}}{6.6} \quad (5.5)$$

$$Skewness = \frac{\phi_{16} + \phi_{84} - 2\phi_{50}}{2(\phi_{84} - \phi_{16})} + \frac{\phi_{5} + \phi_{95} - 2\phi_{50}}{2(\phi_{95} - \phi_{5})} \quad (5.6)$$

$$Kurtosis = \frac{\phi_{95} + \phi_{5}}{2.44\,(\phi_{75} - \phi_{25})} \quad (5.7)$$

where, for example, ϕ_{50} is the 50th percentile of the grain-size distribution plotted on the ϕ-scale (Figure 5.2b). Note that different percentiles are also used in these formulae. This is because the 16th/84th and 5th/95th percentiles approximate to the ± 2 and ± 3 standard deviations from the mean value of a normal distribution, respectively. These formulae are still widely used when the method of sediment analysis does not yield a complete grain-size distribution, because the calculation does not rely too heavily on the fine and coarse tails of the distribution. Where a significant fraction of the material is very fine (> 4ϕ), however, a laser granulometer or settling tube method is more appropriate for the sample. When a complete grain-size distribution is available, the statistical descriptors are calculated using the more accurate method of **moment measures** (Pettijohn *et al.*, 1987)

$$Mean = \bar{x} = \frac{\sum_{i=1}^{n} f_i M_\phi}{100} \quad (5.8)$$

$$Sorting = \sigma = \sqrt{\frac{\sum_{i=1}^{n} f_i (M_\phi - \bar{x})^2}{100}} \quad (5.9)$$

$$Skewness = \sum_{i=1}^{n} \frac{f_i (M_\phi - \bar{x})^3}{100\sigma^3} \quad (5.10)$$

$$Kurtosis = \sum_{i=1}^{n} \frac{f_i (M_\phi - \bar{x})^4}{100} \quad (5.11)$$

where n is the number of data points, f_i is percentage of grains (or percentage of total weight of grains) in each size interval and M_ϕ is the midpoint of each size interval in phi units.

The **mean** (1st-moment) is the most common value used to represent sediment grain size, although if the distribution is highly skewed then the median or mode may be better representative. The Udden-Wentworth Scheme is a classification scheme based on mean grain size (Table 5.1), but there are also classification schemes for the other moment measures (Table 5.2). The **sorting** (2nd-moment) is controlled in part by the range of grain sizes present at the sediment source, as well as processes operating during transport and deposition. For example, rapidly-deposited sediment is often poorly sorted, whereas frequently reworked sediment tends to be well sorted. The **skewness** (3rd-moment) is an indicator of the symmetry of the grain size distribution (Figure 5.3a). For a normal distribution, the measure of skewness is zero.

Negative skewness means that there are more coarse grains than expected in a log-normally distributed sample, and positive skewness means that there are more fine grains than expected. Skewness can arise from the mixing of sediments from different sources, but can also be indicative of sorting during transport and deposition. For example, beach sands are typically negatively skewed, because continual agitation by waves can resuspend the finer particles, leaving an apparent excess of coarser sizes in the bed sediment. There is often a significant shell component in beach sands that also negatively skews the distribution. The **kurtosis** (4th-moment) is an indicator of the peaked shape of the distribution (Figure 5.3b). Distributions with a high peak and low range are termed leptokurtic, whereas distributions with a subdued peak but a wider range are termed platykurtic. These moment measures in combination describe the totality of the grain size distribution of a sediment sample, and can often be used to distinguish its genetic origin or depositional environment (Allen, 1985). Other physical properties including grain mineralogy, shape and density also affect how individual grains behave in response to flow conditions.

Table 5.2 Descriptors for sediment sorting, skewness and kurtosis as defined by Folk and Ward (1957).

Sorting (ϕ-scale)		Skewness (ϕ-scale)		Kurtosis (ϕ-scale)	
< 0.35	Very well sorted	> +0.30	Strongly fine-skewed	< 0.67	Very platykurtic
0.35 to 0.50	Well sorted	+0.30 to +0.10	Fine-skewed	0.37 to 0.90	Platykurtic
0.50 to 0.71	Moderately well sorted	+0.10 to −0.10	Nearly-symmetrical	0.90 to 1.11	Mesokurtic
0.71 to 1.00	Moderately sorted	-0.10 to −0.30	Coarse-skewed	1.11 to 1.50	Leptokurtic
1.00 to 2.00	Poorly sorted	< −0.30	Strongly coarse-skewed	1.50 to 3.00	Very leptokurtic
> 2.00	Very poorly sorted				

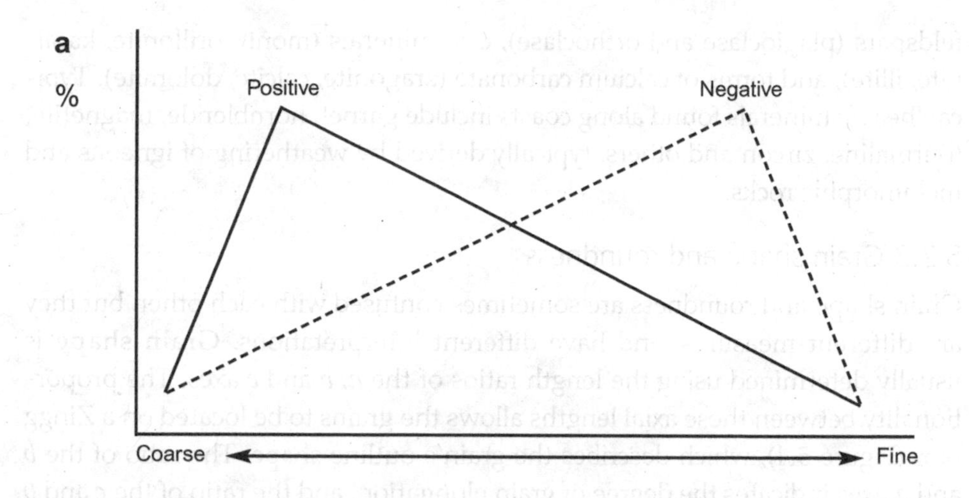

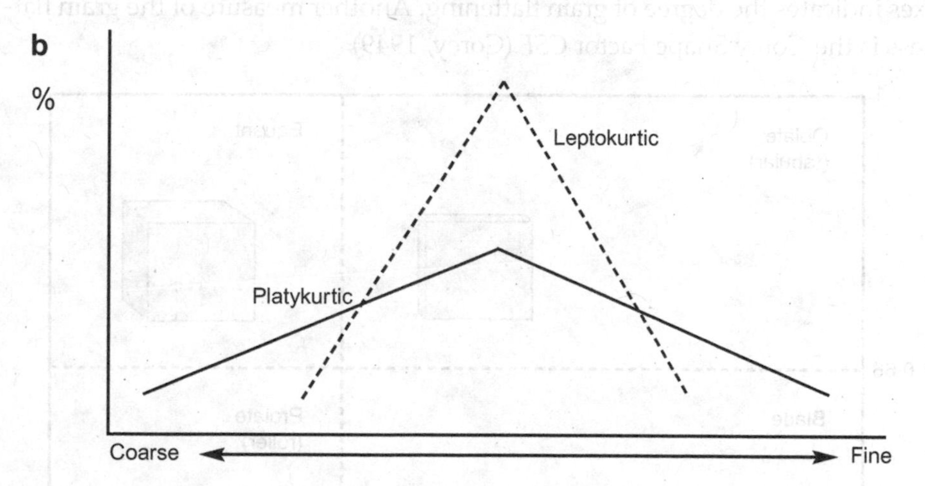

Figure 5.3 Illustration of the (a) skewness and (b) kurtosis of typical coarse and fine sediment samples.

5.2.2 Grain mass and density

The mass of a grain affects its inertia with respect to the forces applied to it by a moving fluid. A grain's **mass** is equal to the product of its volume and density. The volume is clearly related to the grain size. The volume of spherical grains, for example, increases as the grain-diameter cubed. The **density** of a grain is its mass per unit volume, and is largely determined by mineralogy. We can distinguish between 'light' and 'heavy' minerals, with a somewhat arbitrary dividing line at a density of 2900 kg m^{-3}, because a similar separation often occurs in the environment. For example, alternating layers of light and dark sand seen on some beaches are usually concentrations of light and heavy mineral types, respectively. Typical 'light' minerals common along coasts include quartz, feldspars (plagioclase and orthoclase), clay minerals (montmorillonite, kaolinite, illite), and forms of calcium carbonate (aragonite, calcite, dolomite). Typical 'heavy' minerals found along coasts include garnet, hornblende, magnetite, tourmaline, zircon and others, typically derived by weathering of igneous and metamorphic rocks.

5.2.3 Grain shape and roundness

Grain shape and roundness are sometimes confused with each other, but they are different measures, and have different interpretations. **Grain shape** is usually determined using the length ratios of the *a*, *b* and *c* axes. The proportionality between these axial lengths allows the grains to be located on a Zingg plot (Figure 5.4), which describes the grain's outline shape. The ratio of the *b* and *a* axes indicates the degree of grain elongation, and the ratio of the *c* and *b* axes indicates the degree of grain flattening. Another measure of the grain flatness is the Corey Shape Factor *CSF* (Corey, 1949)

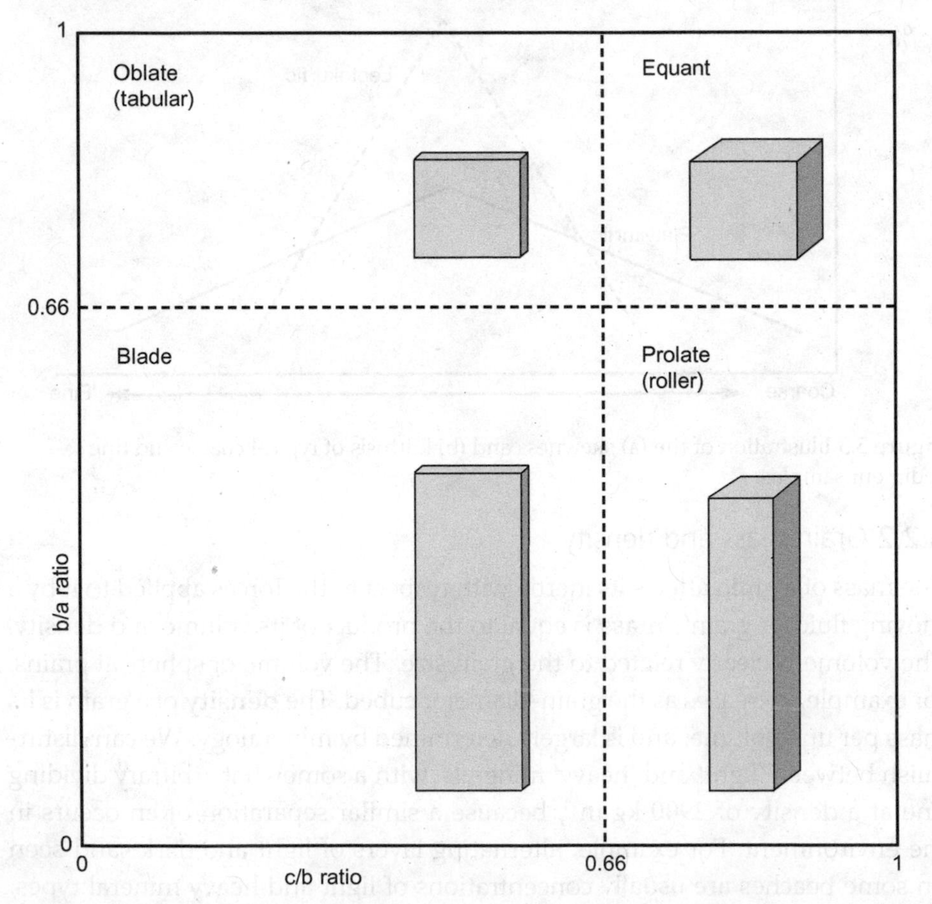

Figure 5.4 Zingg's (1935) classification of grain shape.

$$CSF = \frac{c}{\sqrt{ab}} \tag{5.12}$$

where $CSF = 0$ represents a flat disc and $CSF = 1$ represents a perfect sphere. The shape of the original mineral crystal largely determines the shape of a single grain, although dissolution and abrasion can have a modifying effect. Rock type and structure are also important where grains are of pebble size or larger; for example, slate rocks preferentially form blades where the blade thickness (*c* axis) is determined by the rock's original cleavage. The shape of a grain influences both its entrainment and settling. Flat grains are more difficult to entrain and settle more slowly than spherical grains. Sorting based on shape takes place on gravel beaches, with prolate and equant clasts predominantly found at the base of the beach face and oblate and bladed clasts found towards the top.

Roundness refers to the three-dimensional shape of the grain that considers in particular the roundness of grain corners and protrusions. Roundness can be distinguished from shape by considering the differences between a cubic box and a sphere or ball-shaped grain of the same size as the box. Both these forms have the same *a, b* and *c* axis lengths and therefore the same equant shape, but a box has corners (low roundness) whereas the sphere does not (high roundness) (Figure 5.5). Therefore grain roundness should be considered alongside grain shape. The degree of roundness usually indicates the susceptibility of the grain to chemical weathering and/or the degree of mechanical abrasion it has experienced. Angular grains often indicate resistant minerals, or a depositional site that is close to source. Well-rounded grains indicate either easily weathered minerals, energetic environments where there is active mechanical abrasion, or a depositional site that is far from source. Grain roundness also influences the

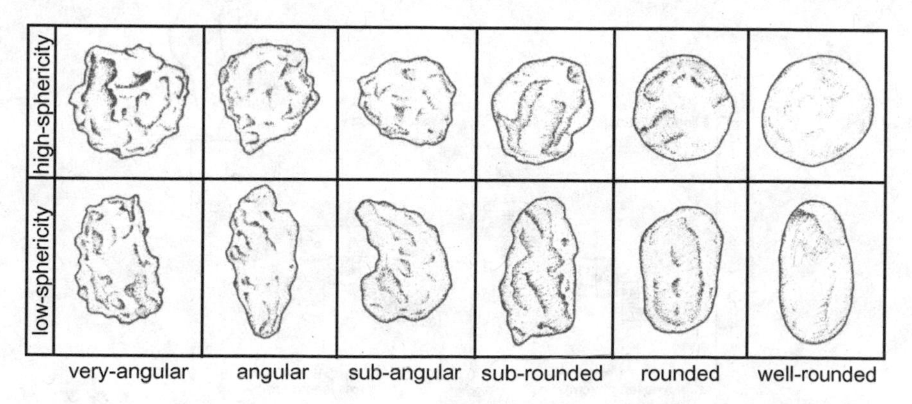

Figure 5.5 Powers' (1953) classification of grain roundness for grains displaying low sphericity and high sphericity (From Tucker, 1995.) (Copyright © 1995 Blackwell Publishers, reproduced with permission.)

sediment's friction angle and packing. Relationships between grain size, shape and roundness are explored in Case Study 5.1.

Case Study 5.1 Grain size and shape variations on a gravel beach

A classic location for examining the size, shape and roundness relationships of clasts is Chesil Beach, southern England. Here, a linear gravel beach backed by a lagoon (The Fleet) extends for 28 km in a northwest–southeast direction. Beach height, beach width and sediment grain size all decrease progressively towards the northwest in the direction of dominant longshore transport (Figure 5.6). Clast lithology is mainly quartzite and flint/chert from the cliffs behind. Although these are relatively hard rock types, the high-energy waves mean that all lithologies are rapidly abraded in the direction of longshore drift. Carr *et al.* (1970) describe how the quartzite and flint/chert clasts change in size and shape at various points along the

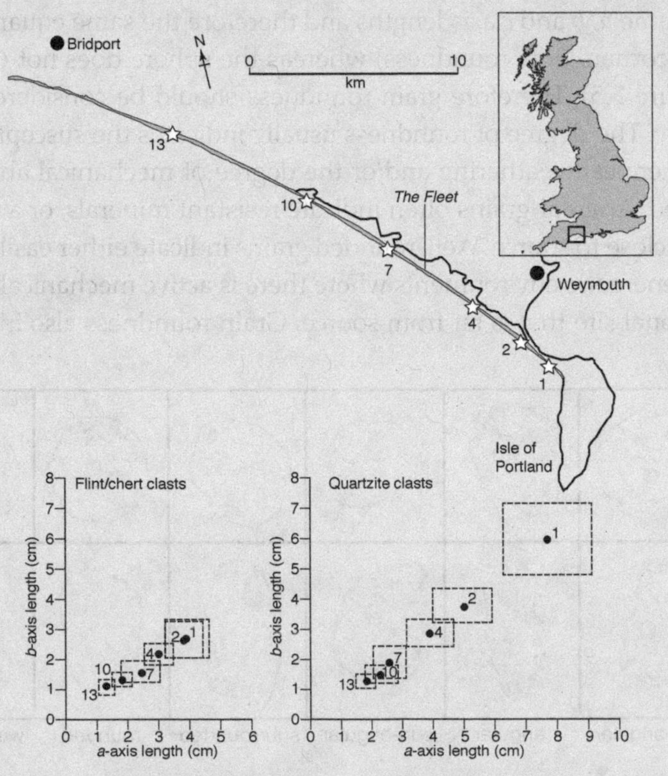

Figure 5.6 Plots of variation in the mean value and range of the *a, b, c* axis dimensions of flint/chert and quartzite clasts from different sites along Chesil Beach, southern England. (Drawn from data from Carr *et al.*, 1970).

beach, plotted in Figure 5.6. There are several important points to note. First, both quartzite and flint/chert clasts get smaller in the direction of longshore drift. This is due to abrasion by waves and during longshore transport. Second, the size range of both lithologies becomes smaller in the direction of longshore drift, in other words the clasts become more uniform. Up-drift (site 1), the quartzite clasts are around 8 cm in diameter and the flint/chert around 4 cm, whereas down-drift (site 13) they are all around 1–2 cm. Third, the much larger diminution of quartzite clasts over the length of the beach compared with the flint/chert clasts shows that the quartzite is more easily eroded, with the erosional products (sand-sized grains) probably transported offshore. Throughout these longshore changes in clast size, it is notable that clast shape does not change at all: all quartzite clasts remain equant and all flint/chert clasts remain bladed.

Clast transport along Chesil Beach is very episodic, both spatially and temporally (Carr, 1971). A key factor controlling longshore drift speed, and therefore the rate of change of grain size, shape and sorting, is the angle of wave approach and wave frequency. This relationship is explored in more detail in Box 5.1. West (2009) maintains an interesting website on the geology of Chesil Beach, available at www.soton.ac.uk/~imw/chespeb.htm

5.2.4 Sediment packing and porosity

The arrangement of deposited grains is referred to as **sediment packing**. Given the wide range of natural grain sizes and shapes that exists, there are many possible packing arrangements. The rate of grain deposition and the direction from which they are deposited can also influence sediment packing. Many sediment grains approximate the shape of a sphere, so it is useful to consider the two packing arrangements possible for uniform spheres (see Allen, 1985). The **cubic** arrangement is where spheres sit directly on top of each other, forming a box-shaped structure with vertical sides. The **rhombohedral** arrangement is where spheres sit in the hollows formed at the intersection of four touching spheres in the layer below them. This packing arrangement forms a rhomboid structure with sloping sides.

Sediment packing is significant because it has implications for bulk-sediment density, sediment concentration and sediment porosity, which are key geotechnical properties of the sediment body as a whole. The **bulk-sediment density** is the total sediment mass divided by the total volume of the packed sediment (i.e. grains plus void spaces). The bulk-sediment density will always be less than the density of constituent grains due to the presence of voids. The **sediment concentration** is the total volume of the grains divided by the total volume of the packed sediment, and can be expressed as a fraction or a percentage. The **sediment porosity** is the volume of void spaces contained within the sediment, which is therefore equal to 100 per cent minus the sediment concentration. For very fine sediments the bulk-sediment density and concentration are controlling factors in their ability to be entrained by waves and currents (Section 5.5.1). Moreover, the sediment porosity is an important factor in beach–groundwater dynamics (see Box 8.2).

In practice, sediment populations have a mix of grain sizes, shapes and mineralogies. This means that their properties and behaviour do not always correspond to theoretical values. In addition, densely packed natural sediments tend towards the upper end of the theoretical ranges for both bulk density and concentration (lower end for porosity), because the smaller grains can occupy the voids between the larger grains. In addition, angular grains can lock together and help support more rounded grains. Indicative values of the sediment concentration and porosity for natural sands are shown in the bottom half of Table 5.3. While the packing arrangement, degree of sediment sorting and grain shape all have a significant effect on sediment concentration and porosity, mean grain size generally does not.

Table 5.3 Measured packing concentrations and porosities of some theoretical and natural materials in water (Allen, 1985). Loose packing was produced by a simple dumping of sediment onto the bed, whereas dense packing was produced by subsequently agitating the sediment. Note that the biogenic *Lithothamnium* sands have an elongated grain shape.

Material	Concentration (%)		Porosity (%)	
Theoretical				
Spheres (cubic packing)	52.0		48.0	
Spheres (rhombohedral packing)	74.0		26.0	
Natural (mean diameter)	Loose	Dense	Loose	Dense
Quartz sands (0.27 mm)	55.8	65.1	44.2	34.9
Quartz sands (1.04 mm)	54.1	62.8	45.9	37.2
Lithothamnium sands (3.2 mm)	58.7	70.3	41.3	58.7

5.2.5 Friction angle of sediment

If one considers a pile of sediment on which grains are accumulating one by one from above, the **friction angle**, also known as the angle of yield, is the maximum slope angle to which the sediment mass will develop prior to the initiation of grain avalanching down the slope (Figure 5.7a). After avalanching, the new slope angle is referred to as the **angle of rest**, which is always less than the friction angle. On a horizontal bed, the friction angle is effectively the angle through which a grain must be rolled in order for it to change its position on the bed, rather than falling back to its original position (Figure 5.7b). Experimental data show that the friction angle depends on grain size, shape, packing arrangement and surface texture of

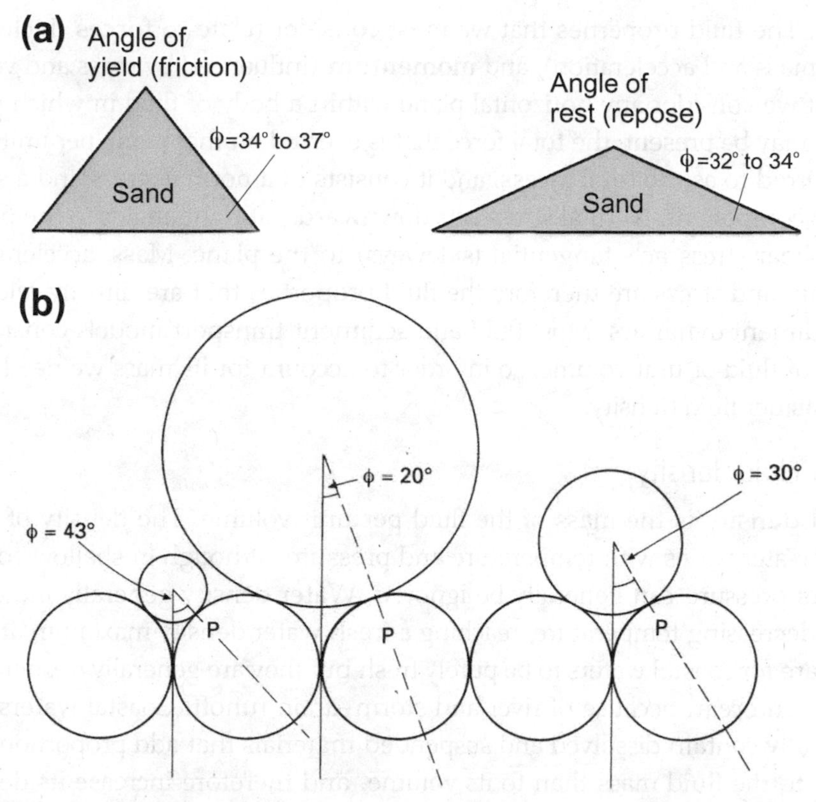

Figure 5.7 – (a) Schematic diagram showing relationships between angle of yield (friction angle) and angle of rest (repose) for a heap of natural sand grains. (b) Schematic diagram showing the role of friction angle in the initiation of sediment motion. The angle enclosed by the vertical line and the line marked P is the angle the grain must pivot through before it can change its position on the bed. This angle is equivalent to the friction angle. For a perfect sphere the pivot angle is 30° if it is resting on a bed of similar sized grains. If the bed is composed of larger or smaller grains, the pivot angle is correspondingly larger and smaller, respectively. (From Pye, 1994.) (Copyright © 1994 Blackwell Publishers, reproduced with permission.)

the grains. Friction angles for natural sand grains are on average 34°–37°, but individual grains can be well outside this range (Pye, 1994). Smaller angles within this range are associated with loosely-packed spherical grains and larger angles with densely-packed angular grains. Friction angles for coarser sediments can reach 45° (Allen, 1985). It is not clear why there is a size-dependence in the experimental data, since the theoretical friction angles for two different sized spheres resting on spheres of similar size are equal. It may be that larger grains tend to have larger roughness elements, increasing the surface friction.

5.3 Fluid properties

Fluid properties can explain how wind, currents and waves can apply forces and impart momentum to sediments in order to initiate and maintain grain movement. The fluid properties that we must consider relate to **forces** (including fluid mass and acceleration), and **momentum** (influenced by mass and velocity). If we consider any horizontal plane within a body of fluid in which sediment may be present, the total force that is exerted on that plane per unit area is referred to as the **total stress**, and it consists of a normal stress and a shear stress component. Normal stress acts downwards, at right angles to the plane, and shear stress acts tangential (sideways) to the plane. Mass, acceleration, velocity and stress are therefore the fluid properties that are directly relevant to sediment dynamics. Most fluid and sediment transport models consider a body of fluid of unit volume, so in order to account for its mass we need only to consider fluid density.

5.3.1 Fluid density

Fluid density is the mass of the fluid per unit volume. The density of pure fresh water varies with temperature and pressure, although in shallow coastal waters pressure can generally be ignored. Water density generally increases with decreasing temperature, reaching a fresh water density maximum at 4°C. It is rare for coastal waters to be purely fresh but they are generally fresher than the open ocean because of river and storm-drain runoff. Coastal waters also generally contain dissolved and suspended materials that add proportionately more to the fluid mass than to its volume, and therefore increase its density over that of fresh water. For example, open ocean waters contain dissolved salts that have a mass of *c.* 35 mg for every 1 kg of water. This standard volume of dissolved solutes is usually denoted as 35 parts per thousand (ppt or ‰) or 35 PSU (practical salinity units). Coastal waters typically contain less dissolved salts due to their dilution by fresh water, so salinity is generally less, in the range 0–35 ppt. Table 5.4 lists some indicative values for water density.

Table 5.4 Typical values for fluid density and molecular viscosity. Note the three orders of magnitude difference in density, and two orders of magnitude difference in viscosity between air and water.

Fluid type	Fluid density (kg m^{-3})	Fluid viscosity (N s m^{-2})
Air		
At 10°C	1.3	1.80×10^{-5}
Water		
Fresh (10°C)	1000	1.06×10^{-3}
Saline (35 ppt at 10°C)	1027	1.40×10^{-3}

5.3.2 Shear stress and viscosity

Most fluids offer some inherent resistance to deformation or flow, and are therefore referred to as viscous fluids. This resistance to deformation is provided by molecular forces within the fluid and is termed the fluid's **molecular viscosity**. The molecular viscosity of water can be measured by an experiment in which a thin body of water is deformed between two smooth plates (Figure 5.8). The bottom plate is kept stationary and the top plate is moved with a velocity u. The layer of fluid in immediate contact with each plate must have the same velocity as the plate, but this fluid velocity decreases with distance away from the moving plate due to internal friction between the water molecules. The result is a linear velocity-gradient within the fluid where the fluid velocity decreases from a maximum within the layer in contact with the top plate, to zero within the layer in contact with the bottom plate (Figure 5.8). At any level in the fluid the normal stress (the vertically-acting force per unit area) is equal to the product of the fluid's mass and the gravitational acceleration. The **shear stress** (the tangential-force per unit area) at any level in the fluid τ can be written as

$$\tau = \mu \frac{du}{dz} \qquad \textbf{(5.13)}$$

where μ is the molecular viscosity of the fluid, and u and z are defined in Figure 5.8. The molecular viscosity provides the link between the shear stress and the velocity-gradient at any level in the fluid. For a given fluid volume the molecular

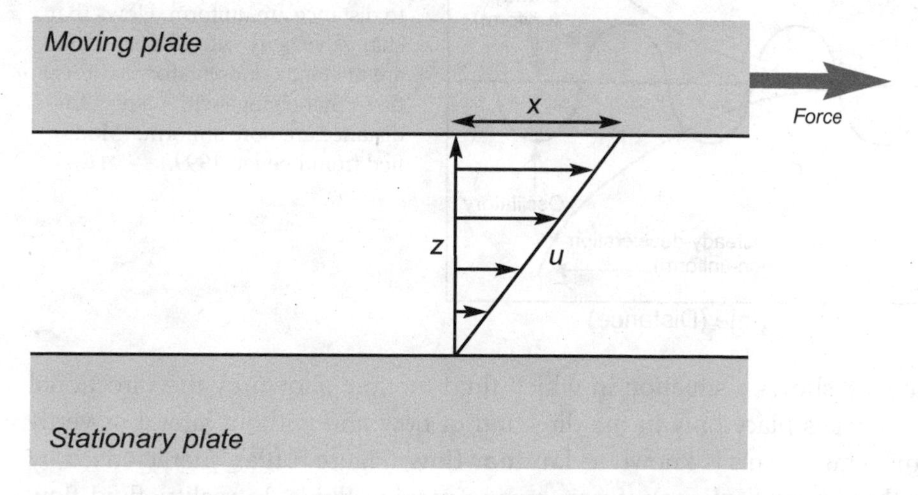

Figure 5.8 Experimental design used to measure the molecular viscosity of a fluid. A thin film of the fluid is sheared between two plates by moving the top plate horizontally and keeping the bottom plate still. The velocity profile through the fluid is represented by arrows whose length is proportional to the fluid velocity. (Modified from Allen, 1994.)

viscosity varies with density, thus the viscosity of water is considerably greater than that of air, and the viscosity of ocean water is greater than that of fresh water (Table 5.4).

5.3.3 Flow velocity, acceleration, laminar and turbulent flow

We must now define specific terms related to fluid motion. The flow velocity u is the distance travelled by a fluid parcel per unit of time. Flow acceleration can be either spatial or temporal. **Spatial acceleration** is the change in flow velocity per unit of distance x (i.e. du/dx), and **temporal acceleration** is the change in flow velocity per unit of time (i.e. du/dt). If the temporal (spatial) acceleration is constant then it produces a uniform **steady flow** in which the velocity is constant through time (space). If there is an increase or decrease in velocity with time (distance) then a non-uniform or **unsteady flow** is produced. If there are rhythmic changes in flow velocity or direction over time, then this produces an **oscillatory flow** (Figure 5.9). For example, if tidal flows are observed over a short enough time period (*c.* 10–15 minutes), they can be approximately steady. If the period of observation is increased beyond one tidal period, then tidal flows are seen to be oscillatory. Patterns and calculations of sediment transport are also related to these flow properties (Section 5.5).

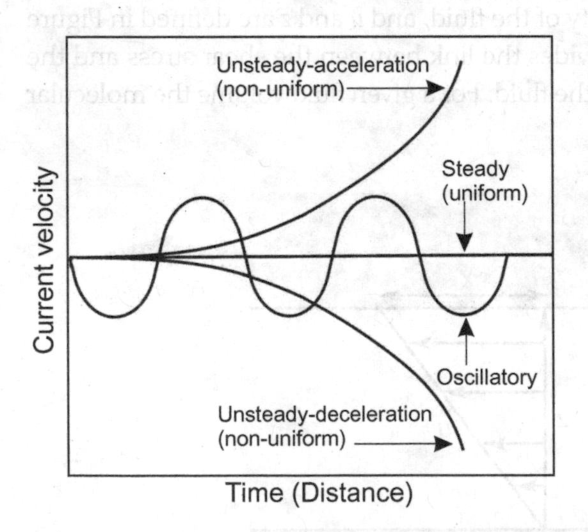

Figure 5.9 Schematic diagram of current velocity records used to define various flow types. Flows that are constant with respect to time are steady, and constant with respect to distance are uniform. Flows that change velocity with respect to time are unsteady or oscillatory, whereas flows that change with respect to distance are non-uniform. (Modified from Leeder, 1999.)

Figure 5.8 shows a situation in which fluid motion, shown by the directional arrows, takes place only in the direction of flow and without lateral or vertical movement. This is known as **laminar flow** (Figure 5.10a), characterised by very thin layers (laminae) of non-mixing moving fluids. In reality, fluid flow is rarely laminar because of frictional effects that retard flow velocity, but is more commonly turbulent. **Turbulent flow** has fluid movement mostly in one

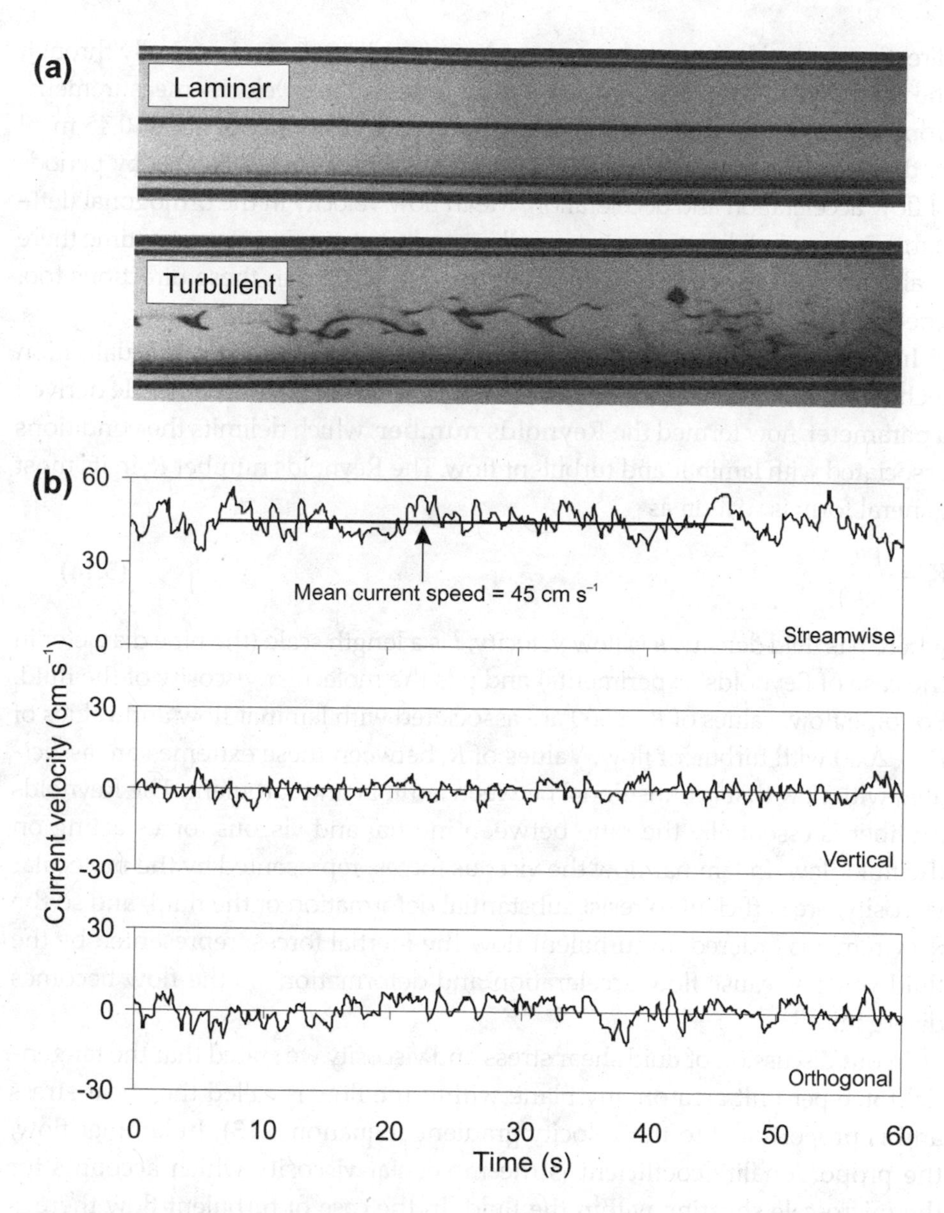

Figure 5.10 (a) Laminar (top panel) and turbulent (bottom panel) pipe flow visualised by a horizontal streak of dye along the centreline of flow. (Photographs by N. Johanneson and C. Lowe. Modified from van Dyke, 1982.) (Copyright © 1982 Parabolic Press, reproduced with permission.) (b) Time series of instantaneous velocity vectors in the streamwise, orthogonal and vertical directions for a turbulent tidal current. Turbulence is indicated by fluctuations in current velocity about the mean. Time series of current velocity for a steady, laminar flow would show horizontal lines without fluctuations.

direction, but with apparently random deviations laterally and vertically through the fluid body. Figure 5.10b shows a time series of flow velocity measurements from a quasi-steady turbulent flow that has a mean velocity of about 0.45 m s^{-1} in the principal streamwise flow direction, but which also fluctuates by periods of flow acceleration and deceleration. Mean flow velocity in the orthogonal (left–right) and vertical directions are typically zero, but at any one instant in time there is also short-lived acceleration/deceleration of the fluid in these directions too. The result is that the fluid parcel swirls downstream in a chaotic way.

In 1883, the physicist Osborne Reynolds (1842–1912) used flow visualisation techniques to demonstrate the nature of laminar and turbulent flow. He derived a parameter, now termed the **Reynolds number**, which delimits the conditions associated with laminar and turbulent flow. The Reynolds number R_e in its most general form is written as

$$R_e = \frac{\rho u l}{\mu} \tag{5.14}$$

where ρ is fluid density, u is flow velocity, l is a length scale (the pipe diameter in the case of Reynolds' experiments) and μ is the molecular viscosity of the fluid. For pipe flow, values of $R_e < 500$ are associated with laminar flow and values of $R_e > 2000$ with turbulent flow. Values of R_e between these extremes are associated with flow that is transitional between laminar and turbulent. The Reynolds number is essentially the ratio between inertial and viscous forces acting on the fluid flow. In laminar flow the viscous forces, represented by the molecular viscosity, are sufficient to resist substantial deformation of the fluid, and so the flow remains ordered. In turbulent flow the inertial forces, represented by the fluid velocity, cause flow acceleration and deformation, so the flow becomes disorganised.

In our discussion of fluid shear stress and viscosity we stated that the tangential force per unit area on any plane within the flow is called the shear stress and is proportional to the velocity gradient (Equation 5.13). In laminar flow, the proportionality coefficient is the molecular viscosity which accounts for the microscale shearing within the fluid. In the case of turbulent flow there is also shear arising from the friction between adjacent fluid eddies with differing momentum. This produces an apparent viscosity that is additional to the molecular viscosity. We call this apparent viscosity due to turbulence the **eddy viscosity** ξ. If Equation 5.13 defines the shear stress within laminar flow, then the shear stress in turbulent flow is defined as

$$\tau = (\mu + \xi)\,\frac{du}{dz} \tag{5.15}$$

Unlike the molecular viscosity, which is constant for a given fluid, the eddy viscosity varies with the flow conditions and increases with increased turbulence. Representative values for eddy viscosity are generally much larger than for molecular viscosity, and indicate the greater mixing and more rapid transfer of momentum that occurs within turbulent flows. This therefore makes turbulent flows more energetic, and potentially more erosive, than laminar flows. The nature of the bed over which the turbulent flow is moving is also important. This factor is now examined.

5.4 Benthic boundary layers

When a fluid is in motion against a boundary such as the sea bed (but also including a coastline or sea wall), friction arises between the two. This friction initially affects only the fluid motion in direct contact with the bed, but over time these effects reach to higher elevations in the flow. The region of fluid that is closest to the bed and influenced by frictional effects is termed the **benthic boundary layer**. Concepts of boundary layer flow are rooted in the work of the German hydrodynamicist Ludwig Prandtl (1875–1953).

Experiments have shown that the mean horizontal velocity within the boundary layer increases from zero at the bed (the no-slip condition) to a maximum value at the top of the boundary layer (Figure 5.8). A plot of mean horizontal flow velocity at increasing elevations above the bed defines a curve that is called the **velocity profile** (Figure 5.11). When a fluid begins to flow over the bed the boundary layer is initially laminar and its thickness grows slowly over time, therefore with distance travelled. In most coastal settings laminar flow is short-lived, and the boundary layer develops through the transitional regime to become a turbulent boundary layer. Momentum transfer from the boundary up through the flow is slow in laminar boundary layers because the fluid viscosity is small (only molecular viscosity), whereas it is rapid in turbulent boundary layers because the viscosity is large (molecular plus eddy viscosity). The enhanced rates of momentum exchange that take place by macroscopic mixing in turbulent boundary layers explain why they can grow in thickness at much faster rates, and why they have much steeper near-bed velocity gradients than laminar boundary layers (Figure 5.11).

In our earlier discussion we described shear stress as the tangential (sideways) component of stress acting on any plane within the fluid. With regard to sediment dynamics the **bed shear stress** is the tangential component of stress occurring on the fluid plane that is in contact with the bed. The bed shear stress is also positively related to the velocity gradient (Equation 5.13). The bed shear stress beneath a turbulent boundary layer is greater than beneath a laminar

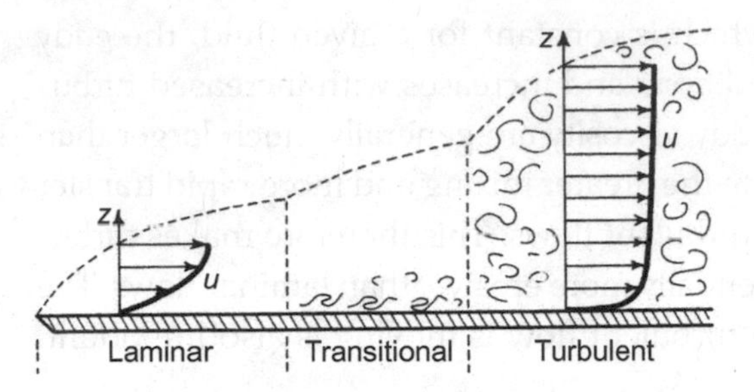

Figure 5.11 Schematic illustration of a growing boundary layer and its transformation from laminar to turbulent flow. Notice that the velocity profile in the turbulent boundary layer is steeper, thus larger current velocities impinge on the bed. (From Allen, 1985.) (Copyright © 1985 George Allen & Unwin, reproduced with permission.)

boundary layer, because of the steeper velocity gradient. This means that eroded products can be lifted from the bed into the flow through turbulent mixing. For a more detailed discussion of benthic boundary layers the interested reader is referred to Allen (1985) and Hsu (2004). Here we focus on methods for calculating bed shear stress in steady and oscillatory boundary layer models.

5.4.1 Boundary layer model for steady flow

Channelised river or tidal flows in deltas and estuaries and wind-driven water currents on the inner shelf are all approximately steady flows when observed over short time periods. In these situations the benthic boundary layer grows with distance travelled by the flow, and if permitted to become fully established, it can occupy the entire water depth. So far we have discussed the benthic boundary layer as a single entity, but in detail it consists of three layers (Figure 5.12a):

- The **bed layer** is generally 1–10 cm thick and consists of up to two sublayers: the buffer sublayer and viscous sublayer. The presence or absence of the viscous sublayer depends on the hydraulic characteristics of the boundary (see below).
- The **logarithmic layer** is where the velocity increases logarithmically with height above the bed. This layer is generally 1–2 m thick.
- The **outer layer** occupies about 85 per cent of the total boundary layer, which may be tens of metres in suitable water depths.

The stated thickness of each layer is only indicative and depends on the steadiness of the flow. The more steady the flow, the thicker the layers will become.

The **viscous sublayer**, when present, is a thin layer of laminar flow (< 1 cm thick) that separates the turbulent buffer sublayer from the bed (Figure 5.12b). It only exists on hydraulically-smooth beds, which are generally composed of mud (silts and clays) or very fine sand. Hydraulically-rough beds with higher friction are generally composed of sediments coarser than fine sand. Coarser grains protrude up through the viscous sublayer and disrupt the laminar flow, thus enabling the turbulent flow in the buffer sublayer to impinge directly onto

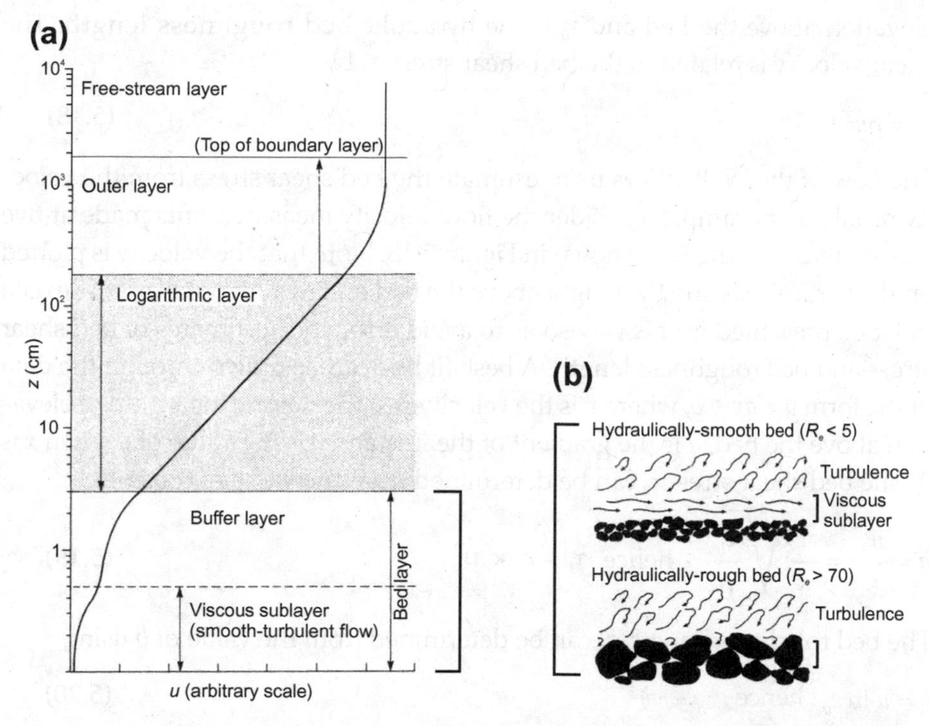

Figure 5.12 (a) Schematic current velocity profile plotted on log-linear axes and showing different layers within the boundary layer. The free-stream layer extends from the top of the boundary layer to the water surface. The vertical dimensions shown are indicative of a tidal current some 10 m deep. (Modified from Wright, 1989.) (b) Schematic diagram showing the relationship between flow and bed conditions in the presence and absence of a viscous sublayer. (Modified from Allen, 1994.)

the bed. Hydraulically-smooth and -rough beds can be distinguished by a variant of the Reynolds number, termed the boundary Reynolds number R_e^*

$$R_e^* = \frac{\rho u_* k'}{\mu} \qquad (5.16)$$

where u_* is the shear velocity and k' is the skin friction roughness length. These two terms are defined below. Hydraulically-smooth boundaries occur when $R_e^* < 5$, hydraulically-rough boundaries occur when $R_e^* > 70$, and transitional conditions occur between values of 5 and 70.

The velocity profile in the logarithmic layer can be described by an equation known as the **Law of the Wall**

$$u = \frac{u_*}{\kappa} \ln\left(\frac{z}{z_0}\right) \qquad (5.17)$$

where u_* is the shear velocity, κ is von Karman's constant (equal to 0.4), z is elevation above the bed and z_0 is the hydraulic **bed roughness length**. The shear velocity is related to the bed shear stress τ_b by

$$\tau_b = \rho u_*^2 \tag{5.18}$$

The Law of the Wall allows us to estimate the bed shear stress from the velocity profile. For example, consider the flow velocity measurements made at five elevations above the bed, shown in Figure 5.13. Note that the velocity is plotted on the vertical axis and the height above the bed on the horizontal axis. The data must be presented in this way so as to avoid erroneous estimates of bed shear stress and bed roughness length. A best-fit line can be drawn through the data of the form $y = ax + b$, where y is the velocity, x is the natural logarithm of elevation above the bed, a is the gradient of the line and b is the value of y when x is 0. The bed shear stress t_b can be determined from the value of α using

$$a = \frac{u_*}{\kappa} = \frac{1}{\kappa}\sqrt{\frac{\tau_b}{\rho}} \quad \text{hence} \quad \tau_b = \rho\,(\kappa\, a)^2 \tag{5.19}$$

The bed roughness length z_0 can be determined from the value of b using

$$b = a \ln z_0 \quad \text{hence} \quad z_0 = e^{-b/a} \tag{5.20}$$

For the example shown in Figure 5.13 the regression coefficients a and b are 0.23 and 1.33, which yields a bed shear stress of 8.69 N m^{-2} and a bed roughness length of 0.0031 m.

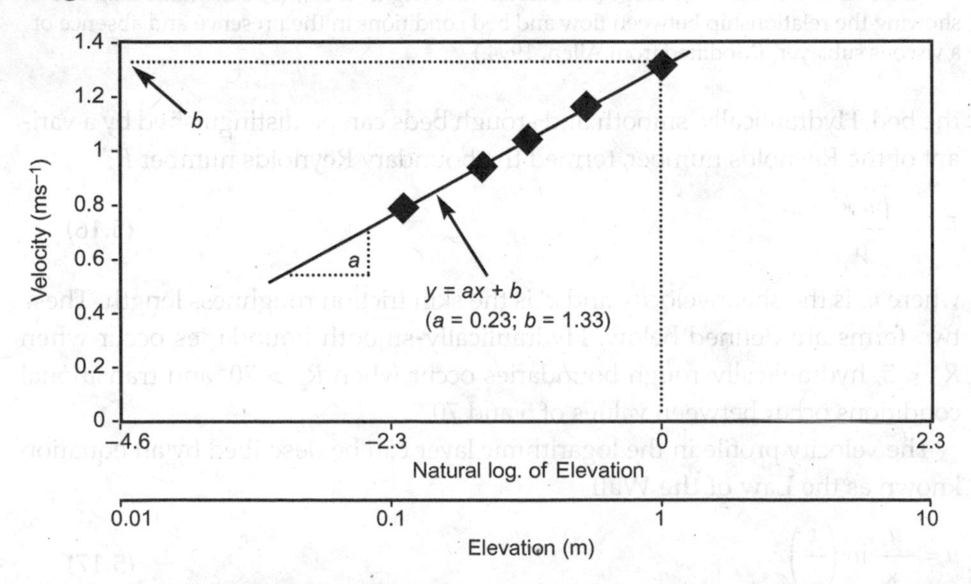

Figure 5.13 Illustration of method for estimating bed shear stress and bed roughness using the Law of the Wall. The velocity measurements at five elevations above the bed are shown as diamonds and the line of best fit through the data is shown as the solid line.

While the method described in Figure 5.13 can estimate bed shear stress and bed roughness, it requires several measurements of flow velocity in the logarithmic layer, which is not always possible. Field investigations of flow over different sea bed types have yielded an alternative method for estimating bed shear stress based on a single velocity measurement

$$\tau_b = \rho C_d u_{100}^2 \quad (5.21)$$

where C_d is the fluid-drag coefficient and u_{100} is the mean horizontal velocity at 100 cm above the bed. Some typical values of the drag coefficient and bed roughness lengths for different sea bed types are listed in Table 5.5.

Table 5.5 Indicative values for z_o and C_d under steady flow over different bed types (Soulsby, 1983).

Bottom type	z_o (mm)	C_d
Mud	0.20	0.0022
Mud/sand	0.70	0.0030
Silt/sand	0.05	0.0016
Sand (unrippled)	0.40	0.0026
Sand (rippled)	6.00	0.0061
Sand/shell	0.30	0.0024
Sand/gravel	0.30	0.0024
Mud/sand/gravel	0.30	0.0024
Gravel	3.00	0.0047

5.4.2 Boundary layer model for oscillatory flow

While steady and oscillatory flows within boundary layers are generally similar, a different method for estimating bed shear stress under waves is required because of the temporal variation in the boundary layer structure under waves. Figure 5.14 shows the boundary layer velocity profile at several phases of the oscillatory wave cycle. During each cycle the horizontal flow velocity accelerates, then decelerates, crosses zero (i.e. changes direction), then accelerates and decelerates again. Consequently, a new boundary layer grows and decays twice during each wave cycle: once on the forward stroke of the wave and once on the backward stroke. Due to the short time periods involved (generally a few seconds), the boundary layer thickness does not reach much more than 10 cm for long-period swell waves, and is considerably smaller for short-period wind waves. The relatively thin boundary layer beneath oscillatory flows means that, for a given free-stream velocity and bed roughness, the bed shear stress in oscillatory flows is always larger than beneath steady flow.

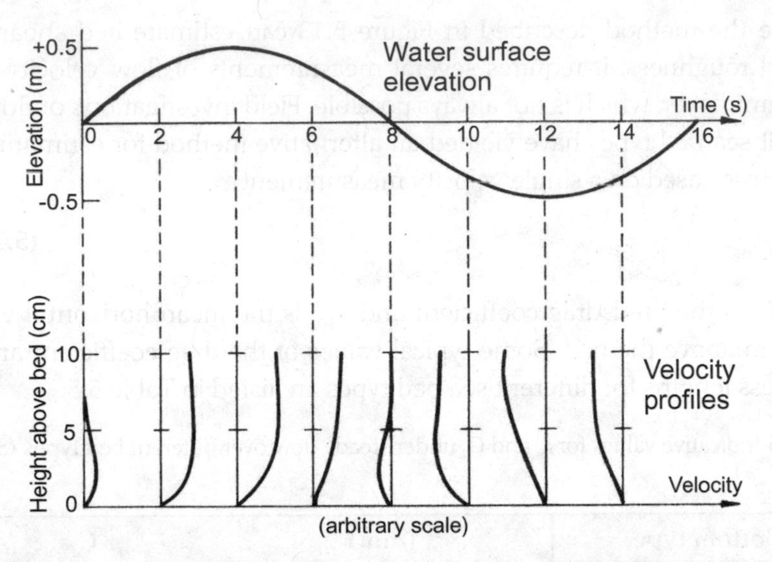

Figure 5.14 Schematic diagram showing current velocity profiles in oscillatory flow. During the first 8 seconds the flow is in the direction of wave travel. The current speed increases and the boundary layer grows in height in the period 0 to 4 seconds. In the period 4 to 8 seconds the current speed decreases and the boundary layer diminishes. A similar pattern occurs during the period 8 to 16 seconds, however, the flow direction has reversed. The vertical dimensions shown are indicative of a swell wave with a height of 1 m and a period of 16 s. (Modified from Sleath, 1984.)

Worked Example 3

The method used to estimate bed shear stress under oscillatory flow first requires a measurement of the maximum horizontal flow velocity at the top of the wave boundary layer u_0. If a measurement is not available then linear wave theory can be used to predict this flow velocity (see Equation 4.13). Next we need to estimate the bed friction factor under waves f_w, which can be obtained from (Nielsen, 1992)

$$f_w = \exp\left[5.5\left(\frac{k_s}{d_o}\right)^{0.2} - 6.3\right] \qquad \textbf{(5.22)}$$

where d_o is the wave orbital diameter, also obtainable from linear wave theory (see Equation 4.12) and k_s is the **Nikuradse roughness length**, which is analogous to the bed roughness length described for steady flows (Section 5.4.1). An empirical recipe is available to estimate the Nikuradse roughness length, but it is not straightforward. There are at least two contributors to the total roughness. The first is the grain (skin friction) contribution k' which is related to the roughness of individual grains making up the bed. The second is the bedform contribution k'' which is related to the roughness of any bedforms present on the bed. In most cases the total Nikuradse roughness length k_s will simply be the sum of the grain and bedform roughness contributions

$$k_s = k' + k'' \quad (5.23)$$

Experimental data suggest the following formulae for estimating each roughness contribution (Nielsen, 1992)

$$k' = D \quad (5.24)$$

and

$$k'' = 8\frac{\eta^2}{\lambda} \quad (5.25)$$

where D is mean grain diameter of the bed sediment, η is bedform height and λ is bedform spacing. Now we can estimate the maximum flow velocity at the top of the boundary layer and the wave friction factor, we can calculate the maximum bed shear stress under waves τ_w using

$$\tau_w = \frac{1}{2}\rho f_w u_o^2 \quad (5.26)$$

Now that we have the tools to calculate the bed shear stress under either steady or oscillatory boundary layers, we can describe the sediment dynamics under these fluid flows.

5.5 Sediment dynamics

The dynamic behaviour of sediment in a moving fluid is strongly determined by sediment grain size. For grain sizes greater than 63 μm the grains are free to behave individually and single-grain properties (i.e. the size of individual grains) are most important. For grain sizes less than 63 μm the sediment grains are cohesive due to electrostatic forces. The dynamic behaviour of cohesive sediment depends less on single-grain properties and more on bulk-sediment properties (e.g. floc size and water content) (see Section 7.4). This distinction in sediment response to fluid flow is most apparent when considering sediment entrainment and deposition mechanisms.

5.5.1 Sediment entrainment and resuspension

A cohesionless grain at rest on a bed of similar grains experiences no acceleration, so all of the forces acting on the grain must be in equilibrium. The forces involved are lift, drag and weight forces. The lift force arises due to the slightly faster flow across the top of the grain compared to the bottom of the grain, which results in a pressure differential. The drag force arises due to the friction between the fluid and the particle, and therefore acts horizontally in the direction of mean flow. The weight force arises from the mass of the grain being acted upon by gravity. In order for the grain to move it must pivot over the adjacent grain through an angle equal to the friction angle. This movement can only occur if the lift and drag forces overcome the weight force. In principle, mathematical expressions can be derived to describe the nature of these forces, which can then be solved for the critical flow conditions required to initiate grain motion. In practice, however, the drag and particularly the lift forces are difficult to quantify due to the variability of flow conditions. For this reason the flow conditions necessary to initiate sediment movement are usually predicted from data from laboratory experiments.

Based on experimental data, Shields (1936) proposed that, when grains begin to move in a steady current, a relationship exists between bed shear stress and grain size. In order to make the data performed under a range of experimental conditions comparable, the **critical bed shear stress** required to initiate grain motion is transformed into a non-dimensional parameter known as the **Shields parameter**

$$\theta_c = \frac{\tau_c}{gD(\rho_s - \rho)} \qquad \textbf{(5.27)}$$

The non-dimensional grain diameter D_* that appears on the horizontal axis in Figure 5.15 is given by

$$D_* = D\left[\frac{\rho^2 g(s-1)}{\mu^2}\right]^{1/3} \qquad \textbf{(5.28)}$$

For non-dimensional grain diameters less than a value of around 10 there is an inverse relationship between grain diameter and the critical Shields parameter, whereby the bed shear stress required to initiate sediment motion increases with decreasing grain diameter (Figure 5.15). This is due to a combination of factors. Silt and clay sized grains are readily compacted to the point that individual grains do not protrude through the viscous sublayer and are therefore not exposed to turbulent flow. This corresponds to hydraulically smooth conditions (Figure 5.12b). Silt and clay grains are also cohesive and therefore experience electrostatic attraction to the bed and to each other which increases in effect as the grains become smaller. This makes these grains difficult to entrain despite their small mass. For non-dimensional grain diameters greater than a value of 10, grain cohesion factors are less important and the grains are sufficiently large

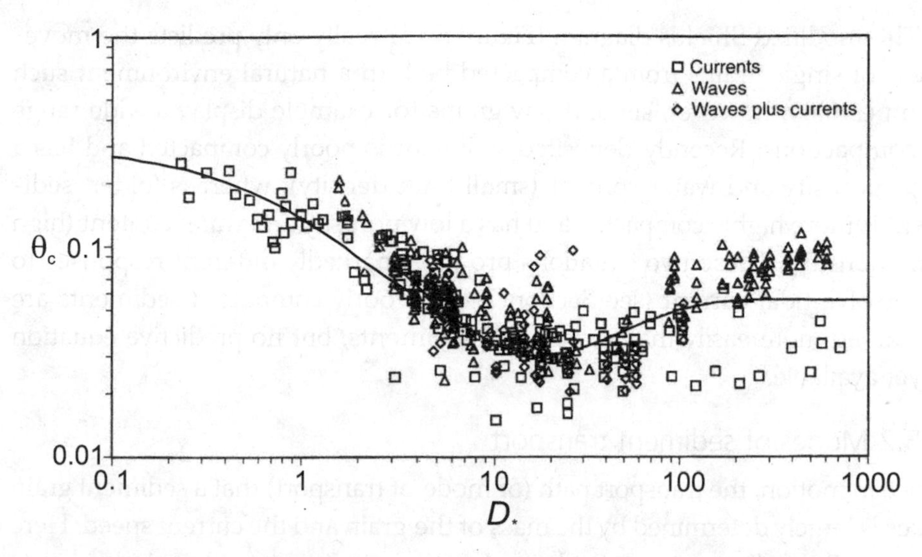

Figure 5.15 Modified Shields diagram showing empirical data and best fit function (Equation 5.29) for predicting the critical Shields parameter necessary to initiate sediment motion. Equation 5.27 can be used to determine the equivalent critical bed shear stress. For the case of waves this approach yields the critical maximum bed shear stress for the wave cycle. (Modified from Soulsby, 1997.)

to break up the viscous sublayer such that turbulent flow impinges directly on the bed. This corresponds to hydraulically-rough conditions (Figure 5.12b). For these larger grains, the weight force is of overriding importance, so there is only a weakly positive relationship between critical bed shear stress and grain size (Figure 5.15).

Based on the data shown in Figure 5.15 the following expression has been proposed to predict the critical Shields parameter (and bed shear stress) required to initiate grain motion

$$\theta_c = \frac{0.30}{1 + 1.2D_*} + 0.055[1 - \exp(-0.02D_*)] \quad \textbf{(5.29)}$$

It is important to note that the data used to produce Figure 5.15 are largely based on experiments involving uniform sediment grains. The poorly sorted sediments commonly found in natural environments may lead to a situation where surface grains are resting on a bed of larger or smaller grains. In the former case the pivot angle through which the grain must move is larger than the case for well sorted sediment (Figure 5.7b). In the latter case the pivot angle is smaller. There is a corresponding effect on the critical bed shear stress required to initiate motion. Equation 5.29 therefore only applies to well-sorted sediments.

The modified Shields diagram (Figure 5.15) really only predicts the movement of single grains from a compacted bed. In a natural environment such as an estuary, however, silt and clay grains for example display a wide range of compactions. Recently-deposited sediment is poorly compacted and has a high porosity and water content (small bulk density), whereas 'older' sediment is more highly compacted and has a low porosity and water content (high bulk density). These two situations produce markedly different responses to an erosive tidal current (see Section 7.4.2). Poorly compacted sediments are entrained more easily than compacted sediments, but no predictive equation is yet available.

5.5.2 Modes of sediment transport

Once in motion, the transport path (or mode of transport) that a sediment grain takes is largely determined by the mass of the grain and the current speed. Here we describe sediment transport by water currents. Transport by wind currents is discussed in Section 9.2. Sediment transport in water has two main modes (Figure 5.16):

- **Bedload** is where grains are supported by either continuous **(traction)** or intermittent contact (**saltation**) with the bed. In the case of traction, grains slide or roll and maintain contact with the bed at all times. This is a relatively slow transport mode and is typical when weak currents are transporting sands or strong currents are transporting pebbles and boulders. In the case of saltation the grains take short hops along the bed in the direction of flow. Saltation is typical when moderate currents are transporting sand or very strong and turbulent currents are transporting gravel and pebbles. Saltation is the most important mechanism of sand transport by wind (see Figure 9.7).

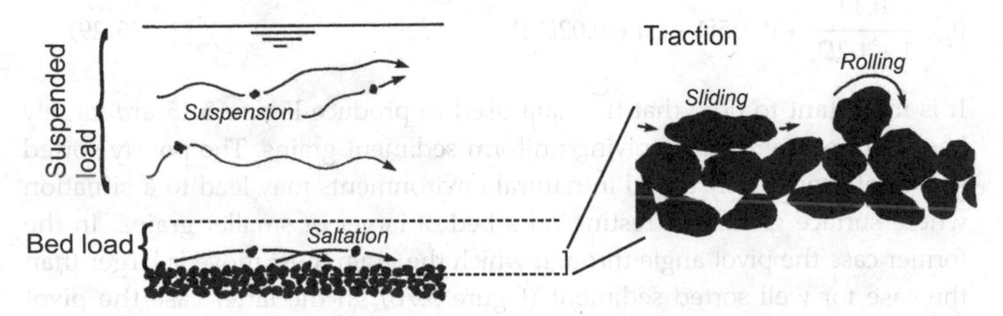

Figure 5.16 Schematic representation of sediment transport modes showing grain paths. Note that bedload includes both saltation and traction. (Modified from Allen, 1994.)

- **Suspended load** is where grains are supported by turbulence within the fluid. The grain paths of suspended load are distinguishable from saltation due to their irregularity, which arises from the grains being buffeted by turbulent eddies within the fluid. Suspension transport is typical when moderate currents are transporting silts or strong currents are transporting sands.

Francis (1973) performed flume experiments to investigate the relationship between sediment transport mode and current velocity. His results can be summarised by introducing the concept of transport stage Ψ

$$\Psi = \frac{u_*}{w_s} \tag{5.30}$$

which is the ratio of the shear velocity u_* to grain settling velocity w_s. The former is a surrogate for all of the forces driving sediment transport, and the latter is a surrogate for all of the forces driving sediment deposition. For low Ψ, only bedload is transported. There is no clear threshold value for Ψ between bedload and suspended load transport modes. As Ψ increases, the

Box 5.1 Life in the slow lane: The transport speeds of large clasts

Sandy sediments are transported across the entire range of transport modes and, in general, the speed at which sand is transported is closely related to the transport stage Ψ (Equation 5.30). This equation shows that sand transport as bedload takes place at relatively slow speeds, and sand transport as suspended load takes place at relatively fast speeds. Gravels and pebbles, however, move almost entirely as bedload and the transport speed of individual clasts depends not only on transport stage, but also on their relationship to the background sediment mass. Based on the ratio of the size of an individual clast to the modal size of the surrounding sediment mass, there are three transport possibilities (Figure 5.17):

- If the ratio is less than half, then the grain movement will be impeded as the clast will lose its surface position and be buried by surrounding sediment; in other words, smaller clasts will be trapped within the interstices of larger ones and will not participate in sediment transport.
- If the ratio is greater than about two, then the clast will have a propensity to move faster than the surrounding sediment mass. This occurs because the clast will project further into the boundary layer, and will therefore experience larger fluid drag forces. The preferential

transport of such larger clasts is termed **overpassing**. As a result these clasts can travel further and faster than those around them.

- Clasts that are much bigger than the modal clast size and so have very large ratios may sink into the sediment (due to their large mass), become immobile, and form a **lag** or **armour** deposit. The armour develops over time as more mobile sediment is transported away leaving the largest clasts behind.

The net result of differential transport speeds is the sediment will become better sorted over time through progressive differential transport of grains of different sizes. In the context of gravel beaches, these processes may lead to segregation of the gravel deposit and the development of graded shorelines.

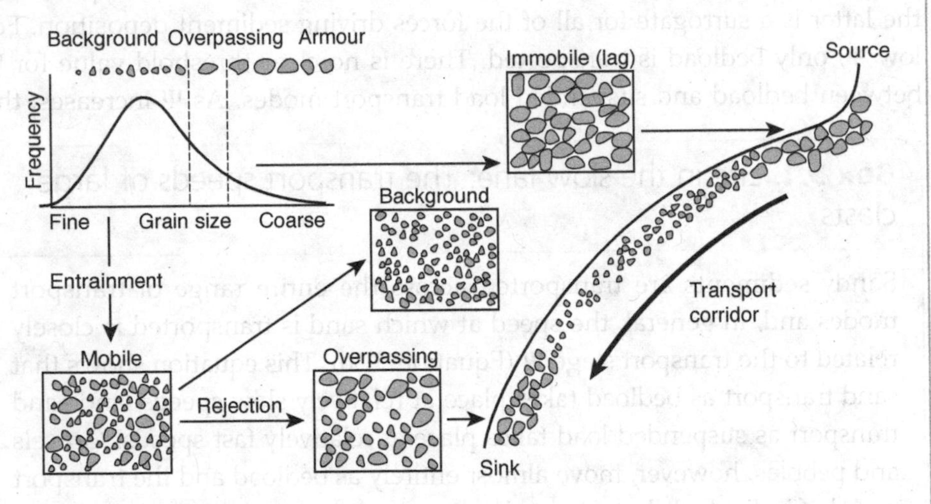

Figure 5.17 Schematic diagram demonstrating the importance of the clast size distribution to the relative speeds that individual clasts are transported. (Modified from Orford *et al.*, 1991.)

amount of material transported as bedload declines and the proportion travelling as suspended load rises. Moreover, the speed at which the grains are transported also increases. For high Ψ, almost all transport is as suspended load moving at a speed that is close to the speed of the fluid. Francis' (1973) experimental data are based on sand- and granule-sized material. A discussion of the factors controlling the speed of coarser sediment transport can be found in Box 5.1.

5.5.3 Models for calculating the transport rate

The **sediment transport rate** q can be defined as the mass of sediment transported per unit of cross-sectional area of flow per unit time. Consider a unit width of fluid that extends from the stationary bed level to the height of the highest saltating grains. The bedload transport rate q_b is the product of the mass of sediment in that cross-sectional area of fluid and the grain velocity, which is equivalent to

$$q_b = \int_{0}^{z=a} u_g(z)C(z)dz \qquad \textbf{(5.31)}$$

where u_g is the grain velocity as a function of height above the bed z, C is the concentration of grains as a function of height above the bed, and a is height of the bedload layer. Even though Equation 5.31 is theoretically valid, the bedload layer is typically only a few centimetres high and so it is difficult to measure grain velocities and concentrations directly. An alternative approach is therefore required. From our discussion of transport stage we know that sediment will be transported farther if the shear velocity (and bed shear stress) increase. It is therefore reasonable to expect that the sediment transport rate is proportional to the bed shear stress exerted by the flow, or more precisely the excess bed shear stress above the critical amount required to initiate grain motion. Field and laboratory experiments indicate that the bedload transport rate can be predicted using an equation of the form (e.g. Meyer-Peter and Muller, 1948)

$$q_b = A(\tau_b - \tau_c)^{1.5} \qquad \textbf{(5.32)}$$

where τ_b is bed shear stress, τ_c is the critical bed shear stress and A is a proportionality coefficient that depends on sediment properties.

The suspended load transport rate can also be determined using Equation 5.31, except that integration occurs over the entire water column height. Typically the only data available to predict suspended sediment transport rate is mean current velocity.

An alternative approach to the sediment transport problem was proposed by Bagnold (1963, 1966), who equated a transporting current to a machine and the sediment transport to work done by that machine. In this approach, termed the **energetics approach**, the work done (sediment transport rate) is proportional to the power of the machine (transporting current). For bedload transport Bagnold proposed

$$q_b = \frac{\tau_b u e_b}{\tan\phi - \tan\beta} \qquad \textbf{(5.33)}$$

where $\tau_b u$ is current power, e_b is the bedload efficiency factor, ϕ is the friction angle of the sediment and β is the slope angle of the bed. Bedload efficiency is smaller than a value of one because the current is not 100 per cent efficient – some of the power is lost due to frictional dissipation as heat and by grain–grain interaction. For suspended load transport Bagnold proposed

$$q_s = \frac{\tau_o u e_s}{(w_s/u) - \tan\beta} \tag{5.34}$$

where e_s is the suspended load efficiency factor. The transport efficiency factors depend on both flow conditions and sediment properties. For equivalent sediment properties, waves are generally more efficient at transporting sediment than steady currents.

In order to apply any of the sediment transport models to a practical situation that involves steady currents or waves, we need measurements of the bed slope, grain size, grain settling velocity and current speed. The bed shear stress is estimated using the method described in Section 5.4. Situations that involve combined waves and steady currents, however, are more complex. Nonlinear interactions between the waves and current produce a combined boundary layer that is not well described by either of the boundary layer models that we have presented. Combined wave-current boundary layer models exist, but they are beyond our scope here. In combined flows waves are mostly responsible for suspending sediment, and steady currents for transporting sediment.

5.5.4 Sediment deposition of non-cohesive sediments (large grains)

Sediment deposition involves the settling of grains towards the bed, from either the bedload or suspended load. In the case of bedload, when the bed shear stress and fluid turbulence are insufficient to keep the sediment moving, the grains cease horizontal movement and rapidly come to rest. Bed properties can be important in trapping sediment grains or inhibiting their movement. In the case of suspended load, the grains must settle a longer distance vertically through the fluid before coming to rest. Sediment will begin to settle when the upward acting fluid forces are insufficient to overcome the downward acting weight force on the grain.

When a single suspended grain starts to be deposited, it will initially experience a period of acceleration towards the bed. This acceleration is short-lived, however, because the grain's acceleration decreases to zero when it reaches its terminal (maximum) **fall velocity**. The grain falls through the fluid due to the downward acting weight force, but when the grain is travelling at its terminal fall velocity, the weight force is matched by a combined buoyancy and fluid-drag force. In the case of a spherical grain in a stationary body of water, the balance of forces can be expressed by

$$\underbrace{\frac{4}{3}\pi r^3 \rho_s g}_{\text{downward-acting weight force}} = \underbrace{\frac{4}{3}\pi r^3 \rho g}_{\text{upward-acting buoyancy force}} + \underbrace{\frac{1}{2} C_d \rho \pi r^2 w_s^2}_{\text{upward-acting fluid-drag force}} \tag{5.35}$$

The weight force is the product of the sphere's volume, its density and gravitational acceleration. The buoyancy force is the product of the sphere's volume, water density and gravitational acceleration. The fluid-drag force is the product of the drag coefficient, water density, the sphere's surface area and its velocity squared. The half at the front of the right-hand side of Equation 5.35 is there because the drag force is only acting on the leading hemisphere of the grain as it falls through the water.

The balance of forces in Equation 5.35 can be rearranged to yield an expression for the terminal fall velocity of the sphere

$$w_s = \sqrt{\frac{8gr}{3C_d} \frac{(\rho_s - \rho)}{\rho}} \tag{5.36}$$

If a spherical grain with a known diameter and density is settling in a fluid with a known density, then Equation 5.36 can be used to calculate the fall velocity, provided the drag coefficient is also known. For Reynolds numbers that are appropriate for small grains, the flow around the grain as it falls through the fluid is laminar and the drag coefficient can be estimated using

$$C_d = \frac{24}{R_G} \tag{5.37}$$

where R_G is the grain's Reynolds number in which the grain settling velocity and the grain diameter are consistent with those used in Equation 5.14. For larger Reynolds numbers, which may relate to larger settling velocities or larger grains in a less viscous fluid, prediction of the drag coefficient is less straightforward. Equations 5.36 and 5.37 constitute **Stokes Law of settling** and are strictly valid only for spheres settling in a fluid with a grain Reynolds number less than 20. In practice, for grains settling in water, this restricts Stokes Law to grain sizes less than 0.15 mm diameter (i.e. very fine sand, silt or clay). Stokes Law is not valid in air, due to the air's low viscosity and the larger density difference between sediment grains and air.

In the case of grains larger than very fine sand, the flow separation around the grains as they settle through the fluid complicates the drag force and leads to unpredictable behaviour. Moreover, the grain shape and roundness can modify the drag and therefore the fall velocity, the effects being larger for larger grains. For this reason the fall velocity is usually measured directly (see Lewis and McConchie, 1994, for a description), or calculated empirically. Soulsby (1997) has analysed data for typical sands and proposes the following formula to estimate the grain settling velocity

$$w_s = \frac{\mu}{\rho D} \left[\sqrt{(10.36^2 + 1.049\, D_*^3)} - 10.36\right] \qquad \textbf{(5.38)}$$

where D_* is the dimensionless grain size given by Equation 5.28. The settling velocity of sand grains increases directly with grain size and grain density, and increases inversely with fluid density and fluid viscosity.

5.5.5 Sediment deposition of cohesive sediments (small grains)

We now consider the settling of cohesive sediments comprising silt and clay with grain diameters smaller than 0.062 mm or 4ϕ. Stokes Law is appropriate for calculating the settling velocity of fine particles if they remain dispersed, but in coastal waters they generally combine to form **flocs**, which are composites of several particles loosely held together. Several flocs may also combine to produce an aggregate (Figure 5.18). The process of **flocculation** arises from the electrical charges present on individual particles. The face of a clay platelet has a slight negative charge and the edge a slight positive charge. When two platelets come into close proximity, the face of one particle and the edge of the other are electrostatically attracted. The probability of the particles coming together in fresh water is low, however, because the negatively charged faces of the two particles, which have a much larger surface area than the edges, will tend to repel the particles from one another. In the case of seawater the probability of effective attraction increases. Seawater is a strong electrolyte which helps neutralise the negatively-charged faces, thus facilitating electrostatic attraction. In order for flocculation in seawater to be significant, individual particles need to be brought very close together, by either Brownian motion within a concentrated suspension, differential settling of variously sized particles, and/or by fluid turbulence. Eisma (1993) discusses flocculation in some detail.

Flocs have the combined mass of their component particles, and so the effect of flocculation is to considerably increase their fall velocity. Since collisions of individual clay particles are a necessary precursor to flocculation,

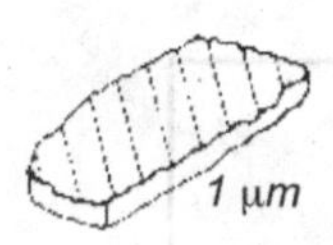

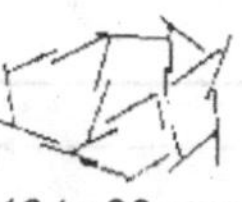

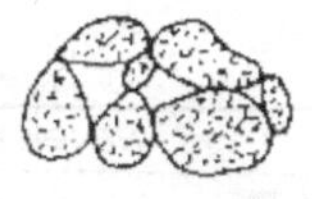

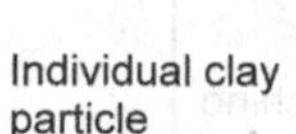

Figure 5.18 Indicative dimensions of a clay particle, a floc and an aggregate. (Modified from Eisma, 1993.)

the more concentrated the particle suspension the greater likelihood there is for flocculation. There is often a clear relationship between settling velocity and suspension concentration, although this relationship is not consistent from one site to another or over time. For example, Uncles *et al.* (2010) discuss how floc concentration and settling patterns vary within the Tamar estuary, southwest England. Spatially, the zone of highest suspended sediment concentration migrates up and down the estuary with tidal state, and temporally most large flocs (200–500 µm) settle around high and low tides when water velocities are lowest (Section 7.4). Local factors such as water biochemistry and the presence/absence of organic films on grains and faecal pellets are also important in flocculation. These factors are difficult to account for and have hindered attempts to develop a predictive equation for the settling velocity of cohesive sediment.

5.6 Bedforms

The generic term **bedform** refers to any upstanding morphological feature that is composed of unconsolidated sediments. The bedforms that we are interested in are those that make quasi-regular patterns on the sea floor as a result of bedload and, to a lesser extent, suspended load transport. Flow conditions control bedform morphology, and in turn bedforms locally modify both flow conditions and sediment transport modes as the bedform evolves. Bedforms increase bed roughness and therefore friction, and so reduce wave and current energy.

5.6.1 Dimensions of bedforms and relation to flow conditions

The nomenclature, flow pattern and sediment transport over a typical asymmetric bedform is shown in Figure 5.19a. As the flow rises over the stoss (or up-flow) side of the bedform it accelerates due to compression of the flow over the bedform crest. Flow acceleration increases sediment transport up the stoss side and towards the brink point of the crest. As the flow reaches the crest it separates from the bed and recirculates in the form of an eddy in the lee of the bedform. The sediment transported to the crest is piled to a slope angle greater

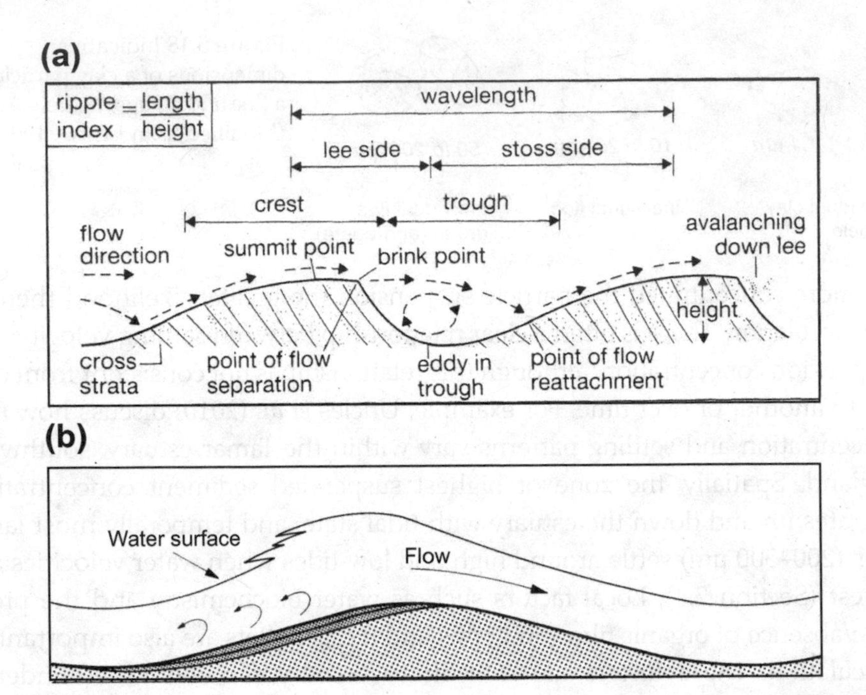

Figure 5.19 Schematic diagram showing nomenclature, flow and sediment transport pattern over bedforms in unidirectional flow: (a) ripple or dune (From Tucker, 1995.) (Copyright © 1995 Blackwell Publishers, reproduced with permission.); and (b) antidune. (From Reineck and Singh, 1980.) (Copyright © 1980 Springer-Verlag, reproduced with permission.)

than its yield angle and it consequently avalanches downslope. This cycle of sediment erosion from the stoss slope and deposition on the lee slope results in the bedform migrating in the direction of flow. The relationship between the bedform and the flow is closely coupled, with small changes in one causing a corresponding change in the other.

This example applies to unidirectional flow, which is typical of nearshore currents outside of the surf zone. Both waves and tides, however, are oscillatory flows with succeeding forward and backward components. The pattern for oscillatory flow over asymmetric wave ripples is shown in Figure 5.20. In this case, the relatively strong onshore stroke of the wave forms an eddy or vortex on the lee side of the wave ripple. Provided the onshore flow persists, this eddy remains trapped in the lee of the ripple, but when the flow reverses, the eddy is thrown upwards off the bottom and a small cloud of suspended sediment generated by the eddy is ejected into the water column above the ripple (see Box 5.2 for further discussion). The sediment cloud is moved seaward by the offshore stroke of the wave. Because the offshore stroke is relatively weak, no eddy forms in the lee of the ripple. The net suspended sediment transport

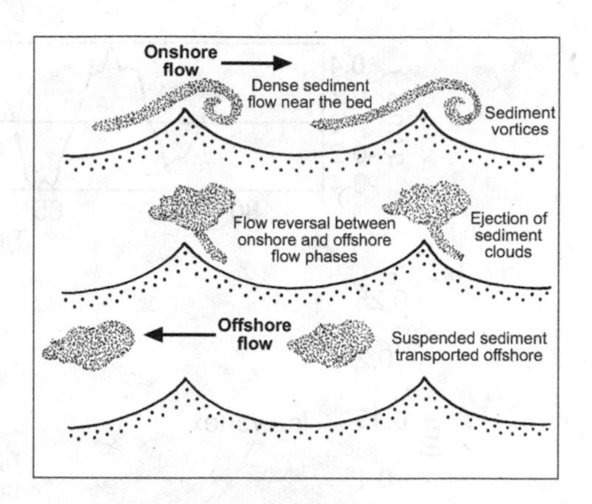

Figure 5.20 Schematic diagram showing suspended sediment transport by onshore asymmetric wave motion over sharp-crested ripples. See text for explanation.

during one wave cycle is therefore seaward, which produces the asymmetry in the ripple shape. If the orbital velocity-magnitude is symmetrical, then the wave ripples are symmetrical and a vortex (with associated sediment cloud) occurs on both sides of the ripple crest each wave cycle (Figure 5.20). This means that, in practice, ripple geometry can be very variable.

Box 5.2 Acoustic-visualisation of sediment suspension over wave ripples

Recent technology has improved our ability to investigate near-bed sediment suspensions beneath waves and currents. One such instrument available is the acoustic back-scatter sensor (ABS), which consists of a sound (acoustic) source and a receiver or hydrophone. When the acoustic signal that is emitted by the ABS encounters a sediment suspension, the signal scatters in all directions. The ABS measures the intensity of the signal that is scattered back towards the instrument. The acoustic backscatter strength is then calibrated against known suspended sediment concentrations to yield a suspension concentration 'map' like the one shown in Figure 5.21.

Figure 5.21 shows a slice through the suspension concentration that is parallel to the direction of wave travel, so that the wave ripple is seen in cross-section. In this example, waves were travelling from left to right, and at this particular time the seaward stroke of the wave was decelerating to zero (top panel). A vortex containing high concentrations (*c.* 1.5 g l^{-1}) of suspended sediment is clearly seen on the lee of the ripple crest. Above the ripple crest is also seen another vortex of suspended sediment that was ejected during an earlier wave cycle. The suspension concentration in the latter is smaller (*c.* 0.25 g l^{-1}) due to some deposition since ejection. Although only a few experiments have produced this type of

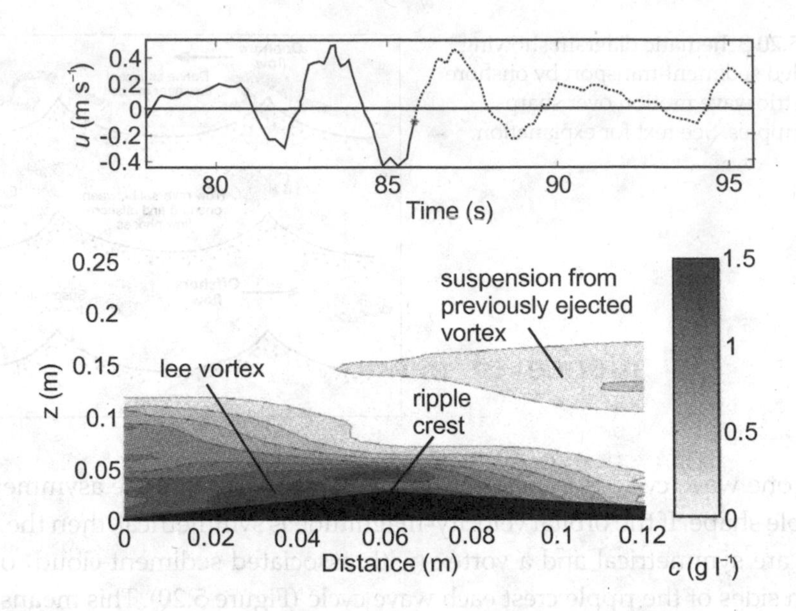

Figure 5.21 Top panel shows the time series of oscillatory flow velocity above the ripple crest. The flow velocity history that is responsible for the suspension pattern shown in the lower panel is marked in bold. The bottom panel shows a snap-shot in time of the suspension concentrations over the ripple crest. See text for explanation. (From Villard and Osborne, 2002.) (Copyright 2002 International Association of Sedimentologists, reproduced with permission.)

acoustic imagery to date (e.g. Kuhnle and Wren, 2009), the ABS has already advanced our understanding of suspension dynamics. A variant of the ABS system is the acoustic Doppler current profiler (ADCP) which uses different sound frequencies (the Doppler shift principle) and in three dimensions, thereby allowing for spatial analysis of variations in suspended sediment concentration in response to variations in flow velocity and direction (e.g. Kostaschuk *et al.*, 2005). These instruments are helping to revolutionise our understanding of fine-sediment transport processes.

Most bedforms fall into the categories of ripples, dunes or antidunes. These bedform types can be distinguished based on their dimensions (Table 5.6). **Ripples** and **dunes** can be further described as either two-dimensional or three-dimensional based on the plan view of their crest line. Straight-crested bedforms are two-dimensional, and sinuous to linguoid bedforms are three-dimensional (Figure 5.22). **Antidunes** are low relief bedforms that have heights similar to ripples, but lengths similar to dunes. They are restricted to rapid, shallow flows that have a wavy water surface that matches the shape of the bedform beneath. This type of flow is called supercritical flow. Antidunes are so named because they have the potential to migrate in the opposite direction to the flow; hence the steepest slope is on the upstream side of the dune and the gentlest slope on the downstream side. Migration is achieved by scour on the upstream side by the breaking surface wave as it slowly migrates upstream (Figure 5.19b). The distinction between flow regimes can be determined using the **Froude Number** F_r

$$F_r = \frac{u}{\sqrt{gh}} \tag{5.39}$$

Subcritical flow occurs when $F_r < 1$ and supercritical flow occurs when $F_r > 1$. Ripples and dunes are the products of subcritical flow and antidunes the product of supercritical flow.

Table 5.6 Dimensions of bedforms and associated flow conditions (after Reineck and Singh, 1980; Sleath, 1984; Nielsen, 1992).

	Unidirectional flow			Oscillatory flow	
	Ripples	*Dunes*	*Antidunes*	*Rolling-grain ripples*	*Vortex ripples*
Length (spacing)	0.1–0.2 m	0.6–30 m	0.1–1 m	0.02–1 m	0.02–1 m
Height	< 0.06 m	0.06–1.5 m	0.01–0.1 m	A few mm	A few cm
Ripple index	8–15	>15	Not applicable	> 10	4–10
Typical flow velocity	Low	Moderate	High	Low and high	Moderate
Typical flow depth	> a few cm	A few dm	A few cm to dm	Up to $\frac{1}{2}$ the wavelength	Up to $\frac{1}{2}$ the wavelength
Typical grain size	0.03–0.6 mm	> 0.3 mm	All sand and gravel	All sand	All sand

While ripples and dunes can be distinguished solely on their dimensions, they also show morphodynamic differences. The size of ripples and dunes increases with flow velocity (and therefore bed shear stress), but they are also influenced by flow depth, sediment type and sediment supply. This means that, on mobile sandy substrates and under energetic flow conditions, dunes can be very large.

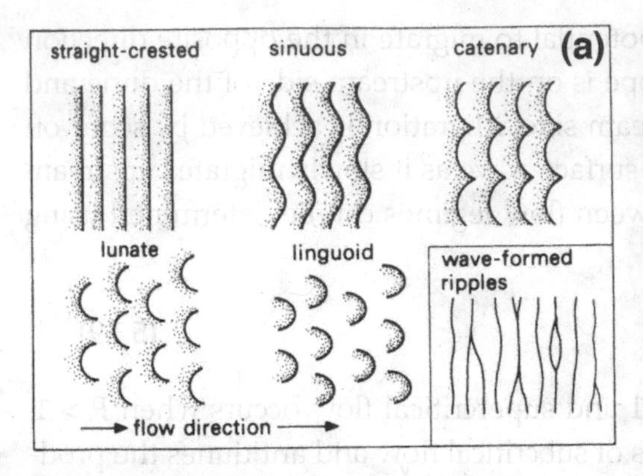

Figure 5.22 – (a) Schematic diagram showing plan view of crest lines for two-dimensional and three-dimensional bedforms. (From Tucker, 1995.) (Copyright © 1995 Blackwell Publishers, reproduced with permission.) (b) Photo of symmetric ripples found in the intertidal zone. (Photo: J. Knight.)

For example, in the tidally-dominated Irish Sea, sea-floor dunes are typically 20–35 m high in water depths of 50–100 m (van Landeghem *et al.*, 2009). With very strong tidal currents in the South China Sea, bedforms are in the region of 2–10 m high in water depths of 20–50 m (Kubicki, 2008).

For a given flow velocity and water depth, larger grain sizes yield larger ripples and dunes. This is true for ripples in sediment up to 0.6 mm diameter. Ripples do not exist beyond this grain size because here the bed is hydraulically rough for all flow velocities capable of transporting sediment, precluding the existence of a viscous sublayer at the base of the boundary layer. The implication is that ripples are scaled to the bed layer thickness, whereas dunes are scaled to the entire flow depth.

5.6.2 Bedform stability fields

Flow conditions change more quickly than bedforms do, particularly dunes where a substantial volume of sediment must be moved to change their size and shape. It is not uncommon, therefore, for bedforms to be in disequilibrium with flow conditions, especially under unsteady tidal flows. The flow conditions under each bedform type is stable and can be identified on a bedform stability diagram (Figure 5.23a). An analogous diagram exists for oscillatory flow over wave ripples (Figure 5.23b). Only two stability fields exist in this case: one for vortex ripples and one for rolling grain (low-steepness) ripples. The former are dominated by suspended sediment transport and the latter by bedload transport.

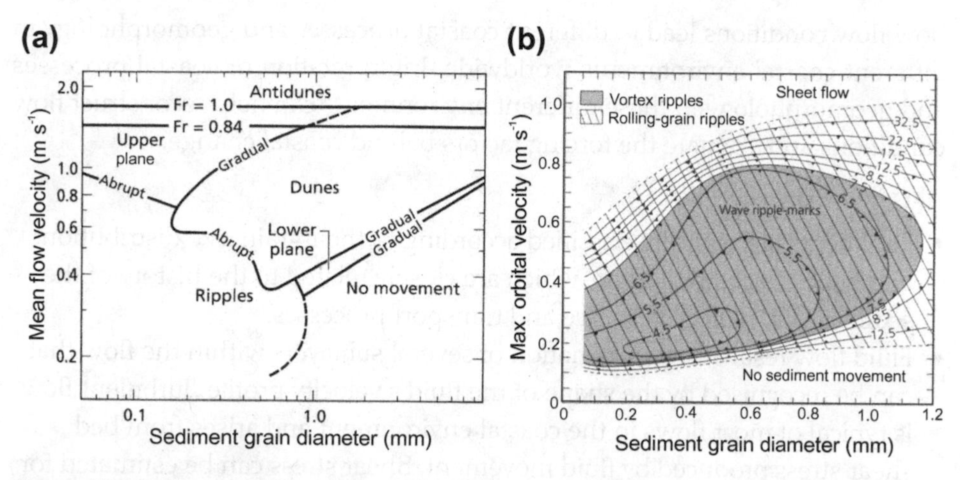

Figure 5.23 Bedform stability diagram for (a) unidirectional flow and (b) oscillatory flow beneath wind waves and swell. (Unidirectional flow modified from Southard and Boguchwal, 1990.) (Copyright © 1990 Society for Sedimentary Petrology, reproduced with permission.) (Oscillatory flow modified from P.A. Allen, 1997 after J.R.L. Allen, 1985.) (Copyright © 1997 Blackwell Publishers, reproduced with permission.)

5.6.3 Interpreting the bedform record along coasts

The flow conditions experienced at a single location change over long time scales by sea-level rise and global climate change, and on shorter time scales by interannual climate variability such as El Niño, sediment supply and anthropogenic factors. This may mean that bedforms observed in nearshore and coastal locations today were not formed under today's flow regime, but instead reflect the flow regime of some time in the past. The bedforms are termed **relict** where they do not reflect today's conditions, and **moribund** where they do reflect today's conditions, but where the bedforms are no longer active. For example, very large, relict tidal sand banks are present at 110–180 m water depth on the continental shelf in the Celtic Sea (Belderson *et al.*, 1986). Present-day tidal currents are not strong enough at this depth to generate these bedforms, and the sand banks were formed during the last deglaciation when sea level was 100 m lower than present. Later sea-level rise drowned the sand banks, removing them from the reach of tidal processes.

Bedforms around coastal margins are composed of sediments of different physical properties, composition, source and age. The bedforms themselves form spatial patterns that reflect an episodic and dynamic history of partial reworking, reshaping, erosion and deposition in response to variations in flow conditions and sediment supply throughout the Holocene (and sometimes earlier), and sea level. Understanding these processes and patterns is a considerable challenge for coastal scientists. The remainder of this book focuses on how flow conditions lead to different coastal processes and geomorphology in different coastal environments worldwide. Interpretation of coastal processes and geomorphology in these different environments can help reconstruct flow conditions and evaluate the forcing factors behind coastal change.

SUMMARY

- Sediments are typically classified according to their grain size distribution and grain shape/roundness, which are closely related to the history of the sediment, including its source and transport processes.
- Fluid flow results in the formation of several sublayers within the flow that can be recognised by the shape of the fluid's velocity profile. Turbulent flow is typical of most flows in the coastal environment and arises from bed shear stress produced by fluid movement. Shear stress can be estimated for steady flow using the Law of the Wall, and can be estimated for oscillatory flow based on the bed roughness and flow velocity at the top of the wave boundary layer.
- The critical bed shear stress required to entrain a sediment grain from its bed can be predicted from a modified Shields diagram. For fine sediments the critical bed shear stress increases with decreasing grain size, due to the fact that these sediments form hydraulically smooth beds and the grains are cohesive. For coarser sediments the critical bed shear stress increases with grain size, reflecting the importance of the weight force. The situation is more complicated in the case of poorly sorted sediments or cohesive sediments with a high water content.
- Sediment is transported as either bedload or suspended load. In bedload the grains are supported by continuous (traction) or intermittent (saltation) contact with the bed. In suspended load the grains are supported by fluid momentum related to turbulence.
- Settling of cohesionless (sandy) sediment is governed by grain properties, whereas settling of cohesive (fine) sediment is governed by the suspension concentration and development of flocs. When deposited, sediments can accumulate and organise into bedforms which are classified according to their dimensions and flow regimes. Current and wave ripples are small-scale bedforms; dunes and antidunes are large-scale bedforms. Ripples and dunes are the products of subcritical flow whereas antidunes are the product of supercritical flow. Ripples occur in relation to hydraulically smooth beds and dunes to hydraulically rough beds.

Reflective questions

These questions are designed to test your comprehension of material covered in this chapter. Suggested answers to these questions can be found on this book's website.

5a. Describe the role of rock type in influencing grain size, shape and roundness.

5b. Explain how water temperature and salinity influence fluid density and viscosity.

5c. The benthic boundary layer is an important control on shear stress and fluid velocity. Draw a diagram that shows the major differences in velocity profile between (1) a flat clay bed, and (2) an irregular gravel bed.

5d. Describe how bedform type and geometry can be used to reconstruct fluid flow conditions.

Further reading

Allen, J.R.L., 1984. *Sedimentary Structures: Their Character and Physical Basis*. Elsevier, Amsterdam. (The most comprehensive work available on bedforms and the physical processes responsible for them.)

Allen, J.R.L., 1985. *Principles of Physical Sedimentology*. George Allen and Unwin, London. (A detailed physical account of all aspects of sediment dynamics.)

Hsu, K.J., 2004. *Physics of Sedimentology*, (2nd edn). Springer, Berlin. (This is a graduate-level text that explores in detail the physical processes of fluid flow and sediment movement.)

Pye, K., 1994. Properties of sediment particles. In: K. Pye (ed) *Sediment Transport and Depositional Processes*. Blackwell Scientific Publications, Oxford, 1–24. (Provides a detailed overview of sediment properties.)

CHAPTER 6

FLUVIAL-DOMINATED COASTAL ENVIRONMENTS – DELTAS

AIMS

This chapter discusses the major hydrodynamic and sedimentary processes associated with the formation of river deltas at the coast. Different delta types reflect the interplay between fluvial and coastal processes. Delta plain, delta front and prodelta environments are described with respect to their morphology and hydrodynamic and sedimentary processes. Deltas are sensitive to changes in forcing by internal (autogenic) factors such as river sediment supply, and external (allogenic) factors such as sea level.

6.1 Introduction

Coastal **deltas** are discrete accumulations of terrigenous sediment deposited where rivers enter into the sea. Similar features are also formed where rivers enter large lakes, but these are not considered here. River sediments may also accumulate at the head of coastal embayments if the coastline is drowned (forming a **bay-head delta**), but these are generally controlled by estuarine processes and are discussed in Section 7.2.1. This chapter focuses on situations where the delta is sufficiently large to cause the coastline to prograde. This implies that sediments must be delivered by the river faster than they are dispersed by waves, tides, gravity and ocean currents. For a river to deliver sufficient sediment volumes, its drainage basin usually needs to be large, although high rates of precipitation and catchment denudation are also sufficient conditions. Inman and Nordstrom's (1971) census of major deltas placed 57 per cent along trailing-edge coastlines (e.g. the Amazon River, Brazil), and 34 per cent along coasts fronting marginal seas that are often protected from ocean waves by island arcs (e.g. the Klang River, Malaysia). Many in this latter group are located in the mid to low latitudes with high precipitation and catchment denudation (e.g. the Fly River, Papua New Guinea).

The dynamics of specific river deltas reflect the interplay between the major delta controlling factors (Figure 6.1). A useful context for exploring these morphodynamic processes and controls is through the delta classification

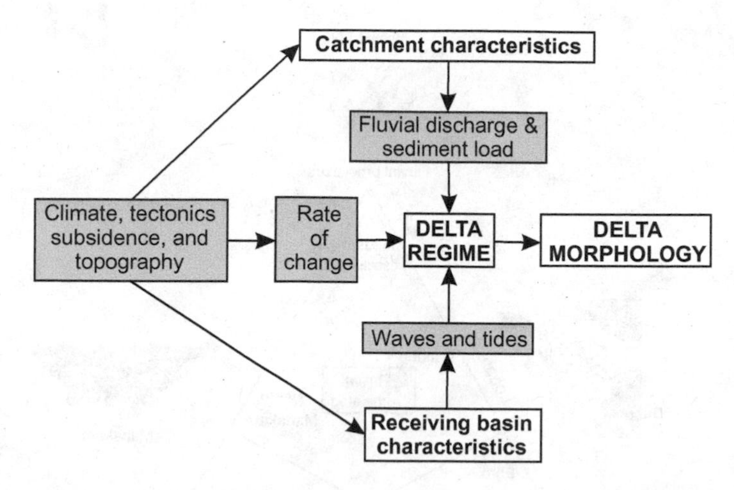

Figure 6.1 Interaction of factors responsible for variations in delta morphology. (Modified from Elliott, 1986.)

scheme proposed by Galloway (1975), because it has widespread acceptance and is reasonably straightforward. Other delta classification schemes are discussed by Reading and Collinson (1996) and Postma (1990).

6.2 Galloway's classification of deltas

Galloway's scheme for delta classification has three end-members (Figure 6.2), which describe the different energy sources that can influence delta morphodynamics:

- **Fluvial-dominated deltas** are characterised by large catchments, high river discharge into protected seas with minimal nearshore wave energy, and a small tidal prism.
- **Wave-dominated deltas** are characterised by relatively high exposure to open-ocean swell.
- **Tide-dominated deltas** are characterised by a tidal prism that is larger than the amount of water supplied by fluvial discharge.

Different river deltas worldwide (e.g. in Figure 6.2) are located in different sectors of the diagram according to the interplay of fluvial, wave and tidal processes. It is therefore the relative, not absolute, magnitudes of these processes that are important. In addition, these named river deltas are of different sizes, ages and locations, and so they have different geological histories.

Fluvial-dominated deltas are the only delta-type that is indisputably a delta. Reading and Collinson (1996) note that many deltas classified as wave-dominated may be better classified as beach-ridge strand plains, since most of their sediment is supplied by littoral drift rather than river processes. Similarly,

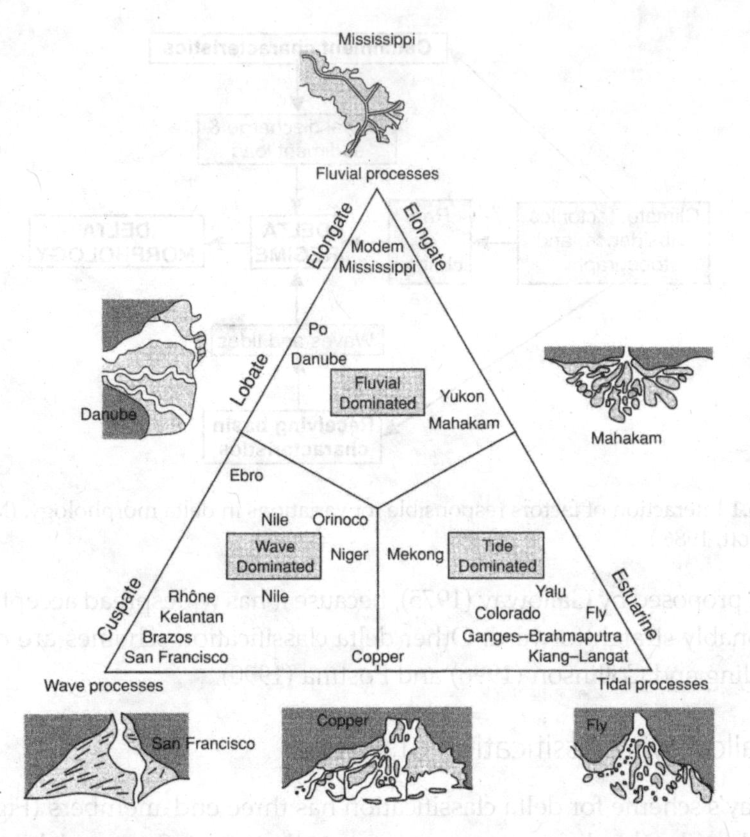

Figure 6.2 Galloway's (1975) classification of fine-grained deltas. The end-member types are fluvial-, tide- and wave-dominated. Specific rivers are named. (From Briggs *et al.*, 1997.) (Copyright © 1997 Taylor & Francis Books, reproduced with permission.)

many tide-dominated deltas could be considered estuaries since they are transgressive systems. Nevertheless, many major deltas experience coastal progradation yet are also modified by waves or tides (e.g. the Niger River, Nigeria, and the Ganges–Brahmaputra river system, Bangladesh). Galloway's scheme therefore has a number of shortcomings but is a useful starting point to consider the relationships between delta morphology and its major hydrodynamic controls. The next sections focus on the morphodynamics of fluvial-dominated deltas, with a brief description of how waves and tides modify delta morphology in particular settings. Tide- and wave-dominated coastal settings are considered in more detail in Chapters 7 and 8, respectively.

While Galloway's scheme focuses only on medium- to fine-grained deltas characteristic of trailing-edge coasts, there are also deltas formed on the tectonically-active leading-edge of continents. These settings have proximal drainage divides, steep hinterland slopes, high rates of denudation, and narrow continental shelves, and as a result produce coarse-grained, steep fan- and braid-type deltas (Colella and Prior, 1990). Coarse-grained deltas are smaller and less common than fine-grained ones, but can be locally important in terms of coastal progradation (Box 6.1).

Box 6.1 Coarse-grained deltas

Along tectonically-active, leading-edge coasts deltas are relatively rare but are typically coarse-grained due to their proximity to the sediment source. Coarse-grained deltas are much steeper slopes than fine-grained deltas, due to the narrow and deep continental shelf that typically fronts leading-edge coasts. Coarse-grained deltas can be classified as fan- or braid-type (Figure 6.3).

(a) Fan delta

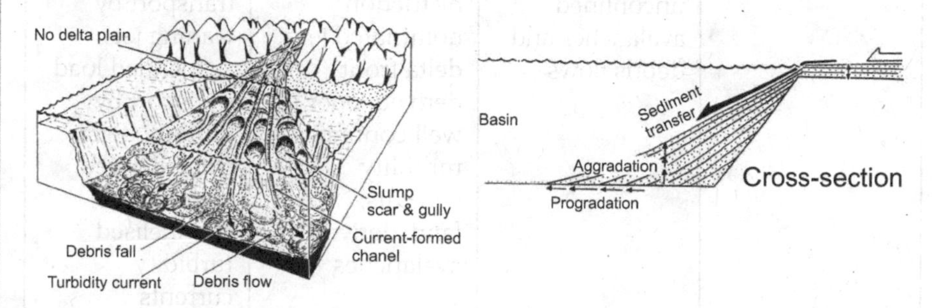

(b) Braid delta

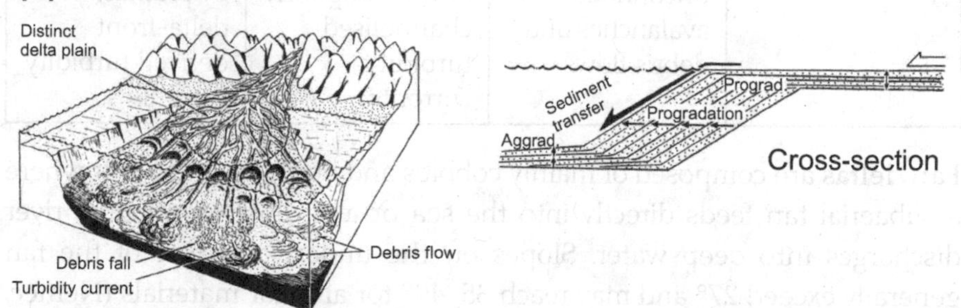

Figure 6.3 Schematic diagram showing oblique and cross-section views of the two types of coarse-grained deltas. (a) Fan deltas are a subaqueous extension of subaerial alluvial fans, thus there is no delta plain and no subaerial progradation. (b) Braid deltas display a distinct delta plain and undergo subaerial progradation. (From Nemec, 1990.) (Copyright © 1990 International Association of Sedimentologists, reproduced with permission.)

Table 6.1 Summary of sediment transport mechanisms for coarse fan and braid deltas, and fine-grained deltas.

Morphological unit	Fan deltas	Braid deltas	Fine-grained deltas
Delta plain	Land-derived slope failure and avalanches; bedload transport by unconfined river flow	Land-derived slope failure and avalanches; bedload transport by poor to well-confined river flow	Bedload and suspended load transport in well-confined channels
Delta front	Land- and delta front-derived, unconfined avalanches and debris flows	Bedload transport by friction dominated jets; delta front-derived, poor to well confined turbidity currents; land-derived avalanches	Bedload and suspended load transport by waning jets; suspended load transport by plumes; delta front-derived slumps and channelised turbidity currents
Prodelta	Land- and delta front-derived, unconfined avalanches and debris flows	Delta front-derived debris flows and poorly channelised turbidity currents	Suspended load transport by plumes; delta-front derived turbidity currents

Fan deltas are composed of mainly cobbles and boulders and occur where a subaerial fan feeds directly into the sea or a bedload-dominated river discharges into deep water. Slopes on the underwater part of the fan generally exceed 27° and may reach 35–40° for angular materials (Nemec, 1990). Almost all sedimentation occurs beneath the water level, hence the delta plain is virtually absent (Figure 6.3a). **Braid deltas** are composed of slightly finer and better sorted sediments delivered by braided rivers that cross a narrow coastal plain before reaching the coast. Offshore slopes are generally shallower, permitting a morphological style of delta plain, delta front and prodelta that is similar to fine-grained deltas (see Section 6.3).

An important difference between coarse- and fine-grained deltas is the mode of sediment transfer (Table 6.1). For coarse-grained deltas, sediment delivery is predominantly by episodic gravity-driven debris avalanches and flows, with some poorly confined river flow during flash floods. The cascade of flows from top to base of the slope builds the long, steep delta front. In contrast with fine-grained deltas, the coarsest and finest materials are generally well mixed across the delta front, even to the deepest water depths. Horizontal progradation and vertical aggradation are restricted to the seaward margin of the delta front, thus there is little change in coastline position unless the basin becomes sufficiently shallow to permit the development of a braid delta.

On braid deltas (Figure 6.3b) both gravity and fluvial transport occur across the delta plain. The upper delta plain receives material via avalanching and debris flows, usually triggered by heavy precipitation. This material is transported to the delta front as bedload in braided streams. At the delta front streams immediately deposit their bedload at the top of the subaqueous slope. When the top of the slope becomes oversteepened, material travels down the delta front by gravity in the form of unconfined debris flows, while finer material is transported as poorly confined debris flows and turbidity currents (see also Box 6.2). Cycles of fan- and braid-delta activity can be triggered by variations in river sediment supply, discharge and climate (e.g. van Dijk *et al.*, 2009).

6.3 Delta morphology

Both fine and coarse-grained deltas generally have a similar internal and plan-view structure, and morphological elements (Figure 6.4). This results from the common role that the river catchment and receiving basin play with respect to the rate and dynamics of sediment accumulation, its changes through time, and the delta regime (i.e. river-, wave-, or tide-dominated). The different permutations of these factors lead to differences in delta morphology. For example, consider two deltas that have the same catchment characteristics, sediment supply and receiving basin characteristics, but different basin subsidence rates. The slowly subsiding delta will prograde farther seaward with a thin coastal sediment wedge, whereas the rapidly subsiding delta will have a thicker sediment wedge, due to the constantly-increasing accommodation space at the river mouth. The two deltas will also develop different entrance conditions. Because the delta with the slower subsidence rate progrades the farthest, it will be the first to become exposed to open-ocean waves and tides, while the other

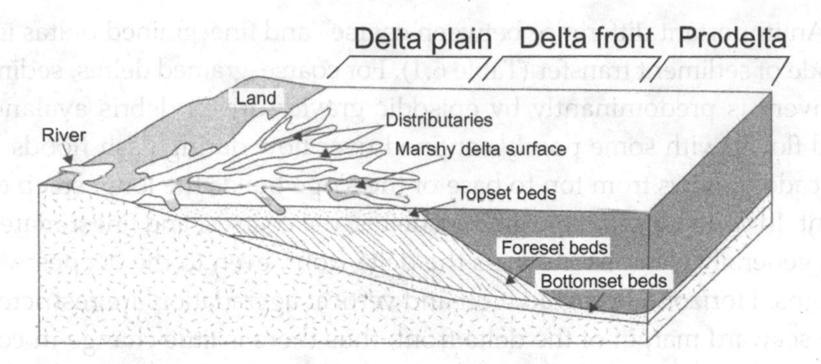

Figure 6.4 The three generic morphostratigraphic units found in all fine-grained deltas: delta plain (topset beds), delta front (foreset beds) and prodelta (bottomset beds). (Modified from Haslett, 2000.)

remains fluvially-dominated. The rates, magnitudes and interplay of these controls (Figure 6.1) therefore determine the resultant delta morphology and its evolution over time and space.

Although the detailed morphology of individual deltas varies, the sediment body of almost all delta types comprises three morphological units. These are the delta plain, delta front and prodelta (Figure 6.4). The **delta plain** is the sedimentary platform that corresponds to the most recent coastal progradation and aggradation by the incoming river. This river sediment causes the **delta front** to prograde seaward through deposition of coarsest sediments closest to the coast at the edge of the coastal plain, and finer sediments farther seaward on the **prodelta**. As the delta front progrades horizontally the delta plain aggrades vertically. The resulting sediment distribution through the delta wedge is characterised by fining in the seaward direction and coarsening in the vertical direction. The sedimentary architecture of most deltas comprises three major morphostratigraphic units termed topsets, foresets and bottomsets, which are closely related to processes operating on the delta plain, delta front and prodelta, respectively (Figure 6.4). We now describe the morphological and sedimentary properties of the delta plain, delta front and prodelta, and the morphostratigraphic units associated with them.

6.3.1 Delta plain

The delta plain represents the seaward termination of the incoming river, and as such it is formed in the same way and is morphologically similar to a river floodplain. River water and sediment is brought to the delta plain by single or multiple channels. The number of these **distributary channels** active on the plain at any one time is mostly related to the slope of the plain and the grain size of the sediment load. Rivers carrying coarse sediment across a steep delta plain have several simultaneously active distributary channels, whereas rivers carrying fine sediment across a gently-sloped delta plain have relatively few, although there may be numerous abandoned channels. Distributary channels behave much like fluvial channels with predominantly unidirectional flow on the upper delta plain, but may have reversing tidal flows on the lower plain and behave more like estuarine channels. Morphologically, the **upper delta plain** includes fluvial features such as abandoned channels, levees, floodplains, lakes and swamps between the active channels, all of which are fed sediment by overbank flow during floods. In contrast, the **lower delta plain** includes estuarine features such as brackish lagoons, tidal flats, marshes and relict beach ridges. The surface slope of the delta plain is generally very gently seaward-dipping, and this slope is always maintained whether the delta front is prograding or aggrading. In cross-section, delta plain sediments form a morphostratigraphic unit known as **topsets** (Figure 6.4) which is generally tabular in geometry and formed by active erosion, deposition and migration of delta plain channels. The distributary channels are relatively high-energy environments, so they tend to contain the coarsest sediment load. The interdistributary areas, however, are low-energy environments with fine sediment deposited from suspension. Due to the migration and abandonment of distributary channels associated with delta switching (Section 6.6), the distribution of coarse and fine sediment on the delta plain can be complex, and coarse and fine topset beds can be juxtaposed by overlapping and channelised cut-and-fill structures. For topsets to aggrade upwards, the basin must also subside, otherwise the distributary channels will continually erode the delta plain. The thickest topset beds are therefore associated with the highest rates of basin subsidence, which is caused by isostatic compensation for sediment loading, and sediment compaction.

6.3.2 Delta front

The delta front is the most active part of the delta system and represents the point where fluvial, wave and tidal processes meet. It is also where the unique nature of each delta is most apparent. Most of the sediment transported through active distributary channels is deposited at the delta front. As river water exits the distributary mouth it expands and decelerates, thereby losing its competency to transport sediment. Sediment deposition forms a **distributary mouth bar**. The size, shape and position of the bar depend on the hydrodynamic behaviour of the river water as it leaves the distributary mouth, plus any modification caused by waves and/or tides. Different hydrodynamic processes at the delta front are discussed in Section 6.5.

As sediment is deposited at the delta front it cascades down the delta slope to form **foresets** (Figure 6.4). Foreset beds are generally rectilinear to slightly concave-up and have a slope angle that is close to the angle of repose of the sediment, determined by its grain size. The most seaward foreset corresponds to the present extent of the active delta. As more sediment is transported to the delta front edge, new foresets are built and the delta progrades, with delta plain topset beds unconformably overlying older delta front foresets. The rate of sediment supply to the delta front, sediment grain size distribution and sorting, and the amount of water incorporated within the sediment cascading by mass wasting down the delta front are all important factors controlling foreset thickness (and so the rate of delta progradation) and therefore delta front morphology (Postma, 1990).

6.3.3 Prodelta

The prodelta (literally, the area in front of the delta) refers to the deeper water and lower energy region located seaward of the delta front. The large distance (sometimes several kilometres) between the prodelta and the distributary mouth of the delta plain means that only the finest sediments reach this zone, generally fine silts and clays. These sediments settle from turbulent suspensions carried into the prodelta environment by either jets of river water effluxing straight into the sea from the delta plain, or from buoyant plumes. The lack of bedload transport into the prodelta area and a generally horizontal to gently-sloping bed mean that sediment layers here tend to aggrade vertically to produce relatively flat-lying and thin **bottomsets** (Figure 6.4). The bottomsets are progressively buried as foresets prograde across them. Not all delta systems include a prodelta, because they tend to form in deeper water below wave base and where fine sediment disperses beyond the delta front. The prodelta is best preserved if sediments are not disturbed by tides or by mass wasting on the delta front.

Relationships between the delta plain, delta front and prodelta can be evaluated with respect to their hydrodynamic processes and internal sedimentary architecture. Many sandy or coarser deltas often have a distinctive tripartite (or three-part) structure comprising topsets, foresets and bottomsets that, together, is referred to as a **Gilbert-type** delta (Postma, 1990). This delta type is best developed where coarse sediments accumulate in relatively deep water, which is typical of steep, leading-edge coastlines such as Norway and Chile (Box 6.1). The internal structure of finer-grained deltas developed on shallower coastal plains is more complex because the incoming rivers tend to have more numerous and more dynamic distributary channels, the sediments accumulate on shallower slopes, and erosional reworking of the sediments is much more common. These finer-grained and shallower deltas are known generically as **Salisbury-type** deltas (Salisbury, 1892).

6.4 The role of mass wasting and gravity

A key process affecting the delta front and prodelta is **mass wasting**, which is a general term that refers to the movement of bed sediment downslope under gravity. The presence of mass wasting can be readily identified from the morphology and sediments of the delta front and prodelta. If mass wasting is not important, delta front foresets will be straight and uniform with no variations in thickness, and the fronting prodelta surface will be smooth and featureless. If mass wasting is important, however, then slumps, depressions and erosive channels and gullies can disrupt the surface of both these zones as a result of **flow transformation**, which describes how the physical properties and flow mechanisms of the sediments change while in motion. For example, an intact rotational slump or submarine slide can develop downslope into a disaggregated turbidity current that can erode a channel down the delta front and across the prodelta (Box 6.2). Such channels are slightly sinuous, up to several kilometres long, up to 20 m deep and up to 1500 m wide.

Box 6.2 Gravity-driven turbidity currents

In Chapter 5 we discussed the sediment transport mechanics associated with fluid drag over the sea bed. There are additional sediment transport modes whose mechanics are based on the downslope component of the sediment weight force, a product of mass and gravity. These gravity-driven transport modes include soil creep, slides, debris flows and turbidity currents. As an example, we discuss the mechanics of turbidity currents.

Turbidity currents are negatively buoyant, highly-concentrated, near-bed suspensions (typically of sand, silt and clay) that flow downslope under their own weight (Figure 6.5). There is no motion of the overlying water necessary to provide drag – gravity does all of the work. Turbidity currents characteristically consist of a head and a body. The body typically travels 16 per cent faster than the head and so is constantly feeding the head region

Figure 6.5 Turbidity current produced in the laboratory displaying the characteristic head and body regions. (From Allen, 1994.) (Copyright © 1994 Blackwell Publishers, reproduced with permission.)

with fluid and sediment. The velocity of the turbidity current increases with the thickness of the body and with the density difference between the suspension and the surrounding fluid.

As the turbidity current flows downslope, the Kelvin-Helmholtz waves (see Figure 6.11 for discussion) that develop at the head serve to entrain relatively clear fluid, thus reducing the bulk density and negative buoyancy of the suspension. Sediment is also deposited at the base of the turbidity current. In order for a turbidity current to be sustained, the mechanisms that reduce the bulk density of the suspension must be balanced by the body region entraining sediment and continually feeding it to the head. It is this sediment entrainment that is responsible for eroding many of the channels observed on the lower delta front and prodelta. Channels can be carved during only a few transport episodes, because the turbidity current's large bulk density (and hence large mass) generates more momentum than relatively clear currents of similar discharge. Detailed discussion of turbidity currents and other gravity-driven flows can be found in Allen (1985) and Leeder (1999).

Reading and Collinson (1996) identify the following preconditions for mass wasting: (1) a very high sedimentation rate which leads to retention of pore waters, resulting in a low shear strength for the sediment; (2) biodegradation of organic detritus within the sea bed and the production of methane gas, which weakens the sediment; and (3) a shock mechanism (e.g. storm waves or submarine earthquake). The processes of mass wasting and flow transformation are significant in delta environments where sediments are delivered to the delta front or continental shelf edge, for a number of reasons:

- Unconsolidated and often fine-grained and water-saturated sediments have great capacity to travel long distances even on very shallow slopes. This means that changes in sediment supply volume, rate or sediment type to the delta front can trigger unpredictable downstream effects.
- When in transport and through erosion, sediments can actively disturb the bed and lead to the escape of subsurface methane. Degassing of methane from buried organic submarine sediments can destabilise the sea bed – with implications for engineering structures such as oil rigs – and also contributes to global warming.
- If moving fast or in high volume, sediments can disturb the water column sufficiently to cause formation of tsunami waves. For example, off the west coast of Norway rapid sliding of the Storegga submarine landslide at *c.* 7900 years BP caused large tsunami waves to be formed, with a run-up on adjacent coasts in Norway and Scotland of up to 25 m (Smith *et al.*, 2004).

All of these processes are dependent on sediment delivery to the delta front by rivers, and redistribution of the sediment across the delta front and prodelta by gravity and by interaction of river water with waves and tides. These processes are now discussed in order to identify their role in delta front morphodynamics.

6.5 Delta front morphodynamics

Bates (1953) argued that delta-front morphology could be understood in terms of hydrodynamic jets, in other words the flow processes that take place at the river mouth. In a subsequent paper, Wright (1977) built on Bates' foundation to provide a synthesis of river mouth morphodynamics. This section summarises the main points of these two classic papers.

The hydrodynamic behaviour of the river effluent as it enters the receiving basin is controlled by the current speed at the river mouth, the slope of the sea bed seaward of the river mouth, and the vertical density distribution through the water column. Depending on how these factors combine, the flow can behave as either a jet or a plume (Table 6.2). The principal determining factor is the density pattern (Figure 6.6). When the water is well mixed vertically the **isopycnals** (lines of equal density) are nearly vertical and the water density at the surface and the

Table 6.2 Morphodynamic summary of distributary mouth processes.

	Jet		Plume		Tides	Waves
	Axial	*Planar*	*Surface*	*Near-bed*		
River current velocity	Large	Large	Moderate	Moderate	Small	Small
Offshore slope	Steep	Gentle	Relatively steep	Relatively steep	Relatively gentle	Relatively steep
Density conditions	Homopycnal	Homopycnal	Hypopycnal	Hyperpycnal	Homopycnal	Homopycnal
Morphology	Narrow, lunate bar	Broad, radial bar or middle-ground bar and bifurcating channel	Subaqueous levees and distal bar	Gullies and channels from mudslides and turbidity currents	Linear channels and shoals aligned perpendicular to the coast	Swash bars and beach ridges aligned parallel to the coast

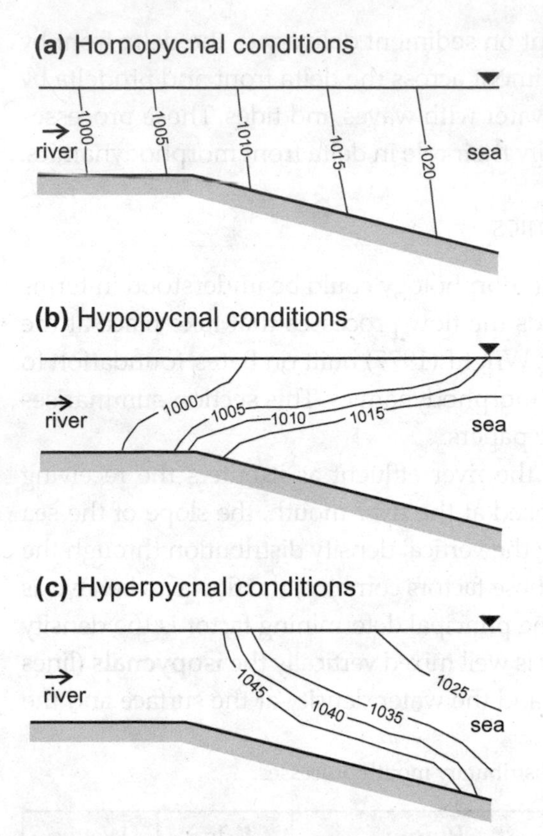

Figure 6.6 Schematic diagram showing three water density patterns: (a) homopycnal conditions, (b) hypopycnal conditions and (c) hyperpycnal conditions. The lines are isopycnals labelled with indicative water densities in kg m^{-3}. Homopycnal conditions are associated with river effluents behaving as jets. Hypopycnal and hyperpycnal conditions are associated with river effluents behaving as positively- and negatively-buoyant plumes, respectively.

bed are similar (i.e. **homopycnal** conditions) (Figure 6.6a). When the water is stratified the isopycnals are sloping, with the least-dense water mass floating on the most-dense water mass. If the river effluent is the least dense, because it is fresher than the basin water, then it extends seaward on the surface and the marine water extends landward along the bed (i.e. **hypopycnal** conditions). If the river effluent is the most dense, because it has high suspended sediment content, then it extends seaward along the bed (i.e. **hyperpycnal** conditions).

Given that the water exiting a river mouth is usually fresh and loaded with sediment and the coastal water is saline, there is always the potential for hypo- or hyperpycnal conditions. Whether this potential is realised or whether mixing and homopycnal conditions prevail depends on the strength of the river current (which will tend to promote mixing when the current is strong) and the density difference between the river and coastal waters (which will tend to promote stratification when the density difference is high). The prevalence of one over the other is represented in the **densimetric Froude Number** F'_r

$$F'_r = \frac{u}{\sqrt{gh'\left(1 - \frac{\rho_R}{\rho_C}\right)}} \qquad (6.1)$$

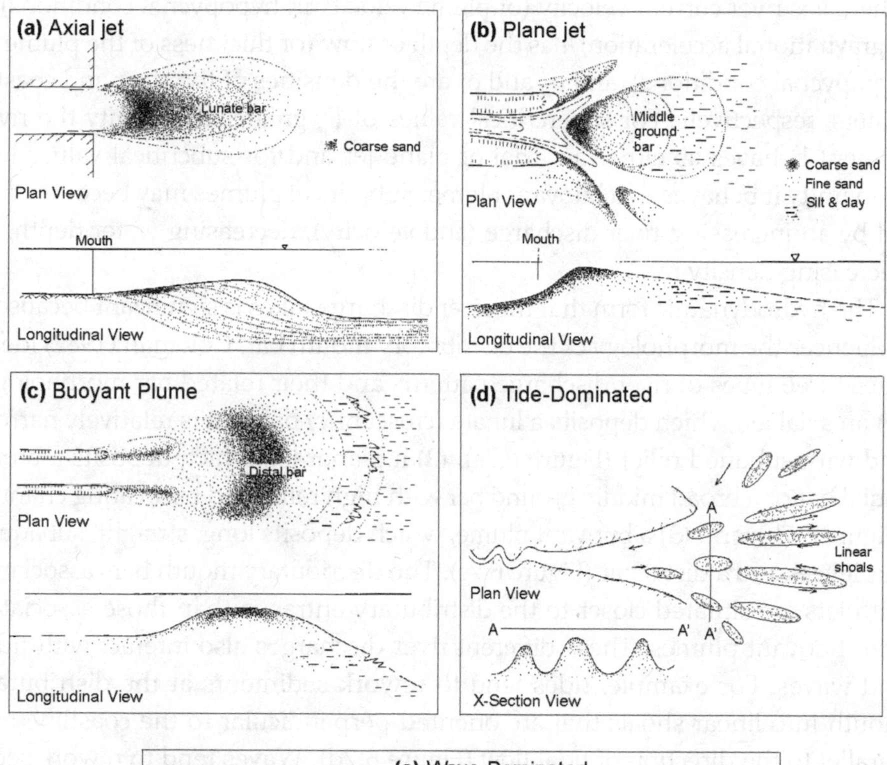

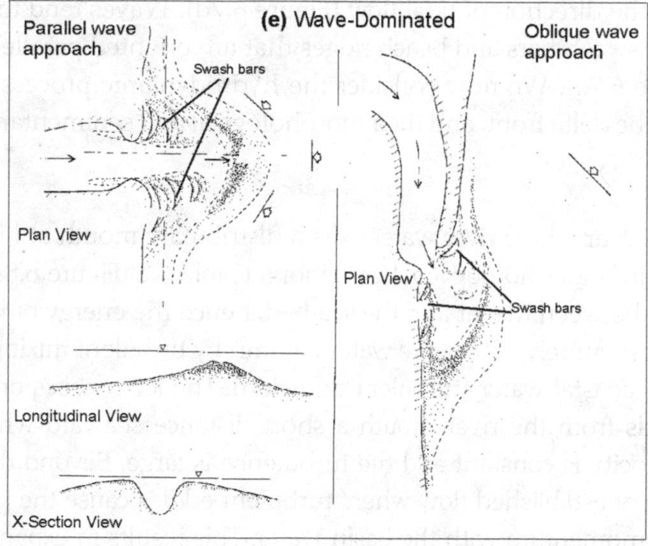

Figure 6.7 Morphology of distributary mouths dominated by: (a) axial jets, (b) plane jets, (c) buoyant plumes, (d) strong tides and (e) strong waves. (From Wright, 1977.) (Copyright © 1977 Geological Society of America, reproduced with permission.)

where u is river current velocity (or plume velocity in hypopycnal conditions), g is gravitational acceleration, h' is the depth of flow (or thickness of the plume in hypopycnal conditions), and ρ_R and ρ_C are the densities of the river and coastal waters, respectively. For supercritical values of F'_r greater than unity the river effluent behaves as either an axial or plane jet and for subcritical values less than unity it behaves as a buoyant plume. Subcritical plumes may become critical by an increasing river discharge (and velocity), decreasing water depth, or decreasing density ratio.

The hydrodynamic form that the river discharge takes is significant because it influences the morphology of the distributory mouth bar. Coleman (1982) identifies three types of river discharge patterns and their related bar morphology: (1) an axial jet, which deposits a lunate (crescentic) bar that is relatively narrow and with subdued relief (Figure 6.7a); (2) a planar jet, which deposits a broad radial bar or a broad middle ground bar with high relief and bifurcating channel (Figure 6.7b); and (3) a buoyant plume, which deposits long, straight, subaqueous levees and a distal bar (Figure 6.7c). The distributary mouth bars associated with jets are situated closer to the distributary entrance than those associated with buoyant plumes. These different river discharges also interact with tides and waves. For example, tides tend to rework sediments at the distributary mouth into linear shoals that are oriented perpendicular to the coastline and parallel to the direction of tidal flow (Figure 6.7d). Waves tend to rework sediments into swash bars and beach ridges that are oriented parallel to the coastline (Figure 6.7e). We now consider the hydrodynamic processes of jets and plumes at the delta front, and their morphological and sedimentary expressions.

6.5.1 Jets

Axial jets occur where river water exits a distributary mouth at high speed into a deep basin under homopycnal conditions (Table 6.2; Figure 6.8a). There is no interaction between the jet and the sea bed, hence the energy of the jet is dissipated almost entirely within the water column by turbulent mixing between the jet and the coastal water (turbulent diffusion). The jet consists of a core region that extends from the river mouth a short distance seaward where the mean current velocity is constant and the turbulence is large. Beyond the core region is the zone of established flow where turbulent eddies cause the jet to exchange water and momentum with the basin water. This results in expansion of the jet when it leaves the constraining channel mouth, deceleration of its velocity, and reduction in the turbulence intensity with distance. The behaviour of the jet with distance from the distributary mouth can be described completely if the current velocity at the mouth and the width of the mouth are known. Figure 6.8a shows that the jet spreads laterally at a roughly constant rate and at an angle of about

12°. The spreading is due to turbulent eddies at the boundary between the jet and basin water. Since the total discharge and momentum flux through any cross-section in the jet must be equal (i.e. conservation of mass and momentum), the cross-sectionally averaged velocity in the jet must decrease with distance from the distributary mouth. If the angle of spread is constant, the spatially-averaged velocity across the section $\bar{u}$ decreases with distance at a constant rate given by

$$\bar{u} = \frac{w}{x \tan\theta} u_o \tag{6.2}$$

where w is width of the distributary mouth, x is distance from the mouth, θ is angle of spread and u_o is velocity at the mouth. In reality, the velocity at the distributary mouth is at a maximum at the centre and a minimum at the margins (Figure 6.8a).

This velocity decrease at the margins is important because it means that the jet can lose its competency to transport sediment and cause sediment deposition in the form of a lunate bar (Figure 6.7a). In cross-section, this bar rises gently to the bar crest and then slopes seaward with steeply-dipping foresets. The lateral extent of the bar is relatively small, consistent with the width of the jet. The coarsest sediment is deposited around the margins of the core region, and sediment fines seaward. Axial jets and lunate bars may occur at newly-created distributary mouths, but the necessary conditions generally do not persist. As

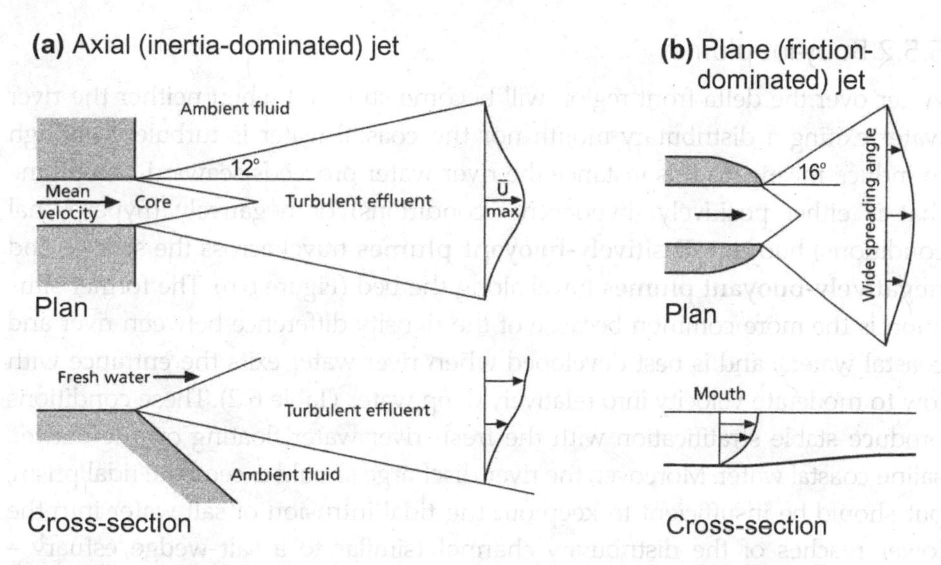

Figure 6.8 Schematic illustration of (a) an axial (inertia-dominated) and (b) a plane (friction-dominated) jet, showing both plan and cross-section views. Note that axial jets do not interact with solid boundaries after exiting the mouth and are free to expand in all directions. Plane jets are strongly influenced by the sea bed. (From Leeder, 1999; modified from Wright, 1977.) (Copyright © 1999 Blackwell Publishers, reproduced with permission.)

deltas develop the receiving basin becomes shallower so the jet is eventually affected by the sea bed and therefore becomes a plane jet.

Plane jets occur where river water exits a distributary mouth at high speed into a shallow basin under homopycnal conditions. The vertical expansion of the jet is restricted by the shallow water depth, and in order to conserve mass and momentum this is compensated for by an increased horizontal spreading angle of about 16°, wider than that of axial jets (Figure 6.8b). In addition to energy dissipation by turbulent diffusion along the sides of the jet, there is also energy dissipation at the base by friction with the sea bed. Due to the greater spreading angle and bed friction, velocities in the jet decrease more rapidly with distance, so neither the core nor the established flow region extend as far seaward as they do in the case of an axial jet. Consequently, sediment deposition is rapid and occurs closer to the entrance. The resulting radial bar slopes rapidly up from the mouth to the bar crest and then slopes gently seaward. Sediment grain size also decreases seaward. However, continued deposition at the bar front results in positive feedback, whereby as the water depth over the bar decreases through sediment accumulation, frictional dissipation of the jet increases causing further sedimentation and shallowing. In order to continue delivering sediment, the jet expands laterally around the bar, forming erosional channels on either side of what is now a middle-ground bar (Figure 6.7b). When the incoming flow splits in this way, the vertical and lateral mixing in each channel is reduced so that flow can continue farther seaward.

6.5.2 Buoyant plumes

Water over the delta front region will become stratified when neither the river water exiting a distributary mouth nor the coastal water is turbulent enough to induce mixing. In this instance the river water proceeds seaward as a plume that is either positively (hypopycnal conditions) or negatively (hyperpycnal conditions) buoyant. **Positively-buoyant plumes** travel across the surface and **negatively-buoyant plumes** travel along the bed (Figure 6.6). The former situation is the more common because of the density difference between river and coastal waters, and is best developed when river water exits the entrance with low to moderate velocity into relatively deep water (Table 6.2). These conditions produce stable stratification with the fresh river water floating on the denser, saline coastal water. Moreover, the river discharge should exceed the tidal prism, but should be insufficient to keep out the tidal intrusion of salt water into the lower reaches of the distributary channel (similar to a salt-wedge estuary – see Section 7.3.1). Negative buoyancy arises if river water is carrying so much suspended sediment that its bulk density is greater than that of the saline coastal water. This usually requires suspended sediment concentrations of the order of 100 g l^{-1}, which is rare. One example is the Huanghe (Yellow) River delta, which

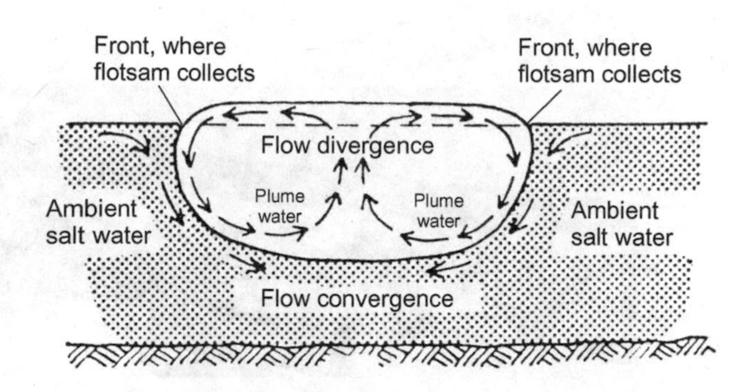

Figure 6.9 Schematic cross-section of a buoyant plume showing circulation in the plane transverse to the flow. The principal flow direction is either into or out of the page. (From Wright, 1977.) (Copyright © 1977 Geological Society of America, reproduced with permission.)

delivers silt from the interior of China to the Gulf of Bohai (Wright *et al.*, 1986). Here we will restrict our discussion to the more common, positively-buoyant plume.

The margins of buoyant plumes are marked by sharp boundaries known as **fronts**, which can be recognised by a colour difference between the plume and the coastal water, or by surface lines of flotsam composed of foam, algae or debris. The flotsam that marks fronts provides a clue as to the nature of flow within the plume (Figure 6.9). Because the plume is positively buoyant and elevated slightly above the coastal water, there is a gentle slope from the centre of the plume out towards its margins. This sloping water surface causes a pressure gradient force and therefore a secondary, lateral surface flow. The diverging surface flow at the centre of the plume draws water vertically upwards and the convergent flow at the margins of the plume drives water downwards, thereby completing the circulation (Figure 6.9). It is the surface flow convergence at the plume margins that accumulates the flotsam. Convergent flow at the base of the plume means that as sediment is transported seaward its lateral dispersion near the bed is limited, so it tends to deposit sediment in linear,

Figure 6.10 Satellite image showing part of the fluvial-dominated Mississippi River delta, USA, displaying the classic 'bird's-foot' morphology. (Modified image from the NASA/Science Photo Library, USA.)

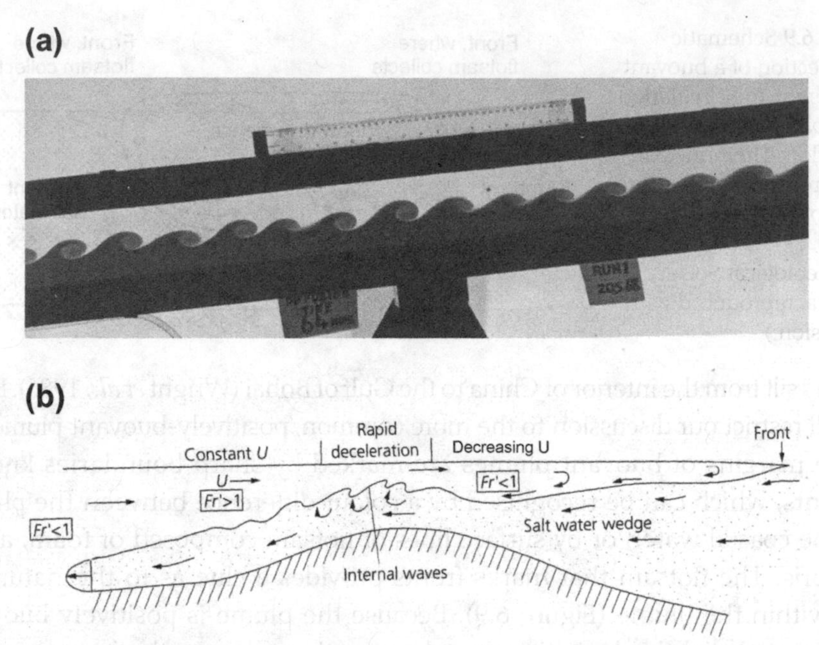

Figure 6.11 (a) Laboratory visualised internal waves produced at the interface between two liquids in relative motion to each other. In this case the waves are Kelvin-Helmholtz waves. (From van Dyke, 1982.) (Copyright 1982 The Parabolic Press, reproduced with permission.) (b) Schematic longitudinal view of a buoyant plume showing the changes in densimetric Froude number as the plume thins in the seaward direction (to the right). When the plume is at its thinnest the Froude number reaches supercritical values and internal waves develop along the base of the plume. (From Allen, 1997; modified from Wright, 1977.) (Copyright © 1997 Blackwell Publishers, reproduced with permission.)

subaqueous levees that grow seawards and extend the length of the distributary channel (Figure 6.7c). Such long, linear levees produce a 'birds-foot' delta front morphology typical of fluvial-dominated deltas such as the Mississippi River delta, USA (Figure 6.10). Here, the river discharge rate varies between 8400 and 28,000 m^3 s^{-1}, whereas mean tide range is only 0.2 m. The predominance of fresh river water, and high annual fine sediment load of 2.1 x 10^8 tonnes, results in the regular occurrence of sediment-laden positively-buoyant plumes which are actively extending the delta front (Milliman and Meade, 1983) (see also Case Study 6.1).

Sediment deposition from positively-buoyant plumes is also responsible for formation of a distal bar located seaward of the subaqueous levees (Figure 6.7c). The mechanics of distal sedimentation from the plume are partly controlled by the changing value of the densimetric Froude number as the plume moves away from the distributary entrance, expanding in width and thinning as it does so. If the plume velocity remains constant, then the densimetric Froude number may eventually reach a value of one (Equation 6.1). If this occurs, then the weight force of the plume is insufficient to counteract the inertial forces driving the plume, so the shear along the density interface between the base of the plume and the coastal water will distort the interface into a series of internal waves (Figure 6.11a). Wright (1977) proposes that these waves promote mixing between the basal plume water and the coastal water, causing the plume to lose mass, momentum and competency to transport sediment. This results in coarser sediment being deposited to form the central part of the distal bar (Figure 6.7c). The plume's deceleration eventually causes the Froude number to drop below unity again, thus suppressing mixing and enabling the plume to continue travelling seaward with the fine sediment load (Figure 6.11b). The fine sediment is progressively deposited beyond the distal bar.

6.5.3 Effects of tides and waves on delta front processes

So far we have considered the fluvial processes by which sediment is carried to and deposited around the delta front region. Tides and waves, however, also exert a strong influence on sediment dynamics and resultant delta front morphology. Tidal processes most strongly affect the morphology and position of bars found adjacent to the river entrance, whereas wave processes most strongly affect the direction and strength of sediment reworking.

Along meso- and macrotidal coasts, the primary effect of the tide on delta front processes is to cause strong mixing and the maintenance of homopycnal conditions. The river discharge typically behaves as a plane jet interacting with strong tidal currents flowing roughly parallel with it (i.e. perpendicular to the coastline). The result is elongated distributary mouth bars separated by tidal channels (Figure 6.7d). The situation is complicated by the fact that flooding tidal currents oppose the river discharge and ebbing currents flow with it. This results in **mutually-evasive sediment transport zones** in which one side of the channel conveys sediment mostly seaward while the other side conveys it mostly landward. In order for the system to still operate as a delta, however, there must be a net sediment flux in the seaward direction in order to achieve coastal progradation (i.e. that volumetric sediment transport on the ebb is greater than on the flood).

Mutually-evasive transport is well developed in the **tide-dominated** Fly River delta, Papua New Guinea (Figure 6.12), which has a relatively large annual average freshwater discharge rate of 6500 $m^3 s^{-1}$ and annual sediment load of 8.5×10^7 tonnes. The tidal discharge rate, however, is about 18 times the river discharge rate. The dominant tidal discharge results in homopycnal, reversing tidal currents that control patterns of sedimentation. Net bedload transport (by combined river and tidal flow) at the entrance of the far northern channel is directed landward on the north-side and seaward on the south-side.

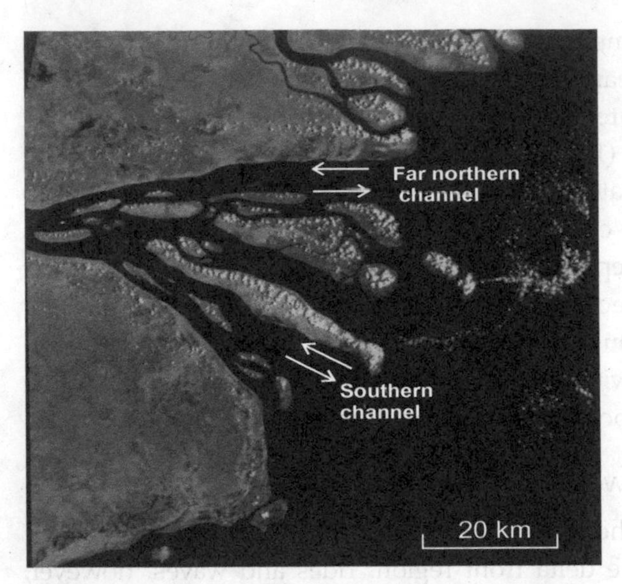

Figure 6.12 Satellite image showing the tide-dominated Fly River delta, Papua New Guinea, displaying a funnel-shaped morphology. Note the linear sand ridges and tidal channels aligned perpendicular to the coast. Arrows indicate direction of net bedload transport at the distributary entrance. (Modified image from Geoscience Australia, reproduced with permission.)

There is a similar mutually-evasive net bedload transport in the southern channel. Despite some sediment being transported landward at specific sites in this system, on the whole, most sediment is being transported seaward to feed an actively prograding delta front (Harris *et al.*, 2004).

When delta systems are connected to a steep shoreface the delta front may be exposed to high-energy waves. Waves also enhance mixing, hence any buoyancy is destroyed and the river effluent behaves as a modified plane jet. Waves encountering the jet experience a shortening of their wavelength and a corresponding increase in steepness. As a result, the waves break farther offshore which widens the surf zone on the delta front, enhancing further mixing and causing wave setup (Section 4.6.5). Wave setup has the effect of reducing the water surface slope through the distributary mouth and therefore reducing the river current velocity, thus the bar tends to be deposited closer to the mouth than it otherwise would be in the absence of waves (Figure 6.7e). The seaward extent of the delta front is also limited by the reworking of delta front sediments by waves in the form of shoreward-migrating swash bars. As these bars weld to the beach, the delta plain progrades as a series of beach ridges and swales. When waves approach with their crest parallel to the coastline they are refracted symmetrically around the distributary mouth bar. When they approach obliquely, however, they cause longshore transport and spit growth, with the river entrance migrating downdrift.

This pattern of wave interaction with delta shoreface sediments is observed in the Shoalhaven River delta, Australia, a typical **wave-dominated** delta (Figure 6.13). The river discharge is highly variable – mean annual discharge is 57 m^3 s^{-1}, but episodic flood events have a discharge of <5000 m^3 s^{-1}. Sediment delivery

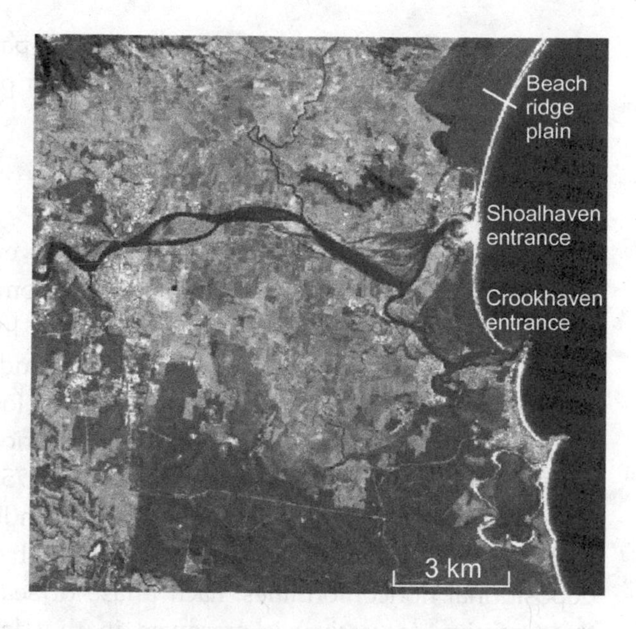

Figure 6.13 Satellite image showing the wave-dominated Shoalhaven River delta, Australia, displaying an arcuate-shaped swash-aligned shoreline. At the time of photography the Shoalhaven entrance was closed. (Modified image from New South Wales Department of Land and Property Information, reproduced with permission.)

to the delta front occurs almost entirely during floods, since the entrance is typically closed for long periods between events due to persistent wave energy pushing delta sediments landward. Wave energy is high because significant wave height exceeds 1.5 m for 50 per cent of the time and exceeds 3 m for 5 per cent of the time (Wright, 1976). Wave energy has shaped the long, arcuate beach to the north of the entrance, which is backed by a wide beach ridge plain composed of delta sediments.

6.6 Climatic controls on delta development

Deltas reflect the interplay of fluvial, tidal and wave processes, and tide-dominated and wave-dominated deltas typically arise where the vigour of the other two processes is subdued due to either tidal state, time of year, or weather conditions. This means that delta behaviour as a whole is strongly climatically influenced. There are two principal ways by which climate–delta system relationships over centennial to millennial time scales can be explored, by focusing on responses to variations in sea level and sediment supply.

6.6.1 Sea-level change and sequence stratigraphy

Mapping the internal architecture of deltas (including identifying prograding and aggrading topsets and downlapping foresets) shows how deltas respond to variations in sea level. This is the concept behind **sequence stratigraphy**. In a classic paper, Posamentier and Vail (1988) show how different scenarios of eustatic sea-level change can result in different patterns of stratigraphic architecture within deltas. The elevation of the delta plain can be related to mean sea-level conditions by Δ_{RSL}

$$\Delta_{RSL} = A - \Delta E - C_N - C_A \pm M \tag{6.3}$$

where A is delta plain aggradation rate, ΔE is eustatic sea-level change, C_N and C_A are natural and accelerated sediment compaction respectively, and M is tectonic and lithospheric changes in Earth's mass. The net result of these interactions is that the delta plain, delta front and prodelta zones can migrate landwards or seawards, or change in width or location over time. For example, Coleman (1988) showed that six major periods of deposition have taken place on the Mississippi River delta in the last 7500 years (Figure 6.14). Sediments deposited over one time period are bundled together in stratigraphic units termed **systems tracts** that are separated by erosional unconformities or depositional nonconformities. Each phase of sea level (highstand, lowstand, transgressive, regressive) is preserved in the delta stratigraphy. Analysis of many cross-sections through the Mississippi River delta shows that different parts of the delta are active at different times (Figure 6.14), and prograde and are then abandoned on an averaged time scale of 1000–2000 years (Coleman, 1988). This is termed delta switching.

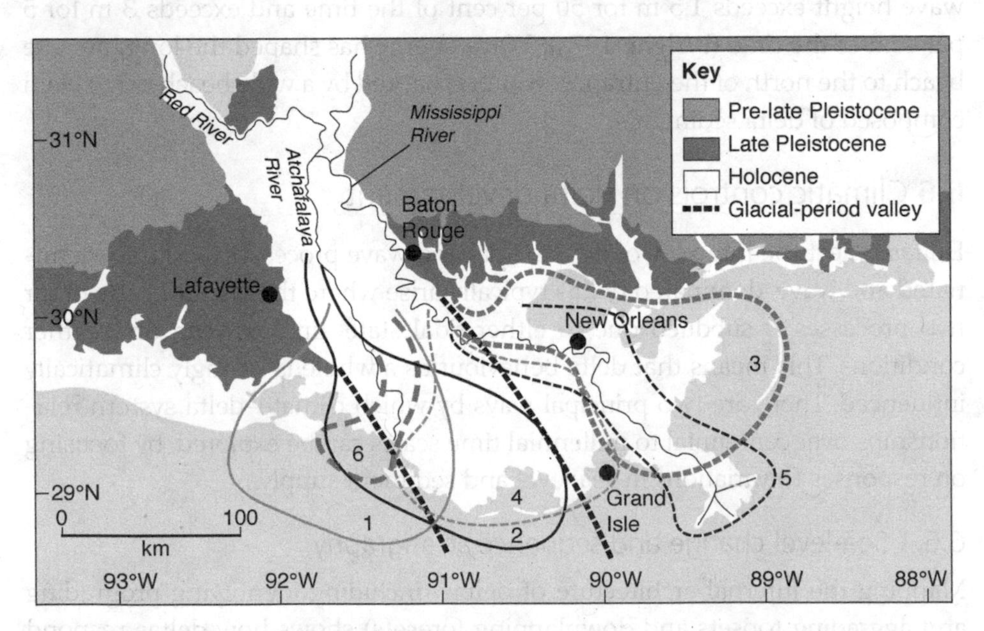

Figure 6.14 Geological map of the Mississippi River delta region showing the major late Pleistocene and Holocene delta lobes (1–6) and their ages. (From Blum and Roberts, 2009.) (Copyright © 2009 Macmillan Publishers Limited, reproduced with permission.)

6.6.2 Delta switching

Presently-active deltas are relatively young features, having developed in the last 7000–10,000 years since sea level reached its present position. Continued river sediment supply is the driving force behind delta progradation, which takes place where proximal accommodation space has been filled up. As a consequence of this sediment excess, the oldest deltas that front coastal valleys that were rapidly infilled first, have undergone several cycles of **delta switching** during the Holocene. Delta switching is where the active region of coastal progradation switches from one location on the delta front to another as a result of the blocking or diversion of distributary channels by fluvial or tidal sediments. The interval between switching events varies between a few hundred to a few thousand years, depending on delta size. Coleman (1982) identifies three types of switching: lobe switching, channel switching and channel extension (Figure 6.15).

Lobe switching occurs where the active area of delta progradation consists of a network of numerous distributary channels. Over time, the active lobe becomes shallow, over-extended and inefficient at conveying sediment all

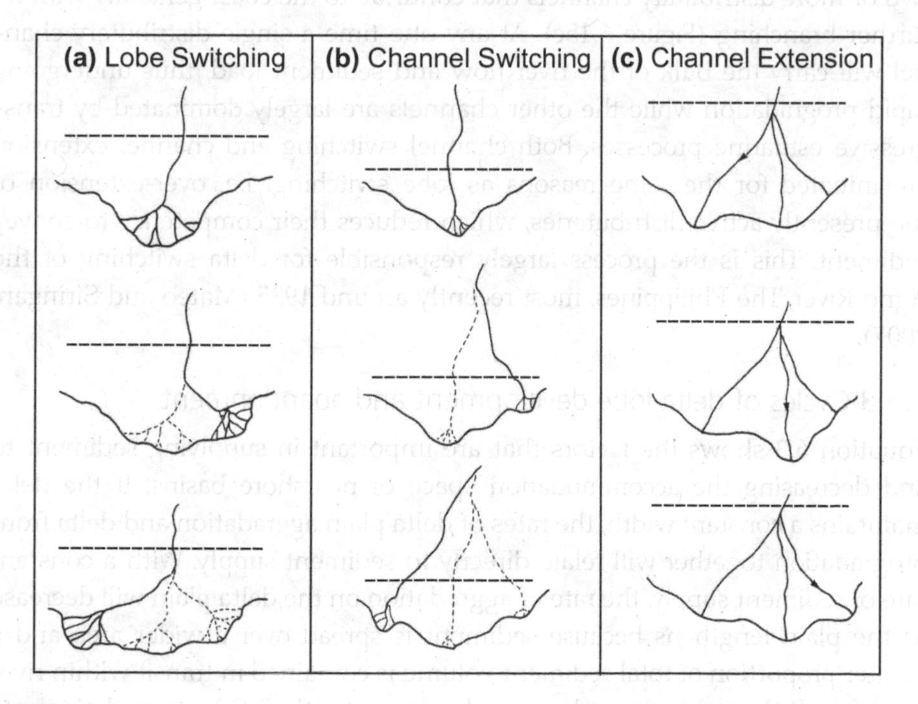

Figure 6.15 Schematic plan view of the three styles of delta switching: (a) lobe switching; (b) channel switching; and (c) channel extension. The border between the fluvial valley and the delta plain is indicated by the dashed, horizontal line. (Modified from Coleman, 1982.)

the way across the delta plain, so the entire network is then abandoned in favour of a more competent channel cut through a shorter, steeper section of the delta plain (Figure 6.15a). When this happens, the first lobe becomes inactive and may undergo erosion by waves and tides. New lobes typically develop in coastal bays between previous lobes, or extend over the top of previously-subsided lobes. Lobe switching is favoured where there are low offshore slopes, and low wave- and tide-energy. This is typical of the River Po delta (Italy) where sediment within previous lobes is reworked by subsequent lobe development (Correggiari *et al.*, 2005). **Channel switching** occurs well upstream, somewhere on the fluvial plain (Figure 6.15b). A new course is cut for the river and a new delta develops. This is seen very clearly by the Atchafalaya River (USA) which has progressively captured some of the Mississippi River's water and sediment since around 1940, and is presently building a new delta lobe (Coleman, 1988) (Figure 6.14). This style of switching produces coastal progradation over a wide area, and is favoured where there are intermediate offshore slopes, and high wave- and tide-energy. **Channel extension** involves the fluvial channel branching on the upper delta plain into two or more distributary channels that continue to the coast generally with no further branching (Figure 6.15c). At any one time a single distributary channel will carry the bulk of the river flow and sediment load, thus undergoing rapid progradation while the other channels are largely dominated by transgressive estuarine processes. Both channel switching and channel extension are initiated for the same reasons as lobe switching; i.e. over-extension of the presently active distributaries, which reduces their competency to convey sediment. This is the process largely responsible for delta switching of the Agno River, The Philippines, most recently around 1935 (Mateo and Siringan, 2007).

6.6.3 Cycles of delta lobe development and abandonment

Equation 6.3 shows the factors that are important in supplying sediment to and decreasing the accommodation space of nearshore basins. If the delta maintains a constant width, the rates of delta plain aggradation and delta front progradation together will relate directly to sediment supply. With a constant rate of sediment supply, the rate of aggradation on the delta plain will decrease as the plain lengthens because sediment is spread over a wider area and a greater proportion of total sediment volume is contained in transit within river channels. If the delta progrades into deeper water then the progradation rate will decrease over time as the foresets get longer. This means that delta lobes will continue to grow bigger indefinitely provided there is no morphodynamic

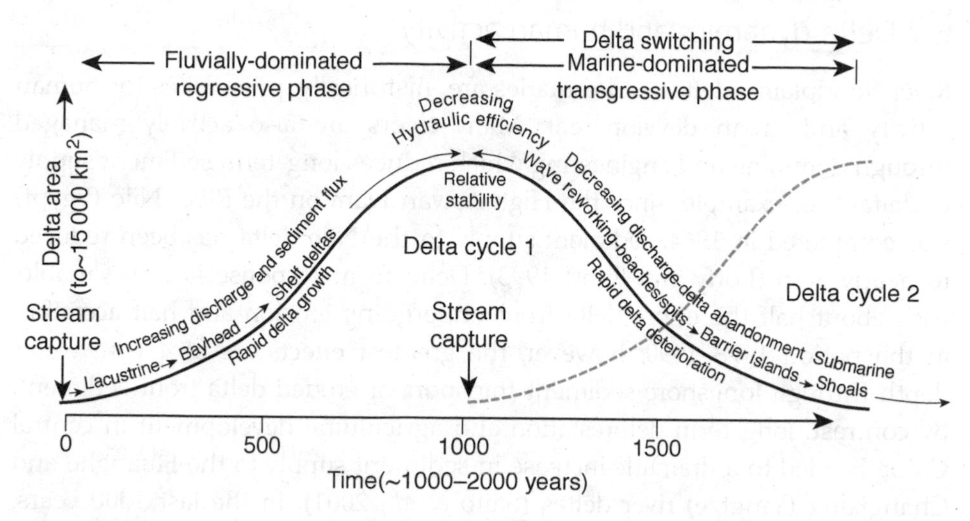

Figure 6.16 Model of the delta cycle, showing the environmental and hydraulic changes that take place as a delta lobe develops. (From Coleman *et al.*, 1998.) (Copyright © 1998 Coastal Education & Research Foundation, reproduced with permission.)

feedback. A model illustrating changes to the delta environment over time as delta lobes develop is shown in Figure 6.16. This shows that delta lobes cease to be active as a result of hydraulic inefficiency within distributary channels, such that channels on the delta plain become infilled by river sediment and channels near the delta front are closed off by tidal flushing or wave-driven longshore transport (Coleman *et al.*, 1998).

What is the time scale over which delta lobes grow and are abandoned? Studies of the Mississippi River suggest that lobes attain a size of around 30,000 km² over a time period of around 100–1500 years, and are then abandoned (Coleman *et al.*, 1998). It is likely that delta lobe size rather than age is more important – the longer the river channels over the delta plain, the more inefficiently those channels are likely to operate. Delta lobe size will depend on the gradient of the coastal hinterland (driving the speed of river transport), sediment supply rate, and accommodation space (i.e. nearshore bathymetry and subsidence). It is also likely that different rivers each operate over a different time scale of delta lobe abandonment, for example, different delta lobes on the Po River (Italy) were formed and abandoned over time scales of 200–500 years (Correggiari *et al.*, 2005).

Worked Example 4

6.7 Delta dynamics and human activity

River floodplains, deltas and estuaries are, historically, prime sites for human activity and urban development. Many rivers are also actively managed through damming and engineering which reduces long-term sediment supply to deltas. For example, since the High Aswan Dam on the River Nile (Egypt) was completed in 1964, sediment supply to the Nile delta has been reduced to nearly zero (Lotfy and Frihy, 1993). Delta front response is very variable, with about half the entire delta front undergoing erosion and half accretion in the period 1978–1990; however, the greatest effects are at 4–6 m water depth through longshore sediment transport of eroded delta front sediment. By contrast, long-term deforestation and agricultural development in central China has led to a dramatic increase in sediment supply to the Huanghe and Changjiang (Yangtze) river deltas (Saito *et al.*, 2001). In the last 1000 years, the Huanghe's sediment yield has increased by an order of magnitude, leading to distributary channel migration, lobe switching and very rapid delta plain progradation (up to 80 km kyr^{1}). Human activity within river catchments and on deltas themselves can radically alter delta plain and delta front dynamics in often unpredictable ways, irrespective of climate forcing. This means that delta morphodynamics is a sensitive indicator of the interplay between geomorphological, sedimentary, anthropogenic and climatic factors. This is explored in more detail in Case Study 6.1.

Case Study 6.1 Deltas on the edge: The changing Mississippi

Deltas worldwide are coming under pressure by a combination of:

- sea-level rise
- land subsidence by sediment compaction, groundwater, oil and gas extraction
- reduced seaward sediment supply from rivers
- increased urbanisation and engineering.

The Mississippi River delta is an example of these factors at work. This river is the largest in North America and its deltaic sediments have attained considerable thickness (35 km) along the northern coast of the Gulf of Mexico, with multiple phases of delta lobe deposition during the Holocene (Figure 6.14). River management and engineering within the catchment, however, has prevented natural channel migration and decreased the sediment volume reaching the delta mouth. For example, large dams on the tributary Missouri River decreased the total Mississippi sediment load by around a third from the 1950s to 1970s (Rossi *et al.*, 2009).

The result is that the Mississippi delta plain and delta front are now sediment-starved. Blum and Roberts (2009) calculated that an extra 18–24 billion tonnes of extra river sediment is needed in order to maintain the delta, given present regional sea-level rise, which is three times the rate that the delta plain is presently aggrading. This makes the delta region much more vulnerable to coastal flooding from both sea-level rise and extreme hurricanes.

Hurricane Katrina (23–30 August 2005), that made landfall over the Mississippi River delta region, was associated with very strong onshore winds, a very deep low-pressure eye and high surge waves that reached up to 10 m above normal levels across the coastline of Louisiana (Fritz *et al.*, 2007) (see also Case Study 3.2). Barrier islands flanking the margins of the delta are directly sustained by delta sediments, but had been in long-term decline due to decreased sediment supply from the delta. This meant the islands were less able to protect the delta front from incoming hurricane-generated waves.

SUMMARY

- The largest fine-grained deltas commonly occur along tectonically-passive trailing edge coasts with large catchment areas, high annual rainfall, high denudation rates, and low wave- and tide-energy. Rapid subsidence will produce the thickest deltaic wedge and slow subsidence will produce the largest deltaic wedge. Coarse-grained deltas occur on tectonically-active leading-edge coastlines and are composed of gravels that are moved by gravity-driven processes into steeper and deeper receiving basins.
- Deltas typically consist of three morphostratigraphic units: the delta plain with topset beds, the delta front with foreset beds and the prodelta with bottomset beds. The delta front is the region of active coastal progradation and displays the greatest morphodynamic variability due to the nature of the river discharge and any modifications caused by tides and waves. Delta front morphology of fine-grained deltas includes subaerial and subaqueous levees and distributary mouth bars.
- When homopycnal conditions occur and the densimetric Froude number is much greater than unity, the river effluent behaves as a jet. When hypopycnal conditions occur and the densimetric Froude number is less than unity, the river effluent behaves as a positively-buoyant plume. In special circumstances hyperpycnal plumes occur and the river effluent behaves as a negatively-buoyant plume. Jets create deltaic bars situated

close to the distributary mouth and plumes create deltaic bars that are located distally. Tides cause deltaic bars to become elongated in the seaward direction. Waves cause deltaic bars to migrate shorewards to become beach ridges.

- Variations in climate (including sea level), wave regime and river sediment supply can lead to distributary channel migration and delta lobe switching, and can contribute to an acceleration of delta erosion. The role of human activity in modifying catchment processes, stabilising river channels and reducing river sediment supply can have profound effects on delta morphodynamics.

Reflective questions

These questions are designed to test your comprehension of material covered in this chapter. Suggested answers to these questions can be found on this book's website.

6a. Describe some of the advantages and disadvantages of Galloway's deltas classification scheme.

6b. Explain the relationship between river discharge characteristics and the morphology and behaviour of distributary mouth bars.

6c. Describe how the delta plain, delta front and prodelta environments are likely to differ between coarse and fine deltas.

6d. Outline some of the ways in which human activity can influence delta morphodynamic processes.

Further reading

Colella, A. and **Prior**, D.B. (eds) 1990. *Coarse-Grained Deltas*. IAHS, Special Publication No. 10. Blackwell Science, Oxford. (A useful compilation of papers on coarse-grained deltas.)

Reading, H.G. and **Collinson**, J.D., 1996. Clastic coasts. In: Reading, H.G. (ed) *Sedimentary Environments: Processes, Facies and Stratigraphy*. Blackwell Science, Oxford, 154–231. (A valuable synthesis of the extensive literature on delta morphology, sedimentology and stratigraphy.)

Stone, G.W. and **Donley**, J.C. (eds) 1998. 'The world deltas symposium: A tribute to James Plummer Morgan (1919–1995).' Special Thematic Section: *Journal of Coastal Research*, 14, 695–916. (A very useful collection of papers covering many aspect of delta geomorphology.)

Wright, L.D., 1977. 'Sediment transport and deposition at river mouths: A synthesis.' *Geological Society of America Bulletin*, 88, 857–868. (A classic paper describing the relationship between effluent dynamics and morphology of distributary-mouth bars.)

CHAPTER 7
TIDE-DOMINATED COASTAL ENVIRONMENTS – ESTUARIES

AIMS

This chapter presents the major properties of wave-dominated and tide-dominated estuary types with respect to their hydrodynamic processes and resultant landforms. Features present within the intertidal zone are first described, and then estuary hydrodynamic processes are discussed, including the interactions of fresh river and saline marine water. Tidal and sediment dynamics within estuaries are described in detail because they control patterns of water circulation and deposition of fine sediments.

7.1 Introduction

Dalrymple *et al.* (1992, p.1132) define an **estuary** as 'the seaward portion of a drowned valley system which receives sediment from both fluvial and marine sources and which contains facies influenced by tide, wave and fluvial processes.' Potter *et al.* (2010, p.497) adopt a slightly different definition, being a 'body of water that is either permanently or periodically open to the sea and which receives at least periodic discharge from a river(s)'. Modern estuaries associated with today's rivers first developed when coastal river valleys were flooded at the end of the postglacial marine transgression. Sea levels stabilised around 6000 years BP, and since that time estuaries have been infilling. Estuaries that received a large sediment influx relative to their accommodation space have completely infilled and are now prograding as coastal deltas (Chapter 6). Estuaries that received a more modest sediment supply still have accommodation space available and are continuing to infill. Sediment can enter an estuary from both the land (river) and the sea (waves and tides). The fundamental difference between deltas and estuaries is that, averaged over many years, the net sediment transport in deltas is seaward whereas in estuaries it is landward. Deltas are progradational systems that are presently extending the coastline, whereas estuaries occupy coastal embayments that are presently infilling.

The possible combinations of coastal processes (river, waves and tides), together with additional processes such as flocculation that are uniquely estuarine, result in a variety of estuary types. As estuaries infill their valleys, an evolving morphodynamic regime further increases this variety. Not surprisingly then, estuaries are among the most complex coastal environments. This chapter presents a summary of estuarine morphology, and discusses major estuarine processes that are not dealt with elsewhere in this book.

7.2 Estuary morphology

In an important paper, Dalrymple *et al.* (1992) propose that most estuaries can be divided into three zones, termed the inner zone, central zone and outer zone, which have different energy regimes, sediment types and morphologies (Figure 7.1). River processes are most significant at the estuary head (landward end or inner zone) and diminish towards the estuary entrance (seaward end or outer zone). The significance of marine processes has the opposite spatial pattern. Considering fluvial and marine processes together, the energy regime in the **inner zone** is river-dominated, in the **outer zone** it is marine-dominated (waves and tides), and in the **central zone** it is mixed (both tide and river processes). The effect of waves is mostly restricted to the seaward margin of the outer zone. The inner and outer zones are the most energetic and are predominantly zones of sediment transfer. Dalrymple *et al.* (1992) argue that for estuaries to be infilling, the averaged net bedload transport direction must be landward in the outer zone. This, combined with the net seaward-directed transport in the inner zone, results in sediment convergence in the central zone which is therefore a sediment sink. The inner and outer zones generally contain the coarsest sediment and the central zone the finest sediment.

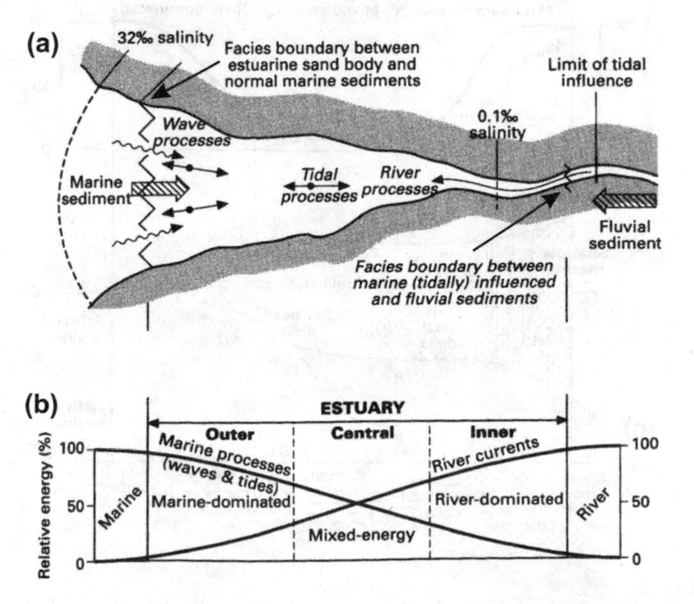

Figure 7.1 (a) Plan view of an estuary showing facies and hydraulic boundaries. (b) Chart showing the change in energy regime along the estuary axis, i.e. the changing mix of wave, tide and river processes. (From Dalrymple *et al.*, 1992.) (Copyright © 1992 Society for Sedimentary Geology, reproduced with permission.)

7.2.1 Estuary classification

The physical properties of estuaries vary dramatically depending on palaeovalley configuration, entrance conditions and degree of infilling. The first is determined by geological inheritance, the second by the predominance of waves or tides at the coastline, and the third by the amount of available sediment. There are many estuary classification schemes, but here we use the one proposed by Dalrymple *et al.* (1992) which considers both morphology and processes. The two end-members of the scheme are wave- and tide-dominated estuaries, each of which has characteristic energy regimes and morphology. The terms wave- and tide-dominated are based on assessment of the relative energy levels of waves and tides in the outer zone of the estuary, which may experience spatial and temporal variability. In reality, most estuaries may show only some elements of the archetypes discussed here, and palaeovalley configuration and sediment supply may be of greater importance.

The outer zone of a **wave-dominated estuary** consists of a **barrier system** that may be subject to washover during storms and a **tidal inlet** (Figure 7.2). Barriers are considered in more detail in Chapter 8. Deltas associated with the jet-like tidal flow through the narrow inlet develop both outside **(ebb-tide delta)** and inside **(flood-tide delta)** of the estuary (Figure 7.3). The outer zone is dominated by wave processes, but their effects decline rapidly with distance from the coast, due to wave breaking over both the barrier and tidal deltas.

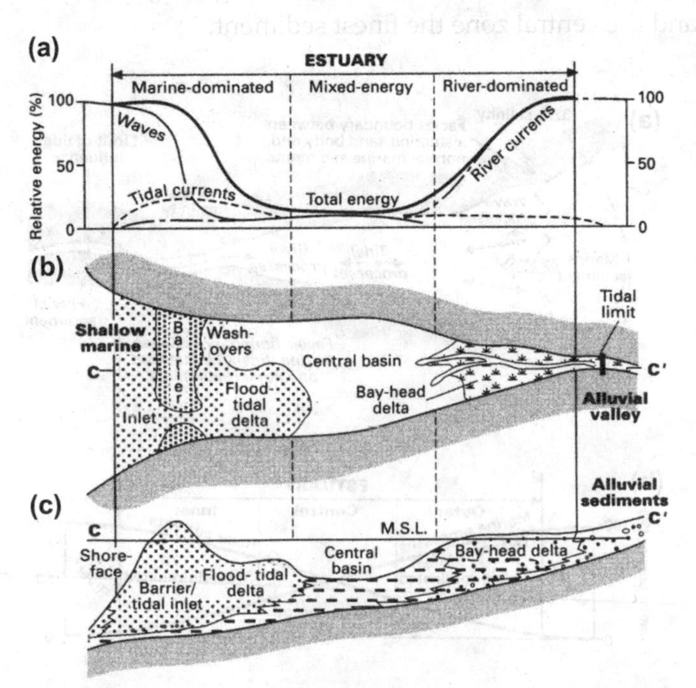

Figure 7.2 (a) Chart showing the change in energy regime along the axis of a wave-dominated estuary. (b) Plan view of the estuary showing positions of principal morphological features. (c) Section view along the estuary axis showing stratigraphy. (From Dalrymple *et al.*, 1992.) (Copyright © 1992 Society for Sedimentary Geology, reproduced with permission.)

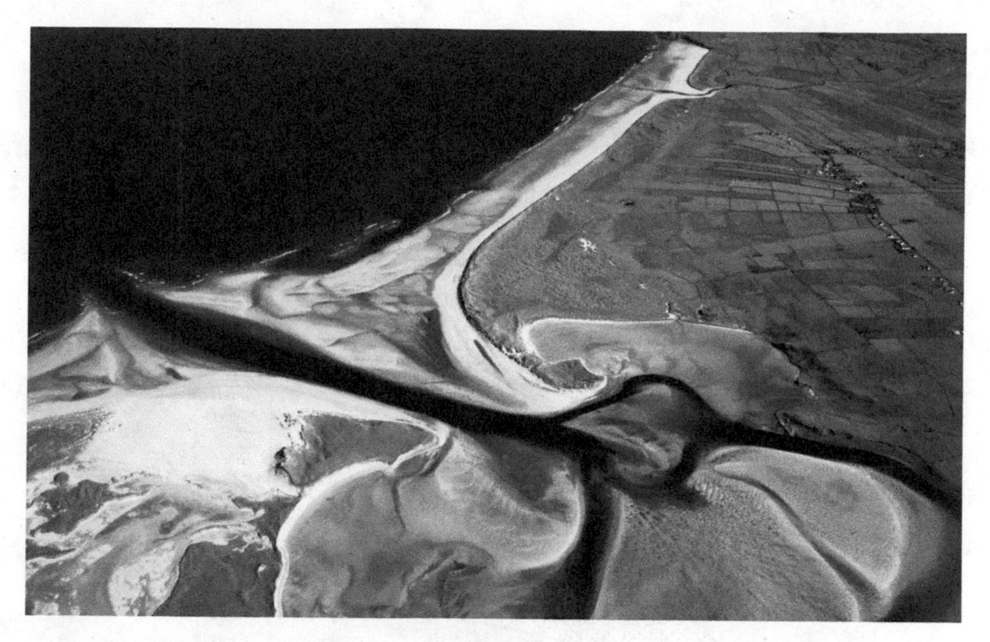

Figure 7.3 Oblique air photo of an estuary mouth in western Ireland showing the prominent flood-tide delta. (Image copyright © 2010 Irish Marine Institute, reproduced with permission.)

There is also a rapid decline in tidal energy with distance from the coast due to the restricted inlet cross-section, but some tidal energy persists into the estuary to shape the flood-tide delta. The outer zone is often composed of medium-size marine sands. A distinguishing characteristic of wave-dominated estuaries is the very low energy level in the central zone. If the estuary is relatively young with high accommodation space, a deep central **mud basin** accumulates the finest sediments, analogous to a prodelta environment. If the estuary is mature, then the central zone is infilled and dominated by salt marshes or mangrove flats composed of predominantly muddy sediments. The inner zone consists of a bay-head delta deposited by the river flow exiting the alluvial valley, typically as a plane jet. The morphological features previously described for the plain and front environments of fluvial-dominated deltas (Section 6.3) can also be recognised in bay-head deltas within wave-dominated estuaries, albeit at a generally smaller-scale.

The outer zone of a **tide-dominated estuary** consists of **linear sand bars** that delineate **tidal channels** (Figure 7.4). The appearance is similar to a tide-dominated delta but these bars do not extend beyond the drowned valley, since the system is infilling a coastal embayment rather than building out the coastline. This outer zone is dominated by tidal processes, and tidal energy often increases landward due to tidal shoaling (Section 7.3.2). Although waves are

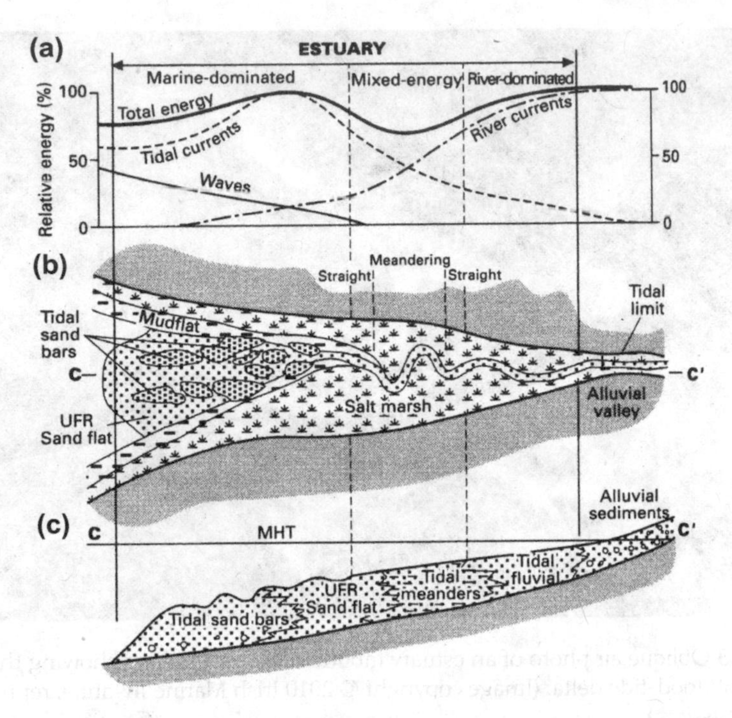

Figure 7.4 (a) Chart showing the change in energy regime along the axis of a tide-dominated estuary. (b) Plan view of the estuary showing positions of principal morphological features. (c) Section view along the estuary axis showing stratigraphy. (From Dalrymple *et al.*, 1992.) (Copyright © 1992 Society for Sedimentary Geology, reproduced with permission.)

subordinate, their influence can sometimes extend farther into a tide-dominated estuary, simply because the entrances of these estuaries are more open than wave-dominated estuaries. Due to high-energy tidal currents in the outer zone, the sediment is predominantly sand. Wave energy in the central basin is insignificant, whereas tidal energy is still relatively high, hence the central zone of tide-dominated estuaries is more energetic than that in wave-dominated estuaries. The relatively straight multiple channels of the outer zone reduce to a single **meandering tidal channel** in the central zone. This channel is influenced by river processes' high discharge periods, but is tide-dominated for most of the time and includes extensive intertidal morphology including salt marsh or mangrove flats. The central zone is a sink for fine sediment, but sediment accumulates here not because of a rapid decline in energy, but due to the role of gravitational circulation and tidal distortion on cohesive sediment dynamics (described in Section 7.4). Due to the persistent tidal energy in the central and inner zones, fluvial sediments become progressively mixed with estuarine sediments and there is no discrete fluvial delta. The pattern of straight channels in the inner and outer zones and a meandering channel in the central zone is characteristic of tide-dominated estuaries.

7.2.2 Tidal channels and the intertidal zone

Up to this point we have only considered energy gradients and their control on morphology along the estuary axis. At any position along this axis, there is also an energy gradient towards the estuary lateral margins. This is mostly due to the relationship between bed elevation and tidal water levels. For example, consider the channel cross-section shown in Figure 7.5, noting that maximum tidal current velocities generally occur at mid tide, and lowest velocities at slack water around high and low tide. That part of the channel below mean low water is the **subtidal zone**, which is almost always submerged and conveys water at all tidal stages. The subtidal channel is the most energetic zone, because it conveys both maximum flood and ebb currents. That part of the channel situated between mean low water and mean high water is the **intertidal zone**, which is alternately submerged and exposed during every tidal cycle. The lower intertidal zone is submerged during maximum tidal currents, but water is shallower and current velocities are correspondingly smaller so its energy level is moderate compared to the subtidal zone. The upper intertidal zone is not submerged for long time periods and is never exposed to the maximum tidal currents, due to its elevation above mid-tide, so its energy level is low. That part

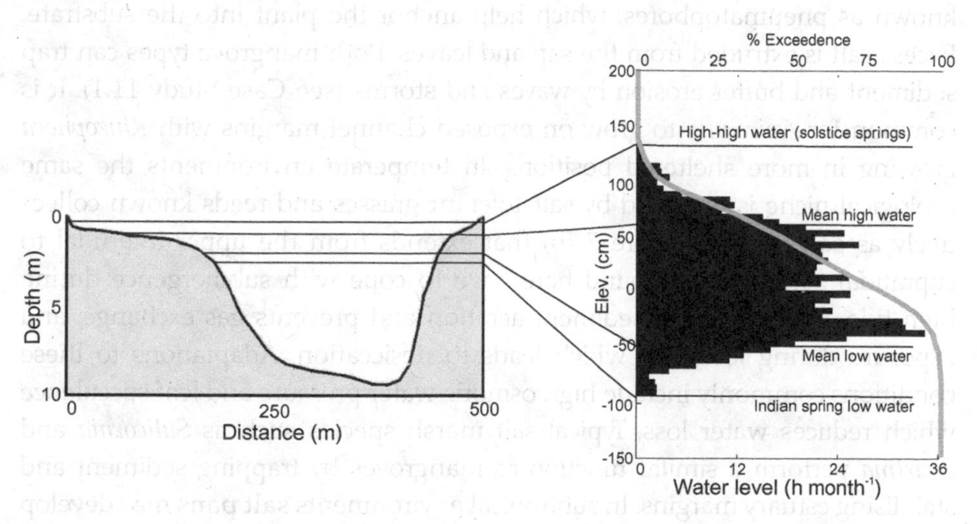

Figure 7.5 Tidal patterns at Gentlemans Halt, Hawkesbury River, New South Wales, Australia. Left panel shows a cross-section of the estuary channel. Right panel shows a histogram (black bars) and percentage-exceedence curve (grey line). The histogram represents the number of hours per month that the water level is located at each elevation in the intertidal zone. The percentage-exceedence curve shows the percentage of time in the month that each elevation in the intertidal zone is submerged.

of the channel situated above mean high water is the **supratidal zone**, which is a low energy environment only submerged during spring high tides and is therefore mostly exposed.

While tidal currents are the most important energy source in estuaries, small wind waves breaking in shallow water along channel margins can cause considerable erosion. At Gentlemans Halt (Australia), for example, the water level is most often located just above mean low water and just below mean high water (Figure 7.5). These two locations on the intertidal profile therefore experience the effects of wind waves most often, and so are marked by local increases in profile steepness due to erosion. Similarly, the lower intertidal zone is most often exposed to scour by maximal tidal currents and is therefore steeper than the upper intertidal zone.

In most estuaries the lower intertidal zone is devoid of vegetation, due to high bed shear stress preventing seedlings taking anchor in the sediment. The upper intertidal zone is less energetic, and in tropical and subtropical environments is frequently colonised by **mangroves** (Figures 7.6 and 7.7a), which are halophytic (salt-tolerant) tree and shrub species that have developed growth habits and physiological strategies to survive in saline or brackish conditions. *Rhizophora* species have prop or aerial roots that provide a large surface area for oxygen absorption, and salt water is excluded by high osmotic pressure within the plant's body. *Avicennia* species have vertical, above-ground roots known as pneumatophores, which help anchor the plant into the substrate. Excess salt is extruded from the sap and leaves. Both mangrove types can trap sediment and buffer erosion by waves and storms (see Case Study 11.1). It is common for *Avicennia* to grow on exposed channel margins with *Rhizophora* growing in more sheltered positions. In temperate environments the same ecological niche is occupied by salt-tolerant grasses and reeds known collectively as **salt marsh** (Figure 7.7b) that extends from the upper intertidal to supratidal zones. Plants found here have to cope with submergence during high tide, which restricts sediment aeration and prevents gas exchange, and exposure during low tide, which leads to desiccation. Adaptations to these conditions commonly include high osmotic water pressure and leaf succulence which reduces water loss. Typical salt marsh species such as *Salicornia* and *Spartina* perform a similar function as mangroves by trapping sediment and stabilising estuary margins. In subtropical environments **salt pans** may develop if localised salt concentrations become too high for even the most salt-tolerant species to survive.

The upper intertidal or supratidal zones may be drained by small **intertidal creeks** that feed into the main estuary channel (Figure 7.6). Due to their elevation these creeks are dry for a significant part of the tidal cycle. At high water,

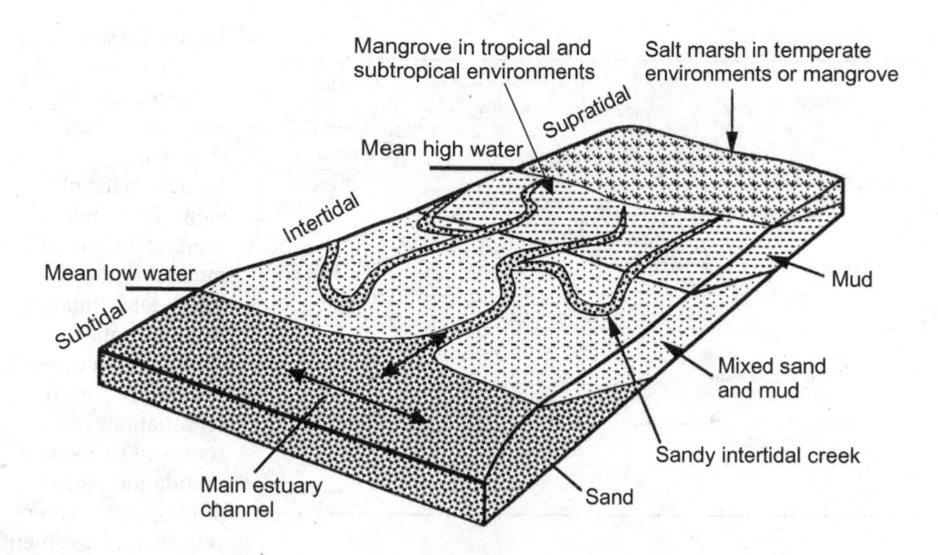

Figure 7.6 Diagram showing typical morphology and sediment distribution for the intertidal margin of an estuary channel. (Modified from Boggs, 1995.)

when the creeks are full (or overflowing on spring tides), the tidal current is usually slack. The strongest currents occur late in the falling tide, which occurs because the water level in the main channel drops more quickly than the water can drain from the tidal flat, resulting in a strong hydraulic gradient which can drive currents with peak velocities of up to 1 m s^{-1} (French *et al.*, 1993). Current speeds across the tidal flat and marsh, outside of the channels, rarely exceed a few cm s^{-1}. The sudden change in flow conditions as the water level moves from the channel confines and onto the intertidal platform results in a decrease in suspended sediment concentration, deposition rate and grain size with distance from the channel. Sediment grain size typically grades from mixed

Figure 7.7 (a) In subtropical and tropical environments, the upper intertidal and supratidal zone may contain mangrove species like these in the Hawkesbury River, New South Wales, Australia. (Photo: M.G. Hughes.) (b) In temperate environments, the supratidal zone may consist of salt marsh such as in this example from Gibralter Point, southwest England. (Photo: G. Masselink.)

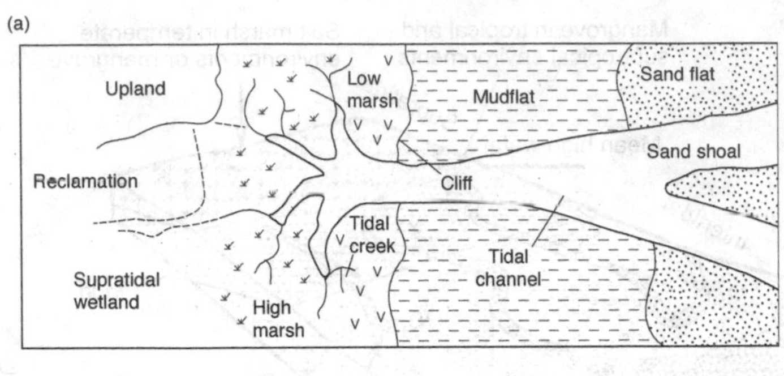

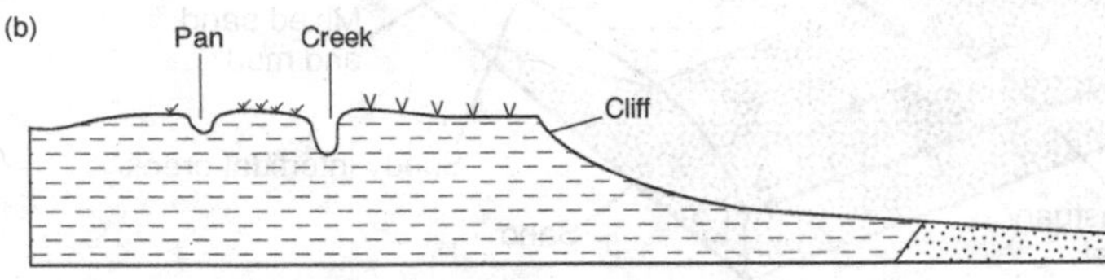

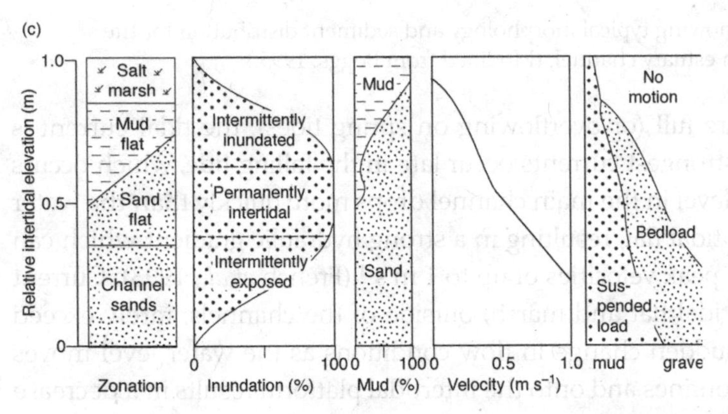

Figure 7.8 Variations in sedimentary properties across a typical muddy shoreline. (a) Schematic plan view of low-tide sandflat, intertidal mudflat and upper intertidal saltmarsh. (b) Schematic cross-section across the area in part a). (c) Variations in sediment properties, inundation period, mud content, water velocity and sediment behaviour. (From Woodroffe, 2002. Based on data from the Wash and Bay of Fundy after Amos, 1995.) (Copyright © 2002 Cambridge University Press, reproduced with permission.)

sand and mud to primarily mud farthest from the channel (Figure 7.8). While the sedimentological behaviour shown in Figure 7.8 is generally correct, it does not consider local topography and roughness affecting the flow and which can have a significant effect on the rate of vertical accretion in mangroves and salt marshes (Case Study 7.1).

Case Study 7.1 Salt marsh accretion rates

The potential accelerated rate of sea-level rise in the near future makes salt marshes particularly vulnerable environments to flooding and erosion. The expected response of these environments to sea-level rise depends primarily on the rate of sediment supply. If sediment supply matches the rate of sea-level rise then vertical accretion of the salt marsh surface will maintain the marsh area. If the sediment supply rate is insufficient then the

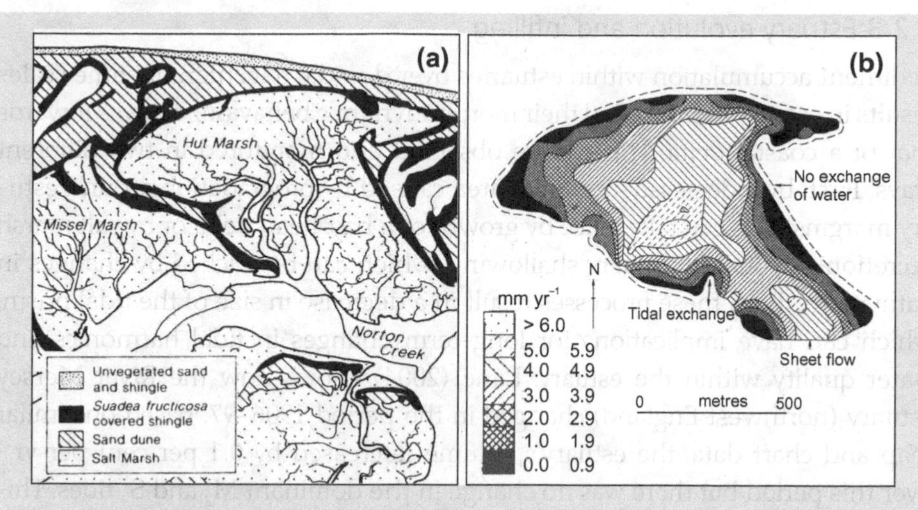

Figure 7.9 (a) Map showing geomorphological setting of Hut Marsh on Scolt Head Island, north Norfolk, England. (b) Map showing five-year mean vertical-accretion rates measured on Hut Marsh (1986–91). (From French *et al.*, 1995.) (Copyright © 1995 Coastal Education and Research Foundation, reproduced with permission.)

marsh will erode and become drowned. The actual response of salt marsh systems will reflect the heterogeneity of the intertidal morphology (Figure 7.8). One of the longest time-series of measurements of salt marsh accretion comes from a macrotidal, back-barrier salt marsh (Hut Marsh) on Scolt Head Island, Norfolk, east England (Figure 7.9).

Across Hut Marsh, French *et al.* (1995) found that there is an inverse relationship between accretion rate and marsh elevation. The north, east and west margins of Hut Marsh are the most elevated, and have accretion rates of 0–3 mm yr^1 (Figure 7.9b). The widest range in accretion rates occurred around the mid-tide level, because accumulation rates consistently higher than average are found along the margins of intertidal creeks. The centre and southeast corner of Hut Marsh display the densest networks of intertidal creeks, and marsh accretion rates in these areas reach 6 mm yr^{-1}. These larger accretion rates result from high suspended sediment concentrations in the channelised flows (*c.* 1000 mg l^{-1}), which spill overbank on spring high tides and settle adjacent to the channel margin. Similar spatial patterns of marsh accretion are seen in a high-resolution LiDAR topographic study from the River Dee estuary, northwest England (Moore *et al.*, 2009).

7.2.3 Estuary evolution and infilling

Sediment accumulation within estuaries over decadal to centennial time scales results in estuary infilling, and their morphodynamic behaviour evolves towards that of a coastal delta. This can be observed and monitored in two different ways. First, by a decrease in estuary area caused by progradation around estuary margins. This can take place by growth of a bay-head delta or by salt marsh accretion. Second, by estuary shallowing, which can be tracked by changes in bathymetry. Both these processes result in a decrease in size of the tidal prism, which can have implications for long-term changes in tidal harmonics and water quality within the estuary. Lane (2004) shows how the River Mersey estuary (northwest England) changed in the period 1906–97. Based on similar map and chart data, the estuary's volume decreased by 0.1 per cent per yr^{-1} over this period but there was no change in the dominant M_2 and S_2 tides. This suggests the estuary is still maintaining hydrodynamic stability despite changes in volume.

Dalrymple *et al.* (1992) believe that, as estuaries infill and their morphodynamic behaviour changes from an estuary to a delta, the meandering tidal channel in the central zone of tide- and mature wave-dominated estuaries also changes. The presence of tight meanders indicates that net bedload transport is landward in the channel seaward of the meanders, so the system is behaving as an estuary. The absence of the meandering zone indicates that the net bedload transport is seawards, so the system is behaving as a delta.

Wave-dominated estuaries infill through seaward progradation of the bay-head delta and landward extension of the flood-tide delta (Figure 7.10). Eventually, fluvial and marine sands bury the central basin muds. As the central basin shrinks, the tidal channel in the outer zone links up with the fluvial channel in the inner zone to produce a meandering tidal channel. At this point in time, from its seaward to landward end, the wave-dominated estuary has the straight-meandering-straight tidal channel that characterises tide-dominated estuaries. From this time on, the evolutionary path of the two estuary types is similar. Tide-dominated estuaries begin by rapidly infilling their outer zone through energetic tidal currents that redistribute sand into broadening and seaward-extending sand ridges. As this process continues the three estuary zones steadily migrate seaward along the drowned valley, progressively infilling it. The position of the meandering tidal channel also migrates seaward, eventually reaching the present coastline where the straightened channel upstream can then convey sediments directly to the coastline and begin building a delta. Infilling of wave-dominated estuaries is also enhanced where increased wave energy at the estuary mouth increases littoral drift and closes off the inlet.

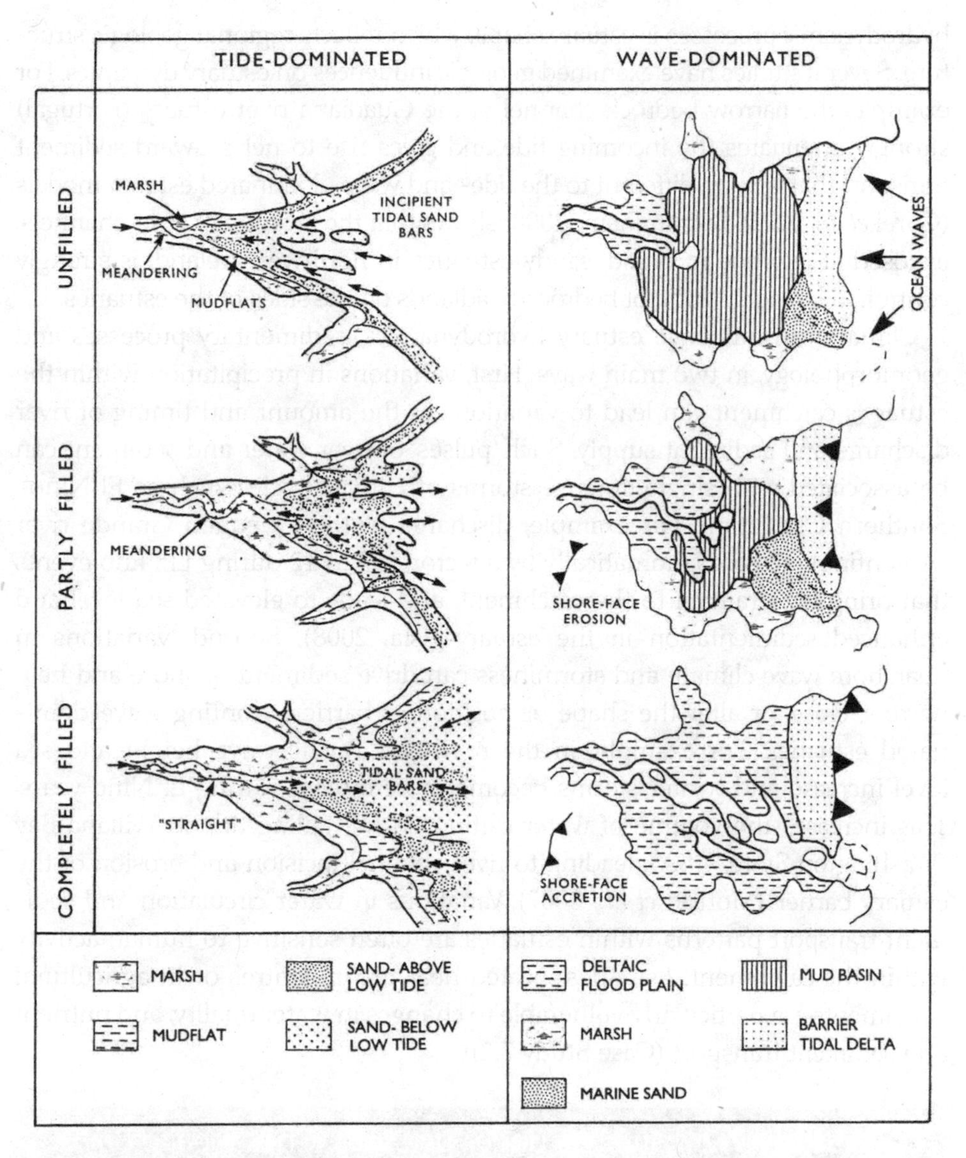

Figure 7.10 From top to bottom, stages in the filling of a wave- and tide-dominated estuary. (From Trenhaile, 1997. Originally modified from Roy *et al.*, 1980 and Dalrymple *et al.*, 1992.) (Copyright © 1997 Oxford University Press, reproduced with permission.)

7.2.4 Controls on estuary evolution and infilling

Patterns of estuary evolution and infilling reflect the role of external forcing, on long time scales, by sea-level change, tectonic subsidence, fluvial sediment supply and sediment compaction; and hydrodynamic processes of tides and marine–fresh water interaction that contribute to sediment transport and deposition within the estuary (described in Section 7.3). An important influence on hydrodynamic processes is estuary shape, which reflects regional geologic structure. Several studies have examined geologic influences on estuary dynamics. For example, the narrow bedrock channel of the Guadiana river estuary (Portugal) strongly attenuates the incoming tide and gives rise to net seaward sediment transport that is quite different to the tide- and wave-dominated estuary models (Garel *et al.*, 2009). Burningham (2008) shows that the location of river channels, exposed at low tide in broad, sandy estuaries in northwest Ireland, is strongly controlled by the position of bedrock headlands on the sides of the estuaries.

Climate can influence estuary hydrodynamics, sedimentary processes and geomorphology, in two main ways. First, variations in precipitation within the estuary's catchment can lead to variations in the amount and timing of river discharge and sediment supply. Such 'pulses' of river water and sediment can be associated with single, intense storms or climate cycles such as El Niño–Southern Oscillation. For example, discharge of the Quequén Grande river (Argentina) increases dramatically by a factor of 3 to 12 during El Niño events that bring high rainfall to the catchment, and leads to elevated sea level and enhanced sedimentation in the estuary (Isla, 2008). Second, variations in nearshore wave climate and storminess can drive sediment onshore and help narrow, close, or alter the shape or position of barriers fronting wave-dominated estuaries. For example, in the northeast Pacific wave height and sea level increase and winter storms become more intense during El Niño years. This increases the volume of water exiting on the falling tide in Willapa Bay (Washington State, USA), leading to river channel incision and erosion of the estuary barrier (Morton *et al.*, 2007). Variations in water circulation and sediment transport patterns within estuaries are often sensitive to human activity within the catchment. Estuaries located near urban centres or in agricultural catchments are particularly vulnerable to changes in water quality, and nutrient and sediment transport (Case Study 7.2).

Case Study 7.2 Estuaries under pressure: Chesapeake Bay

Chesapeake Bay is the largest estuary in the USA and is structurally complex, with large variations in width, a shallow average depth of 6.46 m, and water input from 50 major rivers of which the Susquehanna and the Potomac are the most important. The shallowness of the bay and its highly indented coastline mean that water circulation patterns are very complex. The tidal range of the bay is generally less than 1 m, but tidal currents are strongly flood-dominated. In addition, residual currents during neaps are also around 50 per cent stronger than during springs (Li and Zhong, 2009), and low river discharge during the summer and autumn are associated with water stratification and low water quality.

Although the physical geography and water circulation patterns in Chesapeake Bay have contributed to some significant environmental problems, human activity within the watershed is also important. Large cities including Baltimore and Washington DC contribute waste water, sewage, nutrients and heavy metals into the bay. These impact negatively on aquaculture production and biodiversity.

An important monitor of the vitality of Chesapeake Bay is water quality, which encompasses nutrient and suspended sediment loads, dissolved oxygen content, and chlorophyll level. High nutrient levels in incoming river water causes phytoplankton blooms whose high oxygen demands then lead to low dissolved oxygen content in the bay's water. Based on the abundance and diversity of diatoms and chemical indicators in estuary sediments, Cooper and Brush (1993) show that sedimentation rates, total organic context and anoxic water conditions (where 0 mg O_2 l^{-1}) have changed dramatically since initial European land clearance within the watershed around 1760 AD. Low oxygen content and high nitrogen and phosphorus loads remain a significant issue affecting biodiversity within the bay. In order to address these problems, the Chesapeake Bay Program was set up as an interdisciplinary partnership of bay users, with the aim of increasing water quality and restoring intertidal habitats (www.chesapeakebay.net). This includes regulating water input and water treatment, reducing shoreline erosion, planting aquatic grasses, and encouraging fish migration up tributary rivers.

7.3 Estuary hydrodynamics

A key element of estuary hydrodynamics is the mixing of salt and fresh water masses that are delivered to the estuary by tidal and river flows, respectively. In Chapter 3 we discussed how tides are generated, but this does not explain tidal behaviour inside estuaries. We first consider the factors controlling estuarine mixing, and demonstrate how partial mixing can produce estuarine currents that are additional to tide and river flows. We then describe how channel morphology can influence tidal behaviour.

7.3.1 Stratification, mixing and gravitational circulation

Whether fresh river water and saline coastal water remain segregated or combine in an estuary is determined by the effectiveness of molecular

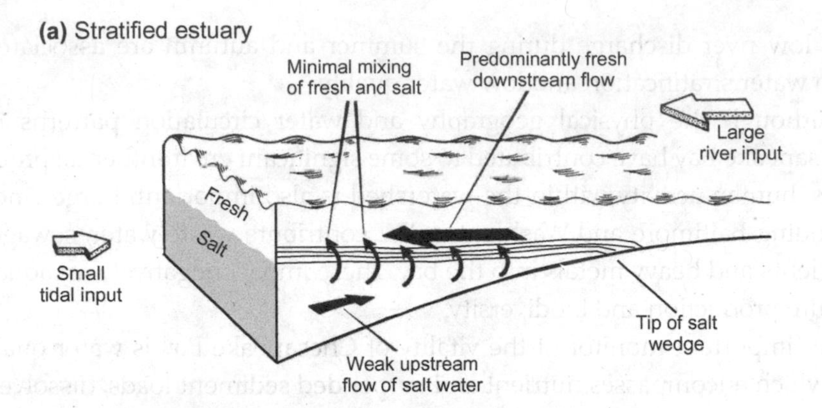

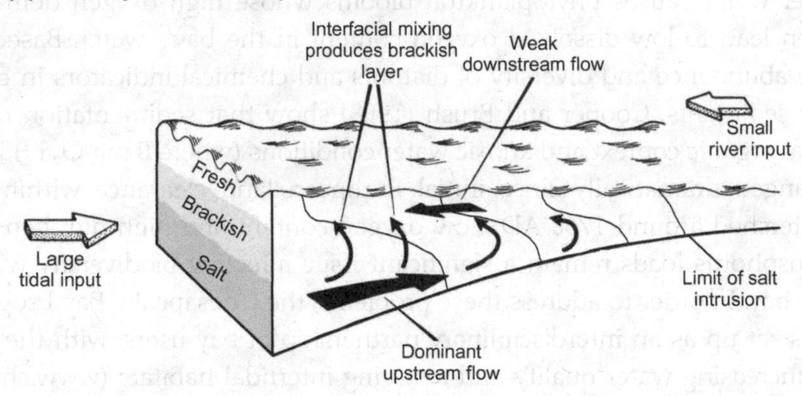

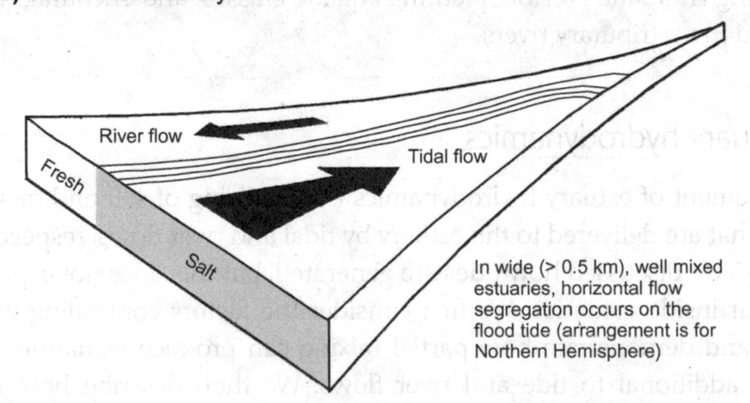

Figure 7.11 Diagram illustrating three main types of estuaries based on density stratification: (a) stratified, (b) partially-mixed and (c) well-mixed estuary. Water masses are indicated by shading on the front face of each block. Salinity contours (arbitrary scale) are indicated by thin lines on the side face of each block in (a) and (b) and on the top face of the block in (c). In (a) and (b) vertical mixing is indicated by thin arrow and non-tidal currents are indicated by thick arrows. (Modified from Pethick, 1984.)

diffusion and turbulent mixing. **Molecular diffusion** refers to the movement of salt molecules from areas of high to low concentration in the salt and fresh water bodies, respectively. The length scale of diffusion mixing is very small. **Turbulent mixing** is far more effective and involves the movement of parcels of fresh water into the salt water body and *vice versa* by eddies, which leads to partially- or well-mixed estuaries. When these processes are ineffective, the two water masses remain segregated and the estuary is stratified.

Stratified estuaries commonly occur along microtidal coasts where there is sufficient river discharge to develop a fresh surface water mass, but the discharge rate is insufficient to completely expel the lower saline water mass from the estuary or generate sufficient turbulence to cause effective mixing (Figure 7.11a). Salinity contours (isohalines) are horizontal and close together, forming a **halocline**. A **salt wedge**, defined on the upper surface by the halocline and on the lower surface by the estuary bed, thins up the estuary where it sits below the fresh river water, and is generally stationary, although it may migrate in response to variations in fluvial discharge or the tide. If it becomes too dynamic, however, turbulence is generated and the stratification breaks down. In stratified estuaries, mixing occurs along the interface between the fresh and salt water mostly by molecular diffusion, and is therefore rather weak, which helps to maintain stability of the stratification.

Partially-mixed estuaries develop if the tidal energy is sufficient to cause increased shear along the halocline and the development of internal waves (e.g. Figure 6.11a). These waves, and the turbulence caused by shearing between river/tidal flow and the estuary channel, results in considerable mixing (Figure 7.11b). Salinity contours dip steeply near the surface and near the bed, marking layers of predominantly fresh and salt water. In the central part of the water column, however, the salinity contours are gently sloping and indicate mixing between the fresh and salt to produce a brackish layer. The salinity decreases up the estuary at both the surface and at depth.

Well-mixed estuaries occur when mixing is so effective that the salinity gradient in the vertical direction vanishes entirely. If well-mixed estuaries are sufficiently broad for the Coriolis force to be effective, the river and sea water may become horizontally segregated (Figure 7.11c). This is best developed on the flood tide when the fresh and marine water masses are flowing in opposite directions. Looking downstream in the Southern (Northern) Hemisphere, the river flow is steered by the Coriolis force towards the left (right) bank and the landward-directed flooding tide is deflected left (right) towards the opposite bank. During the ebb in both hemispheres, both the tide and river flow are deflected towards the same bank.

Case Study 7.3 Spencer Gulf, Australia: An inverse estuary

The small volume of river water draining into the Spencer Gulf, South Australia, means that saline coastal waters in the Gulf are not significantly diluted, and the salt content in the water at the shallow head of the estuary increases due to high evaporation in this arid, subtropical environment. This results in a strong salinity gradient along the estuary axis, with maximum salinities at the head of the estuary (Figure 7.12).

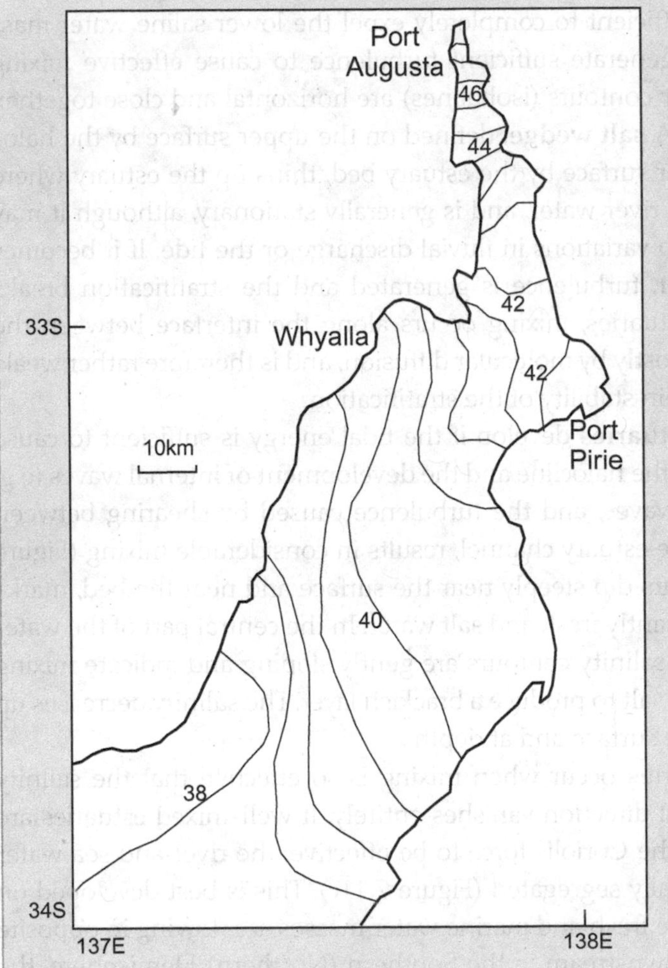

Figure 7.12 Map of Spencer Gulf, South Australia, showing depth-averaged salinity contours in practical salinity units (PSU), March 1984. Ocean water is typically about 35 PSU. Note the increase in salinity northwards toward Port Augusta at the estuary head. (Figure provided by M. Tomczak.)

Given the direct relationship between water salinity and density, a strong salinity gradient corresponds to a strong density gradient. Following our derivation of the pressure-gradient force in Box 3.1, we can determine the hydrostatic pressure P at any given depth as

$$P = \rho g h$$

where ρ is water density, g is gravitational acceleration and h is height of water above the depth of interest. Now consider an estuary with a horizontal water surface, but a difference in water density between the entrance and head of the estuary. The hydrostatic pressure at the entrance P_E and head P_H is, respectively

$$P_E = \rho g h \quad \text{and} \quad P_H = (\rho + \Delta\rho)\, g h$$

where ρ is water density at the estuary entrance and $\Delta\rho$ is the difference in density between the estuary entrance and estuary head. The pressure gradient along the estuary is then

$$\frac{\Delta P}{\Delta x} = \frac{P_H - P_E}{\Delta x} = g h \frac{\Delta \rho}{\Delta x}$$

Along any horizontal plane in the water column there is a pressure-gradient force produced by the density gradient, which drives a current from the area of high pressure to the area of low pressure. In normal estuaries where salinity decreases towards the head of the estuary it will be a landward-directed current. In inverse estuaries where salinity increases towards the head of the estuary, such as the Spencer Gulf, it is a seaward-directed current.

In some estuaries water circulation is relatively weak compared to tidal currents, but is sufficient to influence the net transport of fine suspended sediment. This **gravitational circulation** consists of a near-bed current that is directed landward and a surface current that is directed seaward (Figure 7.11b). The circulation is best developed in partially-mixed estuaries because they have a strong horizontal salinity gradient along the estuary axis. The pressure-gradient force (Box 3.1) is caused by a gradient in water surface elevation, but in partially-mixed estuaries this force is due to a gradient in water density, which drives water from the high pressure (most saline) region at the estuary entrance to the low pressure (least saline) region at the estuary head. This landward flow is generally only apparent near the bed, because the seaward river flow near the surface usually overwhelms it. In arid environments where fresh water discharge into an estuary may be absent for long periods of time and where there is high evaporation, water salinity can actually increase towards the head of the estuary. This is referred to as an **inverse estuary** in which the gravitational circulation is directed seawards (Case Study 7.3).

7.3.2 Tidal dynamics in estuary channels

In Chapter 3, we showed that the tide is a shallow-water wave and that its speed C is given by

$$C = \sqrt{gh} = \frac{L}{T} \tag{7.1}$$

thus

$$L = T\sqrt{gh} \tag{7.2}$$

where g is gravitational acceleration, h is water depth, L is tidal wavelength and T is tidal period. For a semi-diurnal tide with a period of 12.42 hours the tidal wavelengths for various water depths are listed in Table 7.1. The Amazon River, Brazil, has a tide-affected channel length of 850 km and can accommodate several tidal wavelengths (Defant, 1961). High tide therefore occurs at several locations along the estuary length at any one time. Most estuaries, however, can only accommodate a single or a fraction of a single tidal wavelength. For example, the Hawkesbury River, New South Wales, Australia, has a tidal length of 145 km and an average depth of 5 m, so it can accommodate about one half of a tidal wavelength. This means that the time of high tide at the estuary head corresponds with low tide at the estuary mouth.

Table 7.1 Tidal speed and wavelength based on linear theory for shallow water waves. Calculations made using Equations 7.1 and 7.2.

Depth (m)	Tidal speed (m s^{-1})	Tidal wavelength (km)
2	4.4	198
4	6.3	280
6	7.7	343
8	8.9	396
10	9.9	443
15	12.1	542
20	14.0	626
25	15.7	700
30	17.2	767

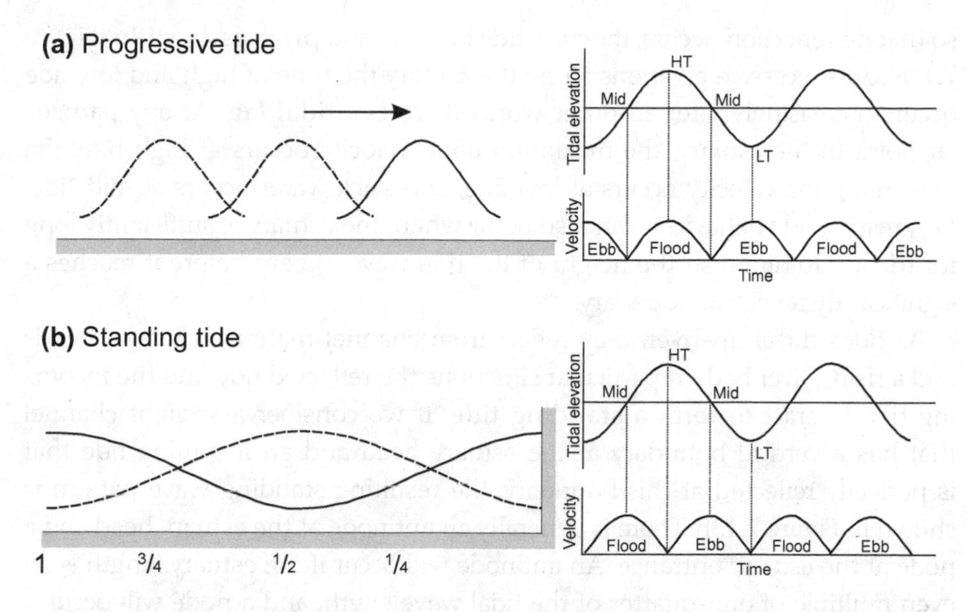

Figure 7.13 Diagram showing the nature of (a) progressive and (b) standing tides. Left-hand panels show the dynamic behaviour of the water level along the estuary. The fractions shown beneath the standing tide indicate the location of the estuary entrance if the estuary length is an integer-multiple of one-quarter the tidal wavelength. Right-hand panels show relationships between tidal water level and current velocity.

Tidal behaviour in long estuary channels (those longer than about one quarter of a tidal wavelength) is quite different to that in short estuaries. In long estuaries, the tide displays progressive and/or standing wave behaviour, whereas in short estuaries a time-varying hydraulic gradient between the tidal water levels outside and inside the estuary produces periodic water level oscillations and currents. In this section we first discuss the tidal dynamics of long estuaries and then the dynamics of short estuaries.

The movement of the tide wave in an estuary controls the vertical change in water level and the horizontal motion of the water column. The vertical rise in water level from low to high tide results from the passage of the wave front, and the fall from high to low results from the passage of the rear of the wave. For roughly half the tidal period, a horizontal current flows in the landward direction and is called the **flood current**. Similarly, for the other half of the tidal period an **ebb current** flows in the seaward direction. Between the flood and ebb currents, during which time the flow direction reverses, the current velocity is at its lowest, and is termed **slack water**. The timing of water level rise and fall relative to the ebb and flood currents depends on whether the tide behaves as a progressive or a standing wave.

If we consider a theoretical tidal channel that is straight and infinitely long, so that no reflection occurs, then the tide behaves as a **progressive tide** (Figure 7.13a). At successive positions along the estuary the time of high and low tide occurs increasingly later, in other words there is a **tidal lag**. At any particular point in the estuary, the maximum flood velocity occurs at high tide, the maximum ebb velocity occurs at low tide, and slack water occurs at mid-tide. Progressive tide behaviour can also occur where the estuary is sufficiently long for friction to diminish the height of the tide wave to zero before it reaches a significantly reflective boundary.

As tides travel up-river, they reflect from channel margins, channel bends and a rising river bed. In particular situations the reflected tide and the incoming tide interact to form a **standing tide**. If we consider a straight channel that has a vertical boundary at the estuary head and an incoming tide that is perfectly reflected at this boundary, the resulting standing wave pattern is shown in Figure 7.13b. There is generally an antinode at the estuary head and a node at the estuary entrance. An antinode will occur if the estuary length is an even multiple of one-quarter of the tidal wavelength, and a node will occur if the estuary length is an odd multiple. Maximum tide range occurs at the antinodes and minimum tide range at the nodes. In the case of a perfect standing tide, the tidal range at the antinodes inside the estuary will be twice the range of the coastal tide outside the estuary. Moreover, the maximum flood velocity occurs at mid-tide on the rising tide, the maximum ebb velocity occurs at mid-tide on the falling tide, and slack water occurs at high and low tide (Figure 7.13b). If the estuary length is one-quarter of the wavelength, then high water occurs simultaneously throughout the estuary, as does low water.

Wright *et al.* (1973) show that a standing tide is responsible for producing the funnel shape typical of many macrotidal estuaries. In order for a channel shape to be stable the system must maximise entropy, or in other words, eliminate any gradients in the sediment transport rate along the channel. A funnel shape, which narrows and increases in tidal range towards the estuary head, maximises entropy. This increasing tidal range is precisely the behaviour that a standing tide displays in an estuary that is one-quarter of the tidal wavelength (Figure 7.13b). It should be no surprise, therefore, that the length of the estuary funnel is usually one-quarter of the tidal wavelength.

Although there are some estuary configurations that favour either progressive or standing tides, in most cases the tide displays a combination of both. Slack water within estuaries commonly occurs 1–2 hours after high or low tide and maximum current velocity occurs shortly after mid-tide. Often the tide behaviour is more progressive near the estuary mouth and more standing near the estuary head, where most reflection occurs.

Estuary channels tend to narrow and shallow with increasing distance from

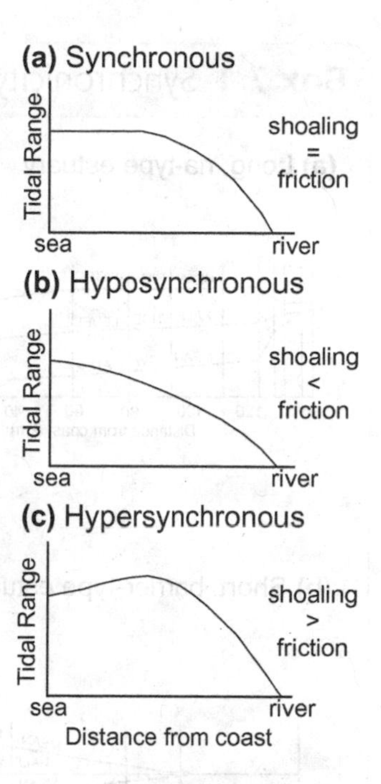

Figure 7.14 Graphs showing tidal range as a function of distance from the coast, which illustrate the definition of: (a) synchronous, (b) hyposynchronous and (c) hypersynchronous estuaries. The relative strength of shoaling versus friction effects is also indicated in each case. The scale on the axes is arbitrary.

the coast. The reduction in water depth up-estuary causes a reduction in the speed and hence the wavelength of the tide (Equation 7.1). This is the same effect as with shoaling wind waves (Section 4.5.2). In order to conserve the energy flux, the energy density must increase so that the wave height, or in this case the tidal range, must increase. Opposing this shoaling and convergence effect is friction between the tide and the estuary channel, which dissipates energy and thus reduces the tidal range. These two competing processes are rarely balanced, and three scenarios are possible (Figure 7.14):

- **Synchronous estuaries** are those where shoaling and friction effects are balanced in the lower estuary so that the tidal range is constant with distance from the coast. Eventually, however, friction becomes overwhelming and the tidal range decreases to zero at the tidal limit (Figure 7.14a).
- **Hyposynchronous estuaries** are those where friction is of overriding importance everywhere and the tidal range decreases throughout the estuary (Figure 7.14b).
- **Hypersynchronous estuaries** are those where shoaling is dominant in the lower estuary and there is an increase in tidal range with distance from the coast. Eventually, however, friction overwhelms the shoaling effect and the tidal range decreases to zero at the tidal limit (Figure 7.14c).

Long estuaries tend to be hypersynchronous and short estuaries hyposynchronous, but the synchronicity can be strongly influenced by estuary entrance conditions (Box. 7.1).

Box 7.1 Synchronicity in New South Wales estuaries

(a) Long, ria-type estuary

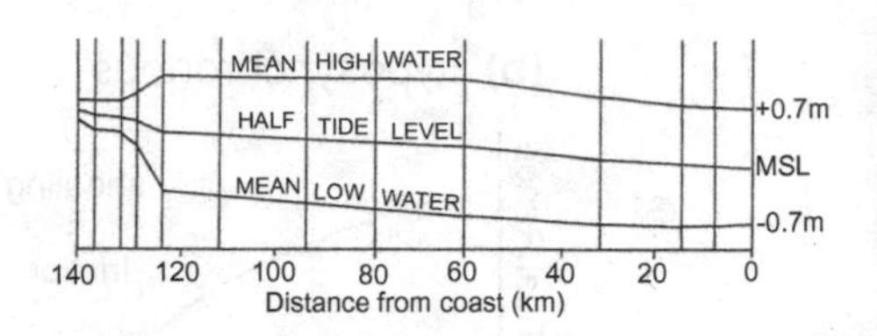

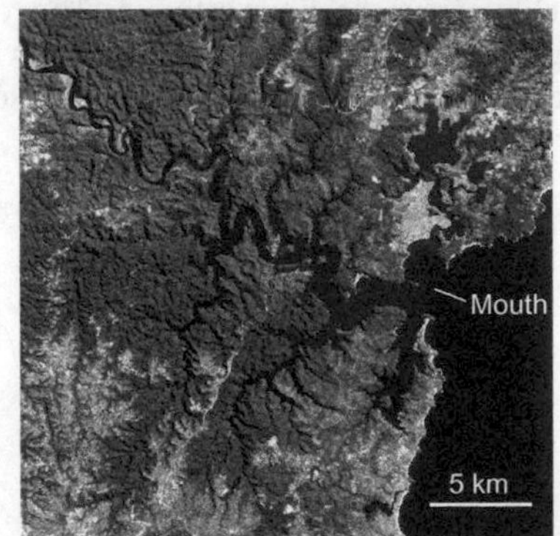

(b) Short, barrier-type estuary

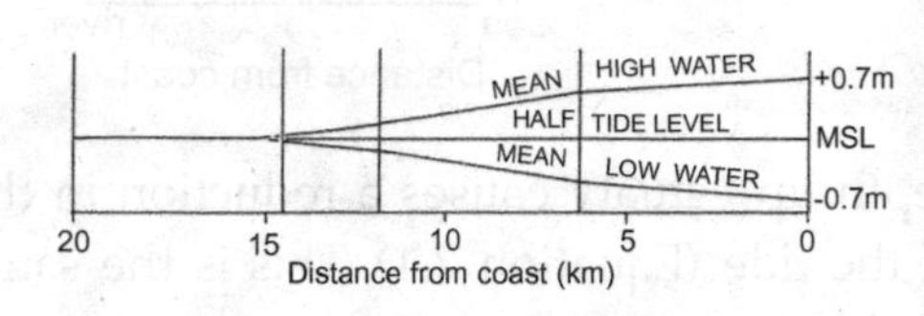

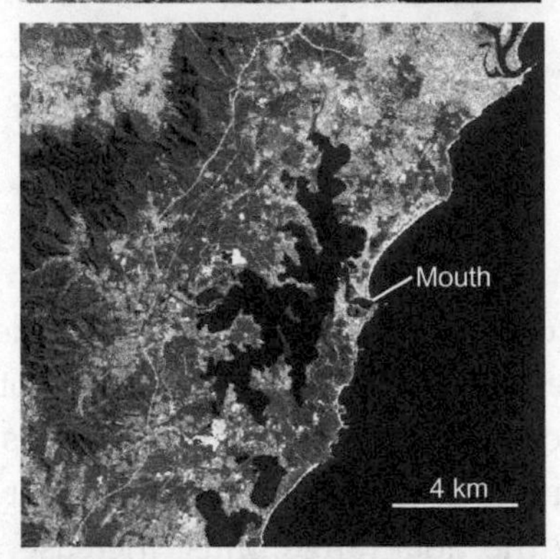

(c) Long, river-type estuary

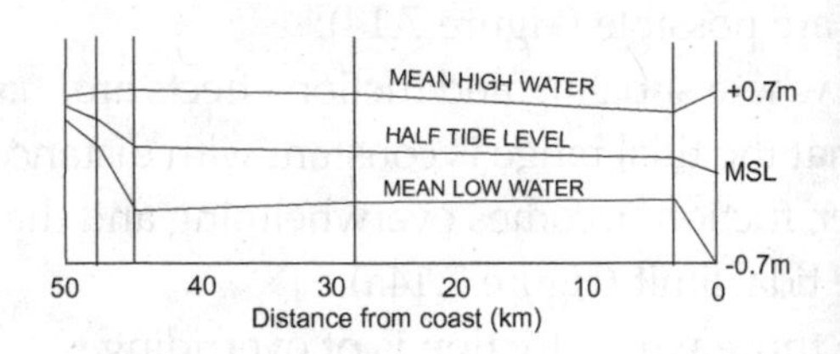

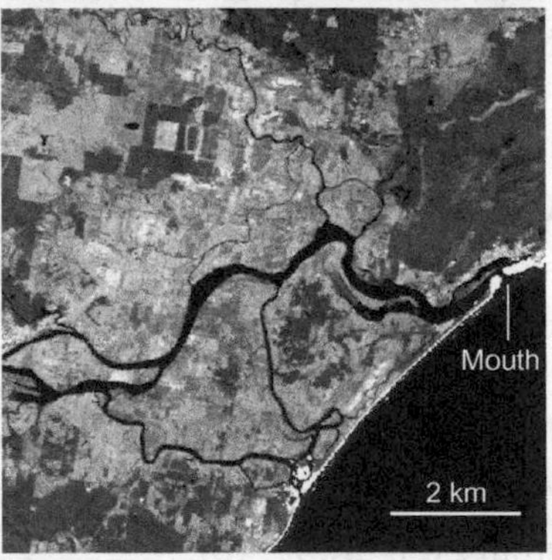

Figure 7.15 Left-hand panels show the elevations of mean high water, half tide level and mean low water as a function of distance from the coast for three estuaries in New South Wales, Australia. Tidal range, and thus synchronicity, is indicated by the change in vertical distance between mean high and low water along the estuaries. Aerial photographs of each estuary, highlighting the differing entrance conditions, are shown in the right-hand panels. The estuaries are: (a) Hawkesbury River, (b) Lake Macquarie and (c) Manning River. (Tidal elevations are modified from New South Wales Government, 1992.) (Satellite images are modified from Geoscience Australia's OzEstuaries database.)

The narrow continental shelf and steep shoreface along the coast of New South Wales (Australia) means that 96 per cent of the offshore wave energy reaches the coastline (Wright, 1976). The modal deepwater significant wave height is 1.5 m, and wave height exceeds 4 m for 1 per cent of the time (Lawson and Abernathy, 1975). Given the energetic wave regime and a microtidal range of only 1.6 m, the estuaries along this coast are all wave-dominated, but the variation in hinterland relief and sediment supply has yielded a variety of estuary entrance conditions.

The flooding of deeply incised river valleys and a low sediment supply have produced long ria-type estuaries with open, hydraulically-efficient, hypersynchronous estuaries (Figure 7.15a). Flooding of broad shallow river valleys and a low sediment supply has produced short barrier-type estuaries with constricted, hydraulically inefficient entrances. Frictional attenuation is of overwhelming importance over the entire short estuary length, and they are hyposynchronous (Figure 7.15b). The flooding of moderately incised river valleys with a large sediment supply has produced long river-type estuaries that also have constricted entrances. These estuaries display mixed synchronicity. Frictional attenuation is significant through the short entrance channel, but once inside the estuary the tidal channel is sufficiently long to permit shoaling to occur (Figure 7.15c).

A common feature of long estuaries is **tidal distortion**, which is usually expressed as a short rising tide and a longer falling tide. Tidal distortion reflects a progressive steepening of the wave front as it travels along the estuary (Figure 7.16a). Off the coast, the tidal range is a small proportion of the water depth so there is little difference in the speeds of the crest and trough. Inside the estuary, however, the tide is a significant proportion of the water depth and the deeper water under the crest travels faster than the shallower water under the trough. This can be demonstrated by re-writing Equation 7.1 as

$$c = \sqrt{g(\bar{h} \pm a)} \qquad \textbf{(7.3)}$$

where $\bar{h}$ is the local tidally-averaged water depth in the estuary and a is tidal amplitude ($-a$ for low tide and $+a$ for high tide). Figure 7.16b shows the speeds of the tide crest and trough and the speed differential as a function of water depth. For water depths of 20 m and more the two speeds are similar and the speed differential tends towards zero, so that the tide is symmetric. In shallower water depths, particularly < 5 m depth, the speed differential increases rapidly with a corresponding increase in tidal distortion. In some cases the distortion becomes so pronounced that the front of the tide is vertical, much like the front

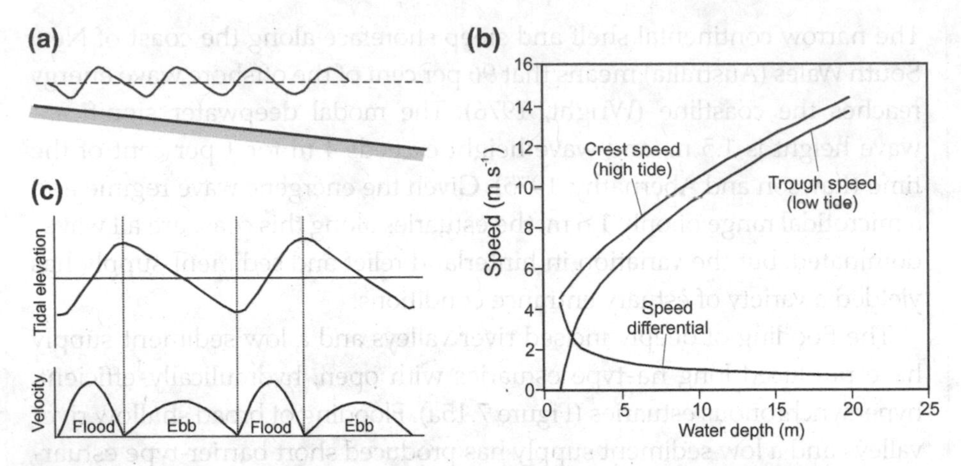

Figure 7.16 (a) Diagram showing the increasing distortion of the tide wave as it progresses along the estuary. (b) Graph showing the velocity of the crest (high tide) and trough (low tide) of a shallow water wave as a function of water depth, calculated using Equation 7.3. The velocity differential between the two increases markedly in very shallow water depths. (c) Graphs showing the relationship between tidal water level and current velocity. Note the velocity-magnitude asymmetry between the dominant flood current and the subordinate ebb current. Scales on the axes are arbitrary.

of a breaking wave in the surf zone, and is termed a **tidal bore**. Because the discharge volume through the channel on the flood tide closely matches the discharge volume on the ebbing tide, the inequality between the flood and ebb durations must produce a velocity-magnitude asymmetry between the tidal currents (Figure 7.16c). In order for the discharge volume to be conserved, the shorter duration of the flood current requires it be of larger magnitude than the ebb current.

Wave-dominated estuaries are often short (less than one-quarter the tidal wavelength) and their entrance channel is restricted, so the tide does not display wave-like behaviour inside the estuary. The periodic rise and fall of the water level in these estuaries results in different water levels inside and outside of the entrance. When the coastal tide outside the estuary is rising it eventually becomes higher than the water inside, so the water surface slopes into the estuary and drives the flood current. The total volume of water entering the estuary on the flooding tide is called the estuary's **tidal prism**, and it raises the water level in the estuary by an amount roughly corresponding to the volume of water divided by the surface area of the estuary. In a similar fashion, when the tide outside the estuary is falling it eventually becomes lower than the water inside, so the water surface slopes out of the estuary and drives the ebb current. The volume of ebbing water then reduces the water level in the estuary by a corresponding amount.

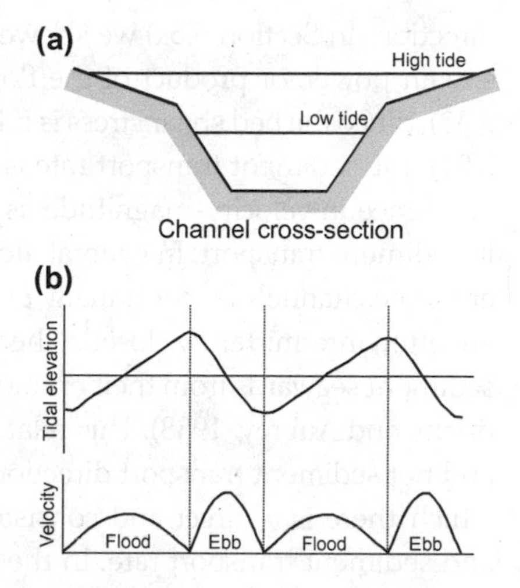

Figure 7.17 (a) Channel cross-section characteristic of short, wave-dominated estuaries in which the volume of water in the intertidal channel is a substantial proportion of the total channel volume. (b) Graphs showing the relationship between tidal water level and current velocity. Note the velocity-magnitude asymmetry between the dominant ebb current and the subordinate flood current. Scales on the axes are arbitrary.

While the coastal tide is the driver of water levels and currents inside short estuaries, it is not the same situation as the tide wave propagating through long estuaries. Nevertheless, tidal distortion and velocity-magnitude asymmetry also occur in short estuaries, although the mechanism is different to long estuaries. Friedrichs and Aubrey (1988) demonstrate that tidal distortion can also develop due to channel shape. Consider a channel that has a large cross-sectional area in the intertidal zone relative to the subtidal zone (Figure 7.17a). Between mid- and high tide, when the maximum flood current velocity is expected to occur, there is inefficient water exchange through the estuary. This hydraulic inefficiency is due to large roughness elements in the upper intertidal zone. As a result the flooding tide is slow to turn and therefore has a longer duration than the ebbing tide (Figure 7.17b). Between mid- and low tide, when the maximum ebb current velocity is expected to occur, almost all of the flow volume is conveyed in the subtidal part of the channel, which is hydraulically efficient. As a result the ebbing tide is quick to turn and therefore has a shorter duration than the flooding tide. If the ebb tide is shorter than the flood tide then conservation of discharge volume demands that it must have the largest velocity-magnitude (Figure 7.17b).

The sense of velocity-magnitude asymmetry can be consistent throughout an estuary or vary along the estuary channel. Estuaries or channel reaches that display a flooding tide that is larger in velocity-magnitude and shorter in duration than the ebbing tide are said to be **flood-dominant**, whereas those that display an ebbing tide that is largest in magnitude and shortest in duration are **ebb-dominant**. Flood- or ebb-dominance often translates directly to net landward or seaward sediment transport, respectively. Velocity-magnitude asymmetry need not be large in order to produce a dominant sediment transport direction. In Section 5.5.3 we showed that sediment transport rate is related to stream power, or product of the flow velocity and bed shear stress (Equation 5.33). Since the bed shear stress is related to the flow velocity squared (Equation 5.21), the sediment transport rate is related to the velocity cubed. Thus a small difference in velocity-magnitude asymmetry can lead to a large net difference in sediment transport. In general, **flood-dominant estuaries** tend to infill their entrance channels by continually pushing coastal sediment landward and thus are often intermittently closed, whereas **ebb-dominant estuaries** tend to flush sediment seawards from their entrance channels and thus are often stable (Friedrichs and Aubrey, 1988). This relationship between flood- or ebb-dominance and net sediment transport direction really only holds for sandy sediments, for which there is a direct and consistent relationship between bed shear stress and sediment transport rate. In the case of finer, cohesive sediments there are other factors that complicate the prediction of net sediment transport based on velocity-magnitude asymmetry.

7.4 The turbidity maximum zone: Cohesive sediment dynamics

In some estuaries there is a well-defined zone in which the suspended sediment concentration (i.e. water turbidity) is on average higher than the waters farther seaward or landward (Figure 7.18). This **turbidity maximum zone** (TMZ) has suspended sediment concentrations of the order of 100 mg l^{-1} in microtidal estuaries such as the Hawkesbury River, Australia (Hughes *et al.*, 1998), and up to 20,000 mg l^{-1} in macrotidal estuaries such as the River Severn, UK (Kirby, 1988). The position of the TMZ varies with tidal stage and river discharge (Figure 7.18).

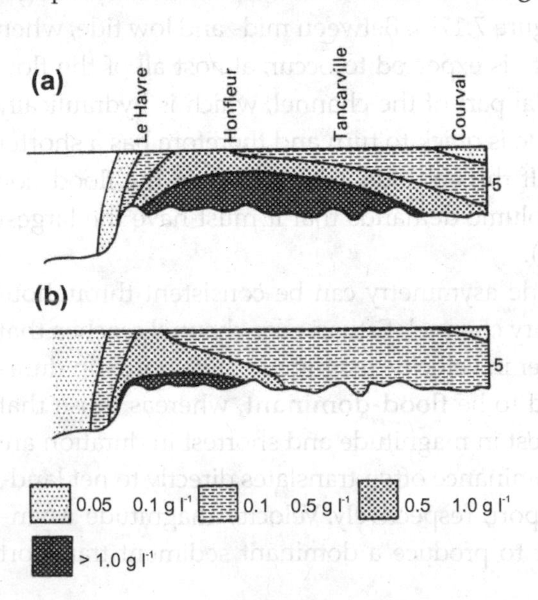

Figure 7.18 Section view along the axis of the Seine Estuary, France, showing the TMZ during two different river discharges: (a) 200 $m^3 s^{-1}$ and (b) 800 $m^3 s^{-1}$. (From Dyer, 1994; modified from Avoine, 1981.) (Copyright © 1994 Blackwell Publishers, reproduced with permission.)

For example, the TMZ in the River Seine (France) moves seaward when the river discharge increases, and oscillates back and forth during the ebb and flood stages of each spring tide (Avoine, 1981). The magnitude of the peak suspended sediment concentration observed in the TMZ is determined by the amount of fine sediment stored in the zone and the strength of the tidal currents available to stir the sediment into suspension.

7.4.1 Generation mechanisms

In partially-mixed estuaries, gravitational circulation results in a weak seaward-directed flow in surface waters and a weak landward-directed flow in bottom waters (Figure 7.19). If the superimposed tidal currents are approximately symmetrical, this gravitational circulation is sufficient to control the net transport of suspended sediment. Fine sediment brought downstream by the river will begin to settle upon reaching the estuary. Nevertheless, once the sediment settles to the bottom waters it will experience a net transport landward over many tide cycles. There is a **null point** in the estuary where the near-bed river current and the near-bed part of the gravitational circulation are roughly equal in their competency to transport sediment. Since the currents are equal but opposite in direction, this null point is where the sediment carried landward by the gravitational circulation will accumulate. In reality tidal currents smear the null point along the estuary axis into a null zone, or TMZ, that is roughly the dimension of the **tidal excursion length** (the horizontal distance a parcel of water is moved over a tide cycle). The TMZ is located near the landward limit of salt intrusion, usually where water salinities are in the range 1–5 ppt, and thus insufficient to competently drive the gravitational circulation any farther landward. In macrotidal estuaries the role of tidal distortion in generating a TMZ becomes paramount. In long estuaries, the tidal distortion is such that flood currents display the largest velocities, although they are of shorter duration than ebb currents. The farther landward the tide travels, the more enhanced this velocity-magnitude asymmetry and resultant sediment flux becomes (Figure 7.20). Averaged over many tide cycles, residual sediment transport is landward along the estuary, so the null point in macrotidal estuaries is located farther landward than in microtidal estuaries.

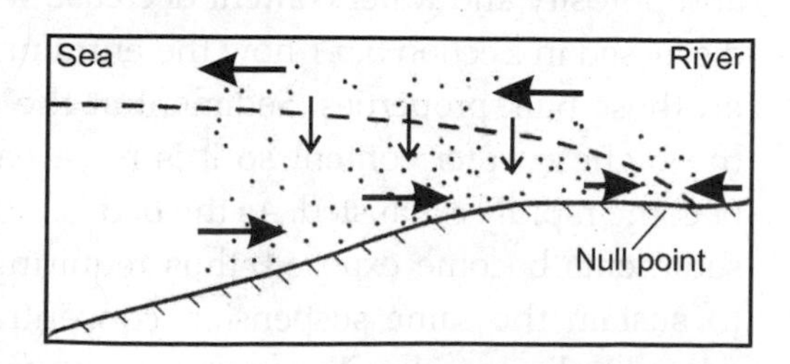

Figure 7.19 Section view of a partially-mixed estuary showing the pattern of non-tidal currents (bold arrows), sediment settling (thin arrows) and the position of the null point for suspended sediment transport. The concentration of suspended sediment is indicated by the concentration of dots. (Modified from Dyer, 1986.)

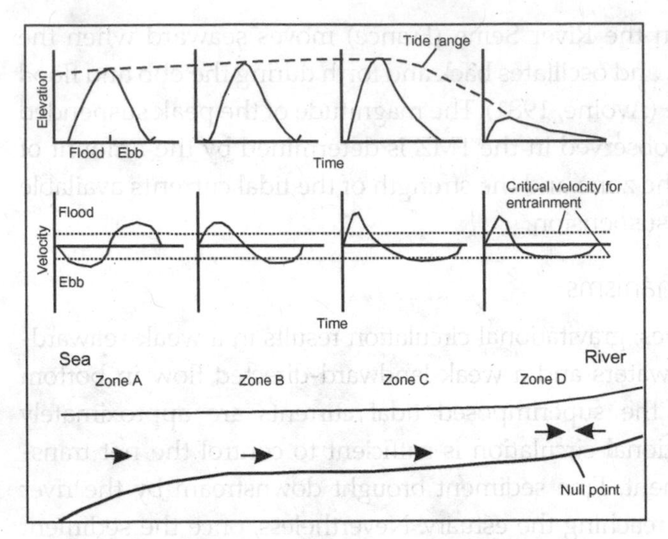

Figure 7.20 Section view of a well-mixed macrotidal estuary showing the change in tidal distortion, velocity-magnitude asymmetry, and net sediment transport direction along the length of the estuary. (Modified from Allen *et al.*, 1980.)

7.4.2 Entrainment and velocity-suspension lags

The TMZ is always most active and suspended sediment concentrations are always largest during spring tides. During neaps, the suspended sediment concentrations usually decline dramatically, and in some cases the turbidity maximum may disappear altogether. Although the link between the magnitude of tidal current velocities and suspended sediment concentration in the TMZ is obvious, it is not straightforward. For example, over a spring–neap tidal cycle in the Tamar Estuary, southwest England, the maximum and minimum suspension concentrations in the TMZ lag behind the timing of maximum and minimum tide range by up to several days (Uncles *et al.*, 1994). Moreover, within a single tide cycle, the maximum and minimum concentrations lag behind the timing of maximum and minimum current velocity by up to several hours. These **velocity-suspension lags** suggest complexity in the sediment entrainment process that goes beyond our discussion in Section 5.5.1, and arises from the cohesive nature of the fine sediments that dominate the TMZ.

Dyer (1998) describes several factors unique to cohesive sediments that can lead to velocity-suspension lags. For example, sediment compaction increases and porosity and water content decrease with depth in a muddy substrate. We discussed in Section 5.5.1 how the entrainment of cohesive sediment depends on these bulk properties. Sediment at the surface is least compacted and has the highest water content so it is most easily entrained, but this source can become rapidly exhausted. As the bed is eroded, more compacted and resistant sediments become exposed, thus requiring increasingly larger shear stresses to sustain the same suspension concentration. This can lead to a situation where sediment already in suspension can be maintained by shear stresses smaller than those required to erode the bed. Another important factor is the low settling velocity of silts and clays. The fact that it may take longer for a suspension to settle than it takes the tide to accelerate, decelerate and reverse direction can lead to complex time-lag relationships between bed shear stress and suspension concentration. In summary, the combined effect of cohesive sediment behaviour and unsteady tidal currents means that the suspension concentrations in the TMZ are rarely in equilibrium with hydrodynamic forcing. This is particularly true of muddy, macrotidal estuaries that have highly-concentrated suspensions (e.g. Allen *et al.*, 1980).

7.4.3 Settling of highly-concentrated suspensions: Fluid muds

When suspended sediment is settling during slack water, near bed concentrations in the TMZ of some estuaries reach levels greater than 10 g l^{-1}. These highly-concentrated near-bed suspensions are termed **fluid mud**, and they have a different settling behaviour to the more common dilute suspensions. When the tidal current goes slack, the suspension goes through a sequence of settling, deposition and consolidation. The suspension may start out being relatively uniform with depth, but eventually an interface develops between clear water above and the suspension below (Figure 7.21). This interface, termed the

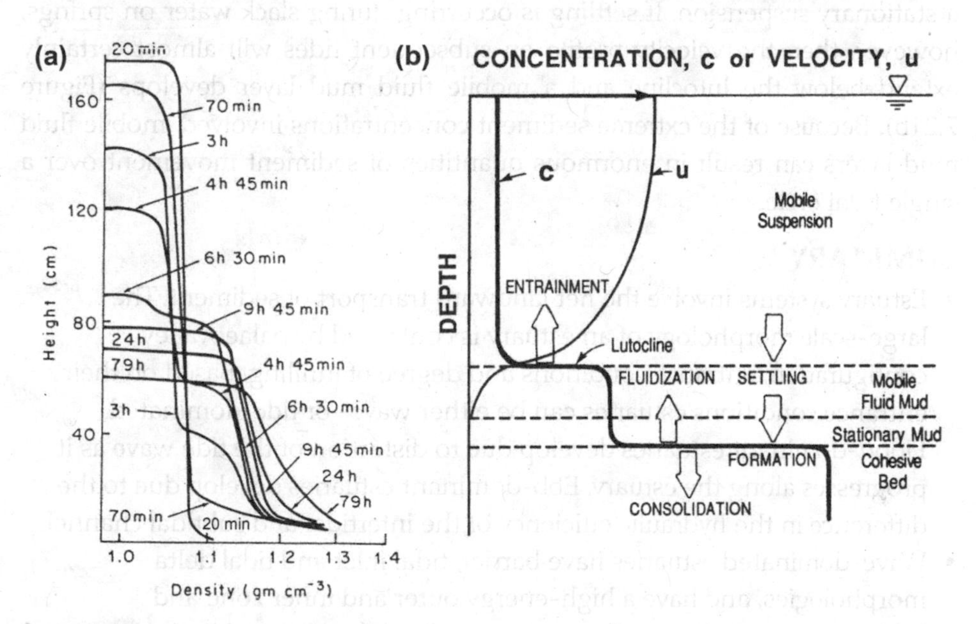

Figure 7.21 (a) Density profiles through time measured in an experimental suspension. (From Been and Sills, 1981.) (Copyright © 1981 Thomas Telford, reproduced with permission.) (b) Diagram showing a near-bed suspension concentration profile and relationship between the current velocity profile and the fluid mud layer. (From Mehta, 1989.) (Copyright © 1989 American Geophysical Union, reproduced with permission.)

lutocline, marks a sharp vertical change in suspended sediment concentration and the fluid's bulk density. Above the lutocline, sediment is free to settle at its single particle or aggregate fall velocity. Below the lutocline, the sediment experiences **hindered settling**, because the concentration of particles is so high that the paths for settling towards the bed are narrow and convoluted, and the upward drag of escaping fluid is significant. The fluid escapes vertically because the lutocline position in the fluid is steadily falling over time so that more sediment is being accommodated into less volume. This increases the pore water pressure causing water to be expelled upwards. At some point the bulk density at the bottom of the suspension is so high that no further settling can take place, and this marks the surface of the depositing bed. As deposition continues from the bottom of the suspension, the bed rises up towards the falling lutocline. When the two meet deposition is complete and the process of consolidation continues through the slow expulsion of pore water.

The time scale for fluid mud to completely settle is long compared with that of the semi-diurnal tide, so the current goes from slack back to maximum velocity before settling is complete. If the settling of fluid mud is taking place as the tide is going into neaps, then the velocity profile during subsequent tides may not penetrate the lutocline and the fluid mud beneath will continue to settle as a stationary suspension. If settling is occurring during slack water on springs, however, then the velocity profile on subsequent tides will almost certainly extend below the lutocline and a mobile fluid mud layer develops (Figure 7.21b). Because of the extreme sediment concentrations involved, mobile fluid mud layers can result in enormous quantities of sediment movement over a single tidal cycle.

SUMMARY

- Estuary systems involve the net landward transport of sediment. The large-scale morphology of an estuary is controlled by palaeovalley configuration, entrance conditions and degree of infilling. Based on their entrance conditions estuaries can be either wave- or tide-dominated. Flood-dominant estuaries develop due to distortion of the tide wave as it progresses along the estuary. Ebb-dominant estuaries develop due to the difference in the hydraulic efficiency of the intertidal and subtidal channel.
- Wave-dominated estuaries have barrier, tidal inlet and tidal delta morphologies, and have a high-energy outer and inner zone and low-energy central zone. The coarsest sediments are found in the inner and outer zones, and finest sediments in the deep central basin.
- Tide-dominated estuaries have linear sand ridges and multiple straight channels with a single meandering channel, broad intertidal mud flats, salt

marshes or mangroves. Outer, central and inner zones are relatively high energy compared with wave-dominated deltas, with coarsest sediments found in the outer and inner zones, and finest sediments in the central zone.

- In long estuaries (greater than one-quarter of the tidal wavelength), the tide displays wave-like behaviour. A lag between the timing of high (and low) tide at the coast and at the head of the estuary is a common element of progressive waves. In short estuaries, a time-varying hydraulic gradient between water levels outside and inside the estuary produces periodic water level oscillations and currents.
- The turbidity maximum zone (TMZ) is where the largest suspended sediment concentrations occur and is best developed in the central zone of long, muddy estuaries. Gravitational circulation is most important in partially-mixed micro/mesotidal estuaries, and velocity-magnitude asymmetry is most important in well-mixed macrotidal estuaries.

Reflective questions

These questions are designed to test your comprehension of material covered in this chapter. Suggested answers to these questions can be found on this book's website.

7a. In your own words, describe the major differences in morphology between wave- and tide-dominated estuaries.

7b. Outline the main sedimentary processes that take place in the intertidal zone of estuaries.

7c. Explain the relationships between the different mixing processes and the development of stratified and well-mixed estuaries.

7d. Consider some of the implications of the tidal prism and velocity-magnitude asymmetry of the flood and ebb tides with respect to sediments and landforms found within estuaries.

Further reading

Dalrymple, R.W., **Zaitlin**, B.A. and **Boyd**, R., 1992. 'Estuarine facies models: Conceptual basis and stratigraphic implications.' *Journal of Sedimentary Petrology*, 62, 1130–1146. (Provides a summary of the major geological and sedimentological aspects of estuaries and the estuary classification scheme used in this chapter.)

Dyer, K.R., 1986. *Coastal and Estuarine Sediment Dynamics*. Wiley, Chichester. (An advanced text on estuaries that includes chapters on estuarine sedimentation and cohesive sediment dynamics.)

Dyer, K.R., 1998. *Estuaries: A Physical Introduction*. Wiley, Chichester. (A comprehensive introduction to tidal dynamics, stratification and mixing in estuaries.)

Prandle, D., 2009. *Estuaries: Dynamics, Mixing, Sedimentation and Morphology*. Cambridge University Press, Cambridge. (An advanced text on estuary dynamics, including monitoring and modelling.)

CHAPTER 8

WAVE-DOMINATED COASTAL ENVIRONMENTS – THE SHOREFACE, BEACHES AND BARRIERS

AIMS

This chapter discusses the processes and landforms associated with the shoreface, beaches and barriers developed on sandy coasts under the dominant influence of waves. First, processes that influence sediment transport on the shoreface are described. Then, wave processes on beaches are discussed with respect to the formation of different beach types, and the formation and dynamics of coastal barriers. 'Hard' and 'soft' coastal engineering strategies can have unanticipated impacts on the workings of coastal systems.

8.1 Introduction

In Chapter 1 we discussed a classification of depositional coastal environments based on the dominance of fluvial, tide or wave processes (Figure 1.7). Chapters 6 and 7 dealt with fluvial and tide-dominated coastal environments, respectively. In this chapter we discuss the shoreface–beach–barrier system which is the main depositional element of wave-dominated coasts (Roy *et al.*, 1994). The shoreface, beach and barrier environments are linked by sediment transport pathways caused mainly by wave processes. In coastal and nearshore environments, however, the capacity for sediment transport is not unlimited, and the volume of transported sediment is strongly influenced by interrelationships between hydrodynamic processes, bathymetry, coastline geometry and sediment supply. As a result, sediment transport tends to take place within **littoral cells**, which refer to a discrete geographic area of the coast and shoreface within which sediment is moved. Boundaries of littoral cells are often delimited at the coast by bedrock headlands, and offshore by sediment bedload partings. Measured rates of shoreface, beach, cliff and dune erosion, and river discharge, enable the coastal sediment budget to be calculated for each littoral cell. For example, in the Santa Barbara littoral cell (California, USA), Patsch and Griggs (2008) use dredging records to estimate sediment transport rates within the cell and to close the cell's sediment budget. Around the Australian coast, Short (2010) identifies 17 littoral cells that correspond to different sets of environmental conditions including wave exposure, tidal regime and coastline type. In our discussion of the shoreface, beaches and barriers it is useful to consider how these environments are interconnected through processes of sediment erosion, transport and deposition.

8.2 The shoreface

The **shoreface** is the upper part of the continental shelf that is affected by contemporary wave processes and extends from the landward limit of wave runup to the seaward depth limit for wave-driven sediment transport (Cowell *et al.*, 1999). The shoreface can be subdivided into an upper and lower section and, although nearshore bars may be present, it usually has a concave-up profile (Figure 8.1). The **upper shoreface**, referred to as the littoral zone by Hallermeier (1981), is the region in which erosion and accretion result in measurable changes in bed elevation during a typical year. The **lower shoreface**, referred to as the shoal zone by Hallermeier (1981), experiences sediment transport processes under typical wave conditions, but significant morphological changes only occur during extreme storm events. The upper and lower shoreface are therefore different morphodynamic regions.

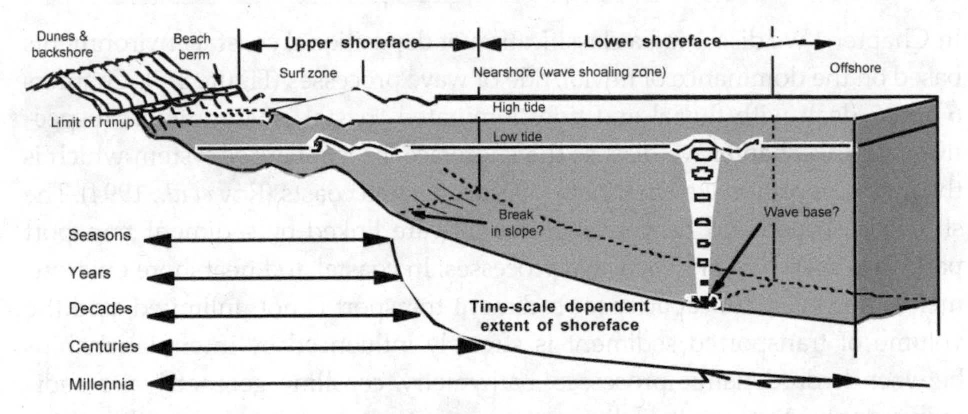

Figure 8.1 Schematic diagram showing the morphological elements of the shoreface profile. The seaward limit of the shoreface is dependent on the time scale of interest. (From Cowell *et al.*, 1999.) (Copyright © 1999 John Wiley & Sons, reproduced with permission.)

8.2.1 Depth of closure

The boundary between the upper and lower shoreface is referred to as the **depth of closure** (or **closure depth**) which can be considered as the seaward limit to significant cross-shore sediment transport. This concept is useful for applications such as estimating sediment budgets, numerical models of coastal change, beach nourishment design and disposal of dredged material.

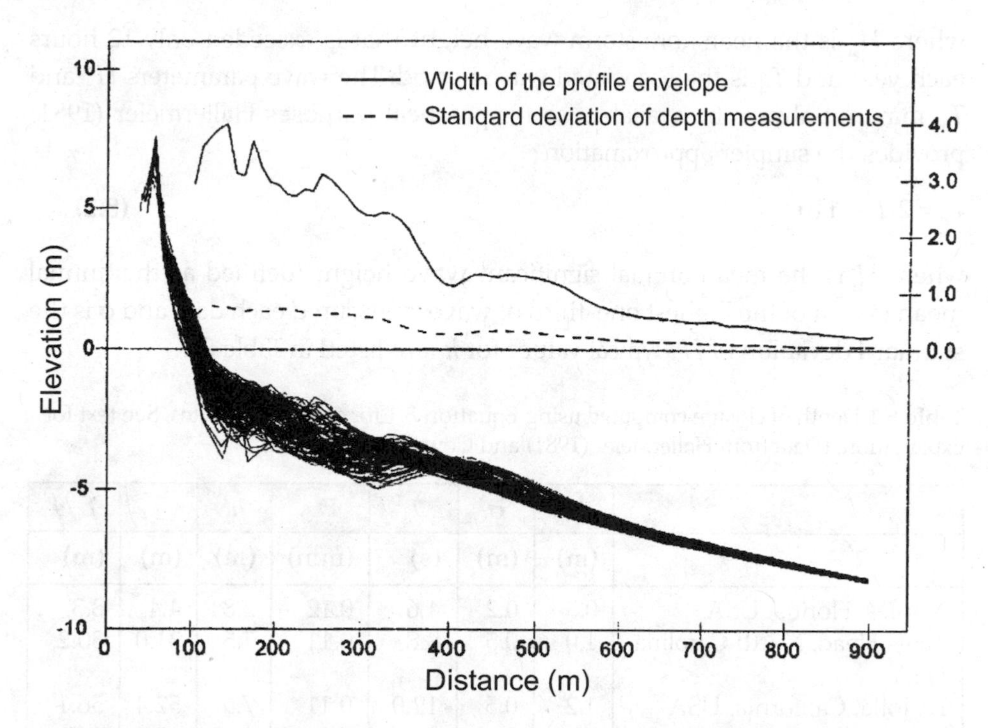

Figure 8.2 The envelope of a shoreface profile bundle surveyed at the Field Research Facility, Duck, North Carolina, USA. The upper part of the graph shows the maximum vertical change in the depth and the standard deviation of the depth variations. (From Komar, 1998.)

The depth of closure can be identified if high-resolution repeat surveys of the shoreface are available. Plotting all cross-shore profiles from one location at different times together creates a profile bundle (Figure 8.2). The **sweep zone** is defined as the area between the minimum and maximum heights of the profile bundle, and the thickness of the sweep zone is an indication of the morphological variability over the studied time span. The morphological variability is greatest in the surf zone region and decreases progressively seaward. The profile bundles converge to a depth at which morphological change is insignificant (within measurement error), which is the closure depth. In practice, this depth varies based on wave height, nearshore geology and sediment supply (Robertson *et al.*, 2008).

High-resolution survey data are often not available and the depth of closure has to be determined indirectly. Hallermeier (1981) uses morphological data from a large number of shorefaces and related observed closure depths to incident wave conditions. He proposes that the depth of closure h_c can be predicted using the annual wave climate according to

$$h_c = 2.28H_{sx} - 68.5\frac{H_{sx}^2}{gT_{sx}^2} \quad (8.1)$$

where H_{sx} is the nearshore storm wave height that is exceeded only 12 hours each year and T_{sx} is the associated wave period. The wave parameters H_{sx} and T_{sx} may not always be available, so for practical purposes Hallermeier (1981) provides the simpler approximation

$$h_c = 2\bar{H}_s + 11\sigma \tag{8.2}$$

where $\bar{H}_s$ is the mean annual significant wave height (defined as the annual mean height of the highest one-third of waves measured each day) and σ is the standard deviation of $\bar{H}_s$. Typical values for h_c are listed in Table 8.1.

Table 8.1 Depth of closure computed using Equation 8.1 for a number of sites. See text for explanation. (Data from Hallermeier (1981) and Cowell *et al.* (1999).)

	$\bar{H}_s$	σ	$\bar{T}_s$	D_{50}	h_c	h_i	$L_o/4$
	(m)	(m)	(s)	(mm)	(m)	(m)	(m)
Naples, Florida, USA	0.3	0.2	4.6	0.12	2.8	4.4	8.3
Nags Head, North Carolina, USA	1.0	0.5	8.8	0.11	7.5	31.0	30.2
La Jolla, California, USA	1.2	0.5	12.0	0.11	7.9	52.4	56.1
Netherlands	1.2	0.8	5.0	0.19	11.2	13.4	9.7
Southeast Australia	1.5	1.2	9.5	0.16	16.2	36.7	35.2

The transition from upper to lower shoreface is generally characterised by a change in sediment characteristics rather than a change in morphology. Upper shoreface sands are usually well-sorted and similar to beach sediments, and tend to fine in a seaward direction (George and Hill, 2008) to a depth similar to h_c, seaward of which coarser, poorly sorted sand may be found. On prograding, deltaic shorefaces, however, the coarse sand facies is usually absent and sandy sediments become finer from the upper to the lower shoreface, grading to clays on the continental shelf (see Section 6.3).

8.2.2 Wave base

The seaward limit of the lower shoreface is less straightforward to define than the depth of closure because it depends on the time scale of interest. The longer the time scale, the greater the potential for extreme wave events to occur, and therefore the farther seaward the lower shoreface extends. Hallermeier (1981) estimates that the limiting depth h_i for significant cross-shore sediment transport of sand by waves in a typical year is given by

$$h_i = (\bar{H}_s - 0.3\sigma)\bar{T}_s\,(g/5000D_{50})^{0.5} \tag{8.3}$$

where $\overline{T}_s$ is the mean annual significant wave period (defined as the annual mean period of the highest one-third of waves measured each day) and D_{50} is the median sediment size (in m) of a sand sample from a water depth of $h = 1.5h_c$. Typical values for h_i are listed in Table 8.1.

One of the difficulties in applying Equation 8.3 is the availability of reliable sediment data and the representativeness of a sample. An alternative approach to defining the seaward limit of the lower shoreface is by identifying the **wave base**, which the water depth beyond which wave action ceases to disturb bed sediment, which corresponds to the transition from deep-water to intermediate-water waves (Chapter 4). Conventionally this depth of wave base occurs where the water depth is equal to half the deep-water wave length ($h = L_o/2$). However, Komar (1976) suggests a limit of $h = L_o/4$. Redefining the wave base in this way provides water depths for the seaward limit of sediment disturbance by waves that are similar to h_i calculated using Equation 8.3 (Table 8.1). However, the wave base approach does not consider wave height or the sediment characteristics that affect bed mobility.

8.2.3 Equilibrium profile

Assuming that the shoreface is in equilibrium with hydrodynamic conditions and that sufficient sediment is available, the concave-up profile of the upper shoreface can be described by a simple function

$$h = Ax^m \tag{8.4}$$

where h is the still-water depth at a horizontal distance x from the shoreline, A is scaling coefficient that controls profile steepness and m is a dimensionless exponent that determines the profile shape. Convex, straight and concave profiles have values of m of larger than 1, equal to 1 and smaller than 1, respectively. Equation 8.4 is used extensively to describe the cross-shore profile of the upper shoreface.

Fitting of Equation 8.4 to natural upper shoreface profiles by Bruun (1954) and Dean (1977, 1991) suggests an average value of $m = 0.67$. However, values for m range from about 0 to 1.4 in a database of 502 beach profiles from the east coast and Gulf coast of the USA (Dean, 1977). According to Cowell *et al.* (1999), the value for m depends on beach type and is relatively low ($m \approx 0.4$) for reflective beaches and high ($m \approx 0.8$) for dissipative beaches. The steepness parameter A has been empirically related by Dean (1987) to the sediment fall velocity w_s

$$A = 0.067w_s^{0.44} \tag{8.5}$$

where w_s is in cm s^{-1}. The larger the sediment size, the greater the value of A and the steeper the beach profile. For sandy sediments, $A = 0.05–0.25$.

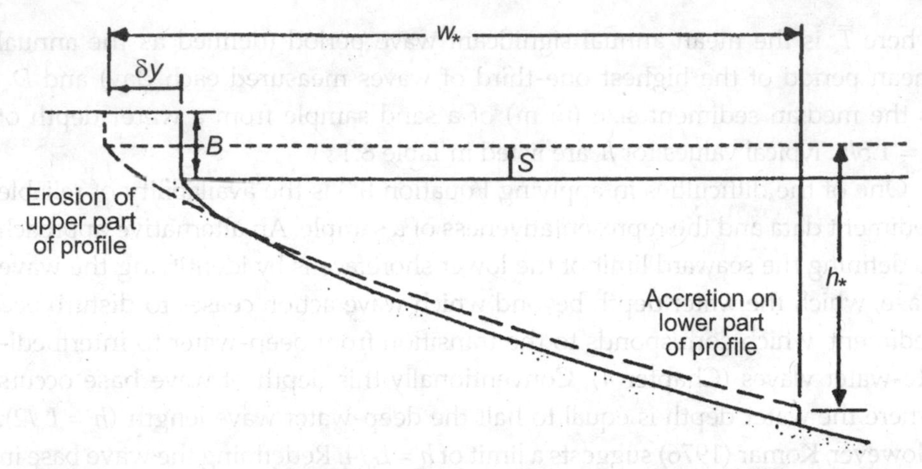

Figure 8.3 Response of shoreface profile to rising sea level according to the Bruun rule. Erosion of the subaerial beach is balanced by deposition on the shoreface. δy = shoreline retreat, S = rise is sea level, w_* = width of the shoreface, h_* = depth of the shoreface and B = height of the subaerial beach. (Modified from Dean, 1991.)

The equilibrium shoreface profile can be used to evaluate whether a particular shoreline is in equilibrium or not (e.g. Kaiser and Frihy, 2009), and to predict shoreline change due to sea-level rise (Figure 8.3). The hypothesis is that when sea level rises, the shoreface profile adjusts by rising and moving landward by coastal erosion, whilst maintaining its equilibrium shape. Bruun (1962) proposes that shoreline retreat δy can be derived by assuming that the amount of erosion on the upper part of the profile is equal to the amount of deposition on the lower part

$$\delta y = -S \frac{w_*}{h_* + B} \tag{8.6}$$

where S is the rise in sea level, w_* is width of the shoreface, h_* is depth of the shoreface (taken as the depth of closure h_c or the wave base h_i) and B is height of the subaerial beach. Equation 8.6 is known as the **Bruun Rule** and is the most widely used approach in predicting the effects of sea-level rise on sandy shorelines (Box 8.1).

Box 8.1 Application and limitations of the Bruun Rule

The Bruun Rule has been widely used by coastal managers, planners and engineers as a predictive tool for coastline response to sea-level rise. The **Bruun Rule** has gained such common usage because:

- it is a simple 'off-the-shelf' model
- it requires little field data as input into the model
- it is a predictive equation that produces a numerical outcome of coastal response to sea-level rise.

There are a number of problems with the Bruun Rule, however:

- It assumes only sea-level rise is responsible for shoreline erosion, whereas longshore sediments transport is at least as important if not more so.
- It assumes that all eroded sediment is redistributed along the profile in the cross-shore direction, in other words, that there is a balanced sediment budget.
- It assumes an instantaneous response to sea-level rise and that an equilibrium profile is always attained.
- It can only be applied to soft-sediment coasts, not rocky or engineered coasts.

Several recent studies have argued that the Bruun Rule should be abandoned because its scientific assumptions are invalid and other more sophisticated models have superseded it (Cooper and Pilkey, 2004). While this is certainly the case, the major problem with the Bruun Rule is not its conceptual basis in physical science, but rather in its application to coastal problems that are more complex than the rule's underlying assumptions. A more useful application of the Bruun Rule is where it is used in combination with other tools, including sediment transport models.

8.2.4 Waves and currents

The seaward limit of the shoreface is defined by the wave base ($L_o/2$ or $L_o/4$), which places the entire shoreface in the shallow- and intermediate-wave region (Chapter 4). In addition to wave-orbital flows, several different currents operate in this region. Surf zone currents operate on the upper shoreface and include longshore wave-driven currents, offshore-directed rip currents and bed return flow. Surf zone currents are discussed in Section 8.3. Tidal currents and wind-driven currents are found on the shoreface seaward of the surf zone.

Tidal currents generally flow in an alongshore direction, and only have an important cross-shore component near estuaries and tidal inlets. Similar to incident wave motion, tidal currents also show flow reversal, albeit on a longer time scale than waves. In microtidal environments, tide-driven currents are generally weak, but on macrotidal shorefaces they may exceed 0.25 m s^{-1}. Usually, flood current velocities are not balanced by those of the ebb and although these currents are generally weak (< 0.25 m s^{-1}) this creates residual tidal currents that contribute significantly to net sediment transport.

Water responds rapidly to wind action and there is a strong tendency in shallow water for surface currents to be aligned with wind direction. However, over most of the shoreface wind-driven surface currents deviate from wind direc-

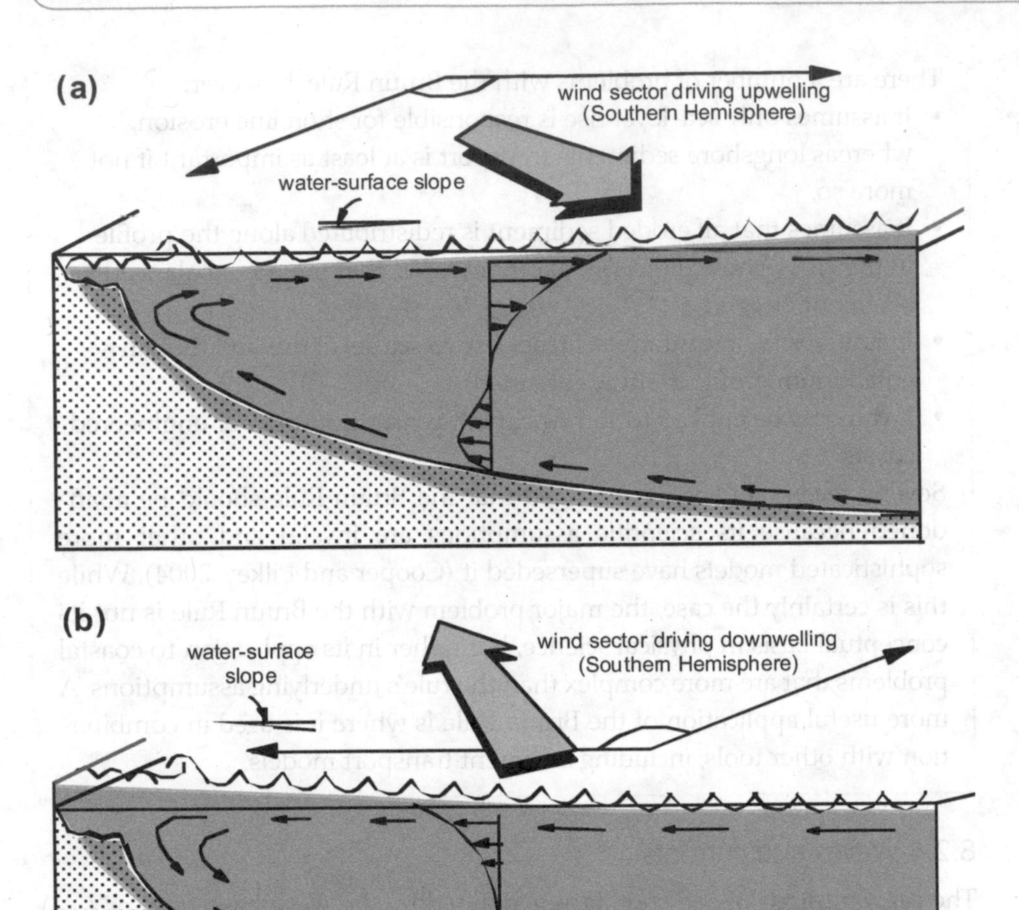

Figure 8.4 Wind-driven currents on the shoreface in the Southern Hemisphere: (a) upwelling with onshore near-bottom flow and b) downwelling with offshore flow near the bed. (From Cowell *et al.*, 1999.) (Copyright © 1999 John Wiley & Sons, reproduced with permission.)

tion by up to 45°, depending on latitude, due to the Coriolis force (Figure 8.4) which is to the right (left) in the Northern (Southern) Hemisphere. If water movement is directed away from the coastline, the water level near the coast will be lowered. This causes coastal **upwelling** whereby the near-bed flow is directed onshore. Conversely, if water movement is directed towards the coastline, the water level near the coast is raised. The elevated coastal water level induces **downwelling** currents that result in offshore flow near the bed. Flow velocities associated with upwelling and downwelling are usually less than 0.25 m s^{-1} (Niedoroda and Swift, 1991). Upwelling and downwelling are discussed in more detail in Section 11.3.

8.2.5 Sediment transport processes

At first glance, it appears that there is only limited potential for net sediment transport on the shoreface because tide- and wind-driven current velocities are generally below the entrainment threshold (Figure 5.16). Wave-orbital velocities are usually strong enough to entrain sediment, but the reversing nature of wave-driven flows reduces their effectiveness in net transport. Three factors contribute to net sediment transport on the shoreface:

- **Wave-current interaction** – Most net sediment transport on the shoreface takes place by waves and currents working in concert. Stresses exerted by wave motion suspend sediments above the bed but may not cause net transport due to the reversing behaviour of the motion. However, if a unidirectional current is superimposed on this to-and-fro motion, net sediment transport will take place in the direction of the current. Since the waves have already supplied the power to put sediment into motion, the unidirectional current can cause net transport no matter how weak the flow.
- **Wave asymmetry** – As waves shoal in shallow water, they become increasingly nonlinear and asymmetric (Section 4.5.3). The change in wave shape is reflected in wave orbital velocities such that the onshore stroke of the wave is stronger but of shorter duration than the offshore stroke. This is termed onshore asymmetry. Sediment transport is related to flow velocity raised to a power of at least 3. Therefore, onshore transport under the wave crest is likely to exceed offshore transport under the wave trough (Figure 8.5), in particular for coarse sediments. These particles tend to be 'nudged' landward during the onshore stroke of the wave, while remaining at rest during the offshore stroke.
- **Bed morphology** – Sediment transport under waves is generally in phase with the wave motion. In other words, maximum onshore (offshore) sediment transport rates coincide with maximum onshore (offshore) flow velocities. In reality, sediment response to the flow is not instantaneous because there are time lags between the flow field and sediment transport patterns, so that maximum flow velocities may not coincide with maximum sediment transport rates. These time lags are significant when suspended sediment transport occurs over a rippled sea bed, and may induce net onshore/offshore transport, despite the presence of symmetrical wave motion (Figures 5.21 and 5.23).

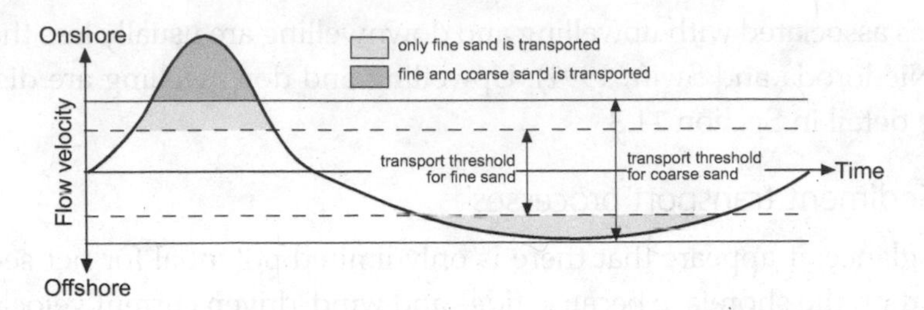

Figure 8.5 The effect of onshore wave asymmetry on sediment transport. Onshore sediment transport, especially of coarse sediment, is promoted because the onshore wave orbital velocity is stronger than the offshore velocity. (Modified from Cowell *et al.*, 1999.)

8.3 Beaches

A **beach** is a wavelain deposit of sand and/or gravel found along marine, lacustrine and estuarine shorelines. The beach relates to that part of the shore profile that extends from the spring low tide level to a landward change in topography such as a cliff or dune field, or where permanent vegetation is established. From a morphodynamic perspective, the part of the nearshore profile that is affected by surf zone processes should also be considered part of the beach. The beach therefore constitutes the upper, concave-up part of the shoreface (Figure 1.1). On most beaches, perturbations to this profile may result from smaller-scale features such as beach cusps, berms and nearshore bars (Figure 8.6), described below. These features give beaches their distinctive morphology and, in combination with differing morphodynamic regimes, allows for the classification of different beach types. Before discussing these smaller-scale features and beach classification, we first introduce the characteristics of wave-driven, nearshore currents which give direction to the movement of beach sediments and shape the beach morphology.

8.3.1 Nearshore currents

In the surf zone, incident waves progressively dissipate their energy as they break. A significant proportion of this energy generates nearshore currents and sediment transport, ultimately forming distinct shoreface morphologies. **Nearshore currents** are due to cross-shore and/or longshore gradients in the mean water surface that arise due to variations in breaker height. This implies that when waves are not breaking, these currents will not be generated. In addition, the intensity of the currents increases with increasing incident wave energy, hence the strongest currents take place during storms and are capable of transporting large quantities of sediment. This is partly due to their often significant flow velocities, but also because sediment entrainment is enhanced by the stirring motion of the breaking waves (Kleinhans and Grasmeijer, 2006).

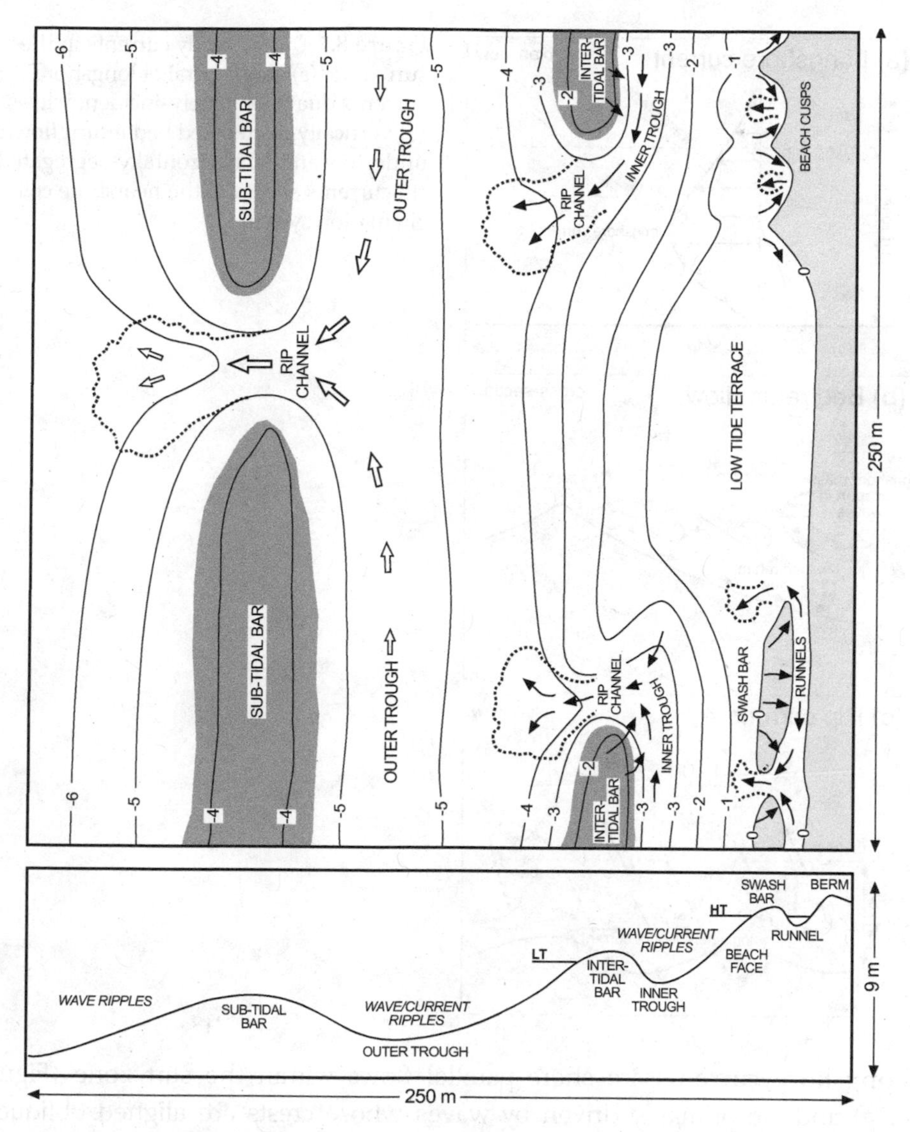

Figure 8.6 Schematic diagram showing dominant morphological features on an idealised beach. The top panel represents a contour plot of the beach morphology (beach to the right), and the bottom panel shows a typical cross-shore beach profile. It is unlikely that all the morphological features shown in the diagram are present at the same time on the same beach. High tide level on this beach is at an elevation of 0 m and the tidal range is 2 m.

There are three types of wave-induced currents which dominate net water movement in the nearshore (Figure 8.7): longshore currents, bed return flow, and rip currents. All current types are quasi-steady, meaning that they flow with a relatively constant velocity for a given set of wave/tide conditions. Longshore currents flow parallel to the shore, while bed return flow and rip currents flow perpendicular to the beach.

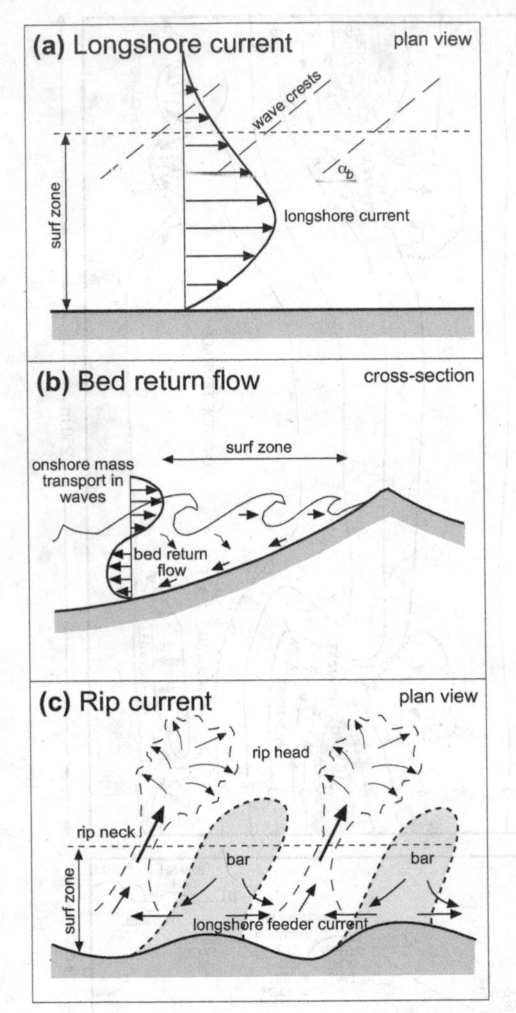

Figure 8.7 Quasi-steady currents in the surf zone: (a) shore-parallel longshore currents due to obliquely-incident waves, (b) vertically-segregated bed return flow or undertow and (c) horizontally-segregated rip currents as part of the nearshore cell circulation system.

Longshore currents are shore-parallel flows within the surf zone (Figure 8.7a) and are primarily driven by waves whose crests are aligned obliquely to the shoreline (e.g. Figure 4.13). Longshore currents may flow in excess of 1 m s^{-1}. On planar beaches, maximum longshore current velocities occur in the mid-surf zone where most of the waves are breaking. On barred beaches, the longshore current is mainly confined to the trough between the bar and the shoreline, or in troughs between multiple bars. The strength of the longshore current increases with the incident wave energy level and with the angle of wave approach. According to Komar and Inman (1970), the longshore current velocity at the mid-surf zone position $\overline{v}_l$ can be given by

$$\overline{v}_l = 1.17\sqrt{gH_b}\,\sin\alpha_b\,\cos\alpha_b \qquad (8.7)$$

where H_b is the root mean square breaker height and α_b is wave angle at the break point. For example, a breaker height of 1 m and breaker angle of 10° will produce a longshore current of 0.63 m s^{-1} at the mid-surf zone position. Longshore currents can be modulated by tidal changes in nearshore water level and may also be affected by alongshore winds. For strong winds blowing in the direction of the longshore current, current velocities may be significantly higher than is predicted using Equation 8.7.

The **bed return flow**, or **undertow**, is an offshore-directed flow near the bed. The current is part of a vertically-segregated circulation pattern characterised by onshore flow in the upper part of the water column and seaward flow near the bottom (Figure 8.7b). Bed return velocities are typically 0.1–0.3 m s^{-1}, but under extreme wave conditions may reach up to 0.5 m s^{-1}. On planar beaches, maximum bed return velocities occur around the mid-surf zone position, but on barred beaches the strongest flows are at, or slightly landward of, the bar crest. The bed return flow is fed by the water carried to the shore by breakers and bores, but a more precise explanation involves the set-up gradient across the surf zone produced by wave energy dissipation (Svendsen, 1984). The set-up produces a seaward-directed pressure gradient of water. This pressure gradient is, on average, balanced by the momentum of the waves directed to the shore (Box 4.3). However, close to the bed, the offshore-directed pressure gradient is greater than the onshore-directed force due to the wave momentum. This results in a net offshore-directed force on near-bed water particles, which drives the bed return current.

Rip currents are strong, narrow currents of 0.5–1 m s^{-1} that flow seaward through the surf zone in channels, and present a significant coastal hazard (Figure 8.7c). They are an integral part of the nearshore cell circulation system and consist of: (1) onshore transport of water between rip currents; (2) longshore **feeder currents** that are fully contained within the surf zone and carry water into the rip; (3) a fast-flowing **rip neck** that extends from the confluence of two opposing feeder currents and transports water seaward through the surf zone; and (4) the **rip head**, which is a region of decreasing velocity and flow expansion seaward of the surf zone. Maximum current velocities occur in the rip neck and may reach up to 2 m s^{-1} under extreme storm conditions when 'mega-rips' form (Short, 1985). Rips develop best at low tide when rip velocities are generally higher than during high tide (Figure 8.8). Rip cell circulation is usually associated with and controlled by nearshore bar/rip morphology (MacMahan *et al.*, 2006). Wave energy dissipation and set-up is larger across the barred areas than in the rip channels. This results in a longshore pressure gradient from bar to rip that drives feeder currents and therefore the cell circulation. A recent field experiment using a Global Positioning System (GPS) to track drifters has mapping rip current patterns (Austin *et al.*, 2010). The GPS drifter

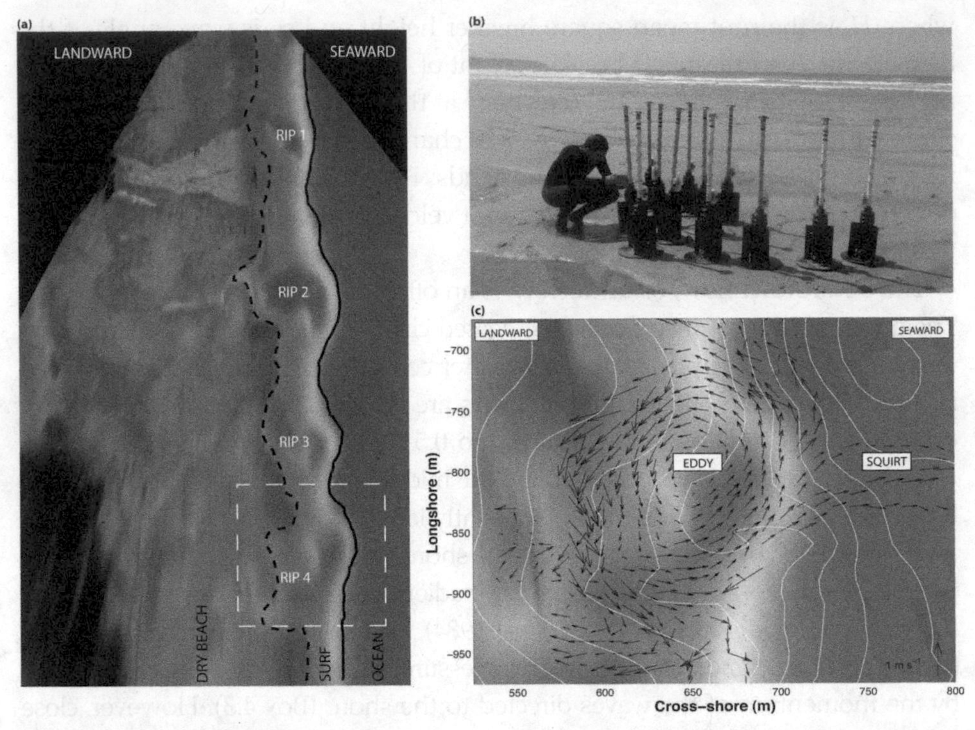

Figure 8.8 (a) Perranporth beach in Cornwall, southwest England, showing distinct rip channels around the low tide level beach. The dashed and solid black lines represent the shoreline position and the seaward edge of the surf zone, respectively. (b) GPS drifters used to obtain rip flow pattern. (c) Rip circulation pattern obtained from a two-hour deployment of twelve GPS drifters in the area boxed in part a. The black arrows represent flow patterns obtained from at least five drifter tracks and grey arrows represent single drifter tracks. The sinuous white band represents the region of wave breaking and the contours show bathymetric elevation. (Source: Images courtesy of M. Austin.)

data suggest that large rotating eddies are present within the surf zone, with only the occasional drifter exiting the surf zone through the rip neck.

The three wave-induced current systems do not occur separately. When waves break obliquely to the shoreline, longshore currents as well as cell circulation with rip currents may be present. In addition, bed return flow is always present under breaking waves, although its role is limited if there is pronounced cell circulation.

8.3.2 Swash morphology

The **swash zone** is the upper part of the beach that is alternately wet and dry on the time scale of individual waves (Figure 8.9). Sediment transport in the swash zone takes place by wave uprush (onshore flow) and backwash (offshore flow). Field measurements demonstrate that swash motion is asymmetric, such that backwash is not simply the reverse of the uprush (Hughes *et al.*, 1997). Generally, onshore flow velocities during uprush are larger, but of shorter duration,

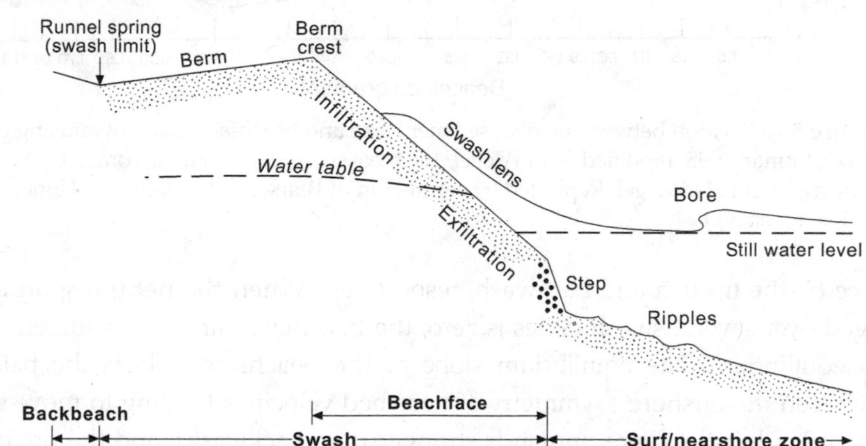

Figure 8.9 (a) Freshly deposited berm on Warriewood Beach, New South Wales, Australia. (Photo: A.D. Short.) (b) Schematic diagram of swash zone and beachface morphology showing terminology and principal processes. (From Hughes and Turner, 1999.) (Copyright © 1999 John Wiley & Sons, reproduced with permission.)

than offshore velocities during backwash. Maximum onshore velocities occur at the start of the uprush and then decrease, whereas offshore velocities increase to a maximum at the end of the backwash. Quasi-steady currents are absent from the swash zone because it is alternately wet and dry. There are also interactions between swash flow and the beach groundwater table, demonstrated by water infiltration into the beach and exfiltration out of the beach. Typical morphological features associated with swash motion include the beachface, berm, beach step and beach cusps.

The **beachface** is the planar, relatively steep upper part of the beach profile subject to swash processes. Sediment is transported up and down the beach-

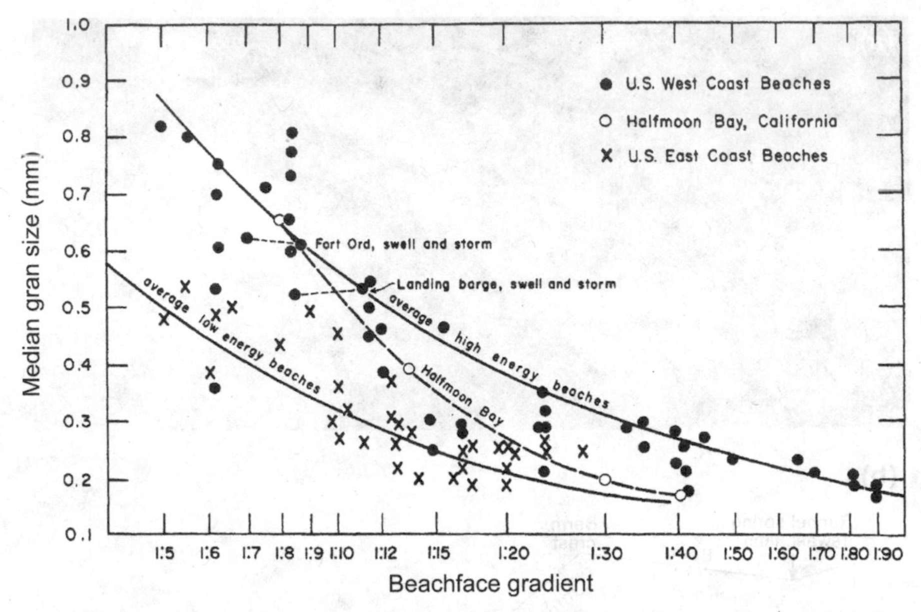

Figure 8.10 Relation between median sediment size and beachface gradient and angle. (From Komar, 1998; modified from Wiegel, 1964; based on data from Bascom, 1951.) (Copyright © 1964 Wiegel. Reprinted by permission of Pearson Education Inc., Upper Saddle River, NJ.)

face by the uprush and backwash, respectively. When the net transport averaged over several swash cycles is zero, the beachface can be considered to be in equilibrium. The equilibrium slope of the beachface reflects the balance between the onshore asymmetry of near-bed velocities tending to move sediment landward (where uprush is stronger than backwash), and the opposing force of gravity acting to move sediment seaward. Onshore swash asymmetry arises mainly from energy losses due to bed friction and infiltration of water

Figure 8.11 (a) Steep boulder beach in County Kerry, Ireland. (b) Shallow sandy beach in Cornwall, southwest England. (Photos: J. Knight.)

into coarse beaches. The equilibrium beachface gradient is positively correlated with sediment grain size (Figures 8.10 and 8.11) and gravel beach gradients steeper than 1/5 are not uncommon (Reis and Gama, 2010).

The **berm** is the nearly planar section landward of the beachface and is an accretionary feature that results from sediment accumulation at the landward extreme of wave influence and is separated from the beachface by the berm crest (Figure 8.9). The berm crest is most distinct on coarse beaches where the beachface is typically steeper, and gravel beaches may have multiple berms at different elevations. On fine-grained beaches, the berm may be indistinct due to similar backbeach and beachface gradients. According to Takeda and Sunamura (1982), the berm height Z_{berm} can be predicted using

$$Z_{\text{berm}} = 0.125 H_b^{5/8} (gT^2)^{3/8} \tag{8.8}$$

The larger the wave height and/or wave period, the larger the vertical wave runup and the higher the berm.

Berms and beachfaces are highly dynamic and respond rapidly to changing wave conditions (Figure 8.12). Berm/beachface accretion and formation of a convex beach profile occur under low-energy swell conditions. Storm conditions, however, result in berm/beachface erosion, development of a concave or scarped beach profile, and formation of a **storm beach**, which is morphologically similar to a berm but formed under different wave conditions. Vertical accretion of the berm requires overtopping of the berm crest, followed by sediment deposition on the landward side. However, erosion of the berm also requires overtopping of the berm crest. Threshold conditions determining berm erosion and accretion are discussed in Section 8.3.5.

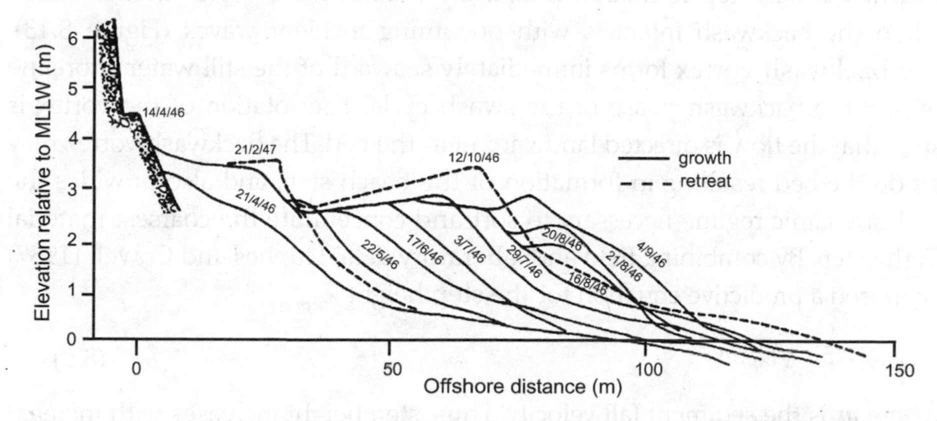

Figure 8.12 – The progressive growth and retreat of the berm at Carmel, California, USA, from 1946–1947. MLLW refers to mean lowest low water. (From Komar, 1998; modified from Wiegel, 1964; based on data from Bascom, 1953.) (Copyright © 1964 Wiegel. Reprinted by permission of Pearson Education Inc., Upper Saddle River, NJ.)

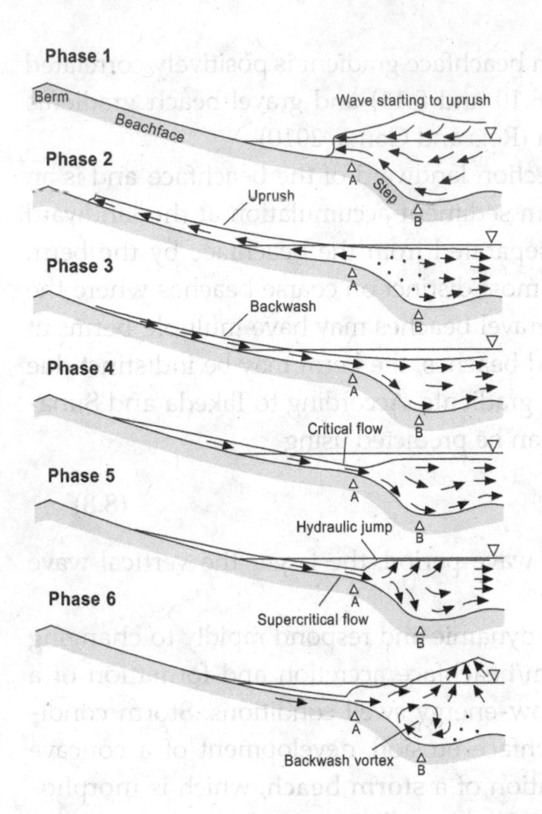

Figure 8.13 Characteristic swash flow patterns over the beachface and step during uprush (phases 1 and 2) and backwash (phases 3 to 6). The length of the arrows shows the relative flow velocity. (From Hughes and Turner, 1999; modified from Larson and Sunamura, 1993.) (Copyright © 1999 John Wiley & Sons, reproduced with permission.)

The **beach step** is a small, submerged scarp located at the base of the beachface and can range in height from several centimetres to over a metre (Figure 8.9). Beach steps generally comprise the coarsest material found on a beach, therefore their morphology is most pronounced on steep coarse sand and gravel beaches. Beach step formation is causally linked to the vortex that develops when the backwash interacts with oncoming incident waves (Figure 8.13). The **backwash vortex** forms immediately seaward of the still water shoreline during the backwash phase of the swash cycle. The rotation of the vortex is such that the flow is directed landward near the bed. The backwash vortex may erode the bed resulting in formation of the beach step, and also provides the hydrodynamic regime necessary to sort and concentrate the coarsest material in the step. By combining field and laboratory data, Hughes and Cowell (1987) proposed a predictive equation for the step height Z_{step}

$$Z_{\text{step}} = 0.55\ \sqrt{H_b T w_s} \tag{8.9}$$

where w_s is the sediment fall velocity. Thus, step height increases with increasing wave height, period and sediment size.

Beach cusps are rhythmic shoreline features formed by swash action and may develop on sand or gravel beaches. They are typically (quasi-) regularly spaced

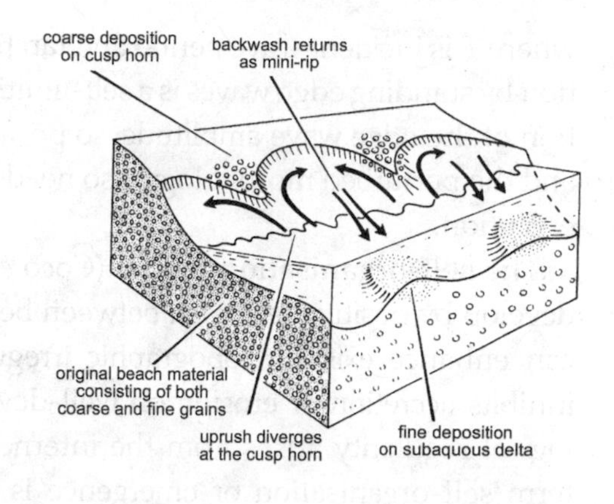

Figure 8.14 Schematic diagram showing beach cusp morphology. Arrows show wave uprush and backwash. (From Pethick, 1984.) (Copyright © 1984 Edward Arnold, reproduced with permission.)

and have a crescentic planform (Figure 8.14). Beach cusp morphology comprises gentle-gradient, seaward-facing embayments separated by steep-gradient, seaward-pointing horns. Cusp spacing (the distance between consecutive horns) is 10–50 cm and related to the horizontal extent of the swash motion (see Box 1.2). This swash motion is distinctly three-dimensional. Wave uprush is deflected from the horns into the embayments where the backwash flows offshore as mini-rips. On cusp horns, the wave uprush is diverted into the two adjacent embayments and so is considerably stronger than the backwash. As a result, the horn margins have relatively steep gradients due to preferential onshore sediment transport. In the embayments, the backwash volume is larger than the uprush volume due to the contributions from the two adjacent horns. Consequently, offshore transport dominates, resulting in gentle beachface gradients.

Although three-dimensional swash flow circulation may help explain small to large cusp evolution, it cannot account for cusp formation on a planar beach. Two theoretical models can help explain these features: standing edge waves and self-organisation. Field evidence supporting these models remains inconclusive.

The **standing edge wave model** (Guza and Inman, 1975) suggests that swash from incident waves is superimposed upon standing edge waves to produce a systematic longshore variation in swash height which results in a regular erosional perturbation (edge waves are a special class of waves that travel along the beach, rather than towards the beach (see Section 4.6.4)). This theory suggests that cusp embayments occur at edge wave antinodes, whereas cusp horns occur at edge wave nodes. The distance between horns is equal to half the wavelength of the prevailing edge wave. Although different types of edge waves may occur on beaches, the sub-harmonic edge wave, which has a period twice that of the incident waves, is implicated in the formation of beach cusps and predicts a beach cusp spacing λ of

$$\lambda = \frac{g}{\pi} T^2 \tan \beta \tag{8.10}$$

where T is incident wave period and $\tan \beta$ is beach gradient. Beach cusp formation by standing edge waves is a self-limiting process because it results in reduction of the edge wave amplitude, so positive feedback between incident waves and the perturbed morphology also needs to be invoked to explain further cusp evolution.

The **self-organisation model** (Coco *et al.*, 2001) suggests that beach cusps develop by positive feedback between beach morphology and swash flow that can enhance existing topographic irregularities, and negative feedback that inhibits accretion or erosion on well-developed cusps (see Box 1.2). Morphological regularity arises from the internal dynamics of the system, hence the term 'self-organisation' or 'emergence' is used. According to the self-organisation model, cusp spacing λ is proportional to the horizontal extent of the swash motion S according to

$$\lambda = fS \tag{8.11}$$

where the constant of proportionality f is *c.* 1.5.

8.3.3 Surf zone morphology

Water motion in the surf zone is dominated by breaking incident waves, standing infragravity waves, and quasi-steady currents. These cause sediment transport at a range of frequencies and in different directions. Surf zone morphology can be planar but usually bars and troughs are present. **Nearshore bars** are an

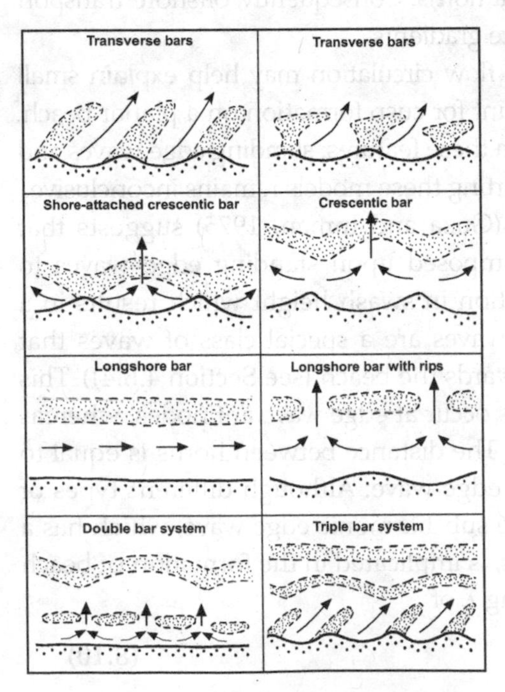

Figure 8.15 Examples of different types of nearshore bar morphology.

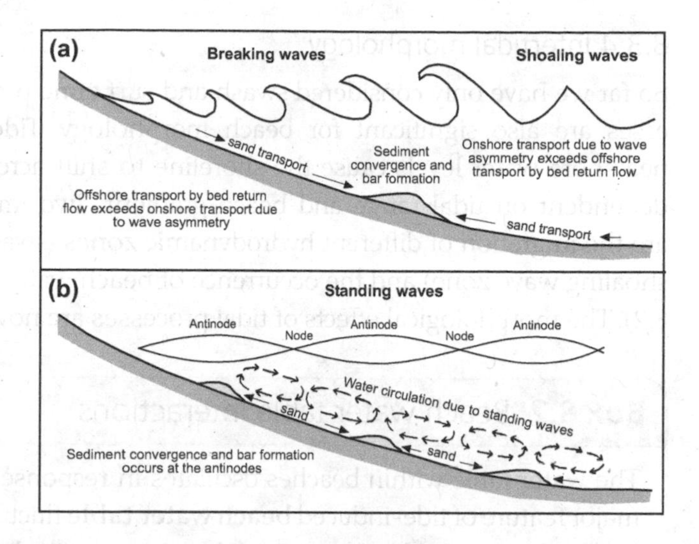

Figure 8.16 Two models of bar formation: (a) breakpoint mechanism; (b) standing wave model. (Modified from Komar, 1998.)

expression of hydrodynamic and sediment transport gradients, with bars (net deposition) forming by sediment convergence and troughs (net erosion) by sediment divergence. Nearshore bar morphology can include transverse bars, crescentic bars and longshore bars (Figure 8.15). Multiple bars may also be present.

Nearshore bars generally form by convergence near the breakpoint as a result of onshore sediment transport outside the surf zone due to wave asymmetry, and offshore transport in the surf zone due to bed return flow (Figure 8.16a; Aagaard *et al.*, 2008). However, rhythmic and multi-bar formation has also been attributed to sediment transport associated with standing infragravity waves (Figure 8.16b; Holman and Bowen, 1982) in which bars form at antinodal positions of infragravity waves if sediment transport is predominantly by bedload, and at nodal positions if suspension is the main transport mechanism.

Nearshore bars are highly dynamic and can migrate onshore or offshore in response to changing wave energy level, tidal range, bar size and water depth over the bar (Ruessink *et al.*, 2009). Offshore **bar migration** is often observed during high wave events when sediment transport by bed return flow is more important than that caused by the onshore asymmetry of the waves. Onshore bar migration is often observed during low- to moderate-energy wave conditions when the shoreward sediment flux due to wave asymmetry dominates over bed return flow. If nearshore bars merge with the beachface, a **welded bar** is formed. Bar migration rates are typically 1–10 m day^{-1} but may reach up to 30 m day^{-1} under extreme conditions. Offshore bar migration rates are generally higher than onshore rates on the same beach. Nearshore bars may show annual cyclic behaviour with a generation phase just below the low-tide water line, a net seaward migration phase through the nearshore, and finally a destruction phase at the seaward end of the nearshore (Aagaard *et al.*, 2010).

8.3.4 Intertidal morphology

So far we have only considered swash and surf zone processes, but tidal processes are also significant for beach morphology. Tide-induced changes in nearshore water levels cause the shoreline to shift across the intertidal zone, dependent on tidal range and beach gradient. Two important consequences are the migration of different hydrodynamic zones (swash zone, surf zone and shoaling wave zone) and the occurrence of beach water table fluctuations (Box 8.2). The morphological effects of tidal processes are now discussed.

Box 8.2 Beach water table interactions

The water table within beaches oscillates in response to the ocean tide. A major feature of tide-induced **beach water table** fluctuations is their asymmetry, characterised by the water table rising faster than it falls. During the flooding tide, the rising water table will keep up with the rising ocean tide level, but during the ebbing tide the water table falls more slowly than the ocean tide level does, leading to beach water table decoupling. This results in formation of a **seepage face** where groundwater outcrops onto the beachface (Figure 8.17). The intersection of the beach groundwater table with the beach is termed the exit point.

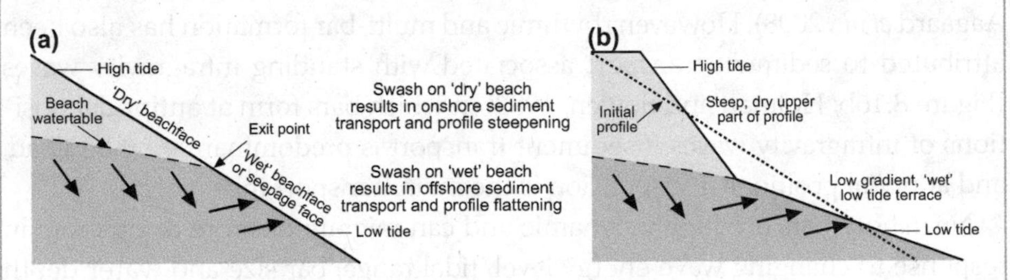

Figure 8.17 (a) Sketch of the exit point and the seepage face of groundwater on a gravel beach. (Modified from Turner, 1993.) (b) Development of a steep upper beach with low-gradient low tide terrace as a result of interactions between swash hydrodynamics and the beach groundwater table.

Water table seepage is dependent on beach gradient and sediment properties which determine the drainage capability of the beach. Low-gradient, fine-grained macrotidal beaches ($D_{50} < 0.3$ mm) do not drain very well and decoupling occurs shortly after high tide. This results in formation of a wide seepage face, occupying most of the intertidal zone. Steep-gradient, coarse-grained macrotidal beaches ($D_{50} > 0.3$ mm) drain efficiently and decoupling occurs closer to low tide, if at all, resulting in a narrow seepage face.

The seepage face defines two morphodynamic regions: (1) an upper intertidal region that alternates between saturated and unsaturated over the tidal cycle; and (2) a lower intertidal region that remains permanently saturated. Turner (1995) postulates that these regions have different sediment transport vectors, with onshore transport occurring over the upper intertidal region and offshore transport over the lower intertidal region. Such sediment divergence may explain the distinct break in slope often observed on large tide range beaches, and the development of a low tide terrace (Figure 8.17b). On the upper profile, steepening is enhanced by swash infiltration on the rising tide, which in turn promotes profile drainage through subsequent tidal cycles. In contrast, offshore transport is promoted across the low tide terrace, maintaining its low gradient.

On tidal beaches, swash, surf zone and shoaling wave processes shift both vertically and horizontally with the rise and fall of the tide. If the tide range is sufficiently large, parts of the intertidal beach profile may be subjected to the entire suite of processes during a single tidal cycle. Masselink and Short (1993) use a numerical model to determine the amount of time, expressed as a percentage of total submergence time, that different parts of the intertidal profile are subjected to swash, surf or shoaling wave processes for different wave-tide conditions over a lunar tidal cycle (Figure 8.18). They find that the relative occurrence of the different hydrodynamic processes is mainly dependent on the ratio between mean spring tide range *MSR* and breaker wave height H_b

$$RTR = \frac{\mathrm{MSR}}{H_b} \tag{8.12}$$

where *RTR* is the **relative tide range**. Swash and surf zone processes prevail around the high tide level, but their relative occurrence decreases with increasing *RTR*. For $RTR \geq 15$, shoaling wave processes dominate over almost the entire intertidal zone.

The model results indicate that swash and surf zone processes are relatively unimportant on beaches with large tidal ranges (Figure 8.18). In addition, stationary water level conditions are only encountered around the high and low tide levels, so there is generally insufficient time for development of berms, beach cusps, beach steps, bars and rip currents. Therefore, the intertidal zone of such beaches is often flat and featureless.

Despite the fact that tide-induced water level fluctuations inhibit bar formation, **intertidal bars** may be present on tidal beaches. Such bars are primarily controlled by swash processes and so are referred to as **swash bars**. However,

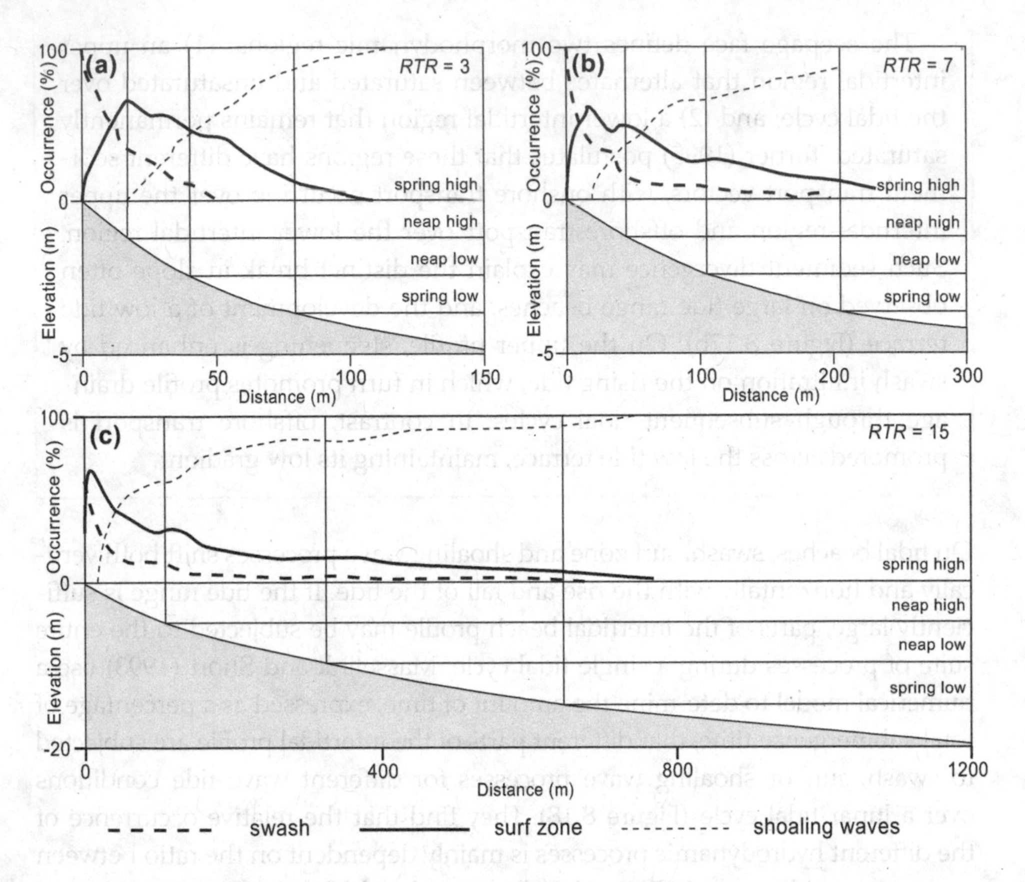

Figure 8.18 The relative occurrence of swash, surf and shoaling wave processes across the beach profile calculated over a lunar tidal cycle for three different values of the relative tide range *RTR*: (a) *RTR* = 3, (b) *RTR* = 7 and (c) *RTR* = 15. (From Masselink and Turner, 1999; modified from Masselink and Short, 1993.) (Copyright © 1999 John Wiley & Sons, reproduced with permission.)

they are also affected by breaking wave processes. Intertidal bars generally migrate onshore under low-energy conditions, but high-energy waves interrupt onshore migration and may flatten the bar within a period of several hours. On macrotidal beaches, numerous intertidal bars may be present which is referred to as **ridge and runnel morphology**. The ridges are somewhat immobile features, but ridges are usually built up under calm conditions and flattened during storms. King (1972) notes that ridge locations roughly correspond to the parts of the intertidal profile where the water level is stationary for the longest time, but van Houwelingen *et al.* (2008) argue that ridges form from a combination of swash, surf and shoaling processes across different parts of the tidal range. Masselink and Anthony (2001) demonstrate that intertidal ridges are largest and most numerous around mean sea level, i.e. where the water level is never stationary and characterised by the largest migration rates.

8.3.5 Seasonal and cyclic beach changes

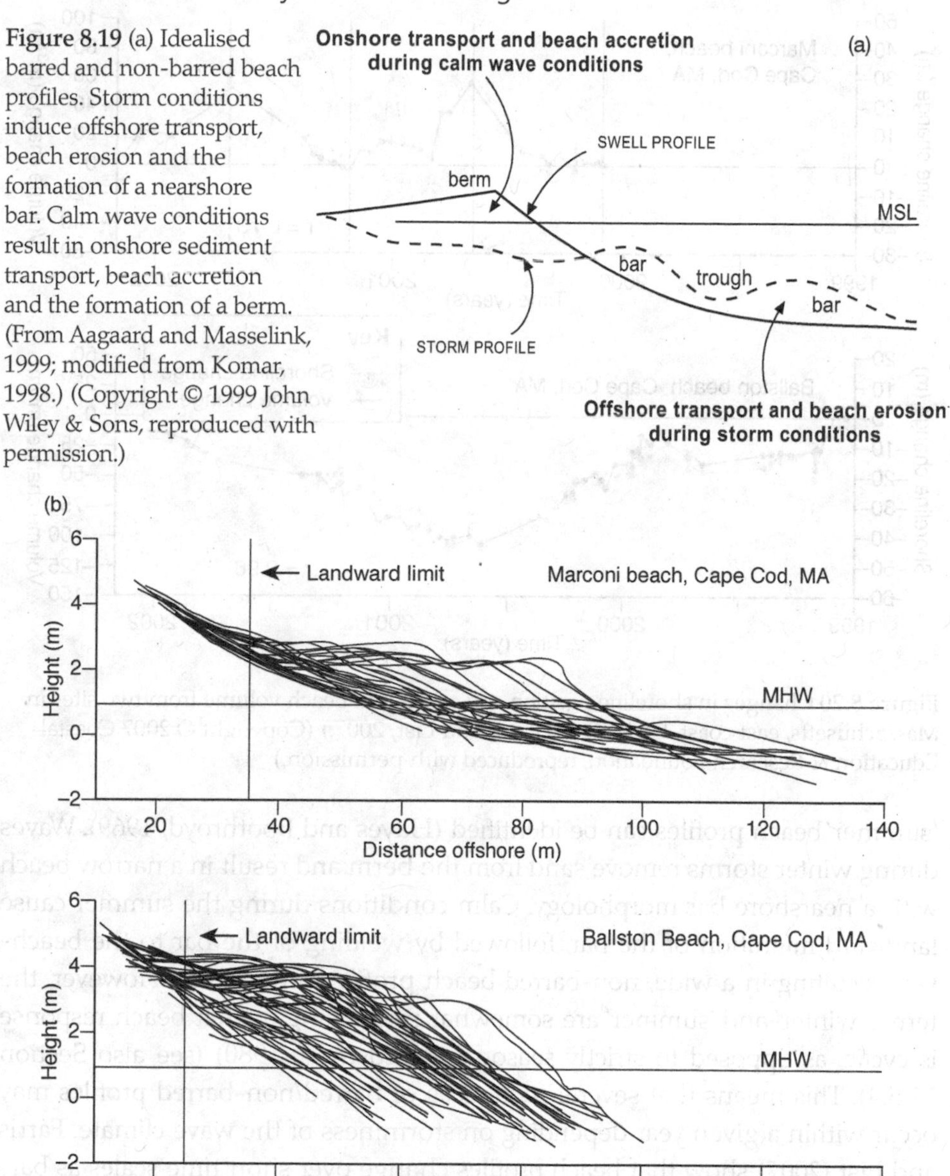

Figure 8.19 (a) Idealised barred and non-barred beach profiles. Storm conditions induce offshore transport, beach erosion and the formation of a nearshore bar. Calm wave conditions result in onshore sediment transport, beach accretion and the formation of a berm. (From Aagaard and Masselink, 1999; modified from Komar, 1998.) (Copyright © 1999 John Wiley & Sons, reproduced with permission.)

Figure 8.19 (b) Beach profiles (surveyed every two weeks) from two sites in Massachusetts, east coast USA. (From Farris and List, 2007.) (Copyright © 2007 Coastal Education & Research Foundation, reproduced with permission.)

Beach morphology responds dynamically to changing wave conditions, in particular the exchange of sediment between the berm and bar (Figure 8.19a). Long-term monitoring on the west coast of the USA shows that 'winter' and

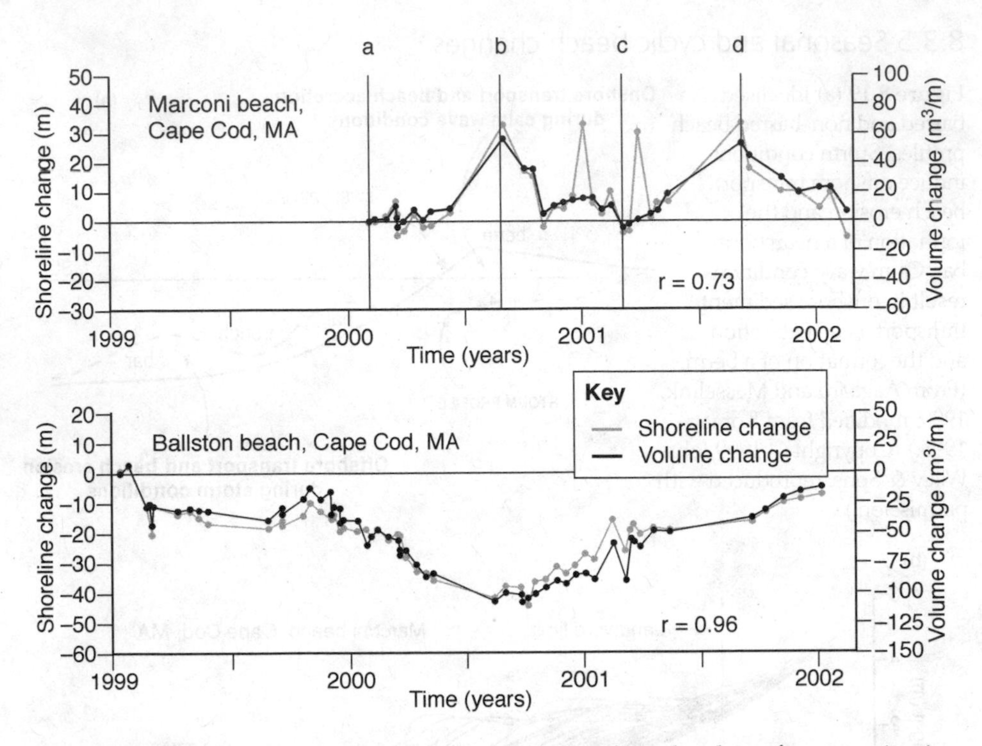

Figure 8.20 Changes in shoreline position and calculated beach volume from two sites in Massachusetts, east coast, USA. (From Farris and List, 2007.) (Copyright © 2007 Coastal Education & Research Foundation, reproduced with permission.)

'summer' beach profiles can be identified (Hayes and Boothroyd, 1969). Waves during winter storms remove sand from the berm and result in a narrow beach with a nearshore bar morphology. Calm conditions during the summer cause landward migration of the bar, followed by welding of the bar to the beach-face resulting in a wide, non-barred beach profile with a berm. However, the terms 'winter' and 'summer' are somewhat misleading in that beach response is cyclic, as opposed to strictly seasonal (Nordstrom, 1980) (see also Section 11.5.4). This means that several sequences of barred/non-barred profiles may occur within a given year depending on storminess of the wave climate. Farris and List (2007) show that beach profiles change over short time scales as bars move onshore and/or new bars develop (Figure 8.19b). They also show that there is a positive relationship (correlation coefficient of 0.71–0.96 for five locations) between shoreline change and beach volume change. Additionally, there is a high degree of seasonal and shorter variability in shoreline position and beach volume (Figure 8.20).

8.3.6 Beach classification

Classification of beaches into distinct groups or morphotypes can provide a useful framework for examining beach morphodynamics and morphological change. For example, sandy beaches can be classified into steep and gentle-gradient beaches based on their overall slope, but these two beach types represent fundamentally different process regimes. On steep beaches, a surf zone is generally absent and incident waves break directly on the beachface. A significant part of incoming wave energy is reflected back from the shoreline and hence these beaches are referred to as **reflective beaches**. On gentle-gradient beaches, the surf zone is wide with multiple lines of spilling breakers, so much incoming wave energy is dissipated during wave breaking process and these beaches are therefore known as **dissipative beaches**. A useful parameter to determine the relative importance of reflection and dissipation is the **surf scaling parameter** ε

$$\varepsilon = \frac{2\pi^2 H_b}{gT^2 \tan^2 \beta} \qquad \textbf{(8.13)}$$

where H_b is breaker height, g is gravity, T is wave period and $\tan \beta$ is beach gradient across the swash/surf zone. Reflective conditions prevail when $\varepsilon < 2.5$ and dissipative conditions when $\varepsilon > 20$ (Guza and Inman, 1975). When $\varepsilon = 2.5$–20 both reflection and dissipation occur and beaches formed under these conditions are termed **intermediate beaches**.

Smaller-scale morphological features superimposed on the beach profile are also strongly related to morphodynamic regime. A steep beachface with beach cusps and/or a beach step is characteristic of reflective beaches. Nearshore bar morphology dissected by rip channels is typical of intermediate beaches. Multi-barred morphology is generally associated with dissipative beaches.

Beach type can to some extent be predicted based on different environmental parameters including the dimensionless fall velocity Ω

$$\Omega = \frac{H_o}{w_s T} \qquad (8.14)$$

where H_o is deep water wave height, T is wave period and w_s is sediment fall velocity. This parameter, originally developed as a predictor for on/offshore sediment transport and bar occurrence, has been implemented in the **Australian beach model** proposed by Wright and Short (1984). This model (Figure 8.21) specifically relates to medium to high wave energy, microtidal sandy coastlines, and identifies six main beach types based on the dimensionless fall velocity:

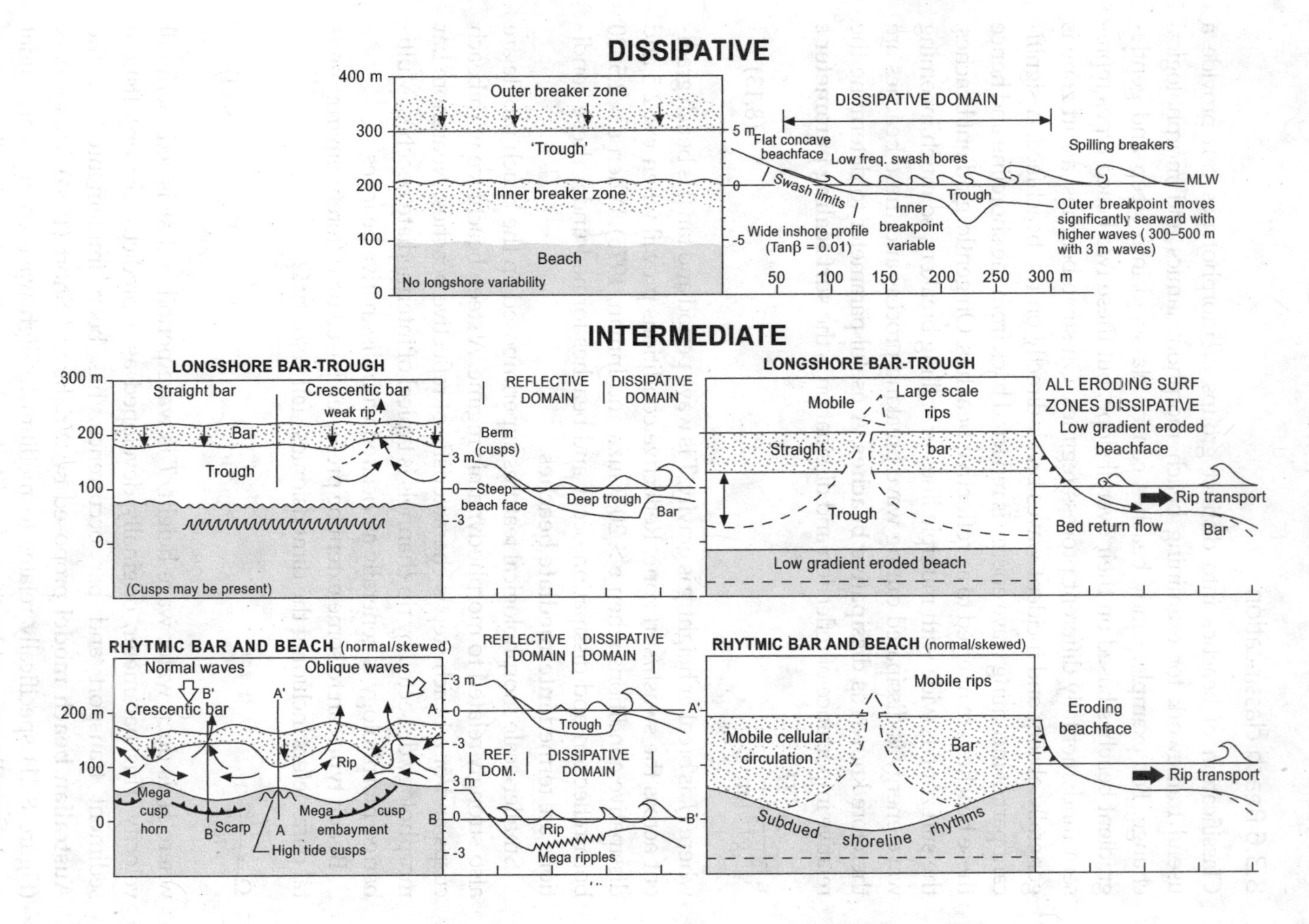

DISSIPATIVE
400 m
300
200
100
0
Outer breaker zone
'Trough'
Inner breaker zone
Beach
No longshore variability
DISSIPATIVE DOMAIN
5 m
0
-5
Flat concave beachface
Low freq. swash bores
Spilling breakers
MLW
Swash limits
Inner breakpoint variable
Trough
Outer breakpoint moves significantly seaward with higher waves (300–500 m with 3 m waves)
Wide inshore profile (Tanβ = 0.01)
50
100
150
200
250
300 m
INTERMEDIATE
LONGSHORE BAR-TROUGH
300 m
200
100
0
Straight bar
Crescentic bar
weak rip
Bar
Trough
(Cusps may be present)
REFLECTIVE DOMAIN
DISSIPATIVE DOMAIN
Berm (cusps)
3 m
0
-3
Steep beach face
Deep trough
Bar
LONGSHORE BAR-TROUGH
Mobile
Large scale rips
Straight
bar
Trough
Low gradient eroded beach
ALL ERODING SURF ZONES DISSIPATIVE
Low gradient eroded beachface
Rip transport
Bed return flow
Bar
RHYTMIC BAR AND BEACH (normal/skewed)
Normal waves
Oblique waves
200 m
100
0
B'
A'
Crescentic bar
A
Rip
Mega cusp horn
B
Scarp
A
Mega cusp embayment
B
High tide cusps
REFLECTIVE DOMAIN
DISSIPATIVE DOMAIN
3 m
0
-3
A'
Trough
REF. DOM.
DISSIPATIVE DOMAIN
3 m
0
-3
B'
Rip
Mega ripples
RHYTMIC BAR AND BEACH (normal/skewed)
Mobile rips
Mobile cellular circulation
Bar
Subdued shoreline rhythms
Eroding beachface
Rip transport

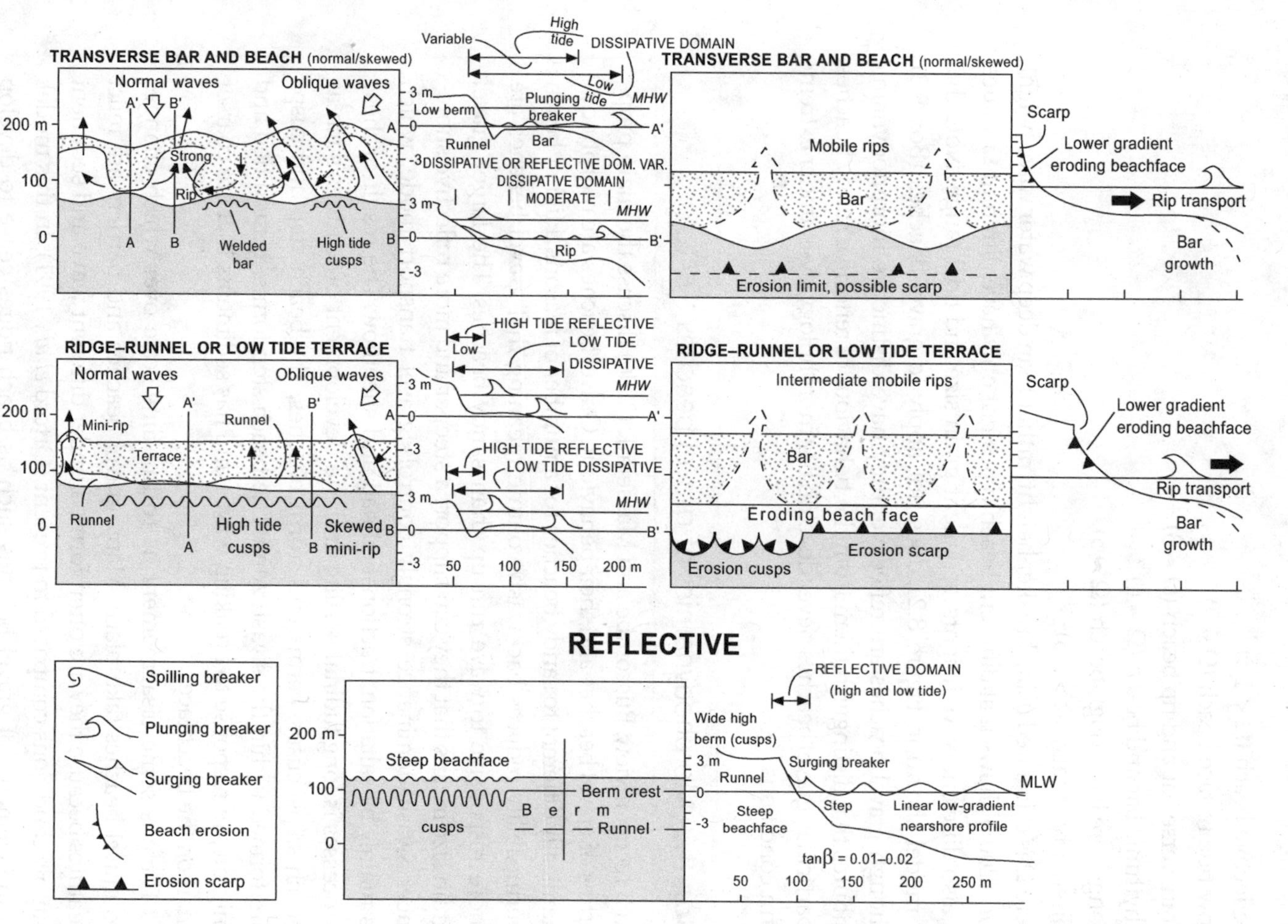

Figure 8.21 Three-dimensional sequence of wave-dominated beach changes for accretionary (left side of diagram, from top to bottom) and erosional (right side of diagram, from bottom to top) wave conditions. (From Short, 1999.) (Copyright © 1999 John Wiley & Sons, reproduced with permission.)

- reflective beach ($\Omega < 1.5$)
- low tide terrace beach ($\Omega \approx 2$)
- transverse bar and rip beach ($\Omega \approx 3$)
- rhythmic bar and beach ($\Omega \approx 4$)
- longshore bar-trough beach ($\Omega \approx 5$)
- dissipative beach ($\Omega > 5.5$)

where Ω is calculated using breaker height, rather than deep water wave height. Short (2006) shows that different beach types not only have different Ω values, but also different wave heights, sediment grain size and tidal range. Note that the model shown in Figure 8.21 does not apply to gravel beaches (Box 8.3). Although gravel beaches are reflective, they have distinctive morphodynamic attributes that distinguish them from the behaviour of reflective sandy beaches. In particular, gravel beaches never develop bar morphology, even under extreme storm conditions.

Box 8.3 Morphodynamics of gravel beaches

In a recent review, Buscombe and Masselink (2006) discuss the main properties of gravel beaches and their behaviour. Gravel beaches are morphodynamically different to sandy beaches for a number of reasons. Higher beach permeability reduces backwash volume meaning that gravel beaches are more responsive to wave run up than sandy beaches. The larger gravel grain size means that they can support a steeper and more reflective shoreface. Swash-dominance strongly controls gravel transport mode, which is mainly by saltation, traction-bedload and sheetflow. The result of these processes is longitudinal sorting and the development of features such as beach steps, cusps, berms and storm beaches. The beach step and cusps are formed within the swash zone. The beach step forms, is modified and migrates in response to breaking waves and has sediments that are coarser than on the beachface.

There is some disagreement as to the time scale over which changes to gravel beaches take place. Many gravel beaches and barriers are relict features because they are often formed under different wave and sediment supply conditions compared to present (Orford *et al.*, 1991). In this model, some elements of gravel beaches such as beach ridges cease to develop when sea level falls. However, shorter time scale changes can also take place in response to variation in wave height during storms, resulting in erosion or migration of berms and the beach step.

Worked Example 5

The beach model can help explain spatial differences between sandy beaches, and predict how beach morphology changes under rising and falling wave conditions. An increase in wave height causes offshore sediment transport and the beach will move towards the dissipative beach state. A decrease in wave height induces onshore sediment transport and the beach will move towards the reflective beach state. The end-members of the beach type continuum (i.e. the reflective and dissipative beaches) are relatively stable over time, whereas the intermediate beach types are the most dynamic.

The effect of tides is not considered in Figure 8.21 but they can be significant, particularly in macrotidal environments. The role of tides increases relative to wave effects with increasing tidal range and with a reduction in incident wave energy level. Quantification of wave *versus* tide effects may be achieved using the relative tide range parameter RTR (Equation 8.12). Tidal effects are relatively insignificant when $RTR < 3$, but dominate sediment transport processes and morphology when $RTR > 15$.

Masselink and Short (1993) formulated a beach classification model that takes into account waves, tides and sediment by combining the relative tide

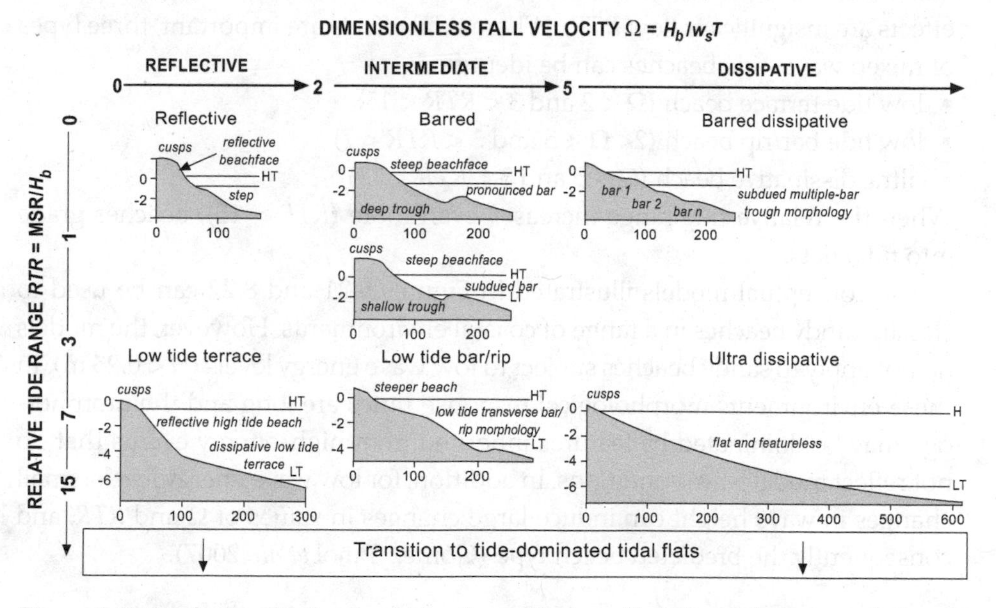

Figure 8.22 Beach classification model for the prediction of beach morphology on the basis of the dimensionless fall velocity ($\Omega = H_b/w_sT$, where H_b is breaker height, w_s is sediment fall velocity of mid-beachface sediment and T is wave period) and the relative tide range (RTR = MSR/H_b, where MSR is the mean spring tide range). (From Masselink and Short, 1993.) (Copyright © 1999 John Wiley & Sons, reproduced with permission.)

Figure 8.23 Examples of: (a) reflective beach, Pearl Beach, Australia (photo: P.J. Cowell); (b) intermediate beach, Palm Beach, Australia (photo: G. Masselink); (c) dissipative beach, Goolwa, Australia (photo: G. Masselink); (d) low tide terrace beach, Westward Ho!, England (photo: G. Masselink); (e) low tide bar-rip beach, Nine Mile Beach, Australia (photo: A.D. Short); and (f) ultra-dissipative beach, Farnborough Beach, Australia (photo: A.D. Short).

range parameter *RTR* and the dimensionless fall velocity Ω (Figures 8.22 and 8.23). Reflective, intermediate and dissipative beach types occur when tidal effects are insignificant (low *RTR*). Where tidal effects are important, three types of mixed wave–tide beaches can be identified:

- low tide terrace beach ($\Omega < 2$ and $3 < RTR < 15$)
- low tide bar/rip beach ($2 < \Omega < 5$ and $3 < RTR < 7$)
- ultra-dissipative beach ($\Omega > 5$ and $3 < RTR < 15$)

When the relative tide range increases even more ($RTR > 15$), beaches grade into tidal flats.

The conceptual models illustrated in Figures 8.21 and 8.22 can be used to classify sandy beaches in a range of coastal environments. However, the models do not apply to sandy beaches subject to low wave energy levels ($H < 0.25$ m). In these environments morphological response times are long and the morphology may be dominated by features inherited from high-energy events that do not reflect modal wave conditions. In addition, for low wave energy levels, small changes in wave height can induce large changes in values of Ω and *RTR*, and consequently the predicted beach type (Gómez-Pujol *et al.*, 2007).

8.4 Barriers and related accumulation features

In the two previous sections we have discussed the morphodynamics of shorefaces and beaches. Collectively, these make up large-scale coastal accumulation features known as **barriers**. This section will discuss how barriers can be described and classified by their planform shape and cross-sectional morphology and stratigraphy. We first discuss the role of longshore sediment transport on the morphology of wave-dominated coasts.

8.4.1 Longshore sediment transport processes

Sediment transport by longshore currents is important from a sediment budget point of view because the net removal of material results in coastal erosion, while the net influx of material results in accretion. Knowledge of longshore transport rates is therefore very useful for understanding coastal change.

The rate of **longshore sediment transport** or **littoral drift** can be predicted using the CERC equation (CERC, 1984)

$$Q_l = KP_l = K\frac{\rho H_b^2 \sqrt{gh_b}}{16(\rho_s - \rho)(1-a)} 2 \sin\alpha_b \qquad \textbf{(8.15)}$$

where Q_l is the volumetric transport rate in $m^3\ day^{-1}$, K is a constant, P_l is the longshore component of the wave energy flux, ρ is density of sea water, H_b is breaking wave height, g is gravity, h_b is water depth where the waves break, ρ_s is density of sediment, and α is porosity index. Typical net littoral drift rates are on the order of 100,000 $m^3\ yr^{-1}$. Due to the often reversing nature of longshore currents, the total littoral drift rate is likely to exceed the net rate many times. In calculating drift rate or volume for real beaches, however, there are many complicating factors including changes in wave climate, coastal engineering structures, and variations in sediment supply. In addition, there is uncertainty in the correct value of K – the original CERC equation uses a value of 0.39, but values down to 0.2 have also been used on sandy beaches (Bayram *et al.*, 2007) and 0.054 for a gravel beach (Ruiz de Alegria-Arzaburu and Masselink, 2010).

The relation between coastal morphology and longshore transport is illustrated by the predictive model of May and Tanner (1973) which can be used to explain the development of an **equilibrium bay** (Figure 8.24). Wave refraction around the headland causes wave convergence on the headland and wave divergence in the adjacent bay (see Figure 4.15). This results in a longshore variation in wave approach angle and energy level, with wave energy decreasing progressively from headland to bay. The largest wave angles are found at the transition from headland to bay, while smaller wave angles occur at the headland and in the bay. From the longshore variation in wave energy (or wave height) and wave angle, the change in the longshore wave energy flux P_l can be computed. The largest value of P_l occurs somewhere between the headland and where the wave angle is maximum. Subsequently, the longshore variation in the littoral drift rate Q_l, which is proportional to P_l, can be determined.

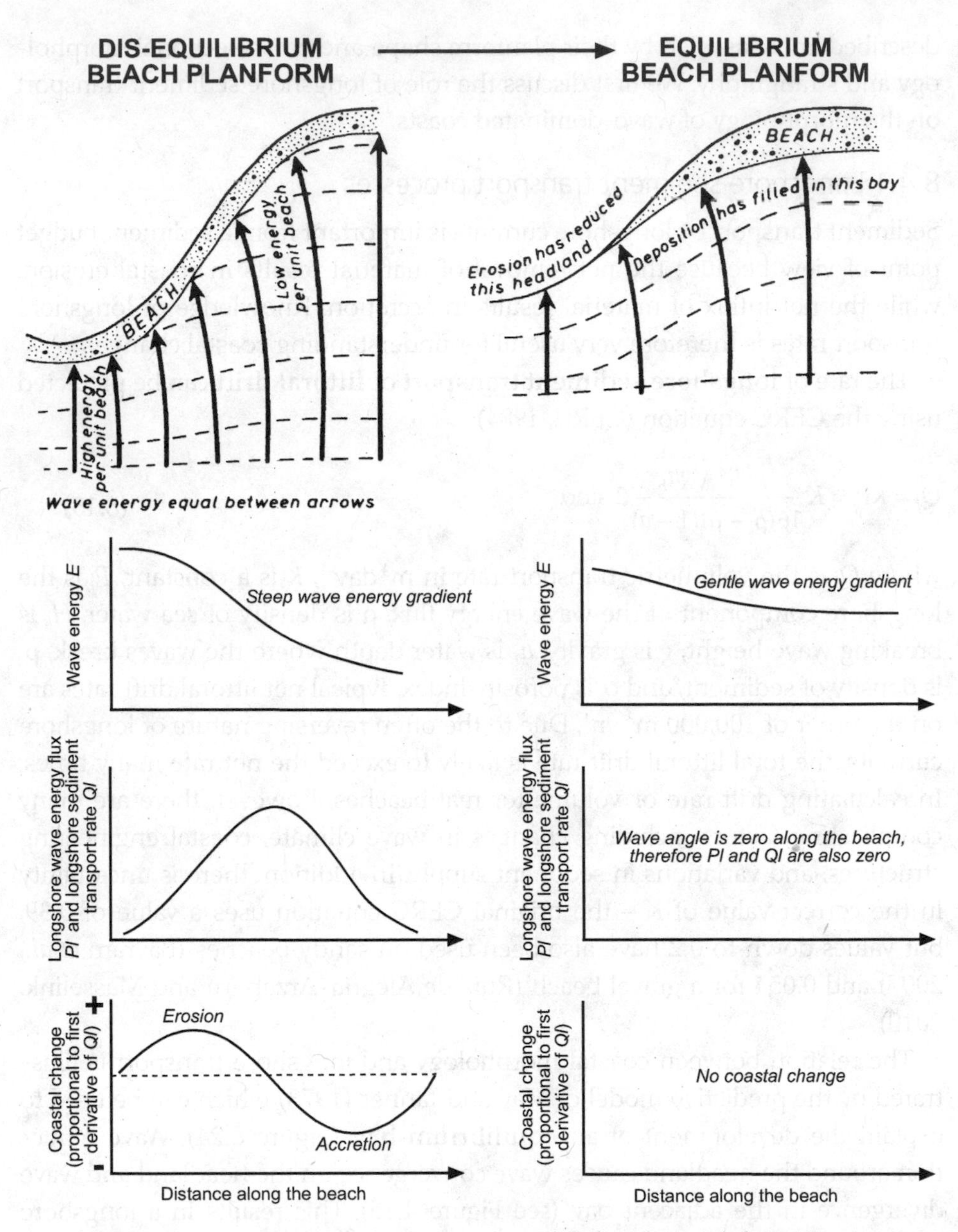

Figure 8.24 Relation between wave refraction, littoral drift and coastal change in a headland-bay system. (Modified from Pethick, 1984; modified from May and Tanner, 1973.)

From the longshore variation in Q_l we can now predict changes in the coastline. If the littoral drift rate increases along a stretch of coast, progressively more sediment will be entrained. This sediment will have to be provided by erosional processes so that, over long time scales, the coastline is expected to retreat. Similarly, if the littoral drift rate decreases, sediment will progressively be deposited, causing coastal progradation. Therefore, the rate of change in the littoral drift rate is proportional to the first derivative of the longshore gradient of Q_l (i.e. dQ_l/dx) and controls coastline changes. In our example of the headland/bay system, the coastline from the headland to where P_l is maximum exhibits a progressive increase in the littoral drift rate and hence will experience progressive erosion. The coastline from where P_l is maximum to the bay is characterised by a progressive decrease in the littoral drift rate and will therefore undergo progressive deposition. This analysis shows that deposition occurs in the low-energy zones of the bay and erosion at the headlands. These processes will continue until the morphology is such that everywhere along the coast the wave angles are zero, meaning that coastline geometry has reached equilibrium with respect to wave approach, and littoral drift ceases. However, coastlines in the real world are rarely in equilibrium, even though they may appear to be so over some spatial and/or temporal scales.

8.4.2 Planform shape of barriers

A large variety of barrier planforms exists, but it is first useful to distinguish between two fundamentally different types of coastline:

- **Swash-aligned coasts** are oriented parallel to the crests of the prevailing incident waves (Figure 8.25a). They are closed littoral cells in terms of longshore sediment transport, and net littoral drift rates are zero.
- **Drift-aligned coasts** are oriented obliquely to the crest of the prevailing waves (Figure 8.25b) and are open littoral cells in terms of longshore sediment transport because sediment is in transit through them, and they are therefore sensitive to changes in drift rates.

The planform shape of many coasts is a combination of drift and swash-alignment. In addition, along any coastal stretch there may be alternate swash- and drift-aligned sections depending on coastal geology and local sediment supply.

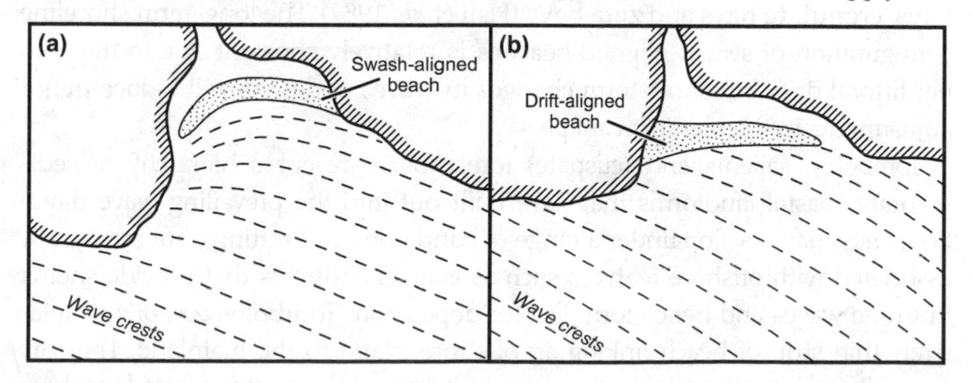

Figure 8.25 Two main types of coastal alignment: (a) swash-aligned and (b) drift-aligned. (Modified from Davies, 1980.)

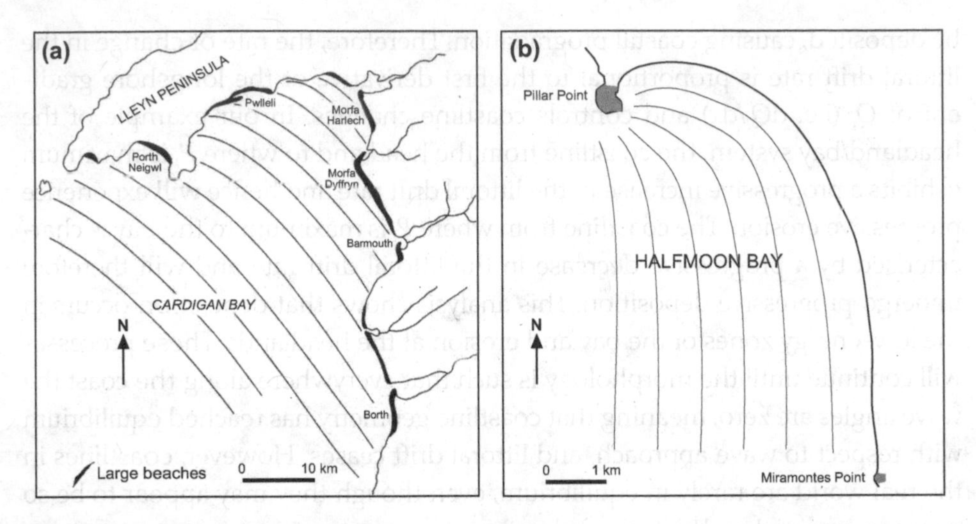

Figure 8.26 Examples of swash-aligned coastlines. (a) Wave refraction pattern for southwesterly waves and orientation of large beaches within Cardigan Bay, Wales. All large beaches are oriented parallel to the refracted wave crests. Note that the spits at the end of the beaches at Pwllheli and Morfa Dyffryn are not swash-aligned. (From Hansom, 1988.) (Copyright © 1988 Cambridge University Press, reproduced with permission.) (b) Wave refraction pattern and shoreline curvature of Halfmoon Bay, California. (Modified from Yasso, 1965.)

Embayed beaches are generally swash-aligned because there is little or no additional sediment supplied to the beach from adjacent headlands and essentially no sediment losses. Therefore, their equilibrium planform shape depends solely on wave conditions, in particular the pattern of refraction and diffraction (Section 4.5). The planform shape is curved with the shoreline aligned with the crests of incoming waves (Figure 8.26a). Large headlands can block the littoral drift, so beaches found down-drift from them are also usually swash-aligned (Figure 8.26b). Their shape can be described by a logarithmic spiral, and beaches conforming to such a curved shape have been termed **log-spiral bays, crenulate bays** and **zeta bays** (Hsu *et al.*, 1987). The long-term shoreline configuration of swash-aligned beaches is relatively constant due to the zero net littoral drift, but short-term changes in wave conditions will induce minor adjustments in the planform shape.

Tombolos, salients and (cuspate) forelands represent a hierarchy of sedimentary coastal landforms that have built out into the prevailing wave direction. They may develop under a range of conditions and settings, but are usually associated with offshore features such as islands and reefs that provide shelter from the waves and hence form loci for deposition. **Tombolos** comprise a relatively thin strip of beach linking an offshore island to the mainland. They are generally swash-aligned and have resulted from the convergence of longshore currents and sediment transport into the shadow zone behind the island due to wave refraction and diffraction. Tombolo development is strongly dependent on the alongshore length of the island I and its distance to the mainland J. Sunamura and Mizuno (1987) found that tombolos develop when $J/I < 1.5$, whereas the island does not exert a significant influence on the coast if $J/I > 3.5$. For $J/I = 1.5–3.5$, a salient forms in the lee of the island. The terms 'salient' and (cuspate) 'foreland' have been used rather inconsistently in the coastal literature. **Salients** are relatively inconspicuous protuberances behind submerged reefs or offshore islands, whereas **forelands** extend a considerable distance out from coasts where two dominant swells are in opposition (Figure 8.27).

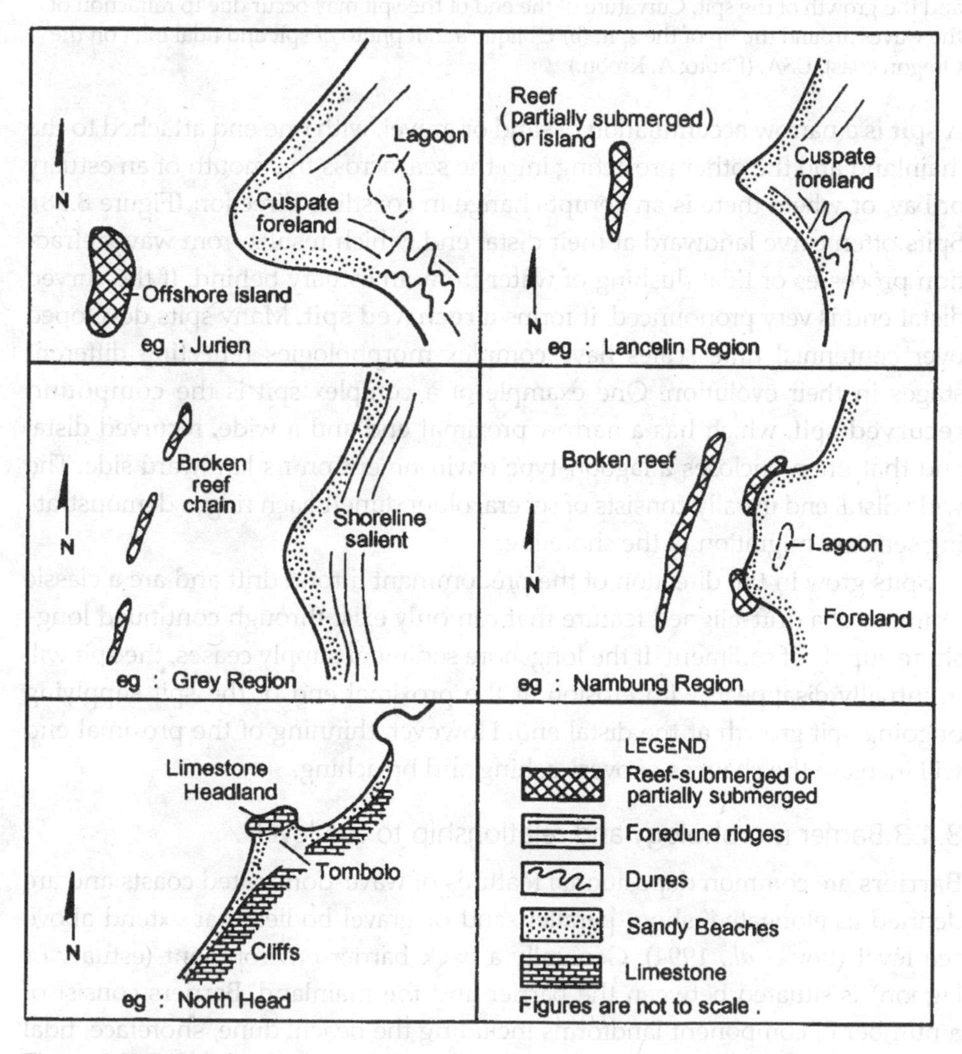

Figure 8.27 Types of tombolos, salient and forelands along the central west coast of Western Australia. (From Sanderson and Eliot, 1996.) (Copyright © 1996 Coastal Education & Research Foundation, reproduced with permission.)

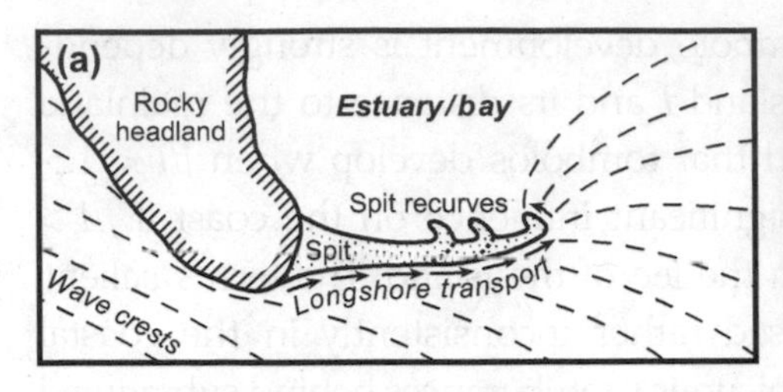

Figure 8.28 (a) Formation of a (recurved) spit. The longshore current transports sediment along the coast. When the current enters deeper water near the tip of the spit, it spreads out and loses its sediment transporting capacity, resulting in the deposition of sediment and the growth of the spit. Curvature of the end of the spit may occur due to refraction of the waves around the tip of the spit. (b) Oblique aerial photo of spit and tidal inlet on the Oregon coast, USA. (Photo: A. Kroon.)

A **spit** is a narrow accumulation of sand or gravel, with one end attached to the mainland and the other projecting into the sea, across the mouth of an estuary or bay, or where there is an abrupt change in coastline direction (Figure 8.28). Spits often curve landward at their distal end, which results from wave refraction processes or tidal flushing of water from an estuary behind. If the curved distal end is very pronounced, it forms a **recurved spit**. Many spits developed over centennial time scales have complex morphologies reflecting different stages in their evolution. One example of a complex spit is the **compound recurved spit**, which has a narrow proximal end and a wide, recurved distal end that often encloses a lagoon-type environment on its landward side. The wide distal end usually consists of several older dune/beach ridges demonstrating seaward migration of the shoreline.

Spits grow in the direction of the predominant littoral drift and are a classic example of a drift-aligned feature that can only exist through continued longshore supply of sediment. If the longshore sediment supply ceases, the spit will eventually disappear, with erosion at the proximal end of the spit supplying ongoing spit growth at the distal end. However, thinning of the proximal end will increase the chances of overwashing and breaching.

8.4.3 Barrier morphology and relationship to sea level

Barriers are common depositional features of wave-dominated coasts and are defined as elongated, shore-parallel sand or gravel bodies that extend above sea level (Roy *et al.*, 1994). Generally a back-barrier environment (estuary or lagoon) is situated between the barrier and the mainland. Barriers consist of a number of component landforms including the beach, dune, shoreface, tidal delta, inlet and washover forms (Figure 8.29a). Together they form a continuum of barrier morphologies, ranging from **barrier islands** with shallow lagoons

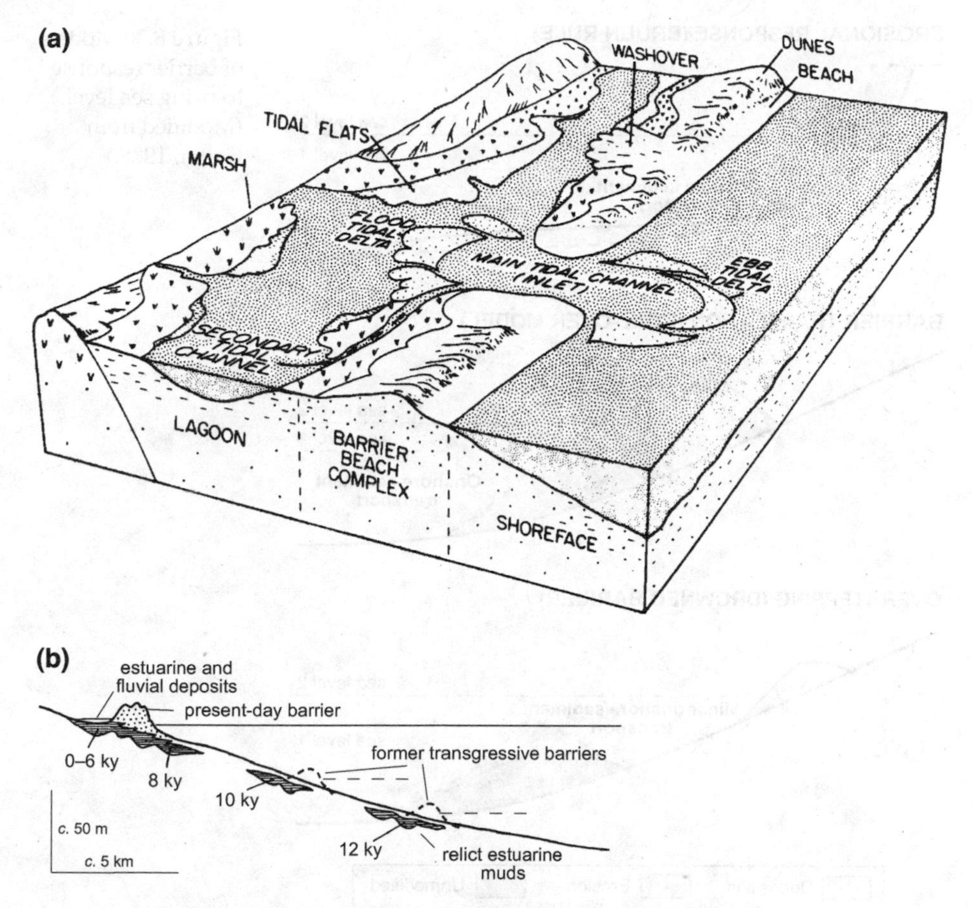

Figure 8.29 (a) Block diagram illustrating the various sub-environments in a barrier island system. (From Reinson, 1984.) (Copyright © 1984 Geological Association of Canada, reproduced with permission.) (b) Coring investigations on the continental shelf of southeastern Australia encounter subsurface relict estuarine sediments indicating formation as the shelf was inundated by the postglacial marine transgression. Estuarine and lagoonal environments only exist in the protection of barriers, which means that these latter features must also have existed at times of lower sea levels. (From Roy *et al.*, 1994.) (Copyright © 1994 Cambridge University Press, reproduced with permission.)

on low-gradient coasts to **mainland beaches** with negligible back-barrier morphologies on steep coasts. Between these end-members are **bay barriers**, which are located within embayments, confined by rocky headlands.

Most contemporary barriers were established during final stages of early Holocene marine transgression, and migrated landward throughout the Holocene under decelerating sea-level rise (Pilkey *et al.*, 2009). Support for the genetic relationship between barrier migration and sea-level rise comes from marine cores that show relict estuarine sediments underlying coarser sands

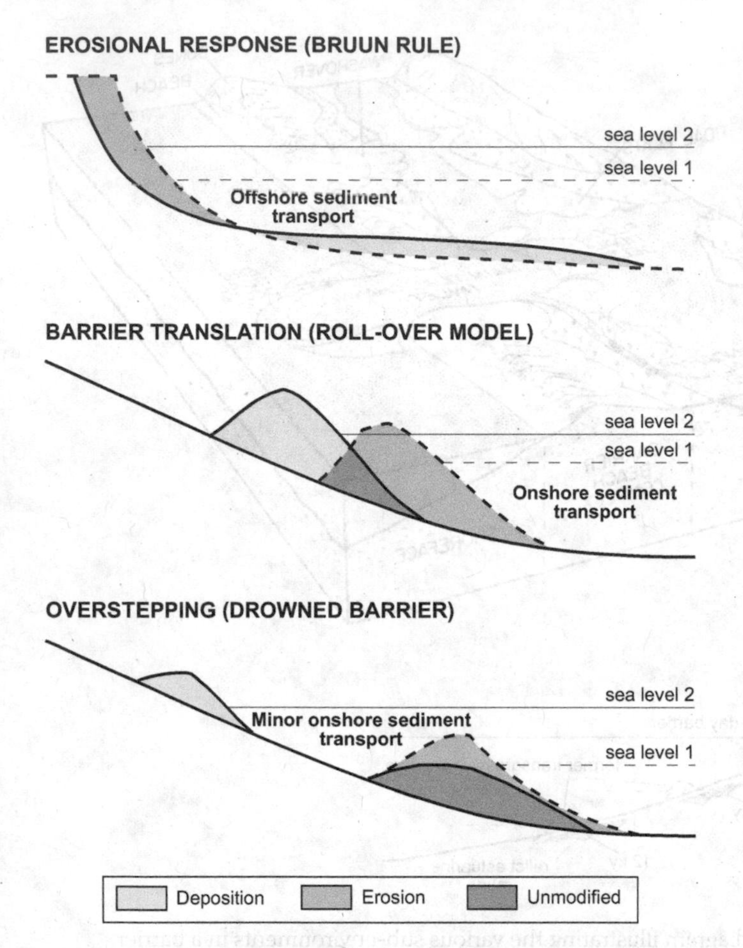

Figure 8.30 Models of barrier response to rising sea level. (Modified from Carter, 1988.)

and gravels that compose the barriers. The absence of open marine sediments behind many barriers indicates that most barriers migrated landward from the outer shelf, closing up their own back-barrier lagoon (Figure 8.29b).

As sea level rises, a barrier may undergo one of three responses (Figure 8.30):

- **Barrier erosion** – The cross-sectional geometry of the shoreface is maintained but gets smaller over time as the entire profile moves landward at a rate that tracks that of sea-level rise. Material eroded from the shoreface is deposited below wave base, hence this model involves offshore transport and the barrier undergoes net sediment loss. This model conforms with the Bruun Rule (Box 8.1).
- **Barrier translation** – The entire barrier migrates landward without loss of material. This is accomplished through erosion of the shoreface and deposition of this sediment behind the barrier in the form of **washover fans**. Barrier translation represents onshore sediment transport and is referred to as the **roll-over model**.

- **Barrier overstepping** – Where sea level rises too fast for the barrier morphology to respond, overstepping of the barrier may occur, leaving it drowned on the sea bed as a relict feature.

Barrier translation is the most common response of all types of barriers to rising sea level (Masetti *et al.*, 2008).

Rates of shoreline recession due to **barrier retreat** depend on a number of factors (Roy *et al.*, 1994). These are:

- **Rate of sea-level rise** – Barrier translation is fairly continuous while sea level is rising rapidly, but may become intermittent and affected by other factors when sea-level rise slows down.
- **Gradient of the underlying substrate** – Barrier retreat rate is most rapid on shallow substrates and slowest on steep substrates.
- **Longshore sediment transport** – Barrier retreat is promoted (inhibited) by a negative (positive) longshore sediment budget, i.e. by net erosion (deposition).
- **Depositional processes in the back-barrier lagoon or estuary** – If back-barrier sedimentation rates are similar to or higher than the rate of sea-level rise, then the barrier will prograde and significantly slow down barrier retreat. The substrate slope steepens over time, inhibiting the rate of translation.

Sand dune development on the barrier island may also affect the rate of barrier retreat. Under rapidly rising sea level, dunes have little time to develop and provide little impediment to storm washover. If larger dunes can develop under slower rates of sea-level rise, the frequency of washover is reduced, slowing the rate of barrier translation. A recent modelling study found that, in order of importance, substrate composition, substrate slope, rate of sea-level rise and sediment supply determine barrier island responses to sea-level rise (Moore *et al.*, 2010). Smaller barrier islands are more sensitive to change than larger ones, and therefore more vulnerable to storms. The role of storms in barrier island morphodynamics is explored in Case Study 8.1.

Case Study 8.1 Hurricane impacts on barrier islands

Barrier islands are sensitive to the strong winds, large waves and elevated sea levels associated with hurricanes, because these conditions are conducive to shoreface erosion and barrier overwashing. Hurricane Ivan affected a 300 km-wide area of the Gulf of Mexico coastline of the USA when it made landfall in September 2004. At landfall it was a Category 3 hurricane with a wind strength of 200 km hr^{-1} and storm surge of up to 4 m.

Several studies have examined the impacts of Hurricane Ivan on the barrier islands of Florida and Alabama, using remote sensing images,

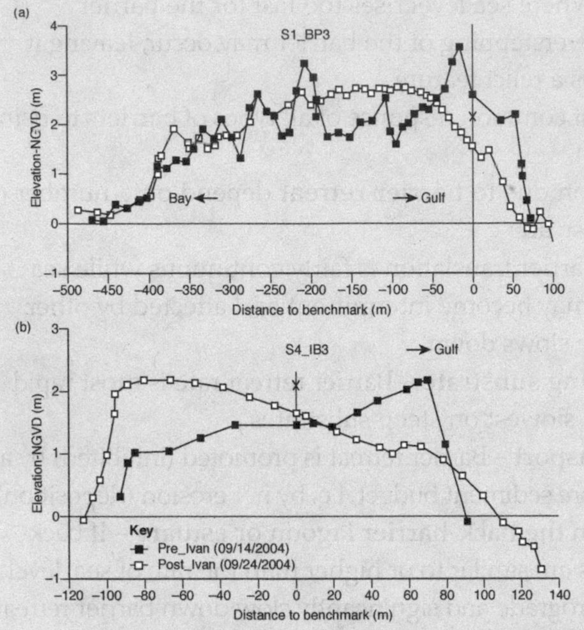

Figure 8.31 Topographic cross-sections over barrier islands at Beasley Park, Santa Rosa Island (top), and Inlet beach (bottom), northwest Florida. (From Wang *et al.*, 2006.) (Copyright © 2006 Coastal Education & Research Foundation, reproduced with permission.)

transects across the islands and field observations of overwashing and sediment transport. For example, on Santa Rosa Island (Florida), Houser and Hamilton (2009) showed that 64 m of shoreline erosion took place during the 2004 hurricane season (including Hurricane Ivan), but that 19 m of progradational recovery had taken place by the following year. An adjacent site, however, showed 30 m of shoreline retreat and no later recovery. The geomorphological impacts of Hurricane Ivan were (Figure 8.31):

- erosional retreat of the barrier island's shoreline by large waves
- deposition of much of this eroded sediment on the landward side of the island by wave overwashing
- reduction in relief of the island as a result of wave overwashing and strong winds
- formation of a relatively flat erosional surface across the island.

High rates of overwashing on barrier islands with subdued relief reduce sediment availability for later recovery. Important controls on the capacity of barrier islands to 'bounce back' after hurricane events therefore include offshore bathymetry, island width, topography and degree of overwashing.

Barriers that migrate landward under the influence of rising sea level are termed **transgressive barriers** (Figure 8.32). Their stratigraphy is such that sediments deposited in seaward locations stratigraphically end up on top of sediments that were deposited in more landward locations (transgressive sequence). Trans-

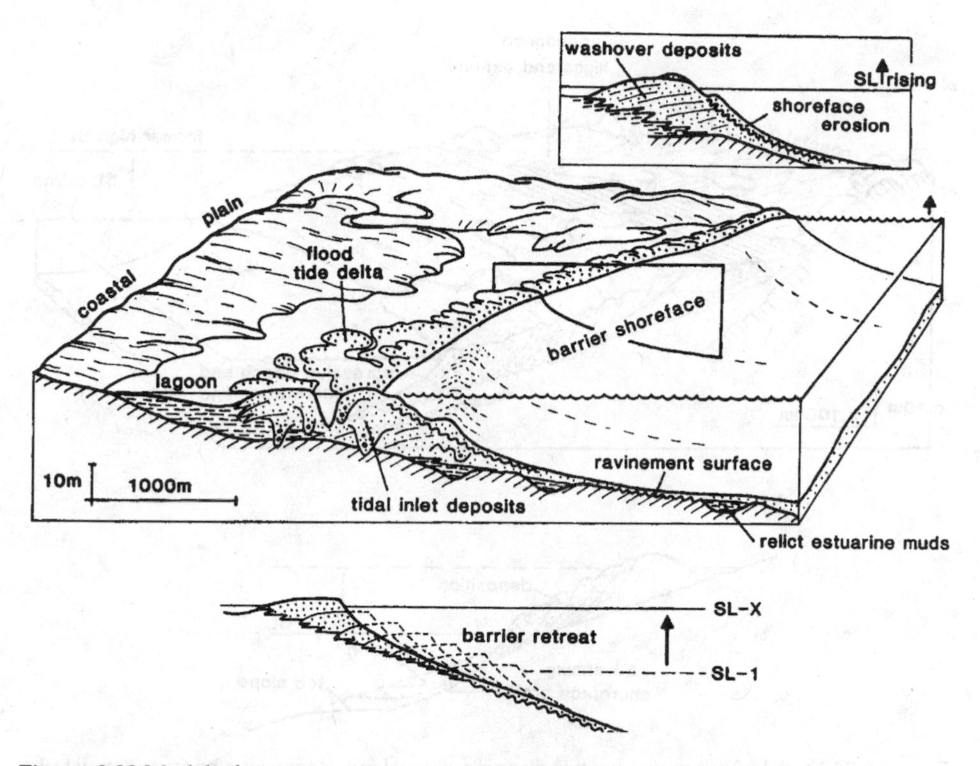

Figure 8.32 Model of transgressive barrier during sea-level rise. Transgressive barriers are almost entirely composed of tidal delta and washover deposits. The barrier migrates into estuarine/lagoonal environments as sea level rises. Landward migration is through a process of erosional shoreface retreat as the barrier adjusts to changing sea level. (From Roy *et al.*, 1994.) (Copyright © 1994 Cambridge University Press, reproduced with permission.)

gressive barriers consist mainly of tidal delta and/or washover deposits, and are underlain by back-barrier estuarine or lagoonal deposits. **Regressive barriers** develop under the influence of falling sea level (Figure 8.33). The stratigraphy of a regressive barrier is such that landward sediments are deposited on top of more seaward ones (regressive sequence). The barrier is often capped by aeolian sand, below which is beach sand, while at still greater depths is sand that accumulated in progressively deeper water, giving way to silt and clay that had been deposited on what was once the continental shelf. Transgressive and regressive barriers result from the interplay between sea level, sediment supply, substrate slope and overwashing.

Many barriers worldwide experienced transgression during the first half of the Holocene when sea level was rising rapidly. **Barrier progradation** then took place under the influence of relatively stable sea level (from *c.* 5000–6000 years BP onwards) and high sediment supply from the shelf. This second phase of Holocene barrier development is often referred to as regressive, but strictly speaking the term should only be used for barriers developed under a falling sea

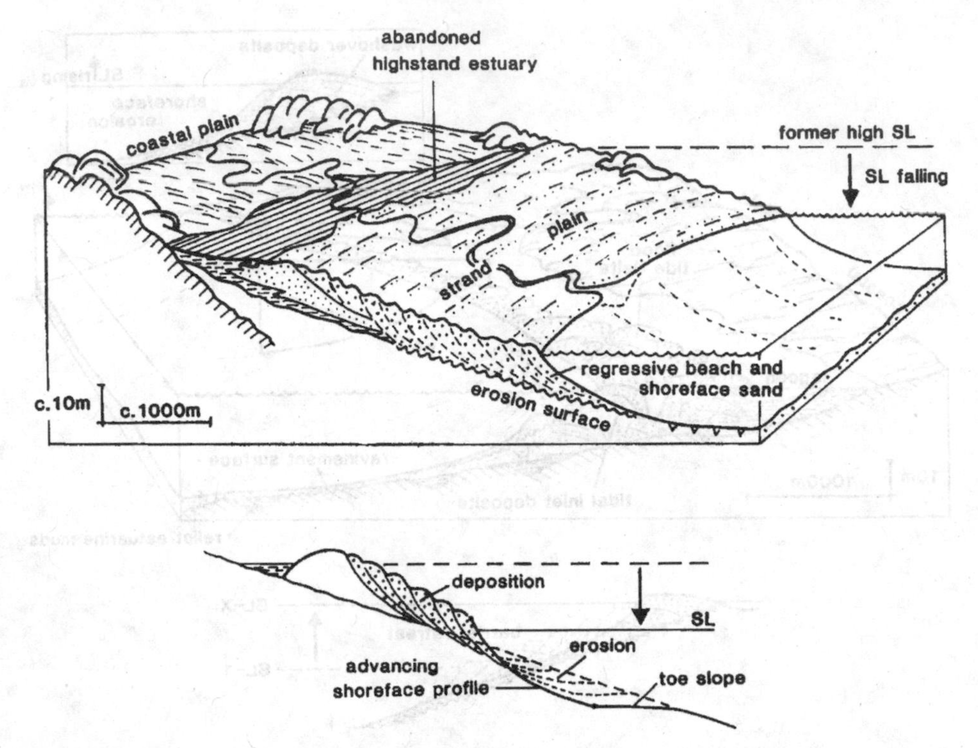

Figure 8.33 Model of regressive barrier during sea-level fall. Regressive barriers consist of a tabular, gently seaward-inclined sand deposit 10–20 m thick with an erosional base. As sea level falls, the inner shelf surface erodes and sand moves onto the shoreface. The surface of the regressive barrier forms a wide strandplain, generally without estuaries. (From Roy *et al.*, 1994.) (Copyright © 1994 Cambridge University Press, reproduced with permission.)

level, hence the term **forced regression** is used. Coastal progradation and forced regression during the second half of the Holocene resulted in the development of coastal **strand plains** of which there are four main types (Figure 8.34):

- **Prograded barrier** – This general term indicates a barrier system that has prograded under the influence of a positive sediment budget, and is not a foredune ridge, beach ridge or chenier plain. Prograded barriers have extensive dune development and may attain elevations of tens of metres above sea level (e.g. coast of the Netherlands).
- **Beach ridge plain** – Beach ridges are wave-deposited, shore-parallel triangular to convex ridges, formed of sand, gravel or shelly sediments. They form by wave and swash processes under fairweather or storm conditions. Beach ridges generally prograde where there is abundant sediment and low offshore gradients, resulting in formation of a beach ridge plain. Dunes may be present on beach ridges, but aeolian processes are secondary to wave and swash processes.

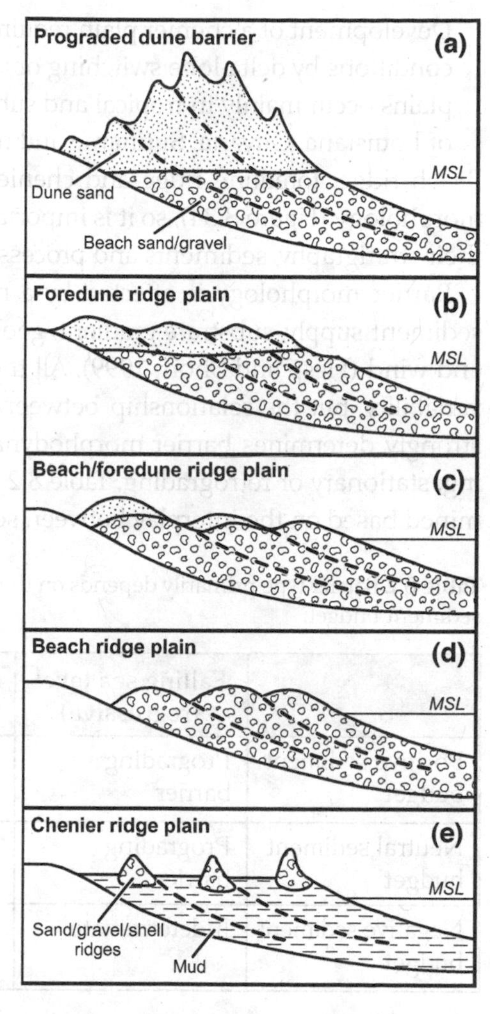

Figure 8.34 Five main types of wave-dominated, coastal plains: (a) prograded dune barrier, (b) foredune ridge plain, (c) beach/foredune ridge plain, (d) beach ridge plain and (e) chenier plain. The different types of coastal plain are similar with respect to their planform and morphology, but can be distinguished on the basis of their stratigraphy.

- **Foredune ridge plain** – Foredunes are shore-parallel ridges formed on the top of the backshore by aeolian processes, and facilitated where the fronting beach progrades. The faster the rate of coastal progradation, the lower and more numerous the foredune ridges are. Dune processes are examined in more detail in Chapter 9.
- **Chenier plain** – A chenier is a coastal ridge composed of sand and shell overlying and surrounded by mudflats or marshes. Chenier plains consist of at least two cheniers separated by a progradational muddy unit. There has been much debate over the formation mechanisms of chenier plain sediments (e.g. Augustinus, 1989). Mud flat deposition occurs during periods of abundant sediment supply, whereas cheniers form during periods of diminished supply, by wave-winnowing which concentrates the sand fraction from the mud, or by sand transported by longshore currents.

Development of a chenier plain requires alternation between these two conditions by delta lobe switching or variations in storminess. Chenier plains occur mainly in tropical and subtropical regions, including the coasts of Louisiana, Guyana, Suriname and northern Australia.

Beach ridge, foredune ridge and chenier plains are similar in planform and morphology (Figure 8.34), so it is important to identify them correctly based on their stratigraphy, sediments and processes.

Barrier morphology is affected by a number of factors including sea level, sediment supply, substrate gradient, geological inheritance, wave energy, tides and wind (Hesp and Short, 1999). All these factors show spatial and temporal variability, but the relationship between sea level and sediment supply most strongly determines barrier morphodynamics and whether they are prograding, stationary or retrograding. Table 8.2 shows how barrier type can be determined based on the interplay between sea-level trend and sediment supply.

Table 8.2 Barrier type primarily depends on the balance between sea-level change and the sediment budget.

	Falling sea level (regressive)	Stationary sea level	Rising sea level (transgressive)
Positive sediment budget	Prograding barrier	Prograding barrier	Indeterminate
Neutral sediment budget	Prograding barrier	Stationary barrier	Retrograding barrier
Negative sediment budget	Indeterminate	Retrograding barrier	Retrograding barrier

8.5 Beaches, barriers and human activity

Human activity strongly influences the morphodynamic behaviour of the shoreface, beaches and barriers because coastal engineering and management practices can modify nearshore wave climate and sediment supply. As a consequence, beaches and barriers are among the most sensitive coastal landforms to human activity, and many coastal management strategies have been developed in order to better control their morphodynamic behaviour, in particular to reduce coastal erosion. The context for our discussion of human activity and coastal management is the range of 'hard' and 'soft' coastal engineering works shown in Figure 8.35. Throughout much of the twentieth century, large timber, concrete or rock structures were considered the best way to combat coastal

management issue	engineered solution(s)	types	description	problems	illustration
cliff erosion	seawalls	vertical wall	a wall constructed out of rock blocks, or bulkheads of wood or steel, or simply semi-vertical mounds of rubble in front of a cliff	rock walls are highly reflective, bulkheads less so. Loose rubble however, absorbs wave energy	seawall; cliff; waves
		curved wall	a concrete constructed concave wall	quite reflective, but the concave structure introduces a dissipative element	
		stepped	a rectilinear stepped hard structure, as gently sloping as possible, often with a curved wave-return wall at the top	the scarps of the steps are reflective, but overall the structure is quite dissipative	wave-return wall; steps
		revetment	a sloping rectilinear armoured structure constructed with less reflective material, such as interlocking blocks (tetrapods), rock-filled gabions, and asphalt	the slope and loose material ensure maximum dissipation of wave energy	armoured blocks
coastal inundation	seawalls	earth banks	a free-standing bank of earth and loose material, often at the landward edge of coastal wetlands	may be susceptible to erosion, and overtopped during extreme high-water events	earth bank
	tidal barriers		barriers built across estuaries with sluice gates that may be closed when threatened by storm surge	extremely costly, and relies on reliable storm surge warning system (*e.g.* Thames Barrier)	
beach stabilization	groynes		shore-normal walls of mainly wood, built across beaches to trap drifting sediment	starve downdrift beaches of sediment	longshore drift; sand; groynes
	beach nourishment		adding sediment to a beach to maintain beach levels and dimensions	sediment is often rapidly removed through erosion and needs regular replenishing; often sourced by dredging coastal waters	
offshore protection	breakwaters		structures situated offshore that intercept waves before they reach the shore. Constructed with concrete and/or rubble	very costly and often suffer damage during storms	waves; protected by breakwater; land
tidal inlet management	jetties		walls built to line the banks of tidal inlets or river outlets in order to stablise the waterway for navigation	the jetties protrude into the sea and promote sediment deposition on the updrift side, but also sediment starvation and erosion on the downdrift side	sand; jetty; erosion; inlet; land; longshore drift

Figure 8.35 Summary of engineered coastal protection works. (From Haslett, 2000.) (Copyright © 2000 Taylor & Francis Books, reproduced with permission.)

erosion. This '**hard engineering**' approach resulted in large sections of coastline being fixed in place. In recent decades, however, there has been increased concern about environmental problems associated with hard engineering, so there has been a shift towards '**soft engineering**' structures that tend to work with coastal processes rather than against them (Cooper and McKenna, 2008).

8.5.1 Management of shoreface dynamics

The major management strategies affecting shoreface dynamics concern reducing nearshore wave intensity through the use of breakwaters, and reducing shoreface erosion through the use of sea walls and revetments.

Breakwaters are shore-parallel structures placed seaward of the shoreline. These are designed to dissipate incident wave energy, reducing the direct impact of storm waves on the shoreface. They are used in harbour design to reduce wave action in the harbour, but when used as a coastal protection

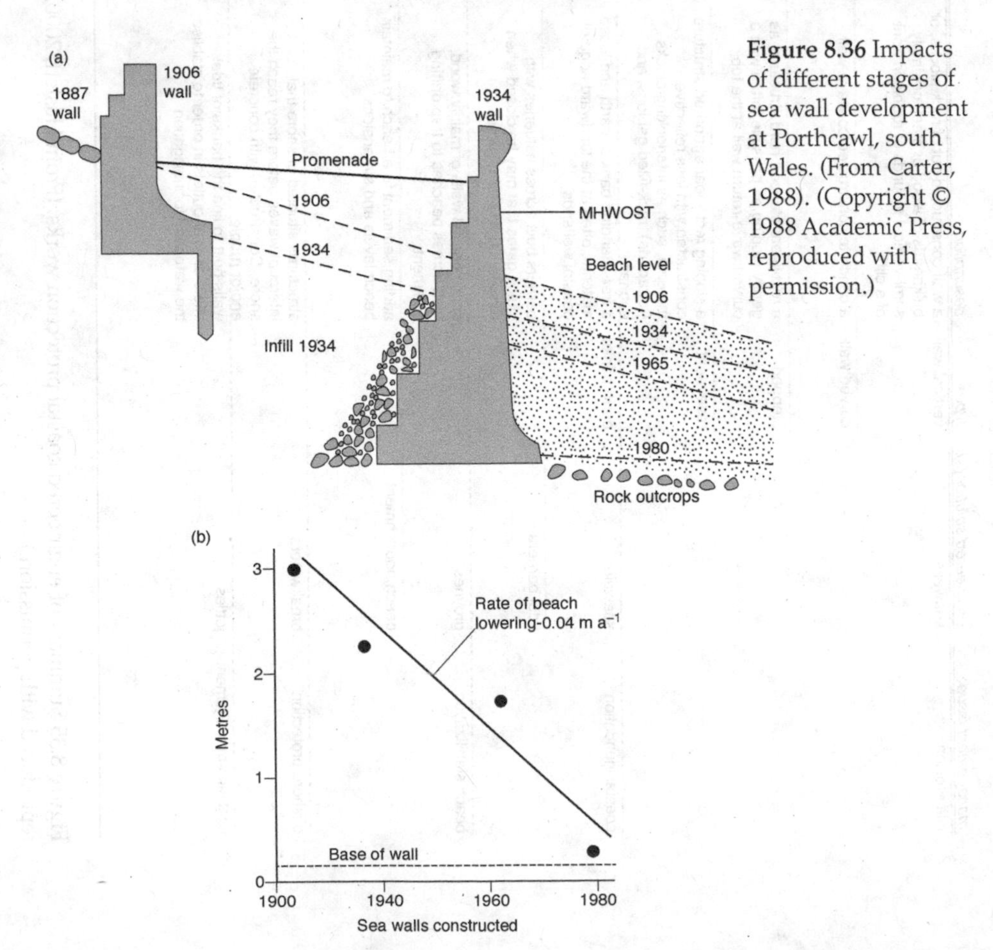

Figure 8.36 Impacts of different stages of sea wall development at Porthcawl, south Wales. (From Carter, 1988). (Copyright © 1988 Academic Press, reproduced with permission.)

structure, breakwaters placed just offshore from the surf zone are referred to as **detached breakwaters**. They are usually built as a series to protect a long stretch of coast. Important design features of a series of breakwaters are the gap distance between individual breakwaters, the length of the breakwaters, and their offshore distance (Pope and Dean, 1986). Submerged breakwaters may also be used with their tops located just below sea level. These work well in microtidal environments but when the tidal range is large, breakwaters are submerged at high tide and exposed at low tide.

Detached breakwaters help shelter the beach area behind them, with calmer water than found on open coasts. Wave shadowing and wave refraction/diffraction patterns contribute to sediment deposition in the lee of the structure, forming salients or tombolos (see Figure 8.27). The type of morphological response depends on breakwater design, with tombolos being promoted by relatively long breakwaters placed at a relatively short distance from the shore (Pope and Dean, 1986). Detached breakwaters do not block the littoral drift (unless tombolos develop), or induce offshore sediment loss. They can also be used as artificial reefs to provide new marine habitats or, more commonly, to increase breaking wave height and so promote conditions suitable for surfing, such as at Boscombe, southern England. Because breakwaters are exposed to extreme wave action, they have high initial costs and need frequent maintenance.

The most commonly used structures for shore and cliff protection are seawalls and revetments built parallel to the shoreline and designed to reduce wave erosion. **Sea walls** are solid structures constructed of concrete, steel sheets or timber. They generally have a vertical face but may also be curved or stepped. **Revetments** are generally smaller and have inclined faces. They may be built from natural stone, known as rip-rap, or concrete armour units. The vertical or concave faces of seawalls reflect the incident wave energy back to sea. However, wave reflection may lead to scouring of the beach in front of the sea wall (Figure 8.36), followed by undermining and eventually collapse of the sea wall. Rip-rap is more permeable and induces less wave reflection than solid, vertical sea walls. Both sea walls and revetments, however, are static features and can impede the exchange of sediment between land and sea. They may accelerate erosion of adjacent, unprotected coastal properties because they affect littoral drift.

8.5.1 Management of beach dynamics

The major management strategies affecting beaches concern reducing the speed of longshore currents or the efficiency with which they induce longshore sediment transport and therefore beach erosion. The most common engineering structures used are groynes.

Groynes are constructed from concrete, rip-rap, steel or timber, and are

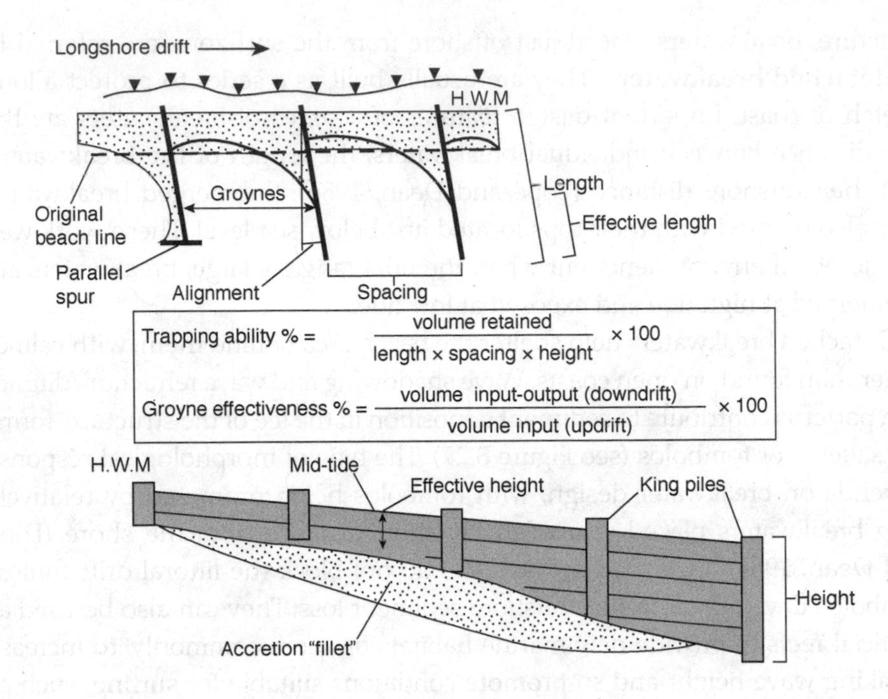

Figure 8.37 Diagram showing the design of a groyne field. (From Carter, 1988.) (Copyright © 1988 Academic Press, reproduced with permission.)

designed to trap a portion of the longshore sediment transport. Groyne construction therefore aims to protect the coast against erosion while maintaining a beach (Schoonees *et al.*, 2006). Groynes are typically perpendicular to the beach and often a series of groynes, referred to as a groyne field, is deployed to protect a large area from erosion (Figure 8.37). An important element in groyne design is the ratio between groyne length and spacing. For sandy beaches a ratio of 1:4 is recommended, while for gravel beaches it is 1:2. Another factor is the length of the groynes relative to the width of the surf zone. Conventionally, the groyne length should be approximately 40–60 per cent of the average surf zone width, allowing the structure to trap some, but not all, of the littoral drift (Komar, 1998). Once the area between two groynes is filled, longshore transport will be able to pass around the end of the structure, which is still in the surf zone.

Groynes are not intended to completely block longshore sediment transport, but in some cases the trapping efficiency of a groyne field is up to 100 per cent, leading to down-drift sediment starvation and enhanced erosion. In some instances the response to this problem has been to extend the groyne field in the down-drift direction, but in doing so enhanced erosion is merely extended to down-coast neighbours. Rip currents also tend to form adjacent to the groynes. During storms these rips may move sediment beyond the surf

Case Study 8.2 Restoration of Miami Beach

Miami Beach is one of the most developed and, in financial terms, one of the most valuable coastlines in the world. The city is built on a narrow barrier island that has a history of dramatic shoreline changes, including opening and closing of tidal inlets. The development of Miami Beach from the 1920s onward occurred against a backdrop of continuing coastal erosion problems. Almost the entire beach frontage was privately owned and this resulted in remedial works being haphazard and inconsistent. By 1965, the Miami shoreline (both ocean and lagoon side) was almost destroyed.

The solution to the coastal erosion problem was beach nourishment. This project took place from 1976 to 1981 and represents the largest beach nourishment scheme undertaken in the USA (Figure 8.38). The nourish-

Figure 8.38 The restoration of Miami Beach by beach nourishment: (a) February 1978 just prior to the project; and (b) October 1981 after completion of the first phase of the project. (From CERC, 1984.) (Reproduced with permission from US Army Corps of Engineers.)

ment project cost at the time more than US$60 million and involved the placement of 13 million m^3 of sand dredged from offshore over a 16 km stretch of coastline (Komar, 1998). The nourishment produced a dry beach 55 m wide at an elevation of 3 m above mean low water. The objective of the project was to provide an expanded recreational beach that also protected the highly-developed beachfront from hurricane waves and storm surge. Additional sand has been added to Miami Beach subsequently, most recently in 1999.

The project is considered to have been successful in that the nourished beach has functioned for more than 30 years and has survived two major hurricanes. The project has also been a success in economical terms. The cost for coastal protection and beach nourishment is US$3 million per year, while the income generated by the Miami Beach economy from visitors is estimated at US$2 billion per year.

zone into water so deep that it cannot readily return to replenish the beach. The use of groynes in coastal protection has declined in recent years, but in some cases it may be the most suitable technique to combat local erosion, particularly if combined with beach nourishment.

Beach nourishment involves the artificial deposition of sediment on the beach or in the nearshore zone to advance the shoreline seaward (Case Study 8.2). Generally, the size of the sand used for nourishment must be equal to or coarser than that of the local sediment, to minimise rapid sediment loss offshore (Castelle *et al.*, 2009). Beach nourishment treats the symptoms of coastal erosion (loss of beach) without addressing the real cause (deficit in the overall sediment budget and/or rising sea level). Therefore, sediment will continue to be lost from the beach and nourishment will have to be repeated. To reduce sediment losses, groynes may be placed at the boundaries of the nourished area.

One of the problems associated with beach nourishment is its public perception. Immediately following nourishment, waves will redistribute the sediment to form a more natural cross-shore profile. The first post-nourishment storm(s) will typically transport sediment from the subaerial beach to the nearshore, resulting in beach narrowing. This is often perceived by the public as a failure of the nourishment scheme, and a waste of money. However, the sediment is not 'lost' because it often remains in the nearshore. Komar (1998) emphasises the need for monitoring of beach nourishment schemes, and for public education as to what should be expected from such projects. Beach nourishment is often a preferred management option because it is relatively cost-efficient and the results appear more natural.

8.5.2 Management and restoration of barrier islands

Like Miami Beach, many barrier islands are areas of high urban development, and the building of properties, roads, bridges and other infrastructure has led to increased pressure on barrier island morphodynamics. This is largely to do with stabilising the ends and front of the barrier island in order to reduce morphodynamic variability and help ensure continuous ship access through barrier inlets to the lagoon or estuary behind. Stabilising barrier island beaches helps buffer coastal erosion and reduces storm impacts landward of the beach. Both of these engineering actions are significant because they lead to marked changes in sediment dynamics, with less sediment passing from one island to another by longshore transport, and more sediment being trapped within or adjacent to inlets (Figure 8.39). As a result, beach sediment supply decreases, necessitating the deployment of groynes and beach nourishment. Over time, therefore, urbanisation of some barrier islands has led to a cycle of net erosion that can result in the entire barrier island being protected by engineering structures.

These issues of development and management of barrier islands are particularly important along the Gulf of Mexico coast of the USA, where the barrier islands are highly urbanised, valued for their biodiversity and wetland habitat,

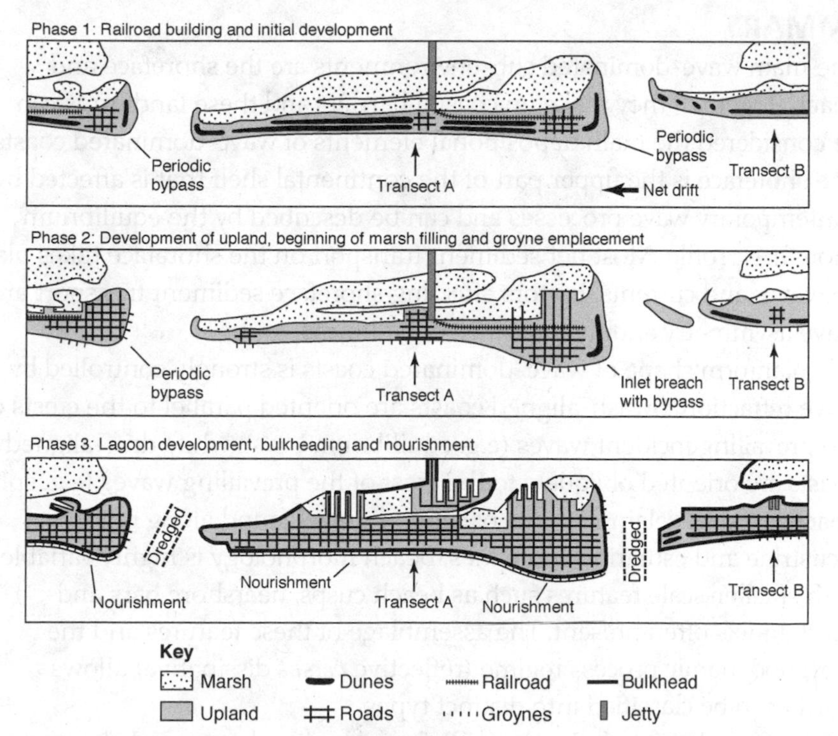

Figure 8.39 Plan view showing stages in development of a barrier island. (From Nordstrom, 1994.) (Copyright © 1994 Cambridge University Press, reproduced with permission.)

and act as a defensive barrier for landfalling hurricanes. As a result, **barrier island restoration** is an important management priority (see Case Study 8.1). The US state of Louisiana is using a range of restoration strategies for the barrier islands that front parts of the Mississippi River delta. For example, the five islands comprising the Isles Dernieres barrier island chain have been undergoing continued coastal erosion and it was predicted that without restoration these islands would have eroded away entirely between 2007 and 2019 (Rodrigue *et al.*, 2008). One of these, Trinity Island, lost one third of its area between 1978 and 1988 as a consequence of sea-level rise and sediment supply changes. The Isles Dernieres Restoration Plan involves increasing the height, width and area of the islands by pumping dredged sediment on land and developing intertidal and supratidal habitats. On Trinity Island in the period 1997–1998, 3.7 million m^3 of sediment was used to build new dune and back-barrier marsh areas. The island increased in height by 0.6–2.4 m, and extended in total shoreline length by 7 km. Cost of the project was $7.6 m. Subsequent monitoring shows that 51 per cent of the new sediment was lost in the period 1998–2006, mainly as a result of hurricane erosion. Biodiversity, however, has been maintained in the new habitats.

SUMMARY

- The main wave-dominated sub-environments are the shoreface and beach. Together, they make up coastal barriers and these landforms can be considered the main depositional elements of wave-dominated coasts. The shoreface is the upper part of the continental shelf that is affected by contemporary wave processes and can be described by the equilibrium shoreface profile. Most net sediment transport on the shoreface takes place by waves and currents. Factors affecting shoreface sediment transport are wave asymmetry and the presence of bedforms.
- The planform shape of wave-dominated coasts is strongly controlled by wave refraction. Swash-aligned coasts are oriented parallel to the crests of the prevailing incident waves (e.g. equilibrium bays), while drift-aligned coasts are oriented obliquely to the crest of the prevailing waves (e.g. spits).
- Beaches are wavelain deposits of sand or gravel found along marine, lacustrine and estuarine shorelines. Beach morphology is highly variable with smaller-scale features such as beach cusps, nearshore bars and rip channels often present. The assemblage of these features and the morphodynamic process regime (reflective *versus* dissipative) allows beaches to be classified into distinct types.
- Barriers are elongated, shore-parallel sand bodies that extend above sea level. Barrier morphology is affected by sea level, sediment supply, substrate

gradient, geological inheritance, wave energy, tides and wind. Sea level and sediment supply are the most important factors influencing barrier evolution and morphology. Depending on the combination of these two factors barriers may be prograding, stationary or retrograding.

- Management of the shoreface, beaches and barriers is based on an understanding of their physical processes. Hard and soft engineering strategies are often used to reduce shoreline erosion and maintain sediment volume. Both strategies are somewhat effective but can be associated with negative consequences, as by reducing longshore sediment transport. Barrier island restoration is an example of how coastal management can be used to reduce hurricane impact, increase biodiversity, and protect urban infrastructure.

Reflective questions

These questions are designed to test your comprehension of material covered in this chapter. Suggested answers to these questions can be found on this book's website.

8a. What are some of the difficulties associated with identifying the depth of closure?

8b. Describe the relationship between beach steepness and nearshore, alongshore and rip currents.

8c. If the rate of sea-level rise is the same, what will be the major differences in barrier island dynamics between a sediment-rich and a sediment-poor coast?

8d. What are the major differences between the impacts of hard and soft engineering works on coastal systems?

Further reading

Komar, P.D., 1998. *Beach Processes and Sedimentation* (2nd edn). Prentice Hall, New Jersey. (Chapters 7–11 of this text provide an advanced review of beach processes and morphology.)

Short, A.D. (ed) 1999. *Handbook of Beach and Shoreface Morphodynamics*. Wiley, Chichester. (This has useful reviews on shoreface, beach and barrier morphodynamics.)

Roy, P.S., **Cowell**, P.J., Ferland, M.A. and **Thom**, B.G., 1994. 'Wave-dominated coasts.' In: R.W.G. Carter and C.D. Woodroffe (eds) *Coastal Evolution*, Cambridge University Press, Cambridge, 87–120. (Comprehensive chapter on the morphodynamics of wave-dominated coasts. The emphasis is on barrier morphology and processes.)

Trenhaile, A.S., 1997. *Coastal Dynamics and Landforms*. Oxford University Press, Oxford. (Chapters 4, 5 and 6 of this text give brief but comprehensive overviews of beaches and barriers.)

CHAPTER 9
COASTAL SAND DUNES

AIMS

This chapter examines the formation processes and landforms associated with coastal sand dunes. Processes of sand transport by wind are first discussed. The landforms that result from wind transport and deposition are then examined, with a particular focus on primary and secondary dunes. The morphodynamic behaviour of dune systems over different temporal scales is then discussed. The chapter concludes by considering coastal dune management.

9.1 Introduction

Coastal sand dunes are accumulations of wind-blown sand usually found in association with sandy beaches and estuaries. They are developed in particular on progradational, sand-rich coasts located on the downwind side of large ocean basins (Figure 9.1) and adjacent to areas that have been glaciated. This combination of factors is important in the formation of sand dunes on the coasts of northwest Europe (in particular the British Isles, Poland, Netherlands, Denmark), and northwest USA (Washington, Oregon, California). The lack of extensive dune development at high latitudes is due to the lack of sandy sediments and the action of frost and sea ice. Their relatively modest development at low latitudes (e.g. northeast Australia, Florida, Brazil) is limited by the effect of dense backbeach vegetation, unsuitable sediments and relatively low wave energy levels. Desert dunes formed inland can also meet the coast, such as in Peru and Namibia. The formation of coastal sand dunes requires both an ample supply of sediment and an energetic wind climate capable of moving the sediment and accumulating it into dune forms. Sand dunes therefore start to accumulate when the competency of the wind for sediment transport decreases. Other local factors contributing to dune development include topography, wave climate, tidal range, sea-level change and presence of intertidal litter.

The growth of coastal sand dunes represents the net accumulation of sediment landward of the beach from where the sediment is derived, and therefore reorganisation of the coastal sediment budget with respect to where sediment is stored (Figure 9.2). Furthermore, once sediments accumulate in sand dunes

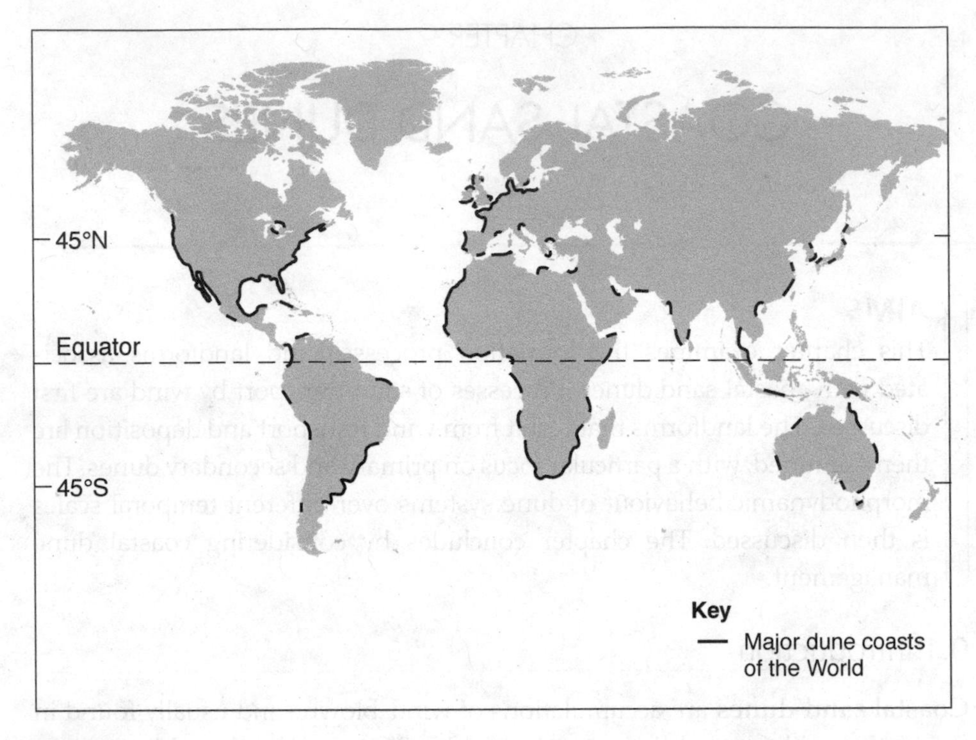

Figure 9.1 Global distribution of major coastal sand dune systems. (From Carter *et al.*, 1990.) (Copyright © 1990 John Wiley & Sons, reproduced with permission.)

they are unlikely to be released back to the beach unless the dune scarp face is eroded by waves. This means that sand dunes often represent a long-term sediment sink. Interactions between beach and dune systems have been examined in a number of studies. For example, Short and Hesp (1982) show that sediment transport from the beach determines the potential size of dunes, with the highest transport rates (and largest dunes) coming from dissipative beaches and lowest rates from reflective beaches (Table 9.1). Psuty (1992) describes different scenarios of beach–dune interaction based on whether sediment is building up or eroding away (Figure 9.3). Changes in beach and dune sediment volumes may reflect external forcing by waves (beach sediment buildup), wind (dune sediment buildup), longshore processes, human activity, or internal (autogenic) factors such as dune vegetation succession. Aagaard *et al.* (2004) use a sediment budget approach to compare changes in beach and nearshore sediment storage with changes in dune sediment storage over time scales of seasons to decades. They show that sediment volume changes on beaches and dunes on the Danish North Sea coast are very similar, in other words that the sediment cell is virtually closed and that sediment merely moves between the onshore and offshore. Beach–dune interactions are discussed in more detail in Section 9.5.

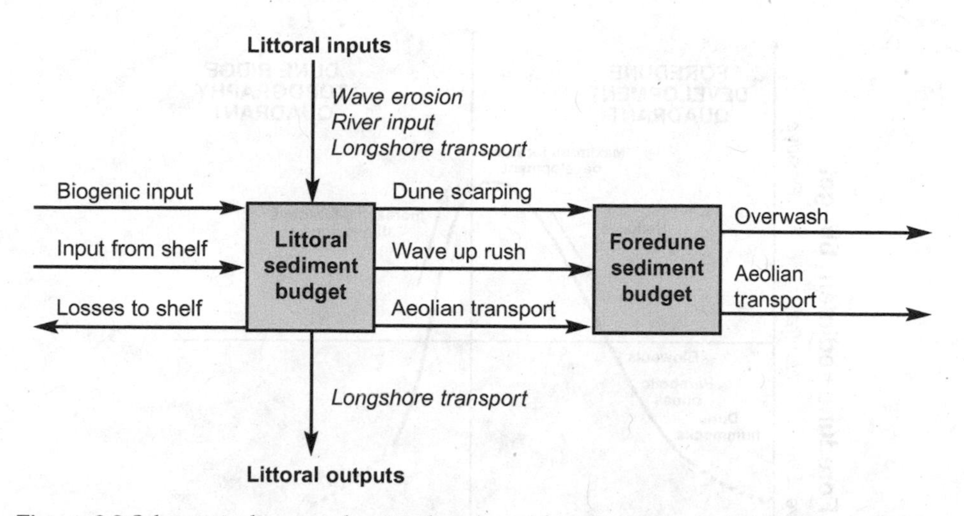

Figure 9.2 Schematic diagram showing beach and foredune sediment interactions. (From Davidson-Arnott, 2010.) (Copyright © 2010 Cambridge University Press, reproduced with permission.)

Coastal dunes help protect the coast from erosion by providing a buffer to extreme waves and winds. Extreme storms result in elevated water levels and beach erosion, and may lead to coastal flooding. However, if a well-developed dune system is present, storm waves will dissipate their energy through dune erosion, reducing the risk of coastal flooding. Any sediment eroded from the dune system and transported offshore will eventually return to the beach under fair weather

Table 9.1 Summary of the Short and Hesp (1982) model relating beach morphodynamic state to dune system dynamics.

Morphodynamic beach state	Frequency: type of dune scarping	Potential aeolian transport: foredune size	Probability of foredune destruction (per 100 years)	Nature of dominant dunes
Dissipative	Low: continuous scarp	High: large	Moderate	Large-scale transgressive dune sheets
Intermediate	Moderate: scarps in rip embayments	Moderate: moderate	Low to high	Parabolic dunes and blowouts
Reflective	High: continuous	Low: small	Low	Foredune scarping and small blowouts

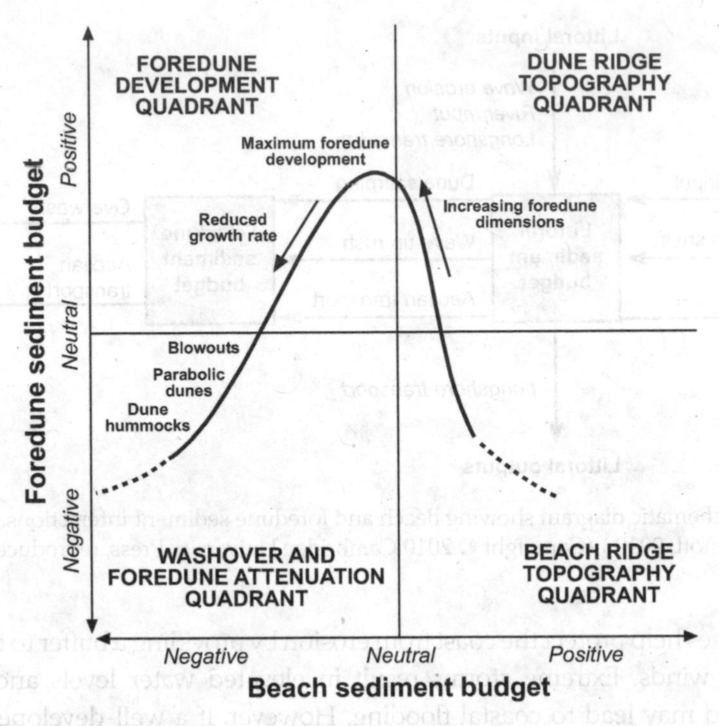

Figure 9.3 Morphologic evolution of the foredune as a function of beach and foredune sediment budget. (Modified from Psuty, 1992.)

conditions, and may act as a sediment source for renewed dune growth. Maintenance of coastal dune systems, their morphological integrity and ecosystems, is therefore an important component of coastal protection and management. This chapter first discusses how coastal sand dunes are formed with respect to aeolian sediment transport, then describes the different dune morphologies.

9.2 Formation of coastal dunes

The formation and dynamics of sand dunes deposited by wind are different to those of dunes of similar morphology deposited by water. Air has lower fluid density than water (1.22 kg m^{-3} compared to 1000 kg m^{-3}), and this difference is significant because it impacts on capacity for sediment transport. For a given grain size, air needs to be moving more quickly than water in order to initiate transport. However, air is more efficient at sediment sorting than water. These factors mean that air- and water-deposited sediments can be distinguished from each other based on their sedimentary properties. Processes of sediment transport by water were discussed in Section 5.5. Here we discuss processes of aeolian sediment transport, but we reproduce the relevant boundary layer equations here (Section 5.4.1) because they are used in our discussion.

9.2.1 Identifying wind flow properties

Wind flow properties including wind speed, turbulence and shear stress change from the land surface upwards in response to variations in topography and vegetation which affect boundary layer roughness. Sediment transport by both water currents and wind arises from the shear stress imposed on the bed by the flow. Wind shear stress τ_0 is related directly to the shear velocity u_*

$$u_* = \sqrt{\frac{\tau_o}{\rho}} \qquad \textbf{(9.1)}$$

where ρ is air density. The shear velocity u_* cannot be measured directly, but can be derived from the vertical velocity profile using the Law of the Wall

$$U_z = \frac{u_*}{\kappa} \ln\left(\frac{z}{z_o}\right) \qquad (9.2)$$

where U_z is the average wind speed at a height z above the ground surface, κ is von Karman's constant (equal to 0.4) and z_o is surface roughness length. This results in a logarithmic velocity profile. Surface roughness also causes drag. The drag coefficient C at height z can be calculated where

$$C_z = \left(\frac{u_*}{U_z}\right)^2 \qquad \textbf{(9.3)}$$

The values of roughness and therefore drag vary over an order of magnitude seasonally in response to vegetation growth. Onshore and offshore winds experience different values of drag due to variations in the roughness of the different surfaces over which they pass. Equation 9.2 can be rearranged and solved for different height values of U_Z, here termed U_{Z1} and U_{Z2}, which can help identify how the shear velocity changes with height above the land surface, where

$$u_* = \frac{\kappa(U_{Z2} - U_{Z1})}{\ln\left(\frac{Z_2}{Z_1}\right)} \qquad \textbf{(9.4)}$$

The relationship between roughness and shear velocity is shown for different coastal environments in Figure 9.4. Tidal flats have the lowest and coastal dunes the highest roughness values.

Wind speed and direction can be measured in the field using a number of different anemometer instruments, usually at a standard height of 1.5 m or 2 m. The most simple cup anemometer comprises three or four cups mounted horizontally, which rotates in proportion to wind speed. The windmill or propeller anemometer comprises a propeller that turns and faces the incoming

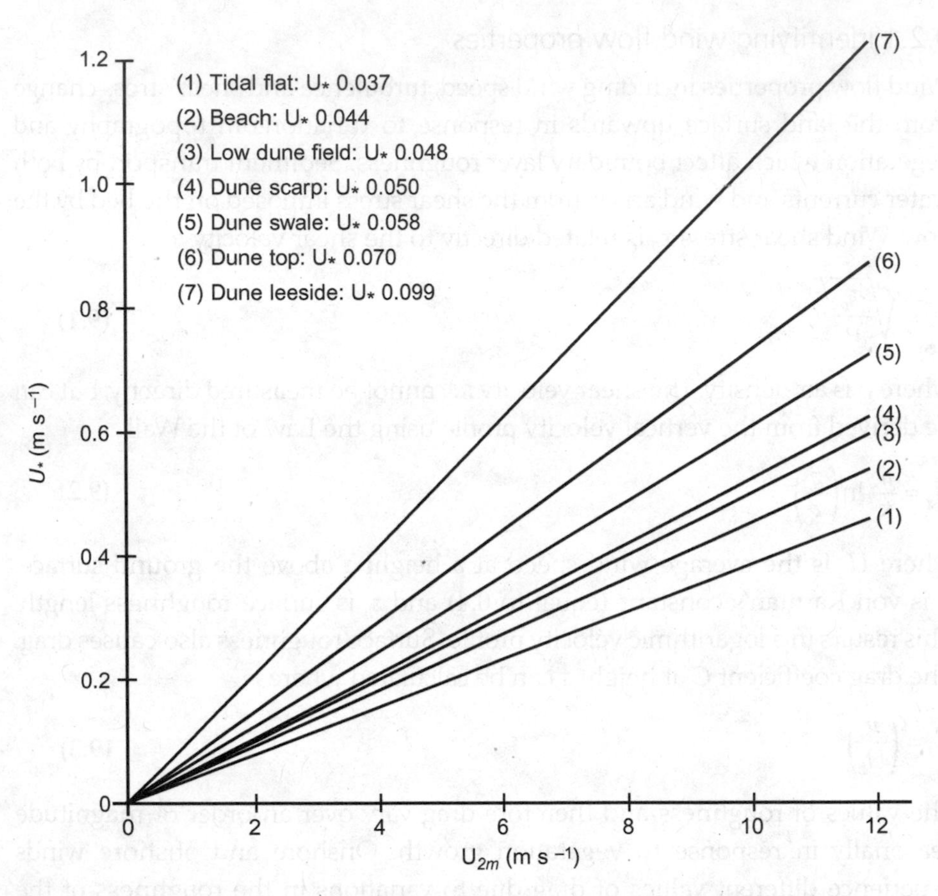

Figure 9.4 Graph showing the relationship between surface roughness Z_0, shear velocity u_* and wind speed at 2 m height U_{z2m}. (Modified from Hsu, 1977.)

wind, with a vane that balances the propeller end and which is aligned in the direction of wind flow. Other fixed instruments use indirect methods to measure wind speed. Hot-wire anemometers are based on the principle that the passage of air over electrically-heated wires will cause cooling of the wires and changes in electrical resistance that are proportional to wind speed. Sonic anemometers measure the time taken for pulses of ultrasonic waves to travel between two pairs of transducers which vary according to wind velocity and air temperature. Wind speeds recorded by an anemometer are standardised according to

$$U_{ZR} = U_{ZM}\left(\frac{Z_R}{Z_M}\right)^n \tag{9.5}$$

where U_{ZR} is wind speed at the reference height Z_R which is usually 10 m, U_{ZM} is wind speed at the anemometer height Z_M, and n is an empirical exponent that varies between 1/11 and 1/7.

Worked Example 6

9.2.2 Getting sediment grains moving

If a sandy surface is exposed to an increasing wind field, aeolian sediment transport will occur, removing sediment from that surface. The general term for this sediment loss is **deflation**. The velocity at which aeolian sediment transport starts to take place is termed the threshold velocity. There are several alternative equations for calculating the **threshold velocity**, but the most important variables involved are wind speed and sediment grain size. A modern understanding of aeolian sediment transport and dynamics comes from the work of Ralph Bagnold (1896–1990), a British army officer who worked on desert dunes in North Africa.

In order to calculate the threshold velocity, we first need to calculate the threshold shear stress u_{*t} for sediment transport. Following Bagnold (1941) this is

$$u_{*t} = A\left[gd\left(\frac{\rho_s - \rho}{\rho}\right)\right]^{0.5} \tag{9.6}$$

where A is an empirical coefficient taken to be a value of 0.1, g is gravity, d is sediment grain size, ρ is air density and ρ_s is sediment density. The threshold wind velocity at height z above the surface (rather than the threshold shear

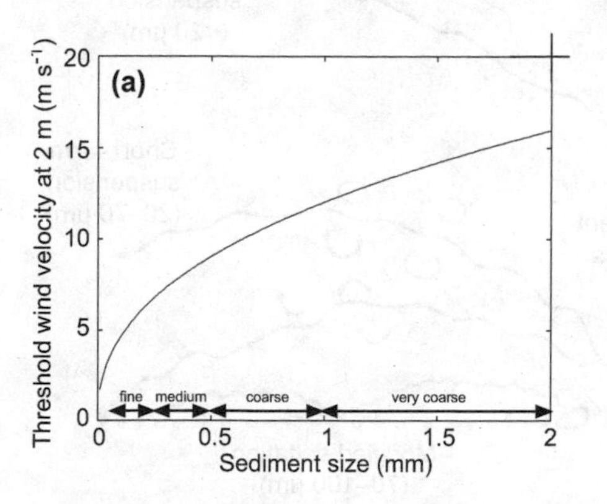

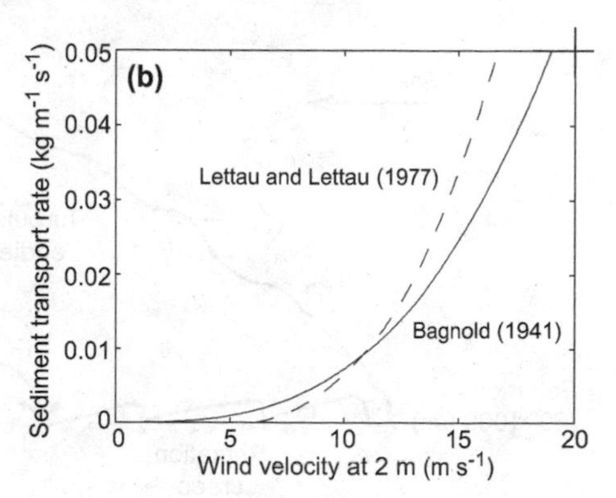

Figure 9.5 (a) Relation between threshold wind velocity (measured 2 m above sand level) and sediment size predicted following Bagnold (1941). (b) Predicted aeolian sediment transport as a function of wind speed (measured 2 m above sand) for a sediment size of 0.2 mm according to Bagnold (1941) and Lettau and Lettau (1977).

velocity) can be derived by inserting the threshold shear velocity u_{*t} into the Law of the Wall (Equation 9.2) and assuming a value for the roughness length z_o. Figure 9.5 shows the threshold wind velocity at 2 m above the sand surface as a function of sediment size derived using Equations 9.2 and 9.6 and assuming $z_o = D/30$. As coastal dunes are commonly made up of fine to medium sand, the typical threshold wind velocity is in the range 4 to 8 m s^{-1}. Thus no sediment is expected to be transported by wind speeds less than 4 m s^{-1}. Threshold velocities are usually measured and/or calculated based upon average wind conditions, but in reality, sediment is usually entrained by peak-rate, intermittent and gusty winds, so averaged threshold conditions may not necessarily be the best approximation for the time-period or conditions under which sediment transport occurs.

9.2.3 Processes of aeolian sediment transport

Once the threshold velocity is exceeded and sand grains are entrained into the windstream, they can potentially travel in one of four distinct ways (Figure 9.6). Grains transported as **creep** roll or slide in a downwind direction but remain in contact with the bed, analogous to traction transport (Section 5.5.2). Grains in **reptation** transport make small 'hops' along the bed whereas those in **saltation** make larger 'leaps'. Grains in **suspension** transport are maintained within the windstream by turbulence. The smaller the grain size, the higher the grain can be lifted into the windstream, which leads to sediment sorting. In terms of coastal dune formation, creep is important in transporting grains up the wind-

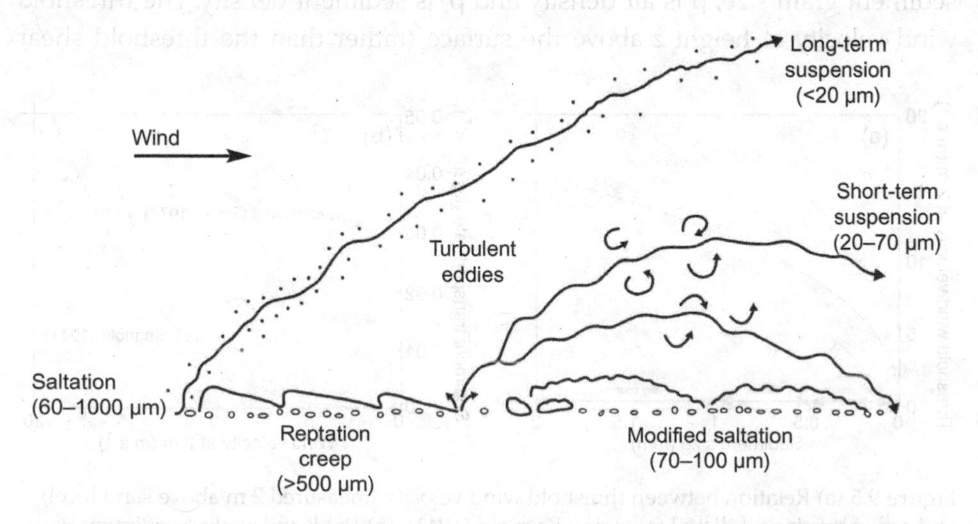

Figure 9.6 Schematic diagram showing the different mechanisms of aeolian sediment transport. (From Davidson-Arnott, 2010.) (Copyright © 2010 Cambridge University Press, reproduced with permission.)

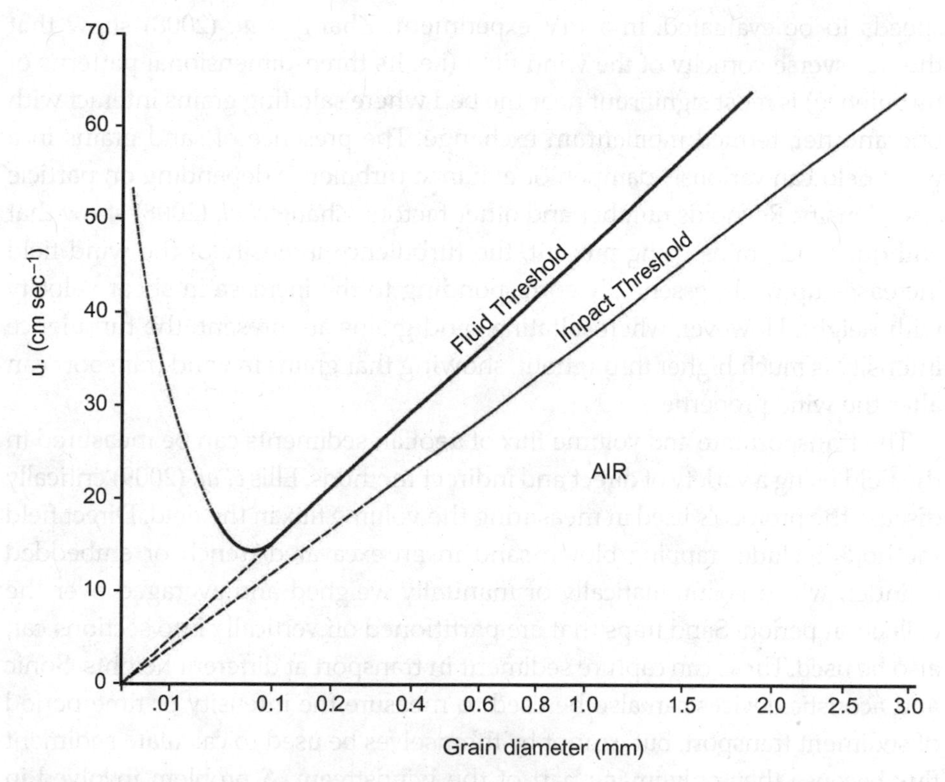

Figure 9.7 Graph showing the relationship between threshold velocity and grain size. (After Bagnold, 1941, from Davidson-Arnott, 2010.) (Copyright © 2010 Cambridge University Press, reproduced with permission.)

ward slope of dunes, but saltation is the most important process, explaining around 95 per cent of total transported sediment volume. Suspension is least important as it only affects the smallest silt and clay particles.

Two threshold shear velocities are recognised: a **fluid threshold** which refers to the velocity required to get sediments moving from rest, and an **impact threshold** which refers to the velocity required to keep moving grains moving by a process of ballistic impact (Figure 9.7). The impact threshold refers to a situation where saltating grains hit the bed, causing resting grains to be ejected into the windstream. The lower impact threshold velocity arises because, once sediment transport starts, the system can experience self-forcing such that transport can continue at around 80 per cent of the fluid threshold velocity.

The dynamic behaviour of sand grains once in motion can be examined in a wind tunnel using a method called **particle image velocimetry** (PIV). This method consists of taking multiple high-speed photographs of grains that are in transport. Comparison of the positions of individual grains on successive images allows the grains' three-dimensional trajectories, interactions and speeds to be evaluated. In a PIV experiment, Zhang *et al.* (2008) show that the transverse vorticity of the wind field (i.e. its three-dimensional patterns of turbulence) is most significant near the bed where saltating grains interact with one another, termed momentum exchange. The presence of sand grains in a wind field can variously dampen or enhance turbulence depending on particle size, density, Reynolds number and other factors. Zhang *et al.* (2008) show that without sand grains being present, the turbulence intensity of the wind field increases upwards, essentially corresponding to the increase in shear velocity with height. However, when saltating sand grains are present, the turbulence intensity is much higher throughout, showing that grains in wind transport can alter the wind properties.

The transport rate and volume flux of aeolian sediments can be measured in the field using a variety of direct and indirect methods. Ellis *et al.* (2009) critically discuss the protocols used in measuring the volume flux in the field. Direct field methods include trapping blown sand in an excavated trench or embedded cylinder, which is automatically or manually weighed and averaged over the collection period. Sand traps that are partitioned off vertically into sections can also be used. These can capture sediment in transport at different heights. Sonic and acoustic devices can also be used to measure the intensity or time period of sediment transport, but cannot by themselves be used to calculate sediment flux because they only image part of the windstream. A problem involved in measuring the sediment flux is that it is difficult to evaluate how the sediment flux may have changed over long time periods. One means by which this can be done is to examine the geomorphological evidence for sediment that is in transit across the beach surface, forming distinctive features called **ventifacts** (Case Study 9.1).

Case Study 9.1 Ventifacts: The geomorphic expression of sand in transit

Ventifacts are boulders or bedrock protrusions that have been abraded by wind-blown sand. They are often found along coasts because of the presence of loose sand, strong winds and absence of stabilising vegetation. Ventifacts are relevant to the understanding of wind transport along coasts because abrasion creates distinctive features on ventifact surfaces from which the direction of wind transport can be inferred. Furthermore, ventifacts are a unique record of blown sand that is in transit from one part of the coastal system (e.g. the beach) to another (e.g. the dunes).

Abrasion generally smoothes and abrades the upwind side of a rock

Figure 9.8 Coastal ventifact developed on a granite boulder. Note the contrast between the abraded windward face and lichen-covered lee side. Wind transport is from bottom to top of the photo. (Photo: J. Knight.)

surface and keeps it free from lichen and weathering products. The abraded side is separated from the non-abraded downwind side by a sharp boundary, usually corresponding to the crest of the boulder, termed a **keel**. The orientation of the keel and facing direction of the abraded side have been used in combination to determine the direction of the greatest abrading wind (Figure 9.8). In northwest Ireland, Knight (2005) showed that across a boulder-strewn sandy beach, inferred abrading wind direction varies by 90° over a distance of 30 m alongshore. The most likely explanation is that boundary layer winds are deflected by the presence of sand dunes and other topographic variations. In addition, the direction of the abrading wind that is inferred from ventifacts can be very different to today's wind climate. This suggests that synoptic wind patterns (or sediment supply) have changed over time. For example, based on sites elsewhere in northwest Ireland where ventifacts have been partially overgrown by recent lichens, Knight and Burningham (2001) suggested that stronger winds during the Little Ice Age led to greater abrasion and ventifact development at this time. This is also the same period when coastal sand dunes were forming in the region.

9.2.4 Aeolian sediment transport rate and volume flux

Numerous empirical and semi-empirical equations have been used to predict **aeolian transport rates**, also know as the **volume flux** (Table 9.2). Chapman (1990) and Sherman *et al.* (1998) evaluate the robustness and predictive capacity of many of these equations. The model of Bagnold (1941) is probably the most widely used

$$q = C\left(\frac{d}{D}\right)^{0.5} \frac{\rho}{g} u_*^3 \qquad (9.7)$$

where q is the sediment transport rate (in kg m^{-1} s^{-1}), C is a constant (see Table 9.2 for values), d is the mean grain diameter and D is a reference grain diameter of 0.25 mm. A major weakness of Equation 9.7 is that it does not include a threshold velocity although, as outlined above, this concept is not without its problems. As a result, Equation 9.7 predicts sediment transport even for below-threshold wind velocities. The model of Lettau and Lettau (1977), however, takes into account the threshold velocity

$$q = C\left(\frac{D}{D_r}\right)^{0.5} \frac{\rho}{g} (u_* - u_{*t}) u_*^2 \qquad (9.8)$$

where C has a value of 4.2 and u_{*t} is given by Equation 9.6. Figure 9.5b compares aeolian sediment transport rates according to Equations 9.7 and 9.8.

Table 9.2 Alternative equations for calculating the mass flux of aeolian transport q. (From Sherman *et al.*, 1998 and Woodroffe, 2002.)

Equation	Source	Comments
$q = C\frac{\rho}{g}\left(\frac{d}{D}\right)^{0.5} U_*^3$	Bagnold, 1941	C = 1.5 for uniform sand, 1.8 for graded sand, 2.8 for mixed sand
$q = C\frac{\rho}{g} U_*^3$	Chepil, 1945	C = 1–3.1
$q = C\frac{\rho}{g}(u_* - u_{*t})(u_* + u_{*t})^2$	Kawamura, 1951	C = 2.78
$q = z\frac{\rho}{g}\left(\frac{d}{D}\right)^{0.75} U_*^3$	Zingg, 1953	z = 0.83
$q = \phi \rho_s g\left(\frac{(\rho_s - \rho)}{\rho} g d^3\right)^{0.5}$	Kadib, 1965	Where ϕ is a transport intensity function based on Einstein's bedload transport model
$q = H\left(\frac{U}{(gd)^{0.5}}\right)^3$	Hsu, 1971	$H = \exp(-0.42 \times 4.91d) \times 10^{-4}$
$q = B\left(\frac{\rho}{g}\right) u^n \left(\frac{d_{50}}{D_{50}}\right)^{0.5} (u_* + u_{*c})^2 (u_* + u_{*c})$	Horikawa *et al.*, 1983	

Figure 9.9 Sand streaks on a beach at Vejers, Denmark. (Photo: M.G. Hughes.)

Practically all aeolian sand transport equations consider the sediment transport rate proportional to the shear velocity cubed ($q \alpha u^3$) (Table 9.2). This implies that, once sediment is moving, a small increase in wind velocity can cause a large increase in the sediment transport rate. For example, only a 25 per cent increase in wind speed will cause a doubling of sediment transport.

Aeolian sand transport equations are generally derived and validated under 'ideal' conditions, and therefore provide an estimate of the maximum transport rate and volume flux that may occur under these conditions. However, the use of time-averaged wind velocity measurements do not account for wind unsteadiness (turbulence and gustiness). Sherman *et al.* (1998) note that a model that performs well in one location may not perform well at another, so the choice of equation used to calculate the volume flux is somewhat subjective. In natural beach and dune systems where there is a grain size range, the smallest grain sizes will be entrained and transported first, leaving the larger grain sizes behind. If wind conditions remain constant, the sediment flux will decrease over time as the volume of sediment suitable for transport is depleted, meaning that such systems become **supply-limited**. In addition, the remaining sediment population will both coarsen and become better sorted, and may result in formation of a coarse **lag**.

Unsteady (turbulent or gusty) wind conditions are the norm rather than the exception in most coastal environments. As a result sand transport is highly episodic. The expression of this episodic transport is the formation of discontinuous and short-lived particle-charged winds, termed **sand streaks** or streams (Figure 9.9). These are sudden 'bursts' of sand transport, usually located within a few tens of centimetres of the beach surface, in response to a single wind gust, which transport sediment over a few tens of metres before redepositing it.

Although the effects of wind gusts are not well known, it is likely that they are geomorphologically significant because they can transport large quantities of sediment over very short time periods.

9.2.5 Factors affecting aeolian sediment transport

Threshold velocities are a useful approximation for the conditions necessary for sediment transport, but other site-specific factors are also important. Measurements of **wind velocity** from anemometers sited at specific heights cannot describe variations in wind velocity or direction at all heights in the windstream, and do not capture episodic gusty events. Most anemometers are also unable to measure how much **turbulence** the windstream has. **Air temperature** and other atmospheric properties are also important because they affect air density. There is an inverse relationship between air temperature and density. For example, at a standard value of air pressure of 1013 hPa, at a temperature of –40 °C air density is 1.515 kg m^{-3}, whereas at a temperature of +15 °C air density is 1.225 kg m^{-3}. Based on Bagnold's equation (Equation 9.6), a grain of 2 mm diameter can be lifted to 2 m height by a wind velocity of 38.4 m s^{-1} if the air temperature is –40 °C, but lifting the same grain to the same height if the air temperature is +15 °C requires a wind velocity of 42.5 m s^{-1}. As a result, colder air can transport sediment much more effectively than warm air.

Land surface **roughness** is also important because it determines shear stress and turbulence which influence lift. Nickling (1978) found that rates of dust transport correlate more strongly with atmospheric turbulence than with wind velocity. Turbulence at the ground surface causes bed erosion, which creates bed relief, which then creates more turbulence. This positive feedback can produce sediment sorting in the direction of wind flow, and is one of the conditions of instability that can lead to the formation of ripples. Features present on the **sand surface** are also significant factors influencing the likelihood of wind transport. In hot coastal environments with high rates of evaporation the sand surface may develop a concentration of salt crystals. These can help 'stick' the sand grains together, increasing the threshold velocity. Gravel and marine shells on sandy beaches can protect the surrounding sand from wind transport. Along estuaries algal mats often develop on fine-sand surfaces, and these perform a similar protective role.

Wet beach surfaces are most commonly exposed during the falling tide. High beach **surface moisture** content, like salt crystals, can 'stick' sand grains together and increase the threshold velocity for entrainment. Beach moisture content nonlinearly affects threshold velocity. Davidson-Arnott *et al.* (2008) measured the low-tide moisture content of a sandy beach on Prince Edward Island, Canada. They found that moisture values vary spatially from around 0.5

to 2 per cent with the mid-beach area tending to be the driest location. Beach topography such as ridges and runnels also influence moisture content. The sediment transport rate on a moist beach is around half of that on a dry beach because of the higher threshold velocity and also because the aeolian transport system is more intermittent. Field observations by Wiggs *et al.* (2004) show that the critical moisture threshold, determining whether sediment transport takes place or not, is around 4–6 per cent, and that a value of 2 per cent had little or no impact on sediment transport rate. This suggests that sediment transport by wind can potentially come from all parts of the beach. They also showed that, initially, moisture level exerted a dominant control on sediment transport rate, but that over time (from minutes to hours) wind velocity became more important after the sand surfaces began to dry out.

Surface moisture also influences the fetch distance over which sediment transport can take place, with moist surface keeping sediment unavailable for entrainment. Here, **fetch distance** refers to the unimpeded length of beach or dune surface over which sediment entrainment and transport can potentially take place. The fetch distance is generally high if the beach has a wide supratidal zone (delimited by the high tide mark), and/or a wide intertidal zone. Beach width, steepness and tidal range are therefore significant influencing factors. However, if the wind blows oblique to the shoreline then beach width is not an accurate measure of fetch distance. Fetch distance is significant, however, because saltation transport systems only become fully established once the wind has blown over a certain beach length. The distance over which saltation becomes established varies with wind angle and beach geometry, as well as in response to sedimentary factors. Where the fetch distance is too short for winds to become fully saturated with sand, the sediment transport can be said to be **fetch-limited**. Udo *et al.* (2008) show that under field conditions the sediment flux can decrease by an order of magnitude if the wind direction (and therefore fetch distance) changes from alongshore to cross-shore.

9.2.6 Aeolian sediment transport and air flow around obstacles

Wind flow is greatly affected by topography over all spatial scales. On the upper beach sand starts to accumulate around variations in microtopography, litter, seaweed and vegetation as a result of alteration of surface shear velocity and therefore sediment transport and deposition patterns. With respect to coastal dunes, flow acceleration on the upwind side and along the flanks of the obstacle is caused by compression of the windstream where lines of equal wind velocity (isovels) converge. The isovels cause increased shear stress towards the apex of the obstacle and may result in scour and formation of blowouts. With a continuous dune ridge and an onshore wind, the air has to be 'lifted' over the

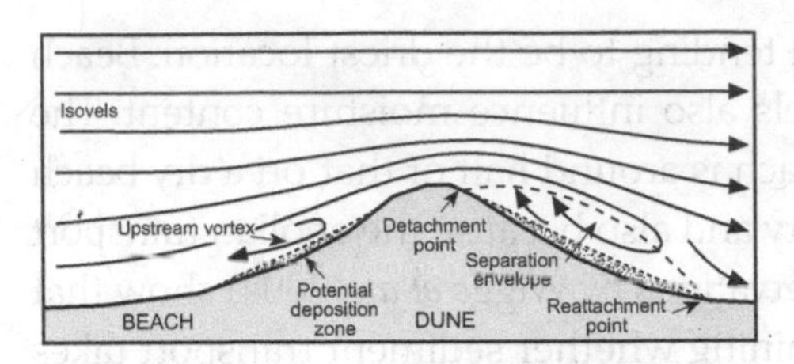

Figure 9.10 Movement of air over a dune as indicated by isovels (lines of equal wind speed). (Modified from Carter, 1988.)

developing dune (Figure 9.10). There may be a zone of flow deceleration on the leeside of the dune as a result of divergence of isovels by flow expansion, leading to sediment deposition here. The zone of this leeside flow separation can extend downwind for a distance five to ten times the obstacle's height. Lynch *et al.* (2010) found that a dune with a height of 11.4 m produced flow separation, whereas a dune with height 6.6 m did not. If a dune ridge is discontinuous and characterised by gaps, wind flow will tend to concentrate on these gaps, causing increased flow velocity and shear stress at these locations. The ensuing scour may widen and deepen the gap, demonstrating positive feedback.

The presence of vegetation is a key factor contributing to the formation of sand dunes. Isolated clumps of vegetation act as small obstacles disturbing the wind flow velocity. Larger stands of vegetation more significantly change the shear velocity profile. The interaction of wind with vegetation is shown in Figure 9.11. The uppermost surface of the vegetation stand acts as a false

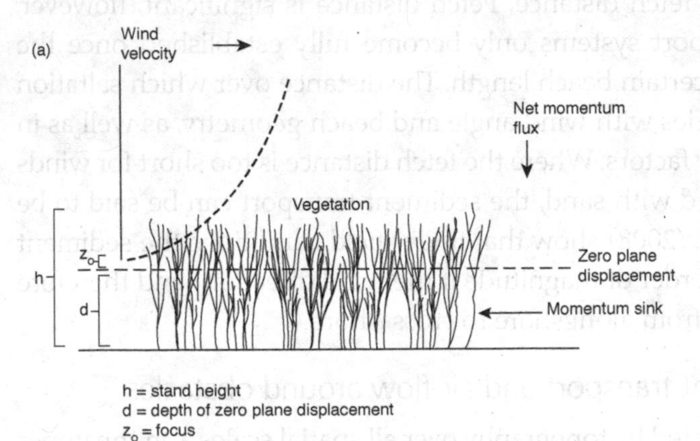

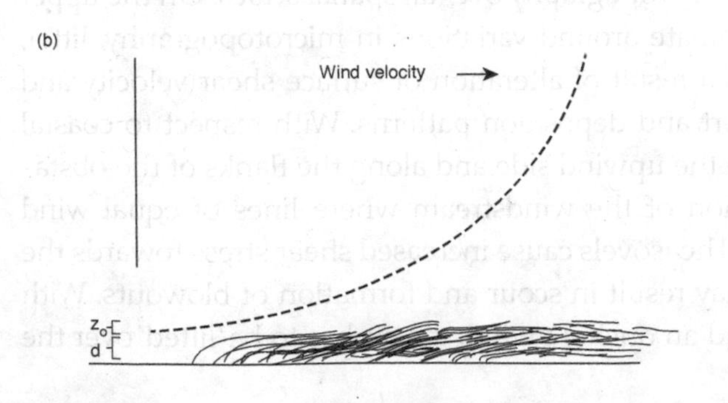

Figure 9.11 Illustration of wind velocity profiles over vegetation stands under (a) normal and (b) high wind speeds. (From Carter, 1988.) (Copyright © 1988 Academic Press, reproduced with permission.)

boundary layer, termed the zero plane displacement, which is located around two-thirds of vegetation stand height. Wind strength within the vegetation stand, however, is very low, facilitating sediment deposition. As wind strength increases the vegetation stand is progressively flattened and the zero plane is located closer to the land surface.

9.3 Coastal dune morphology

Vegetation density and dune development are closely linked through the role of vegetation in decreasing wind velocity and facilitating sediment deposition. Coastal dunes can be broadly classified into two types:

- **Primary dunes** are found closest to the shoreline and are significantly affected by wave processes (e.g. overwashing, storm erosion) and have a dynamic interrelationship with the beach. Primary dunes include the embryo dunes present on the backshore and foredunes that are located on the seaward edge of the dune system.
- **Secondary dunes** are located farther inland and have been dissociated from wave processes through coastal progradation. They include mature foredune ridges, blowouts and transgressive dunes (which includes parabolic dunes).

The morphodynamics of secondary dunes are very similar to those of terrestrial (desert) dunes, but primary dunes are fundamentally different. In fact, Sherman and Bauer (1993) contend that primary dunes are key coastal forms because they are strongly coupled to the nearshore processes on the fronting beach. The primary dune is also geomorphologically conditioned by the colonisation and succession of vegetation assemblages.

9.3.1 Primary dunes

The initial stages of dune development take place on the upper part of the beach, immediately seaward of any pre-existing dunes. The formation of **embryo dunes**, also called incipient dunes, is strongly related to the presence of intertidal litter (including seaweed and flotsam) and pioneer vegetation (Box 9.1). Embryo dunes can also form where pre-existing foredunes are eroded by waves, causing sediment to accumulate at their base (Figure 9.12a). Embryo dunes are low (1–2 m high) vegetated mounds of sand that grow over time, coalescing with each other to form a **foredune** or **foredune ridge**. The growth potential and growth rate from an embryo dune into a foredune depends on interactions between aeolian sediment transport, surface vegetation, and wind conditions. Dune growth over time implies there is sufficient sediment supply, but the direction and rate of dune growth is often limited by vegetation properties, including the speed with which pioneer species can colonise and stabilise the land surface (Box 9.1).

Box 9.1 Colonising an unstable land: Pioneering vegetation

Upper beach and dune environments have unstable, bare sand surfaces with continual sediment movement and reworking. The pioneer vegetation species that are associated with embryo dune development have adapted to succeed in this environment. The precise species involved vary in different parts of the world, but they occupy the same ecological niche and so they tend to have the same ecological characteristics. Key pioneer species on European sand dune systems include *Ammophila arenaria* (European marram grass), *Elymus arenarius* (Lyme grass) and *Euphorbia paralias* (Sea spurge). Pioneer species on North American dune systems include *A. breviligulata* (American beachgrass), *Elymus mollis* (American dunegrass), *Uniola paniculata* (Sea oats) and *Panicum amarum* (Bitter panicgrass). Pioneer species on Australasian dune systems include *Desmoschoenus spiralix* (Golden sand sedge), *Spinifex sericeus* (Beach spinifex) and *Ipomoea pes-caprae* (Goat's foot). Marram grass is a key species often introduced to coastal landscapes in order to stabilise sand dunes and reduce coastal erosion, because it can cope with high rates of sediment deposition of around 0.3–0.5 m yr^{-1}.

These different species have adopted certain physiological strategies in order to survive in coastal dune environments which are nutrient-deficient (particularly nitrogen and phosphorus), dry, unstable, and subject to salt inundation. Spreading (stoloniferous) root systems, typical of *Spinifex* and *Ipomoea*, help stabilise the land surface. Rhizomatous roots, typical of marram grass, are used by plants to synthesise their own nitrogen and so help them survive in a nutrient-poor environment. These roots can also reach downwards to the freshwater table. Rolled leaves, also typical of marram grass, reduce the surface area exposed to wind desiccation and so help conserve the plant's moisture. Succulence or the presence of fleshy leaves is typical of species such as *Euphorbia*. These leaves can store water for long periods, helping the plant cope with desiccation, have small stomata which reduce water loss, and high osmotic water pressure which can exclude salt water.

Foredunes usually have a ridge morphology that is aligned parallel to the coast (Figure 9.12b). Foredune height ranges from 1 m to several tens of metres and they can develop quickly, reaching several metres over a period of 5–10 years (Figure 9.13). Short and Hesp (1982) show that a vegetation cover of 90–100 per cent is associated with linear, beach-parallel foredunes; 45–75 per cent vegetation cover with hummocky dunes that have concave windward faces and areas of leeside

Figure 9.12 Stages of primary dune development. (a) Embryo dunes fronting a foredune ridge, County Donegal, northwest Ireland; (b) successive, linear foredune ridges fronting a beach, Rock, north Cornwall, southwest England. (Photos: J. Knight.)

accumulation; and 0–20 per cent vegetation cover with residual knolls of vegetated dunes, migrating parabolic dunes and transverse blowouts. As foredune ridges develop, by a combination of decreased sediment supply and increased surface stabilisation by plant species, their morphology continues to evolve. As a result, foredunes that are located behind the most seaward dune have irregular profiles and contain overgrown depressions that represent former blowouts. Wind flow patterns over established foredunes are complex, with wind speeds often higher when wind flows oblique to, rather than across, foredunes (Hesp *et al.*, 2005). When winds are obliquely onshore, sediment accumulates at the dune front, whereas when winds are directly onshore sediment is actively transported landward.

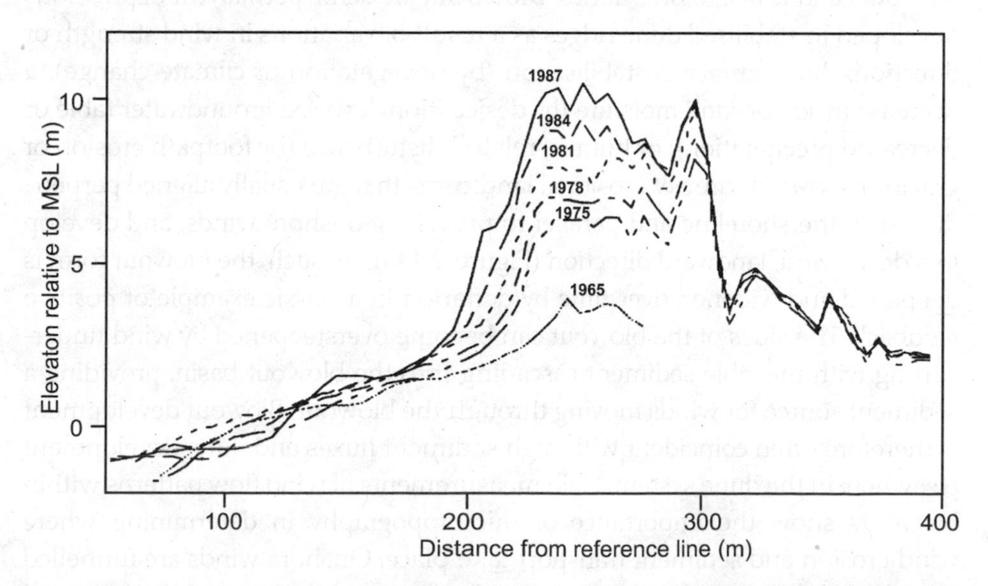

Figure 9.13 Example of foredune growth in the Netherlands. (Modified from Arens, 1994.)

There is also a close relationship between changes in beach width and foredune development. Shoreline progradation provides a sediment source for foredunes, which grow in response. Conversely, foredunes often stop growing (or are eroded) as the beach width narrows by shoreline erosion. At Magilligan Point, Northern Ireland, foredunes grow in response to the development of beach ridges. The foredunes here grow by 6–9 $m^3 m^{-1} yr^{-1}$, equivalent to a vertical growth rate of 0.3–0.7 m yr^{-1} (Carter and Wilson, 1990). Foredune growth continues until sediment supply is cut off as the dunes prograde and a new foredune is developed in front of the older one over a time period of around 35 years (Wilson, 1987). As such, foredunes develop in a seaward direction. Map evidence from Magilligan Point shows that the shoreline and foredunes have prograded by 200 m since the 1950s, so that foredunes have migrated in the direction of the sediment supply.

9.3.2 Secondary dunes

The secondary dune system is found behind the primary dune ridge and includes relict foredune ridges. The age of the dune ridges increases in a landward direction and this is reflected in a concomitant increase in vegetation maturity and soil development, discussed in the following section. Sediment supply to the secondary dunes is limited due to increased distance from source (the beach) and the presence of the primary dune and vegetation cover as an effective barrier to aeolian transport.

There are two main morphological features of secondary coastal dunes – blowouts and transgressive dunes. **Blowouts** are semi-permanent depressions developed in stabilised dune ridges as a result of variations in wind strength or direction; dune surface destabilisation (by devegetation or climate change); a decrease in soil or sand moisture (by desiccation, lowered groundwater table or decreased precipitation); or human-related disturbance (by footpath erosion or grazing). Blowouts are net erosional landforms that are usually aligned perpendicular to the shoreline and parallel to prevailing onshore winds, and develop in a downwind, landward direction (Figure 9.14a). As such, the blowout form is deepened and widened over time by deflation in a classic example of positive feedback. The sides of the blowout can become oversteepened by wind undercutting with unstable sediment cascading into the blowout basin, providing a sediment source for winds moving through the blowout. Blowout development is therefore often coincident with high sediment fluxes and dune development elsewhere in the dune system. Field measurements of wind flow patterns within blowouts show the importance of microtopography in determining where wind erosion and sediment transport take place. Onshore winds are funnelled into the blowout, resulting in flow acceleration and scouring of the base and

Figure 9.14 (a) Multiple blowouts of varying sizes and morphologies in Warnboro Sound, Australia. (b) Small developing transgressive dune field, south of Perth, Australia. (From Hesp, 1999.) (Copyright © 1999 John Wiley & Sons, reproduced with permission.)

walls of the depression (Hugenholtz and Wolfe, 2009). Sediment is often transported towards the landward end of the blowout, either as transgressive dunes within the blowout itself or deposited as fans or ramp dunes at the blowout apex.

Transgressive dunes describe unvegetated, mobile dunes of different forms that are actively migrating landwards or alongshore across a relatively flat surface (Figure 9.14b). Transgressive dune fields are best developed in high wind- and wave-energy environments with an ample sediment supply. Transgressive dunes can range from broad, flat-to-undulating sand sheets, to sand sheets that rise in elevation downwind and terminating in a slip face, termed ramp dunes. Parabolic dunes are common transgressive forms. These are crescent-shaped dunes in which the nose of the dune points downwind and the dune horns point upwind. The dune form migrates downwind as sediment is transported up the windward face of the dune nose and over the crest. Parabolic dunes are generally associated with higher dune migration rates and/or a lower vegetation cover. Bailey and Bristow (2004) examined the migration rate of parabolic dunes at Aberffraw (Anglesey, northwest Wales) (Figure 9.15). Based on air photo evidence from 1940, 1982 and 1993, the dunes migrated inland on average at 1 m yr^{-1}, and vegetation cover increased substantially. The crest of the parabolic dunes also shifted to a more northerly aspect between 1940 and 1993, suggesting that prevailing wind direction changed in this time, from southwesterly to more south-southwesterly.

Other morphological features of secondary dune systems include cliff-top dunes, flat, vegetated plains and interdune slacks. Cliff-top dunes generally take the appearance of secondary rather than primary dune forms and are built on cliff tops rather than directly adjacent to a beach (Case Study 9.2).

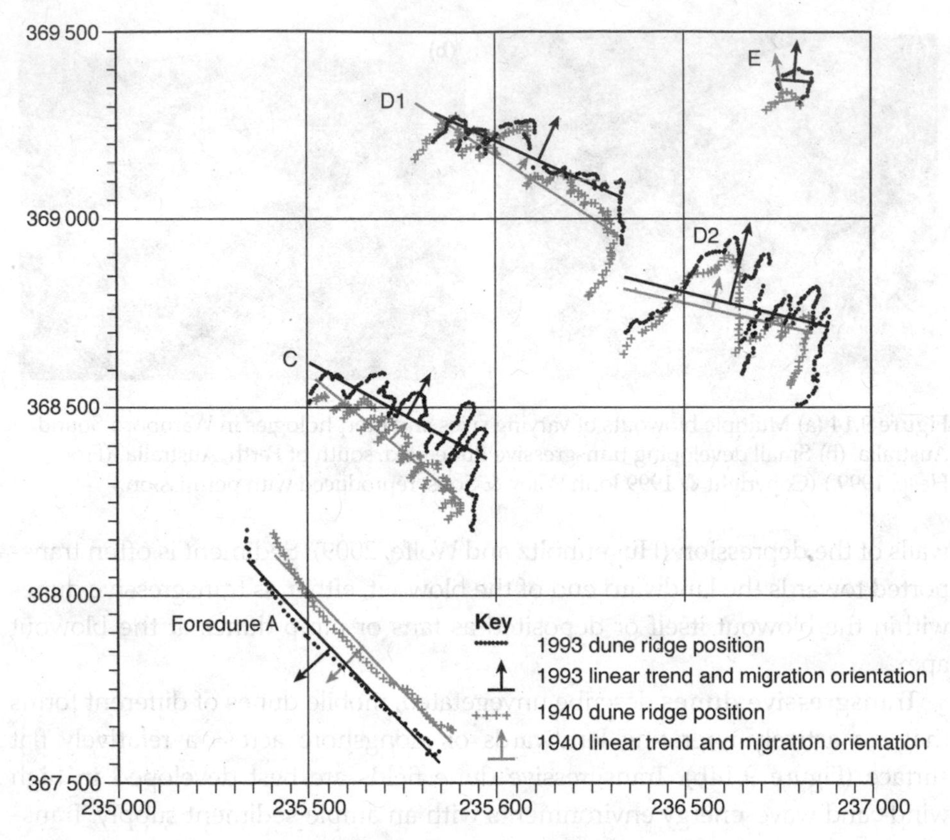

Figure 9.15 Plot of the location of major dune crests at Aberffraw, Anglesey, showing changes in crest line position between 1940 and 1993. Foredune A is moving seaward whereas parabolic dunes C, D and E are moving landward. (From Bailey and Bristow, 2004.) (Copyright © 2004 Elsevier Science, reproduced with permission.)

Flat, vegetated and calcareous plains are often developed along the margins of estuaries and adjacent to high-energy beaches, in particular in western Scotland and Ireland where they are called **machair**. Machair plains are laterally extensive and developed by a combination of deflation down to the level of the water table and washover and other processes that contribute sediment to the plain surface. Machair plains have low vegetation with high grass biodiversity. Larger hollows located between dune ridges, called **interdune slacks**, are also sometimes deflated down to the water table. During winter when the water table rises, ephemeral ponds develop with wet fringing vegetation. Interdune slacks can be distinguished from blowouts on the basis of their size, morphology and position relative to foredunes.

Case Study 9.2 Development of cliff-top dunes

Figure 9.16 Cliff-top dunes developed at Perranporth, north Cornwall, southwest England. (Photo: J. Knight.)

Coastal sand dunes that are located above sea level on the top of bedrock cliffs or platforms, termed cliff-top or perched dunes, are relatively uncommon but have been recorded in southern Australia, Portugal, Denmark, Cornwall (southwest England), Brittany (northwest France) and Ireland. Unlike dunes that accumulate landward of beaches, cliff-top dunes are physically separated from their contemporary sediment source, which is usually a sandy beach located at the cliff foot (Figure 9.16). Jennings (1967) identified a number of scenarios under which cliff-top dunes can form. These are: (1) when dunes migrate over the top of a bedrock ramp which is subsequently eroded away, forming a cliff; (2) when dunes are left stranded on the top of a bedrock cliff or ramp as sea level falls; (3) when a dune system is slowly eroded away as sea level rises; and (4) when dunes migrate seaward towards a pre-existing cliff top. These scenarios suggest that many cliff-top dunes are relict features and started forming in the mid-Holocene when sea-level rise stabilised. Recent studies, however, suggest that some cliff-top dunes have been formed much more recently – for example, in west Jutland (Denmark) cliff-top dunes are located on top of Quaternary sand cliffs. Map and luminescence dating evidence shows that these dunes accumulated in the last 300 years and likely in response to accelerated cliff erosion which provided the sediment source for the dunes (Saye *et al.*, 2006). This evidence shows that sand can be actively blown up the cliff face by strong onshore winds from a sediment source at the cliff base. It does not require sea-level changes (as in Jenning's (1967) model) in order to move the sediment source closer. Cliff-top dunes may therefore be sensitive to changes in sediment availability rather than to changes in sea level.

9.4 Coastal dune evolution

With distance inland, as dune age increases, a number of important changes take place in the dune environment (Figure 9.17). In a landward direction these include changes in sediment mobility, plant succession and soil development. As the land surface becomes vegetated the area of exposed sand decreases, which reduces sediment mobility and promotes further vegetation colonisation. Roots of pioneer species help stabilise the land surface. Pioneer species on embryo and foredunes are succeeded over time in a dune-plant succession that is termed a **psammosere**. On secondary dunes, marram grass is succeeded by perennial grasses including (in the UK) *Festuca rubra* (Red fescue), *Agrostis capillaris* (Common bent grass), *Poa pratensis* (Smooth meadow grass) and *Lotus corniculatus* (Bird's-foot trefoil). On older dunes, perennial grasses are succeeded by shrubby heathland containing *Calluna vulgaris* (Ling heather), and trees including *Quercus* (Oak), *Betula* (Birch), *Fraxinus* (Ash) and *Pinus* (Pine).

Soil development or **pedogenesis** is a key byproduct of plant growth. Dead organic matter (DOM) decomposes on and beneath the ground surface, adding nutrients and altering sediment composition and chemistry. Over time, soil depth increases. Mature dune soils may be some 1.5 m thick (Figure 9.18a). Sediment composition also changes as the level of DOM increases and the soil takes on a darker appearance. The presence of DOM also increases soil moisture retention and nutrient availability. Plant and DOM development cause changes in soil chemistry because they produce organic acids that can leach calcium carbonate fragments, such as marine and land snail shells, from dune soils. As a result, pedogenesis results in an increase in DOM and decrease in $CaCO_3$. Wilson (1987) describes the development of dune soils at Magilligan

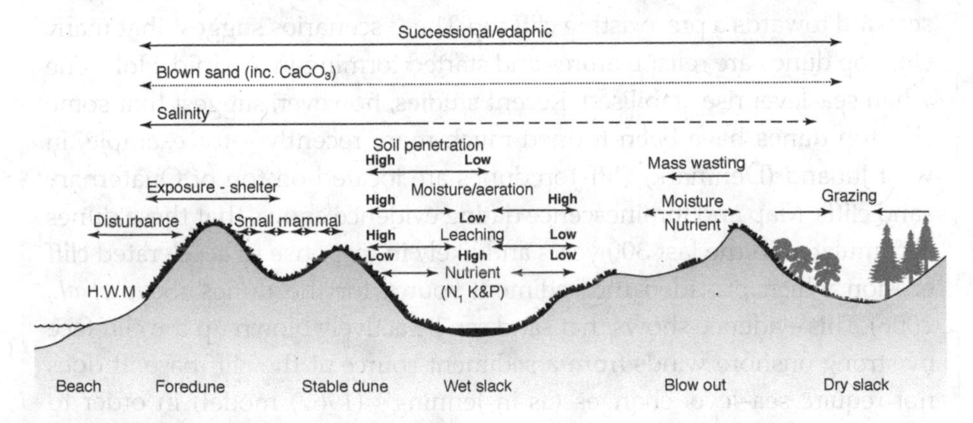

Figure 9.17 Illustration of ecological gradients on coastal dunes, in a transect from the youngest (left) to oldest dunes in the system (right). (From Carter, 1988.) (Copyright © 1988 Academic Press, reproduced with permission.)

Figure 9.18 Sand dune soils, County Donegal, northwest Ireland. (a) Podzol soil; (b) iron pan. (Photos: J. Knight.)

Point, Northern Ireland, by observing soils developed on foredunes of different ages. After only 16 years dune soils have developed a distinctive structure of a pararendzina, a typical light, alluvial-type soil. Levels of DOM increased rapidly from 0.3 per cent (after 9 years) to > 7 per cent (after 31 years). Calcium carbonate content decreased from 12 per cent (after 9 years) to 5 per cent (after 31 years). Soils became more acidic over time (lower pH values) as organic matter increased and $CaCO_3$ content decreased. Well-developed acidic podzol soils (e.g. Figure 9.18a) only developed after some 800 years.

Pedogenesis and development of podzol soils is also accompanied by changes in soil structure caused by water movement through the soil. Leaching of minerals takes place below the organic-rich upper layer of the soil, and these minerals (especially iron) are redeposited at the bottom of the soil profile where they can sometimes form a hard **iron pan** (Figure 9.18b). Iron pans are typical of podzolic, mature dune soils.

9.4.1 Dune evolution and dune stratigraphy

Alternating periods of land surface stability (soil formation) and instability (dune deposition) can lead to complex dune stratigraphies comprising **buried soils**, also termed **palaeosols**, separated by thicknesses of dune sand. Some buried soils may contain tree stumps, shell middens and anthropogenic debris indicative of past human activity on the stable dune surface. For example, the buried soil shown in Figure 9.18a contains charcoal and broken pottery fragments radiocarbon-dated to around 3700 years BP (Bronze Age) (Carter and Wilson, 1991). Using dating and stratigraphic methods, an understanding of the episodic nature of dune evolution can be established. We consider this episodic nature from different dune systems and on four different time scales: 10^4–10^5 (millennial) years, 10^3–10^4 (centennial to millennial) years, 10^2–10^3 (decadal to centennial) years, and > 1 year. These time scales correspond to different types of climate forcing on dunes, from glacial–interglacial cycles to individual storms.

On millennial time scales of glacial–interglacial cycles, sand dunes have evolved on Fraser Island, northeast Australia. Here, sand dunes reach up to 244 m above sea level and have a complex history of development from the mid-Pleistocene onwards, driven by cycles of eustatic sea-level change. Lowstand fluvial deltas during glacial stages provided a sediment source for onshore reworking during postglacial sea-level rise. Glacial–interglacial cycles also caused changes in wind strength and duration that drove dune development (Ward, 2006). Periods of dune development are separated by periods of non-deposition which are marked by palaeosols within the dune succession. Relict parabolic dunes are oriented in the direction of the prevailing southeasterly wind, but are vegetated and stabilised, indicating that they are not active at the present time. However, Levin (2011) showed that there was some dune migration inland in the period 1948–1982 when there were strong tropical cyclones, but since this period decreased cyclone intensity has led to revegetation and a decrease in area of blowouts.

On centennial to millennial time scales, dune sequences along the North Sea coast of Denmark are interbedded with buried soils. Clemmensen *et al.* (2001) described the regional-scale stratigraphy of dune evolution at Vejers, using a combination of radiocarbon dating on buried soils and luminescence dating on the interleaving dune sand. They showed that soils accumulated around 6000 years BP, 4300 years BP, 2600 years BP and 1700–1500 years BP. These periods were followed by renewed sand deposition under a wetter, cooler and more stormy climate, and probably a slightly lower sea level.

On decadal to centennial time scales, dating and stratigraphic evidence shows that the most widespread period of dune development along European coasts took place during the **Little Ice Age**. Intensification of onshore winds during the Little Ice Age period helped promote sediment transport and dune development. For example, Wilson *et al.* (2001) described the chronology of dune development in Northumberland, northeast England, using a combination of dating and sedimentary evidence. They show that sand dunes initially started to form around 4000 years BP when sea level along this coast started to fall due to isostatic factors (Figure 9.19). This initial dune development was transgressive where dunes were anchored on terrestrial sediment (i.e. dunes developed in a landward direction), and regressive where dunes were anchored on marine sediments (i.e. dunes prograded in a seaward direction). Different dune systems along this coastal stretch therefore show different morphodynamic responses to climate forcing on decadal to centennial time scales (Figure 9.19). They also show different temporal responses, with some dunes being active at some time periods, but adjacent dune systems not.

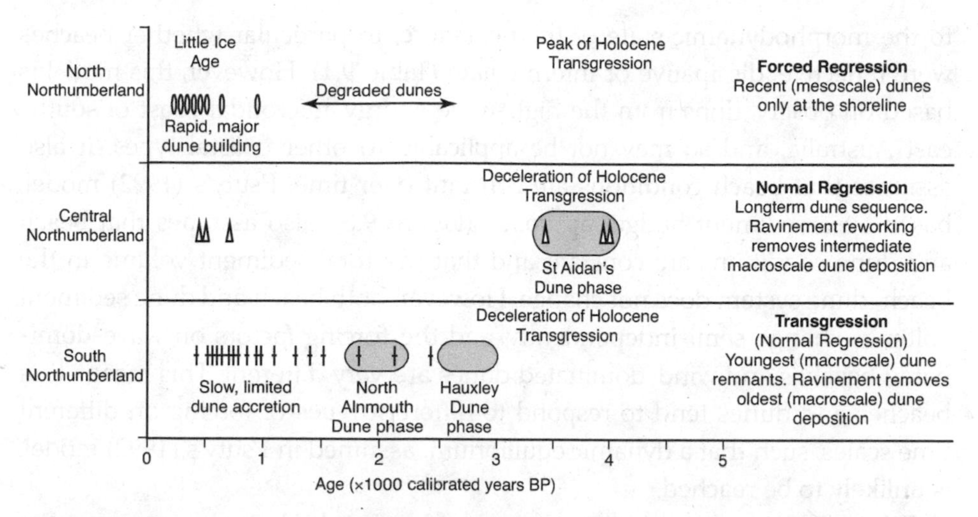

Figure 9.19 Chronology of dune evolution in Northumberland, northeast England, over the last 4000 years. Shaded areas indicate broad periods of dune development. (From Orford *et al.*, 2000.) (Copyright © 2000 Geological Society of London, reproduced with permission.)

On sub-annual (seasonal to event) time scales, dune evolution is also highly variable. Knight *et al.* (2002) compared high-resolution sedimentary data from the Northumberland dunes to the documentary record of historical storms in the North Sea region. Subtle changes in the grain size distribution within sand dune cores can be matched, using an age-depth model derived from luminescence ages on dune sand, with the storm record. Knight *et al.* (2002) argued that historical storms in 1436 and 1530 caused nearshore/beach sediments to be deposited in the dunes by overwashing of large waves. The extensive dune system at Forvie, northeast Scotland, also reflects the power of North Sea storms. It is commonly believed that the village, church and surrounding agricultural land of Forvie was buried by a '9-day storm' in 1413. At Perranporth in north Cornwall, southwest England, a seventh-century church was abandoned following sand engulfment by cliff-top dunes (Figure 9.16) in the eleventh century, and a further church built *c.* 1150 was abandoned in 1804. These events are clear outcomes of climate and event-scale forcing on sensitive dune systems.

9.5 Beach–dune interactions

Traditionally, wave-dominated beaches and wind-dominated coastal dunes have been examined as distinct and separate morphodynamic systems. However, many recent studies show that there are important interactions between beach and dune sub-systems, and that these two environments are strongly coupled and mutually adjusted (Sherman and Bauer, 1993). Short and Hesp (1982) describe how dune morphodynamics are strongly related to the morphodynamic state of the nearshore, in particular whether beaches were reflective, dissipative or intermediate (Table 9.1). However, this model is based on observations from the high-wave energy, microtidal coast of south-east Australia, and so may not be applicable to other coastal types. It also assumes that beach conditions are constant over time. Psuty's (1992) model, based on a sediment budget approach (Figure 9.3), also assumes that beach and dune conditions are constant and that the total sediment volume in the beach–dune system does not change. However, both beach and dune sediment volumes change semi-independently, and the forcing factors on wave-dominated beaches and wind-dominated dunes are very different. This means that beaches and dunes tend to respond to different types of forcing on different time scales, such that a dynamic equilibrium, assumed in Psuty's (1992) model, is unlikely to be reached.

Other more recent studies also consider beach–dune interactions. For example, Peterson *et al.* (2010) show that dunefields on the west coast of the USA are preferentially located adjacent to where longshore transport builds up beach sediment. Variations in longshore transport rate therefore impact on beach width and dune sediment source, but with variable time lags. Saye *et al.* (2005) found that, although eroding dunes are associated with narrow beaches and accreting dunes with wide beaches, this pattern varied between adjacent beaches. Pye and Blott (2008) show that dune erosion rates at Sefton, Lancashire, northwest England, are positively correlated with extreme high tides and show less association with wave regime (which most strongly affects beaches).

Beach–dune interactions are explicitly based on sediment transport from one of these sinks to the other by wind, water and other mechanisms. Sediment transport results in sediment sorting. Beaches and dunes, although composed broadly of the same materials, can often be distinguished on the basis of their grain size population (as well as other properties including grain mineralogy, shape and microfeatures). Examination of the grain size distribution of beach and dune samples using moment measures (Section 5.2.1) has been used in many studies. For example, Friedman (1961) shows that the skewness of dune sands is generally positive (i.e. dune sands are fine) whereas for beach sands skewness is generally negative (i.e. beach sands are coarse). There is less discrimination, however, between samples from within particular environments, such as upper and lower parts of beach profiles. With respect to wind transport, finer sediments can be transported farther inland than coarser sediments, so mean dune grain size decreases landward irrespective of transport mode.

9.6 Dune management

Dune surface stability is critically dependent on vegetation cover, so reduction in vegetation cover, leading to renewed dune activity and reduction of the dune's natural barrier to coastal erosion, is an important management issue. In many environments dunes are an essential part of coastal protection, and maintenance of a 'healthy' dune system is a priority for coastal management. The natural protection of sandy coastlines can be improved by encouraging the growth of dunes in areas where dunes are either absent or discontinuous. With respect to vegetation planting, typically of marram grass but also other species such as *Hippophae rhamnoides* (Sea buckthorn), a number of aspects need to be considered. These include selection of the plant species, timing of planting, planting density, planting width and use of fertiliser. Generally, dune development induced by planting is not as rapid as that due to fencing, but once established, most dunes and dune plants grow steadily.

Dune construction can be achieved through the use of fences (brushwood or wooden palings) at the back of the beach. The fences disrupt the airflow, thereby promoting sediment deposition. CERC (1984) recommends straight fences parallel to the coast with a 'porosity' of 50 per cent, preferably coinciding with the natural dune vegetation or foredune line. Once the fence is at 'full capacity' (i.e. almost buried by sand), dune growth can be maintained by lifting the fences. Fence-built dunes must be stabilised with vegetation or the fence will deteriorate and release the sand. Fence deployment in an area with abundant aeolian sediment transport can achieve rates of vertical dune growth in excess of 1 m yr^{-1} (Figure 9.20). Closely associated with dune construction is **dune stabilisation**, which is directed towards securing bare sand surface in the

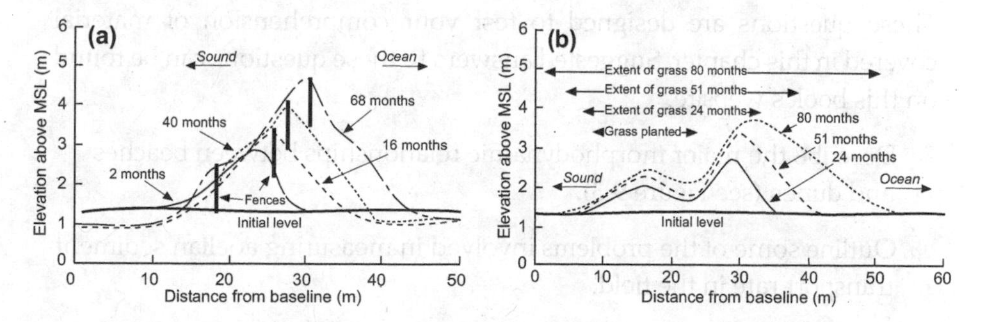

Figure 9.20 Construction of artificial dunes. (a) Sand accumulation by a series of four single-fence lifts, Outer Banks, North Carolina. (b) Sand accumulation following planting with American beachgrass, Ocracoke Island, North Carolina. (Modified from CERC, 1984.)

dunes and repairing gaps in coastal dune ridges. Both fences and vegetation are frequently used to stabilise dunes. Gaps in the dunes may also be filled using earthmoving equipment followed by planting to stabilise the bare sand.

SUMMARY

- Coastal sand dunes represent the net deposition of sand by wind and are important in buffering the effects of coastal erosion. Sediment entrainment takes place when winds increase above the threshold velocity for that grain size, and transport can take place by creep, reptation, saltation and suspension processes. Wind-driven transport is affected by wind velocity, turbulence, air temperature, land surface roughness, surface moisture and fetch distance.
- Primary and secondary coastal dunes can be identified. Primary dunes include embryo dunes and foredunes, which are formed at the seaward edge of the dune system. Secondary dunes include blowouts, transgressive dunes (including parabolic and ramp dunes and sand sheets), machair plains and interdune slacks.
- Coastal dune evolution is strongly linked to vegetation succession and soil development. Dune evolution often reflects climate forcing on different time scales, from individual storm events to glacial–interglacial cycles. Dune–beach interactions can be examined with respect to their sediment budgets, morphological changes and grain size distribution properties.
- Dune management, including dune construction and stabilisation, is an important means by which the coastal zone can be buffered against erosion and other impacts of climate forcing and human activity.

Reflective questions

These questions are designed to test your comprehension of material covered in this chapter. Suggested answers to these questions can be found on this book's website.

9a. Describe the major morphodynamic relationships between beaches and dunes (see Figure 9.3).

9b. Outline some of the problems involved in measuring aoelian sediment transport rate in the field.

9c. How can the morphology of coastal dunes be used to identify prevailing wind direction?

9d. Building from the discussion in Section 9.6, identify other methods by which dune growth can be encouraged.

Further reading

Houston, J.A., **Edmundson**, S.E. and **Rooney**, P.J., 2001. *Coastal Dune Management: Shared Experience of European Conservation Practice*. Liverpool University Press, Liverpool. (This text contains case studies of dune conservation practice.)

Maun, M.A., 2009. *The Biology of Coastal Sand Dunes*. Oxford University Press, Oxford. (This text describes sand dune ecosystems in detail.)

Nordstrom, K.F., **Psuty**, N. and **Carter**, R.W.G. (eds) 1990. *Coastal Dunes: Form and Process*. Wiley, London. (This important text covers many of the topics discussed in this chapter in some detail.)

Trenhaile, A.S., 1997. *Coastal Dynamics and Landforms*. Oxford University Press, Oxford. (Chapter 6 of this text gives an overview of coastal dunes.)

CHAPTER 10

GEOLOGICALLY-CONTROLLED COASTAL ENVIRONMENTS – ROCKY AND CORAL COASTS

AIMS

This chapter discusses the characteristics and behaviour of rocky and coral coasts which are both strongly controlled by geologic processes and patterns. The morphology and geomorphic processes affecting cliffs and shore platforms along rocky coasts are first described, and the relative roles of weathering and wave erosion are considered. The morphology and development of coral reefs, atolls and islands are then examined with respect to their different formational settings, long-term tectonics and ecological processes.

10.1 Introduction

Lithology and geologic processes are a significant control on the physical properties and morphodynamic behaviour of all coastlines. Hard rock fronts around 75 per cent of all coasts globally, and substantially influences adjacent beaches and estuaries by influencing coastline geometry, nearshore bathymetry and sediment supply. Many commentators use the term 'rocky coast' as synonymous with a 'cliffed coast', but geologic patterns influence coastlines in much more subtle ways (Naylor *et al.*, 2010). For this reason we use the term 'geologically-influenced coasts'. Geologic structure, rock type and large-scale tectonic setting are significant geologic properties. Geologic structure includes the orientation of faults, folds and bedding planes. **Concordant** coasts are those where these structures are aligned parallel to the coast and **discordant** coasts where these structures are at an angle to the coast (Figure 10.1). Structural discontinuities act as lines of weakness/resistance that can be exploited by weathering and erosion, most commonly by wave action. Along discordant coasts, differential erosion can lead over time to increasing coastline indentation and development of headland–embayment systems. Large-scale tectonic setting is significant in the formation of leading- and trailing-edge coasts (Section 1.2), and isostatic and coseismic land uplift and depression (Sections 2.3 and 2.4). Tectonic setting is also important in the formation of coral reef islands through the formation and development of volcanic hotspots.

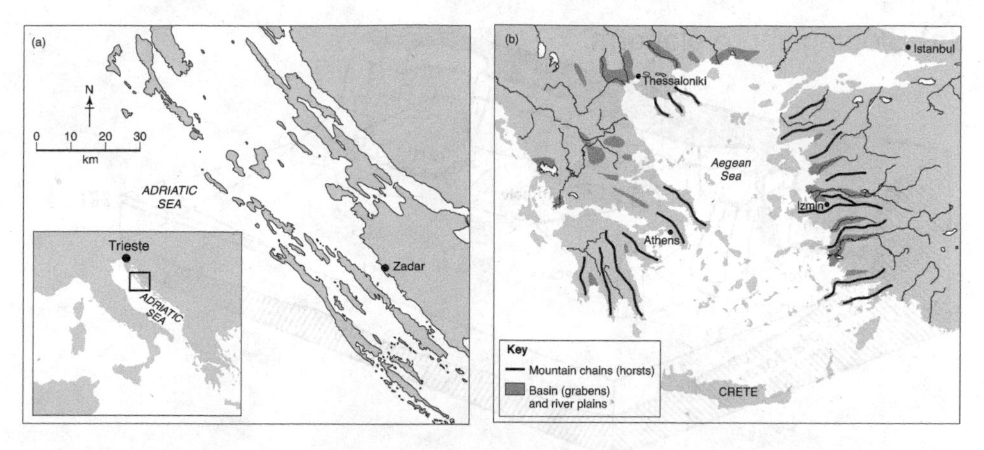

Figure 10.1 Examples of concordant and discordant coasts. (a) Concordant coast along the coastline of Dalmatia, Serbia. (b) Discordant coast of Greece and Turkey around the Aegean Sea. (From Kelletat, 1995.) (Copyright © 1995 Coastal Education & Research Foundation, reproduced with permission.)

This chapter examines the properties, processes and landforms of rocky and coral coasts. These coastal types are grouped together because rock structure, rock type and tectonic setting influence them both. They are also strongly affected by similar weathering processes, and are sensitive to changes in sea level and storminess. Geologically-influenced coasts tend to be viewed as being erosional only, without the presence of depositional features. However, detached fragments of rock and coral can accumulate and influence their dynamics. For this reason, these detached fragments cannot always be described using the same principles that are applied to clay to sand-sized particles as discussed in Chapter 5.

Previously, most studies have considered that rocky and coral coasts have a slow rate of morphological change and long relaxation time. However, many recent studies show that rocky coasts can be dynamic over short time scales and respond to relatively small variations in waves and storminess (see Naylor *et al.*, 2010 for a recent review). Studies also show that coral health, physical structure and ecosystem processes are highly sensitive to changes in storminess, sea surface temperature (SST), oxygen and sediment content (e.g. Hallock, 2005). The discussion followed in this chapter is informed by these recent studies. We start by considering rocky coasts and then turn to coral coasts.

10.2 Rocky coast processes

The morphology and behaviour of rocky coasts reflect the interplay between geologic factors (e.g. structure, rock type, rock hardness, mineralogy) and

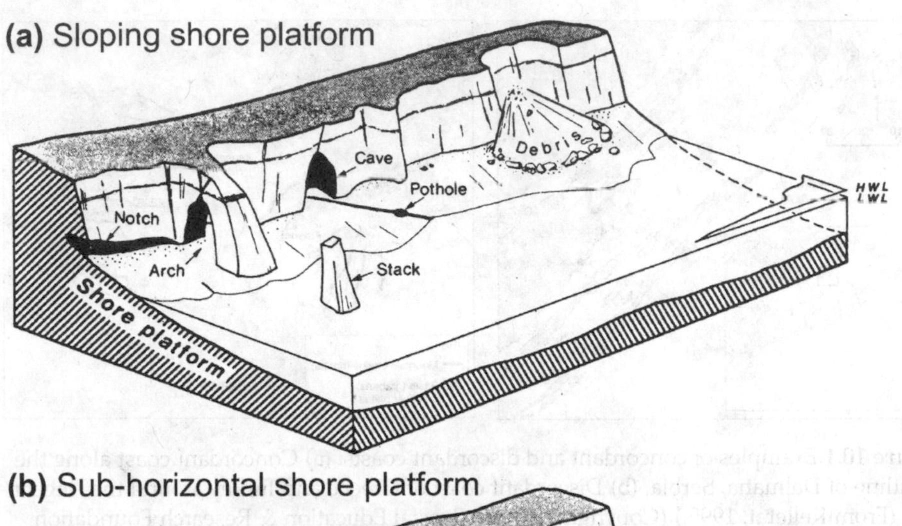

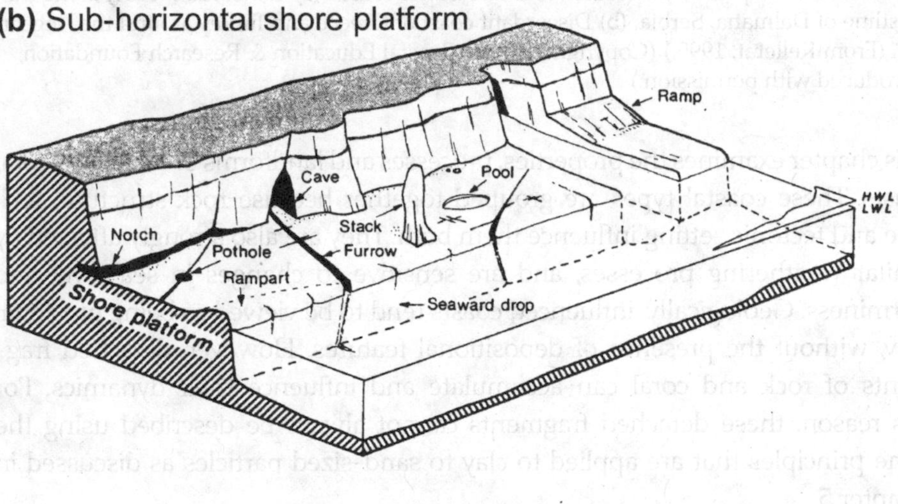

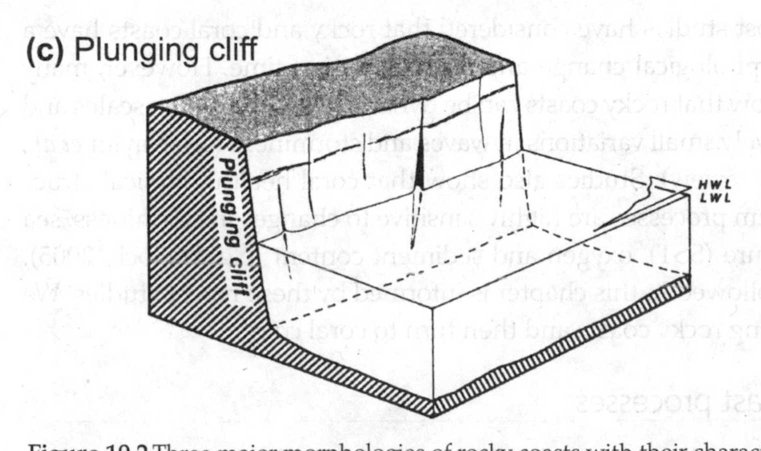

Figure 10.2 Three major morphologies of rocky coasts with their characteristic features: (a) sloping shore platform, (b) sub-horizontal shore platform and (c) plunging cliff. (From Sunamura, 1992.) (Copyright © 1992 John Wiley & Sons, reproduced with permission.)

weathering and erosion processes. Sunamura (1992) categorises rocky coast morphology into three types (Figure 10.2): (a) sloping shore platform, (b) sub-horizontal shore platform and (c) plunging cliff. Although this oversimplifies many field situations, it is useful for our discussion because shoreface steepness and presence/absence of a shore platform largely determine the effectiveness of waves in shoreline erosion.

Rocky coasts are affected by a range of wave, weathering, bioerosion and mass movement processes (for a comprehensive review see Trenhaile, 1987). The different processes often occur concurrently and a key problem is to determine their relative importance and how they are affected by geologic properties (Naylor and Stephenson, 2010). Table 10.1 summarises the major erosional processes that affect rocky coasts, which are then discussed in more detail below.

Table 10.1 Summary of the main erosional processes on rocky coasts.

Process	Description	Typical conditions conducive to the process
Mechanical wave erosion		
Erosion	Removal of loose material by waves	Energetic wave conditions and microtidal tide range
Abrasion	Scouring of rock surfaces by wave-induced flow with mixture of water and sediment	'Soft' rocks, energetic wave conditions, a thin layer of sediment and microtidal tide range
Hydraulic action	Wave-induced pressure variations within the rock causes and widens rock capillaries and cracks	'Weak' rocks, energetic wave conditions and microtidal tide range
Weathering		
Physical weathering	Frost action and cycles of wetting/drying causes and widens rock capillaries and cracks	Sedimentary rocks in cool regions
Salt weathering	Volumetric growth of salt crystals in rock capillaries and cracks widens these capillaries and cracks	Sedimentary rocks in hot and dry regions

Table 10.1 continued

Chemical weathering	A number of chemical processes remove rock material. These processes include hydrolysis, oxidation, hydration and solution	Sedimentary rocks in hot and wet regions
Water-layer levelling	Physical, salt and chemical weathering working together along the edges of rock pools	Sedimentary rocks in areas with high evaporation
Bio-erosion		
Biochemical	Chemical weathering by products of metabolism	Limestone in tropical regions
Biophysical	Physical removal of rock by grazing and boring organisms, expansion of cracks by plant roots	Limestone in tropical regions
Mass movements		
Rock falls and toppling	Rocks falling straight down the face of the cliff	Well-jointed rocks, undercutting of cliff by waves
Slides	Deep-seated failures	Deeply-weathered rock, undercutting of cliff by waves
Flows	Flowing of loose material down a slope	Unconsolidated material, undercutting of cliff by waves

10.2.1 Mechanical wave erosion

Mechanical wave erosion dominates in most swell and storm-wave environments. Under less energetic conditions, waves have limited erosive power but may still be important in removing weathering products from the cliff base. The principal wave effects are surface abrasion due to currents and air pressure variations in rock cracks.

Abrasion is the scouring action of wave-induced currents and includes the sweeping, rolling or dragging of sediment across inclined rock surfaces, and the throwing of coarse material against rock cliffs (attrition). Robinson (1977) found that erosion rates at a cliff base were 15–20 times higher where there was a beach at the foot of the cliff, compared to where there was no beach. Abrasion is most efficient if only a thin layer (< 0.1 m) of sediment is present because a thick layer can protect the underlying rock surface.

Wave breaking on rocks induces air pressure variations in rock capillaries and cracks, a process known as **hydraulic action**. These pressure variations consist of a dynamic (impact of the moving water) and a hydrostatic component (due to the weight of the water column). Breaking waves exert the greatest pressures (particularly plunging breakers), followed by broken and reflected waves (Sunamura, 1992). Hansom *et al.* (2008) used pressure transducers on an artificial cliff in a wave tank to measure pressure changes on rock surfaces caused by waves. If there are standing waves at the cliff base, wave-induced pressure is relatively uniform over the wetted portion of the cliff. However, breaking waves exert a much higher pressure at the height of the wave crest. A rock platform can dissipate wave energy or allow waves to travel as a bore across the platform. In the wave tank experiments, wave impact at the cliff base is measured as an initial pressure increase (at 3 s in the top panel of Figure 10.3), and then a surge of water up the cliff face. On flat cliff tops, the surging water may be pushed high in the air, with the downward force of water causing a peak in pressure on the cliff top (at 8 s in the top panel of Figure 10.3). On sloping cliff tops, the surging water can move landward as irregular bursts of spray (at 6–8 s in the top panel of Figure 10.3).

Loosened rock fragments can be removed by **quarrying**, which leaves angular debris and fresh rock scars along many rocky shorelines. This process of **mechan-**

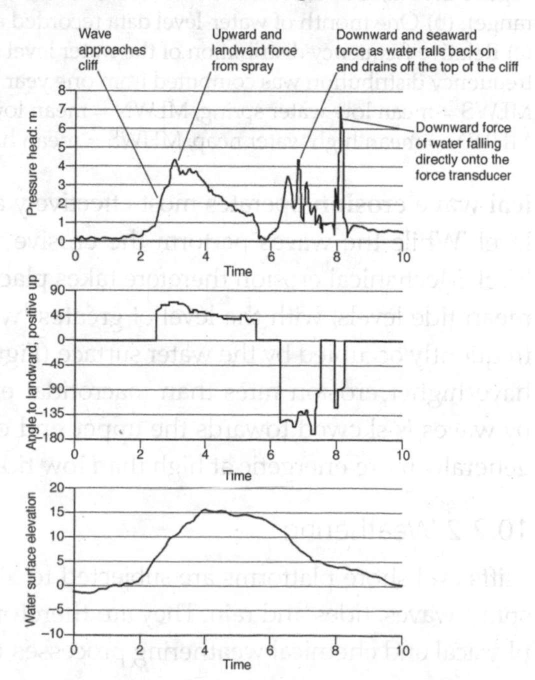

Figure 10.3 Typical patterns of water pressure (top panel), force direction (middle panel) and water surface elevation (bottom panel) in a wave tank experiment, where $H = 14.4$ m and $T = 15$ s. (From Hansom *et al.*, 2008.) (Copyright © 2008 Elsevier Science, reproduced with permission.)

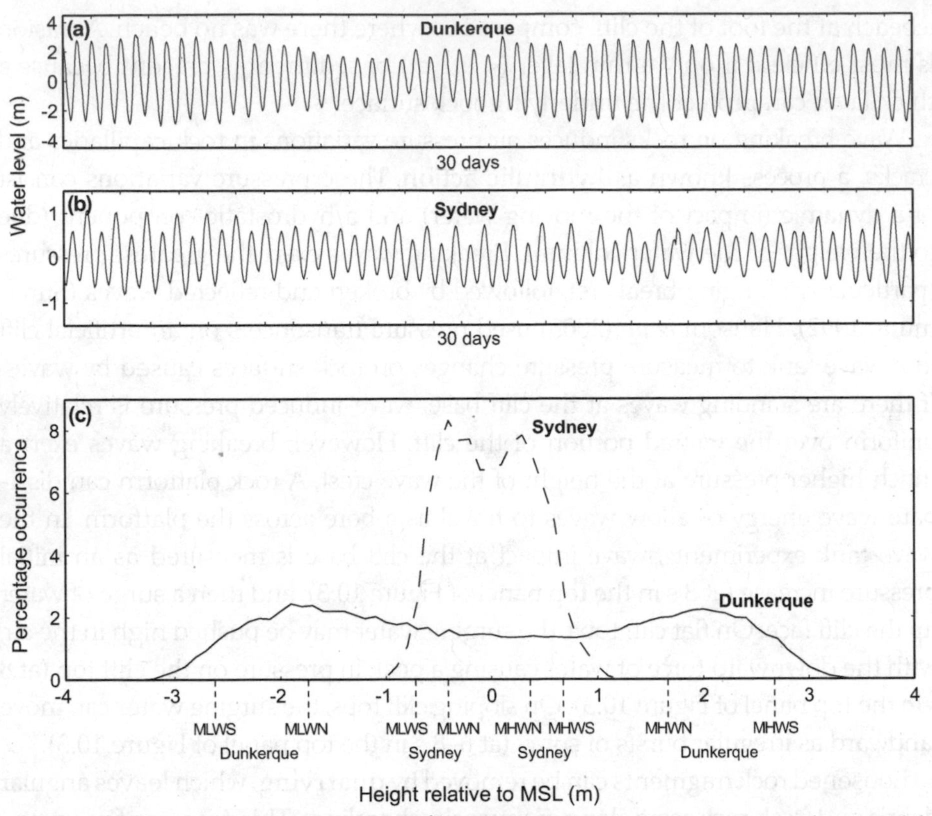

Figure 10.4 (a) One month of water-level data recorded at Dunkerque, France (macrotidal range). (b) One month of water-level data recorded at Sydney, Australia (microtidal range). (c) Relative frequency distribution of the water level for Dunkerque and Sydney. The frequency distribution was computed from one year of data using a bin width of 0.1 m. MLWS = mean low water spring, MLWN = mean low water neap, MSL = mean sea level, MHWN = mean high water neap, MHWS = mean high water spring.

ical wave erosion operates most effectively at, or slightly above, the still water level. While the waves perform the erosive work, tides control the still water level. Mechanical erosion therefore takes place at elevations at and between the mean tide levels, with the level of greatest wear located at the elevations most frequently occupied by the water surface (Figure 10.4). Microtidal environments have higher erosion rates than macrotidal environments. In addition, erosion by waves is skewed towards the upper end of the tidal range because they are generally more energetic at high than low tide and during storms.

10.2.2 Weathering

Cliffs and shore platforms are subjected to alternate wetting and drying by salt spray, waves, tides and rain. They are therefore suitable environments for many physical and chemical weathering processes (Table 10.1). In sheltered areas and on susceptible rocks, **weathering** is an important mechanism of platform breakdown. Like mechanical wave action, weathering works most effectively in, and slightly above, the intertidal zone, and in storm and high-energy wave environments it can be significant above high tide level out of reach of the waves.

Physical weathering can result from frost action and alternate wetting and drying. Rock breakdown is accomplished through the formation and widening of capillaries and cracks. If a frozen rock surface is submerged by the rising tide, the ice within the rock will thaw. Therefore, intertidal rocks may experience two freeze/thaw cycles per day under semi-diurnal tide and sub-zero temperature conditions. The mid-intertidal zone under semi-diurnal tide conditions can experience almost 700 wetting/drying cycles per year. Frost action and wetting/drying cycles can only cause significant weathering if the rock is highly saturated (Trenhaile and Mercan, 1984), which particularly affects fine-grained sedimentary rocks and cold coastal regions.

Salt is deposited on rock surfaces when sea water evaporates. Salt crystals undergo volumetric changes due to adsorption of water, temperature-induced expansion and crystal growth from solution which result in widening of rock capillaries and cracks, and rock weakening (Trenhaile, 1987). Such **salt weathering** therefore operates in a similar manner to frost action, and the efficiency of salt weathering increases with rock permeability and porosity. Conditions favouring salt weathering are high temperatures and low rainfall, therefore salt weathering is most significant in (semi-) arid and Mediterranean climates and on sedimentary rocks.

Chemical weathering usually results from several chemical reactions working together, including hydrolysis, oxidation, hydration and solution. The efficiency of these weathering processes is mainly determined by the amount of water available for the chemical reactions and for the removal of the soluble products. If the weathered product remains in the system, chemical equilibrium may be attained which will inhibit further weathering. This can particularly affect water within rock pools. Chemical weathering is most significant in hot, wet climates, and igneous rocks are generally less susceptible than sedimentary rocks. Cold climates have slow chemical weathering rates, but this may be due to the lack of liquid water rather than low temperatures (Trenhaile, 1987).

Physical, salt and chemical weathering often work together and operate most effectively when and where a rocky substrate experiences frequent wetting and drying cycles, particularly around the margins of rock pools subject to spray or tides. This type of weathering, termed **water-layer levelling** (Matsukura and Matsuoka, 1991), enlarges rock pools and eventually causes them to merge. Water-layer levelling is significant in lowering shore platforms in areas with high evaporation rates, but is less important in cool, wet climates.

10.2.3 Bioerosion

Bioerosion is the removal of rock by organisms. This process is probably most important in tropical regions due to the varied marine biota and abundance of calcareous substrates, but is also effective in the midlatitudes under less vigorous weathering regimes (Spencer and Viles, 2002). Bioerosion is determined by the spatial distribution of marine organisms across the rock surface, which is largely controlled by moisture availability and thus depends on tidal characteristics and wave energy level.

Marine organisms remove rock in various ways. Algae form dense mats on the rock surface and their products of metabolism cause chemical etching of the underlying rock surface. Grazers effectively abrade rock surfaces as they feed on microflora (algae, lichen and fungi) adhering to the rock surface. Chemical or mechanical borers directly remove rock material thereby increasing the area exposed to other weathering processes, but can enhance the rock environment for algal colonisation. Marine flora and fauna can also form organic crusts that protect the underlying rock from wave erosion and weathering. Algal mats can prevent a rock surface from drying out, reducing water-layer levelling.

10.2.4 Mass movements

The steep slopes of rocky coasts suggest that they are potentially unstable and prone to mass movement. A spectrum of **mass movements** can occur on rocky coasts (Figure 10.5), depending primarily on rock lithology and structure. **Rock falls** and **topples** are characteristic of hard, well-jointed rock cliffs that have been undercut by waves. Rock falls and topples often lead to talus at the cliff foot. Rock falls and topples occur most frequently in winter as a result of frost action, rainfall and basal undercutting by storm waves.

Landslides are deep-seated failures that occur when the compressive strength of a rock is exceeded by the load imparted on it. Different landslide types include translational slides, rotational slides and mudslides. Translational slides (or slips) involve movement along a straight plane and are often structurally controlled. Rotational slides fail along a concave-upward surface and typically occur as a result of wave undercutting. A slide will reduce support of the material behind it, causing further sliding and giving rise to multiple rotational landslide systems (Case Study 10.1). Mudslides occur when fine-grained sediment moves on a shear surface. The occurrence of all types of landslides is promoted by: (1) addition of material from an up-slope mass movement, causing an increase in load; (2) steepening of the slope angle due to basal undercutting; and (3) reduction in the compressive strength of the rock due to increase in moisture content and/or weathering. Landslides are therefore often triggered

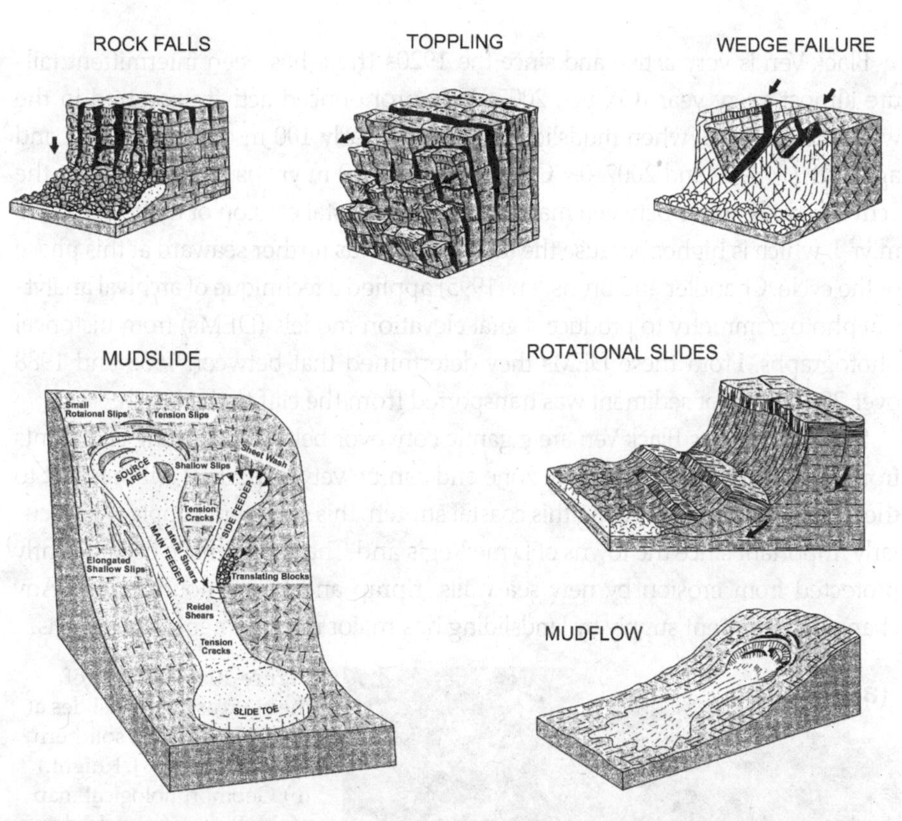

Figure 10.5 Types of mass movement affecting cliffs. (From Brunsden and Goudie, 1997; modified from Allison, 1990.) (Copyright © 1997 Geographical Association, reproduced with permission.)

Case Study 10.1 Black Ven: A coastal landslide system

The cliffs along the coast of Dorset are the highest in southern England and are developed in Liassic (Lower Jurassic) shales, marls and mudstones. The Black Ven landslide complex between Lyme Regis and Charmouth is one of the largest landslide systems in Europe (Cooper, 2007). The uppermost cliffs (145 m high) collapse in single and multiple rotational landslides to form a thick debris-dominated slope (Figure 10.6a). Individual landslide scarp faces lie parallel to each other and to the fronting beach (Figure 10.6b). When very active, flows of rock fragments, debris and mud pour over the landslide benches to merge into broad lobes on the beach. When flow activity slows down, lobe toes are eroded back by the sea to leave residual boulder arcs on the foreshore. Important feedbacks exist between mass movements and nearshore processes. Sediment at the cliff base inhibits further cliff failure, whereas when this sediment is depleted it can lead to an acceleration of cliff erosion. Such coastal landslides therefore undergo cycles of activity.

Black Ven is very active and since the 1920s there has been intermittent failure almost every year (Cooper, 2007). Very pronounced activity occurred in the winter of 1957–58, when mudslides extended nearly 100 m onto the beach, and again in 1968–69 and 2007–08. Cliff retreat of 5–30 m yr^{-1} is characteristic of the active periods, while between major events the coastal erosion of the toe is 15–40 m yr^{-1}, which is higher because the toe lobe extends further seaward at this phase of the cycle. Chandler and Brunsden (1995) applied a technique of archival analytical photogrammetry to produce digital elevation models (DEMs) from historical photographs. From these DEMs they determined that between 1958 and 1988 over 200,000 m^3 of sediment was transported from the cliff to the beach.

Systems such as Black Ven are gigantic conveyor belts transporting sediments from the land into the nearshore zone and can be very important with respect to the littoral drift budget. Along this coastal stretch, this sediment supply is particularly important since the towns of Lyme Regis and Charmouth have been recently protected from erosion by new sea walls, riprap and beach nourishment. Any change in sediment supply by landsliding has major management implications.

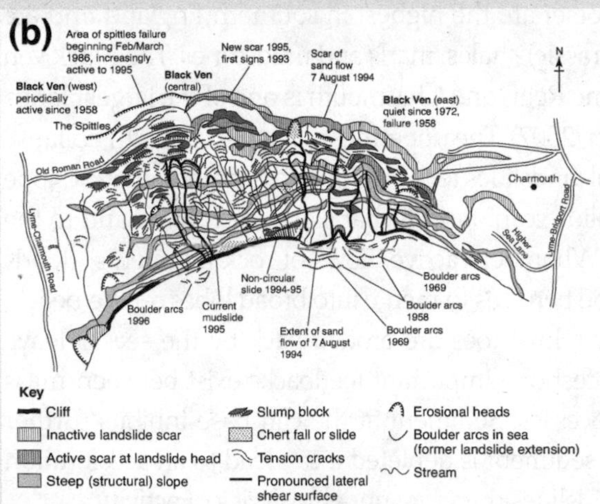

Figure 10.6 (a) Photo of the staircase of landslides at Black Ven, Dorset, southern England (photo: J. Knight.) (b) Geomorphological map of the Black Ven landslides in 1995. (From Cooper, 2007.) (Copyright © 2007 Joint Nature Conservation Committee, reproduced with permission.)

by heavy rainfall and/or increased wave action, and are more common in winter than summer. Many slides develop downslope into **flows** which are mass movements with a high liquid content. Flows of different grain size include debris flows (coarse) and mudflows (fine). Flows take place under exceptional rainfall and groundwater conditions, and contribute to landslide disintegration and sediment release to the fronting beach (Case Study 10.1).

These weathering and erosion processes affect both the cliffs and platforms of rocky coasts (Figure 10.2). We now describe the morphology and behaviour of cliffs and platforms in more detail.

10.3 Cliffs

Coastal **cliffs** are 'steep slopes that border ocean coasts' and occur along *c.* 80 per cent of the world's coastline (Emery and Kuhn, 1982). A wide range of cliff profiles are found in nature (Figure 10.7) which reflects the large number of factors involved in their development, including marine processes, subaerial processes (weathering, mass movements, runoff), rock type (lithology and structure) and sea-level history.

Figure 10.7 Examples of rocky coastlines: (a) stacks at the Twelve Apostles, Victoria, Australia (photo: G. Masselink); (b) rugged coastline of Hartland Quay, Devon, southwest England (photo: G. Masselink); (c) slope-over-wall cliffs at Baggy Point, Exmoor, southwest England (photo: P. Keene); and (d) cliff fronted by shore platform at Boulby, Yorkshire, northeast England (photo: G. Masselink).

10.3.1 Cliff morphology

A simple morphodynamic model illustrates the relative roles of marine and subaerial processes in determining **cliff morphology** (Figure 10.8). Subaerial processes result in sediment movement down the cliff slope to the sea. If the ability of marine processes to remove the debris exceeds debris supply, sediment will not accumulate at the cliff base. In this case, the angle of the cliff profile depends mainly on the structure and lithology of the rock. Vertical cliffs form in massive uniform rocks, whereas the cliff slope will tend to follow the dip of seaward-dipping strata. In either case, the predominance of debris removal over supply will maintain a bare rock face at a constant angle. If the supply of debris exceeds the rate of removal, the material accumulates into a talus slope. The resulting cliff angle will be the angle of repose of the debris. In between these two extremes there will be a range of slope forms, depending on the relative rates of debris supply and removal (Pethick, 1984). The relative roles of marine and subaerial processes also vary over time, with accompanying changes in cliff morphology. For example, following a period of high debris

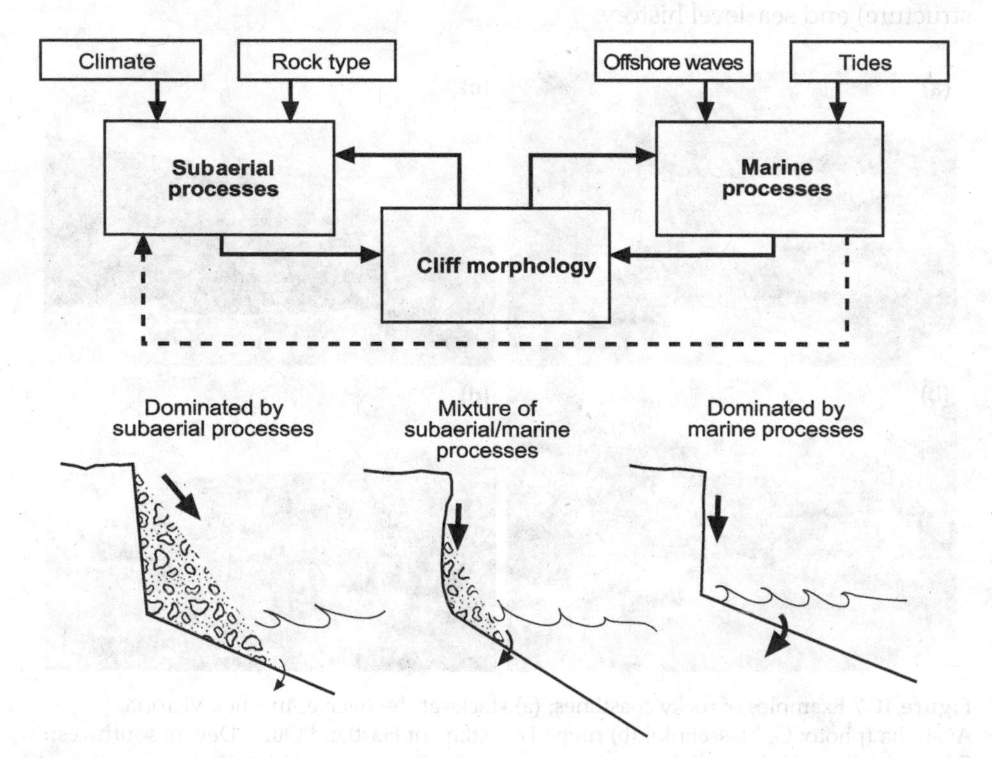

Figure 10.8 Generalised representation of the coastal cliff morphodynamic system. Typical cliff morphologies can be identified based on the relative roles of marine and subaerial processes.

supply, talus at the cliff base may require a significant amount of time before it is removed by wave processes.

Several other models also consider the factors affecting cliff morphology. Hutchinson's (1973) process/form classification of the London Clay cliffs of southeast England shows the interplay of marine and subaerial processes, and is representative of cliff evolution in stiff fissured clay (Figure 10.9):

- Type 1 cliffs occur where the rate of marine erosion is balanced by the rate of sediment supply by shallow landsliding. The slope undergoes parallel retreat and waves only remove the loose material collected at the cliff base. This removal promotes further sliding and a dynamic equilibrium condition may be attained.

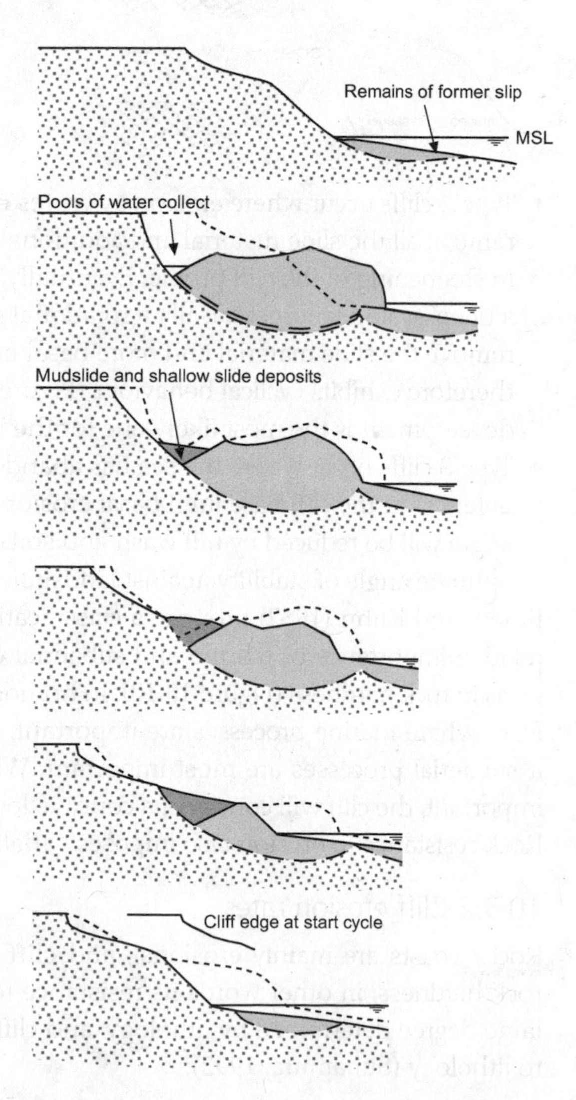

Figure 10.9 Stages in the cyclic behaviour of London Clay cliffs experiencing strong wave erosion. (Modified from Hutchinson, 1973.)

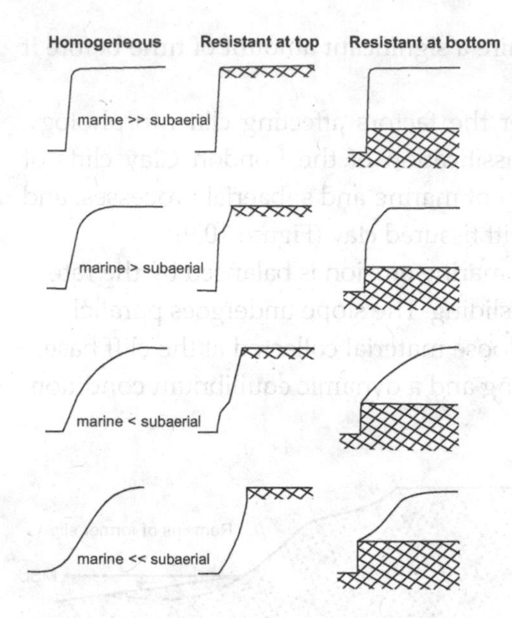

Figure 10.10 Generalised coastal cliff profiles according to variations in rock resistance and the relative efficiency of marine and subaerial processes. It is assumed that the cliffs are cut into horizontally bedded rock and are in steady state equilibrium. The more resistant rock strata are shaded. (Modified from Emery and Kuhn, 1982.)

- Type 2 cliffs occur where erosion by waves exceeds sediment supply. Waves remove all the slide material and additional cliff undercutting results in steepening of the cliff profile. Eventually, this steepening will cause a deep-seated failure, typically a rotational slide. Waves subsequently remove the slide material and more basal erosion occurs. This cliff type therefore exhibits cyclical behaviour (Figure 10.9), and this model for cliff development is the most likely one for the Black Ven landslides.
- Type 3 cliffs occur where the sea has abandoned the cliff or where coastal defences at the cliff base limit basal erosion. After initial sliding, the slope angle will be reduced by hill wash and soil creep until it reaches the ultimate angle of stability against landsliding (*c.* 8° in London Clay).

Emery and Kuhn (1982) propose a classification of marine cliffs based on the relative importance of marine and subaerial erosion, and the effects of variations in rock hardness (Figure 10.10). In homogeneous rocks, steep cliffs tend to form where marine processes are important, whereas convex profiles develop if subaerial processes are most important. Where these processes are equally important, the cliff will consist of a concave lower part and a convex upper part. Rock resistance to erosion also imposes variations on these basic forms.

10.3.2 Cliff erosion rates

Rocky coasts are mainly erosional, and **cliff erosion** is mainly controlled by rock hardness, in other words its resistance to erosion. Rock resistance is to a large degree determined by lithology, and cliff erosion rates are loosely related to lithology (Sunamura, 1992):

- < 0.001 m yr^{-1} for granitic rocks
- 0.001–0.01 m yr^{-1} for limestone
- 0.01–0.1 m yr^{-1} for flysch and shale
- 0.1–1 m yr^{-1} for chalk and Tertiary sedimentary rocks
- 1–10 m yr^{-1} for Quaternary deposits
- > 10 m yr^{-1} for volcanic ejecta

Note that these rates vary over four orders of magnitude, and therefore coasts with different rock types will experience different values of retreat rate and slope sediment supply. It is also useful to translate some of these recession rates to predictions of cliff retreat since sea level reached its present position (*c.* 6000 years BP). Granite coasts could have retreated by about 6 m in 6000 years, while glacial till cliffs along the Holderness coast in eastern England have retreated by almost 3 km just since Roman times (*c.* 2000 years BP).

Other factors are also important. For example, areas of structural weakness due to faulting and fracturing may be focal points of accelerated erosion. Wave exposure is important because they actively erode the cliff base and remove loose debris. The presence of a beach in front of the cliff generally reduces the rate of cliff erosion. Cliff height is significant because low cliffs erode faster than high ones, and less material needs to be removed to accomplish cliff recession. High cliffs, however, produce more debris than low cliffs retreating at the same rate. Coastal engineering structures can lead to the disruption of longshore drift, changing wave exposure, and structures on cliff tops will increase the load, making the cliffs more prone to failure. Cliff evolution and recession are episodic with most morphological change taking place under extreme rainfall and/or wave conditions. Short-term cliff erosion rates should therefore be interpreted carefully because they are often site-specific and unrepresentative of long-term rates.

10.4 Shore platforms

Shore platforms are horizontal or gently sloping intertidal rock surfaces. They have been previously referred to as **wave-cut platforms**, but this term should not be used because it assumes that shore platforms are the result of wave action, which is not always true. Shore platforms are erosional features developed when cliff recession takes place and cliff-foot debris is removed by waves. Shore platform formation is therefore intrinsically linked with cliff erosion, although processes responsible for shore platform lowering are not the same as those that cause cliff recession. The junction between the shore platform and cliff is usually close to the high tide level. Sometimes a high tide beach is present here.

Figure 10.11 Examples of shore platforms: (a) sloping platform in a macrotidal environment (Bude, Cornwall, southwest England) and (b) sub-horizontal platform in a microtidal environment (Wollongong, New South Wales, Australia). (Photos: G. Masselink.)

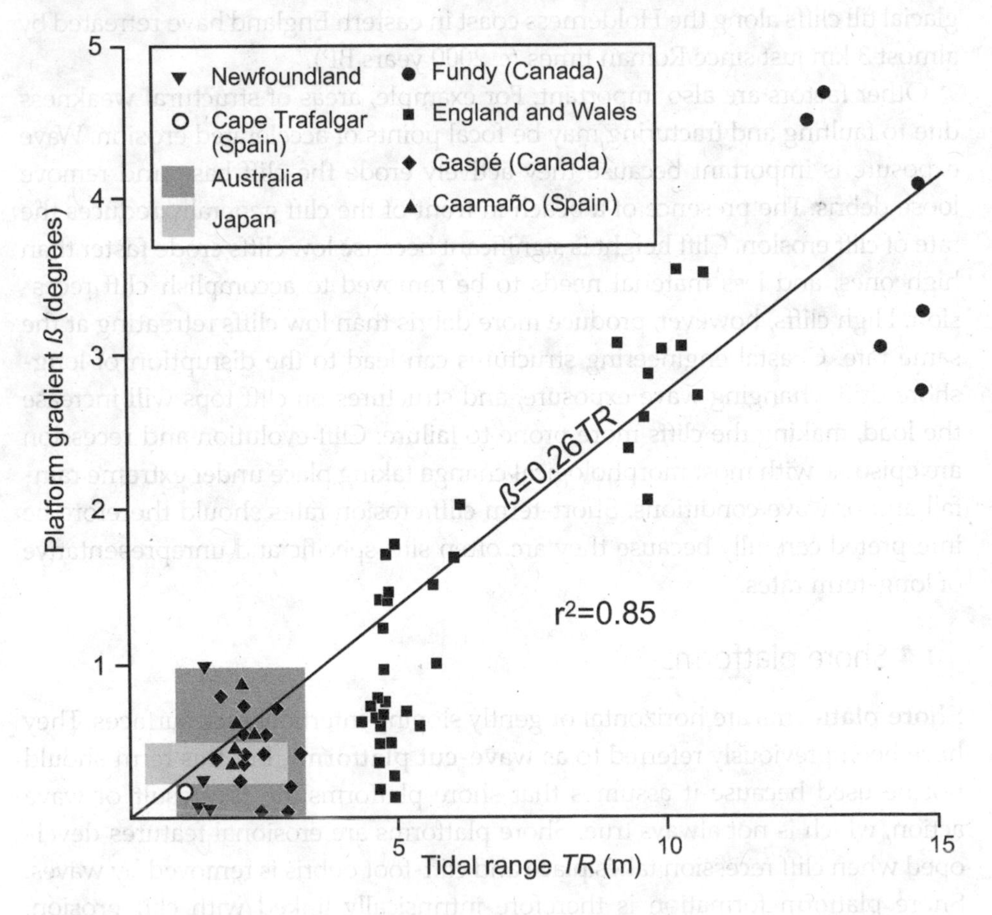

Figure 10.12 Relationship between shore platform gradient β and tidal range *TR*. Each data point represents the regional mean of a large number of platform profiles. The data from Japan and Australia are not included in the regression equation. (From Trenhaile, 1999.) (Copyright © 1999 Coastal Education & Research Foundation, reproduced with permission.)

10.4.1 Shore platform morphology

There are two major morphological types of shore platforms (Figure 10.2):

- **Sloping platforms** (1°–5°) extend from the cliff-platform junction to below low tide level, without any major break of slope or abrupt seaward terminus (Figure 10.11a).
- **Sub-horizontal platforms** extend across the intertidal zone and generally terminate in a low tide cliff or ramp (Figure 10.11b).

According to Trenhaile (1987), sloping platforms are most common in macrotidal environments and sub-horizontal platforms in meso- and microtidal environments. However, Sunamura (1992) believes that different platform types are related to the balance between wave action and rock resistance. He found that sloping platforms develop in relatively weak rocks and under energetic conditions, while sub-horizontal platforms develop in more resistant rocks and less energetic conditions. Shore platforms are absent when the rocks are too resistant and/or the waves are too weak.

Trenhaile (1999) investigated the factors that control platform morphology in south Wales, Gaspé (Canada) and southern Japan. He found a strong, positive relationship between platform gradient and tidal range (Figure 10.12). This relationship may explain the differences between sloping platforms along the macrotidal UK and sub-horizontal platforms along the microtidal coasts of Australasia. Differences in platform slope may reflect Holocene sea-level history and the relative roles of weathering and waves. The morphology of shore platforms associated with headlands also differs from those found within embayments.

10.4.2 Formation of shore platforms: waves versus weathering

Although there has been much recent work on the relative importance of waves and weathering in platform development, this is an old scientific debate. For example, Dana (1849) proposed that platforms are cut by waves at the level of maximum wear, in cliffs that are weathered down to the point at which the rocks are saturated with sea water. Bartrum (1916), however, considered that platforms develop at this saturation level rather than at the level of maximum wear, and that waves only remove weathered material rather than actively cut into the rock. Most theories of platform formation build from this early work.

In reality shore platform processes strongly depend on location, for example, platforms in sheltered environments are unlikely to have been cut by waves. The relative roles of waves and weathering also vary across the platform and with time. To assess the dominant platform-forming processes for a specific shore platform requires an integrated research approach, as is demonstrated in Case Study 10.2.

Case Study 10.2 Shore platforms on Kaikoura Peninsula, New Zealand

Well-developed shore platforms are present on the microtidal Kaikoura Peninsula, South Island, New Zealand (Figure 10.13). These platforms are exposed to an energetic wave climate, where relatively calm seas (< 0.5 m) are interrupted by high-energy storms (> 1.5 m). Stephenson and Kirk (1996) have monitored the platforms on Kaikoura Peninsula over a 20-year period. They used for this purpose the micro-erosion meter (MEM), a device designed to give precise measurements of erosion rates on rock surfaces, and showed an average rate of platform lowering of 1.4 mm yr^{-1}.

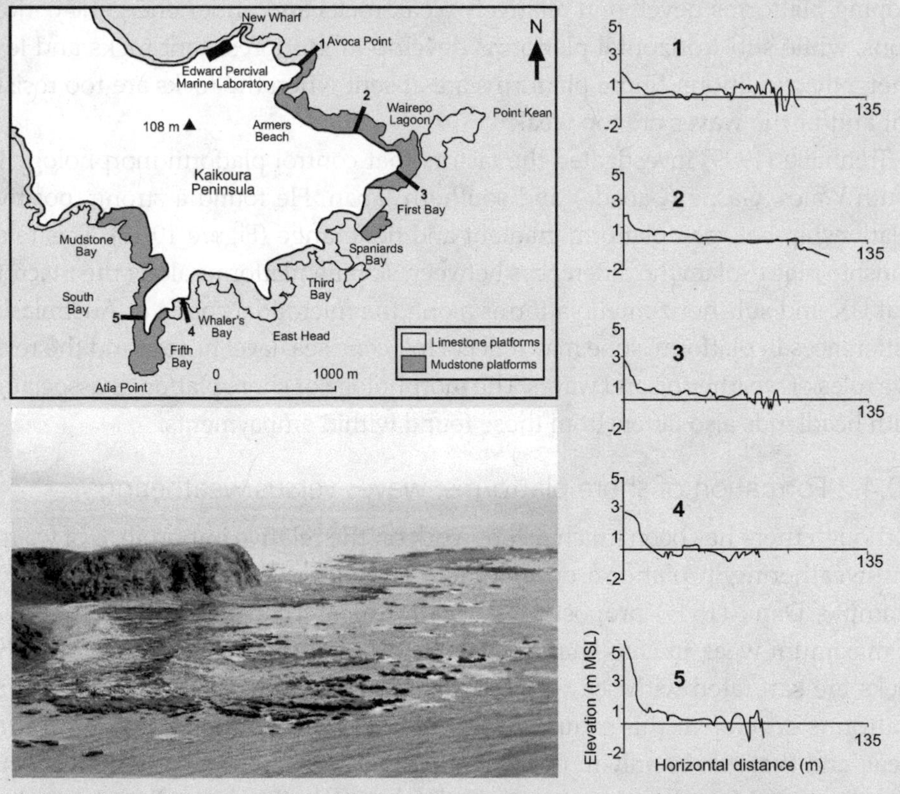

Figure 10.13 Shore platform morphology on Kaikoura Peninsula, South Island, New Zealand. (Modified from Stephenson and Kirk, 2000; photo: W. Stephenson.)

Stephenson and Kirk (2000) found that despite high offshore wave energy, particularly during storms, the amount of wave energy delivered to the platforms on Kaikoura Peninsula is low. This is due to offshore bathymetry, whereby most of the offshore waves break before arriving on the platform surface. It is likely that wave forces here are not capable of causing erosion on the shore platforms and that subaerial weathering plays the dominant role. Evidence of weathering is found in a number of distinctive surface morphologies and rock strength tests. Furthermore, platform erosion rates are strongly correlated to the number of wetting–drying cycles, so it is concluded that the development of shore platforms on Kaikoura Peninsula results mainly from subaerial weathering rather than wave action.

Many field datasets show averaged rates of platform lowering of 0.1–2 mm yr^{-1} (e.g. Trenhaile, 1987), but this rate generally increases as rock hardness decreases. Measurements of rock surface hardness, using the Schmidt hammer rebound method (Goudie, 2006), show that weathering controlled by differential exposure dominates on the lower parts of platforms, whereas rock strength in the upper parts of platforms is lithologically controlled (Chelli *et al.*, 2010). Rates of platform lowering are also linked to cliff recession rates, and field studies suggest that the rate of vertical lowering of the cliff base (top of the shore platform) is 2–5 per cent of the horizontal cliff recession rate.

10.4.3 Shore platform erosion and the generation of boulders

Previous studies have considered that platforms evolve by a combination of cliff recession and platform lowering (e.g. de Lange and Moon, 2005). This evolutionary model assumes that any loosened sediment is rapidly removed by waves and does not accumulate on the platform or interact with cliff, platform or wave processes. However, some recent studies show that platform lowering is not uniform, and that lithological differences and structural discontinuities on the platform exert a strong control on the rate of platform lowering and the size, shape and behaviour of the loosened sediment (e.g. Naylor and Stephenson, 2010). An important outcome of the erosion of sedimentary rock platforms is the generation of detached **boulders**. Many large boulders derived either from shore platforms or rock cliffs are found in locations high on cliff tops or inland that could only have been transported by exceptionally large waves caused by 'mega-storms' or tsunami. Boulders and boulder transport are now examined in detail.

Nott (2003) proposed some equations that describe the significant nearshore wave height required in order to move boulders of a given size by storm and tsunami waves. He proposed three scenarios: where detached boulders are sitting subaerially on a platform surface prior to transport (subaerial boulders); where detached boulders are sitting on a submerged platform surface prior to transport (submerged boulders); and where the platform surface is dissected by joints but where the joint-bounded blocks have not yet been detached from rockhead (joint-bounded blocks). These different boulder/block locations experience different forces of drag, lift, buoyancy and momentum. The predicted height of storm waves at breaking point H_s required to move a boulder/block of *a*, *b*, *c* axial dimensions is, for subaerial boulders

$$H_s \geq \frac{(\rho_s - \rho_w / \rho_w)\,[2a - 4C_m(a/b)(u/g)]}{C_d(ac/b^2) + C_l} \qquad (10.1)$$

for submerged boulders

$$H_s \geq \frac{1/\delta\,(\rho_s - \rho_w / \rho_w)2a}{C_d(ac/b^2) + C_l} \qquad (10.2)$$

and for joint-bounded blocks

$$H_s = \frac{(\rho_s - \rho_w / \rho_w)}{C_l} \qquad (10.3)$$

where ρ_s is boulder density; ρ_w is seawater density; u is instantaneous flow acceleration; g is gravity; C_d is drag coefficient (typically a value of 2 for submerged boulders); C_l is the lift coefficient (typical value of 0.178); and δ is wave-type parameter (which varies as a result of differences in speed between different wave types). Nott (2003) distinguishes between storm and tsunami waves because, as they approach the coast, their wave properties and therefore the combination of forces that they impart on the shore platform change. However, there is debate on the usefulness of Nott's equations because of their large uncertainties and assumptions (e.g. Switzer and Burston, 2010). Nanesena *et al.* (2011) propose variants of Nott's equations that can better illustrate transport in sliding, rolling and saltation modes.

Most commonly, boulders found in association with shore platforms have been detached from the lower intertidal part of the platform and moved to the upper intertidal part of the platform by waves. Boulders are organised during transport into ridges that can be shore-parallel or aligned perpendicular to the shore and oblique to incoming waves. In northwest Ireland, shore-perpendicular ridges are 1.5–2.5 m high, 2–10 m wide and 6.5–14.5 m long, but have a very consistent spacing of *c.* 10 m (Knight *et al.*, 2009). Evidence for very dynamic boulder behaviour comes from tracking the position of individual boulders over time. On the time scale of an individual storm, boulders can move alongshore by several hundred metres, but this transport is highly episodic. Boulder movement over shorter tidal time scales is shown by the presence of knocked and abraded edges, termed bruises, on boulders and the platform surface. Linear bruises on the platform are formed as boulders are dragged down the platform by wave backwash. Despite this evidence for very high boulder mobility even under relatively normal wave conditions, there is a vigorous debate on the relative roles of 'mega-storms' *versus* tsunami, particularly where very large boulders are found at very high elevations (Case Study 10.3).

Case Study 10.3 Megaclasts on the move

Exceptionally large boulders, often called megaclasts, have been recorded along the Atlantic-facing coasts of Scotland and Ireland. Boulders are found up to 50 m above sea level on the Outer Hebrides, Shetland Islands, Orkney Islands (Scotland) and Aran Islands (Ireland) (Scheffers *et al.*, 2009). The size, shape and orientation of the megaclasts can inform on processes of transport. If clasts are rolled into place then *a*-axes are shore-parallel. If clasts are transported by suspension, *a*-axes are parallel to flow and clasts are often imbricated. A clue to the likely transport processes and timing of megaclast emplacement comes from their biological properties and their location. On many megaclasts marine organisms such as barnacles and mussels are still intact in life position. This indicates the intertidal locations from which the megaclasts were derived, and that they were moved recently (last few years–decades). On the Shetland Islands, Hansom and Hall (2009) used radiocarbon dating on peat found beneath the megaclasts, and the same method on marine shells and luminescence dating

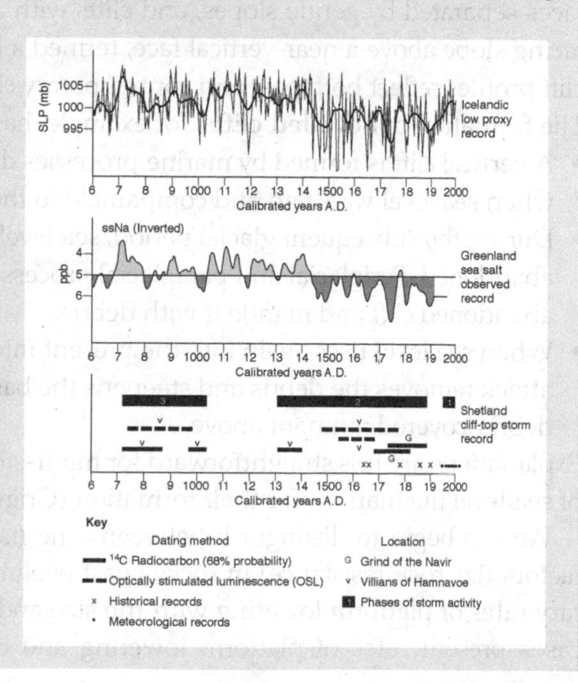

Figure 10.14 A chronology of the cliff-top storm deposits from the Shetland Islands, Scotland. Top panel shows a proxy record of Icelandic low pressure; middle panel shows the sea salt record from the Greenland GISP2 ice core (from Meeker and Mayewski, 2002), indicating the strength of winds off the Atlantic; lower panel indicates dated periods in which boulders were moved or deposited, indicating periods of storminess. (From Hansom and Hall, 2009.) (Copyright © 2009 Elsevier Science, reproduced with permission.)

on sand layers found between the megaclasts. They found that megaclasts were deposited episodically throughout the late Holocene (Figure 10.14). The dated periods of megaclast deposition coincide with peaks in sea salt deposition in Greenland ice cores, indicating a more agitated Atlantic sea surface and stronger zonal winds. In turn, this suggests that the megaclasts were detached and transported by very large storm waves, not tsunami waves. Some of the megaclasts have also trapped modern debris such as plastic buoys and soft drinks cans, indicating that megaclasts are on the move even today.

10.5 The role of inheritance along rocky coasts

Both cliffs and shore platforms have a strong geological inheritance because their morphology is strongly influenced by geologic structure, and morphological patterns tend to persist over long time periods. Apart from geologic structure and evolutionary history, rocky cliffed coasts are also controlled by the balance between debris-supplying and debris-removing processes. In practical terms, it is sometimes difficult to decide whether a cliff is contemporary (i.e. the result of present-day processes), fossil (i.e. the result of past processes) or a combination. Many coastal cliffs are **composite cliffs** consisting of more than one slope element. These include multi-storied cliffs, with two or more steep faces separated by gentle slopes, and cliffs with a convex or straight seaward-facing slope above a near-vertical face, termed a bevelled cliff. Such composite cliff profiles reflect both past and contemporary climatic and sea-level changes. The formation of **bevelled cliffs**, for example, has three stages (Figure 10.15).

- A vertical cliff is formed by marine processes during the last interglacial when sea level was high and comparable to the present day.
- During the subsequent glacial period, sea level falls and the cliff is abandoned. Periglacial and paraglacial processes progressively degrade the abandoned cliff and mantle it with debris.
- When sea level rises again into the present interglacial, renewed wave attack removes the debris and steepens the base of the cliff, leaving the debris-covered remnant above.

Explanations are less straightforward for multi-storied cliffs requiring a number of sea level fluctuations for their formation (Griggs and Trenhaile, 1994).

We can begin to distinguish between time (i.e. inheritance) and geological factors (i.e. rock resistance) in rocky coast evolution by comparing contemporary rates of platform lowering with the size and extent of platforms. In many cases present rates of platform lowering and cliff erosion seem too low to

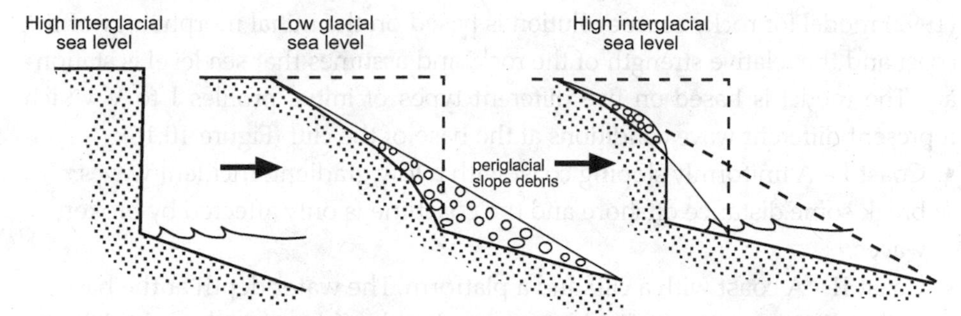

Figure 10.15 Development of a bevelled cliff commonly found in regions that have experienced periglacial conditions during the last glacial. The first stage represents a sea cliff during the last interglacial with sea level close to present. During the subsequent glacial, the abandoned cliff is subjected to periglacial conditions and becomes mantled by weathered material. Marine erosion of the base of the weathered slope occurs during present sea-level conditions, leading to a bevelled cliff. (Modified from Trenhaile, 1987.)

account for the formation of these platforms since sea level reached its present value around 6000 years BP. From this it can be inferred that present wave action is merely modifying older erosional surfaces. However, today's measured rates of change are not necessarily the same as those at an earlier period when platforms were narrower and steeper, and when wave quarrying may have been more significant. Platforms cut in relatively resistant rocks show a strong degree of inheritance, whereas contemporary rates of change in weaker rocks are within the range of values necessary to explain platform development over the last 6000 years. Platforms cut in rocks of intermediate strength may reflect a mix of inherited and contemporary features.

Although abrasion affects platforms that have unimpeded waves and cross-shore currents and loose sandy sediments, most platforms erode at their seaward edge by the removal of boulders by waves. The role of joints and bedding planes on sedimentary rock platforms is therefore critical. Naylor and Stephenson (2010) examined platforms in south Wales and Australia and showed that geologic structures strongly influence rock surface hardness and the processes and spatial patterns of platform erosion. Knight and Burningham (2011) mapped the locations from which boulders were derived from a platform surface. They show that boulders are removed randomly from the lower intertidal zone, meaning that platform erosion is not uniform. It also means that the platform surface in some places is contemporary and has high strength, whereas in other places the surface is old, highly weathered, and has low strength.

10.5.1 Models of rocky coast evolution

Models of rocky coast evolution reflect the interplay between contemporary processes and the role of inheritance, including time and geologic factors. Sunamura's

(1992) model for rocky coast evolution is based on the initial morphology of the coast and the relative strength of the rock, and assumes that sea level is stationary. The model is based on five different types of initial profiles I to V which represent different wave conditions at the base of the cliff (Figure 10.16):

- Coast I – A uniformly sloping coast with a low gradient. Incident waves break some distance offshore and the coastline is only affected by broken waves.
- Coast II – A coast with a cliff and a platform. The water depth at the base of the cliff is zero ($h = 0$). Incident waves break offshore on the subtidal platform and the coastline is only affected by broken waves.
- Coast III – A coast with a cliff and a platform. The water depth at the base of the cliff is between 0 and the depth of wave breaking ($h < h_b$). Wave breaking occurs offshore of the cliff face, but broken waves have more energy than coasts I and II.
- Coast IV – A coast with a cliff and a platform. The water depth at the base of the cliff is equal to the breaker depth ($h = h_b$). Incident waves break directly against the cliff face.
- Coast V – A coast with a plunging cliff. The water depth at the base of the cliff is greater than the breaker depth ($h > h_b$). Incident waves do not break, but are reflected from the cliff face, resulting in standing waves.

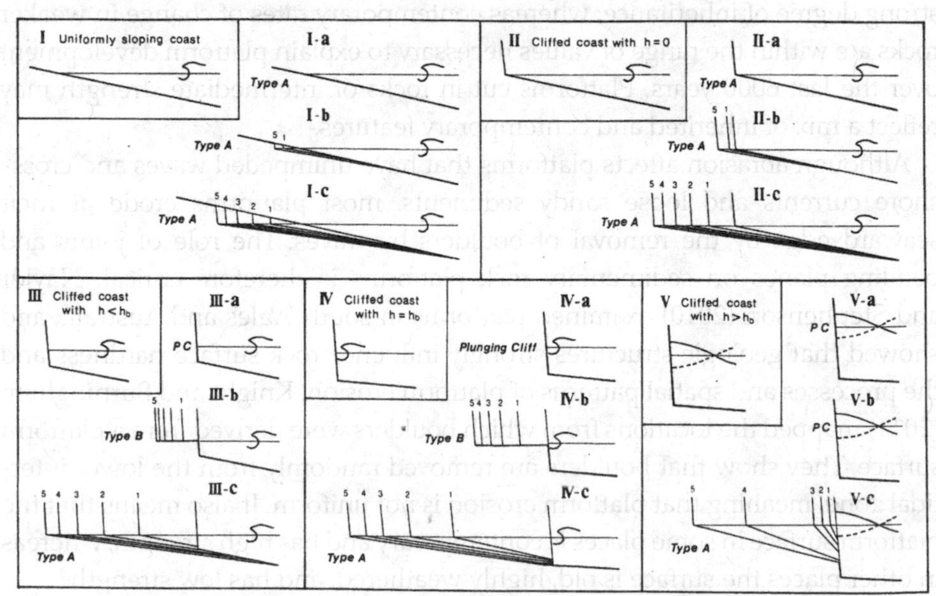

Figure 10.16 Model for rocky coast evolution beginning with five types of initial landform (I through V) with different degrees of rock hardness (**a**, **b** and **c**). 'Type A' refers to sloping shore platforms and 'Type B' indicates sub-horizontal platforms. The water depth at the base of the cliff is indicated by h and the breaker depth is represented by h_b. (From Sunamura, 1992.) (Copyright © 1992 John Wiley & Sons, reproduced with permission.)

The relative resistance of the rock to wave erosion is categorised qualitatively into three classes:

- **a** – very strong and highly resistant to weathering
- **b** – moderately strong and slightly susceptible to weathering
- **c** – very weak and vulnerable to weathering.

The fact that these classes are qualitative, rather than quantitative makes the model difficult to apply. However, Sunamura (1992) attempts to quantify the different strength classes:

$$\text{for a:} \quad \frac{S_c}{\rho g H_l} \geq 590$$

$$\text{for b:} \quad \frac{S_c}{\rho g H_l} = 77 - 590 \qquad \textbf{(10.4)}$$

$$\text{for c:} \quad \frac{S_c}{\rho g H_l} \leq 77$$

where S_c is the compressive strength of the rock (a measure of rock strength), ρ is density of seawater, g is gravity and H_l is the height of the largest waves occurring in the area under consideration. Equation 10.4 represents the first attempt at quantifying the force-balance between wave erosion and rock strength.

The morphological evolution of rocky coasts can now be traced depending on the initial type of coast (I to V) and the relative resistance of the rock to wave erosion (**a** to **c**). Given sufficient time, sloping shore platforms (Type A) develop on coasts **I** and **II**, regardless of the strength of the rock, and on coasts **III** and **IV** if the rock is very weak. However, sub-horizontal shore platforms (Type B) only develop in moderately strong rock on coasts **III** and **IV**. Plunging cliffs form on coasts **III** to **IV** if the rocks are very strong and on coast **V** if the rocks are moderate to very strong.

Models of cliff evolution in unconsolidated rocks such as glacial till deal more explicitly with relationships between cliffline retreat, sediment production and talus slope development. One such model called Soft Cliff and Platform Erosion (SCAPE) describes how cliff erosion leads to beach and shore platform development (Walkden and Hall, 2005). The cliffline retreats over time and gives rise to a sloping platform controlled by negative feedback between wave energy, erosion rate and platform slope. With application to the soft coast of north-east Norfolk, eastern England, output from the SCAPE model was compared to retreat rates from historic maps. This showed that the model was least sensitive to changes in wave height, moderately sensitive to changes in wave direction, and most sensitive to sea-level rise (Dickson *et al.*, 2007).

10.6 Introduction to coral coasts

Coral coasts, which include nearshore coral reefs and coral islands (atolls), are carbonate depositional systems that are related over long time scales to geologic and tectonic processes and climatic variability. The naturalist Charles Darwin (1809–82) was one of the first to discuss the morphological structures and evolution of coral reefs, and to identify relationships between coral islands, tectonics and geology (Darwin, 1842). **Coral reefs** represent the *in situ* production and deposition of calcium carbonate ($CaCO_3$) by biogenic processes. They cover about 2x10^6 km^2 of the tropical oceans and are the largest biologically constructed formations on Earth. Without the presence of organisms, coral reefs would not exist, so they are fundamentally different to the clastic coastal environments considered thus far. Coral coasts represent a delicate balance between biological and physical processes that control reef development and the morphology and dynamics of reef platforms. Here we focus on the geomorphological properties of coral coasts rather than their ecological/biological properties.

To the geomorphologist, coral reefs represent three-dimensional structures that consist of a living reef veneer overlying thick sequences of dead reefs

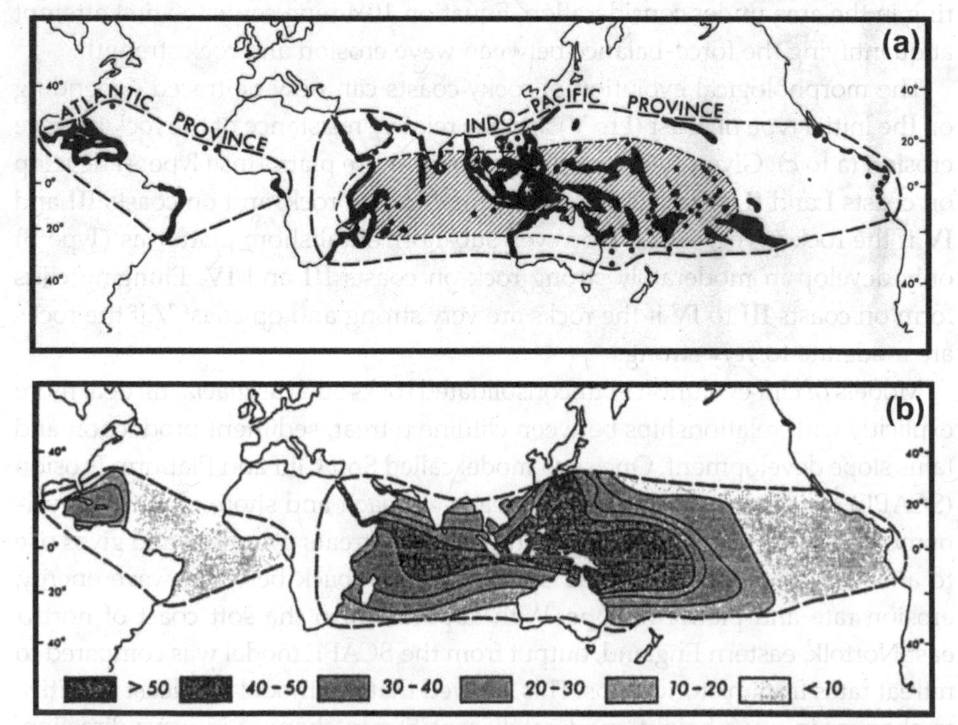

Figure 10.17 Global distribution of: (a) coral reefs and (b) number of reef-building coral genera. The hatched region in (a) includes the area of the most prolific reef development and almost all coral atolls. (From Davies, 1980.)

comprised of cemented $CaCO_3$. Reefs form by the *in situ* constructive abilities of a range of organisms, most notably stony (or scleractinian) corals and other types of corals, coralline and calcareous algae (McCook *et al.*, 2001). Although corals are common throughout the world's oceans, including in deep and cold water, **reef-building corals** (or hermatypic corals) are restricted to tropical and subtropical latitudes, particularly between 30°N and 30°S (Figure 10.17).

10.7 Ecology

10.7.1 Corals: Building blocks of coral reefs

Contrary to common belief, corals are animals, not plants. They are carnivorous suspension feeders that use their tentacles to trap living zooplankton in the water. The basic unit of the coral is the **polyp**, which sits within an external skeleton (exoskeleton) of $CaCO_3$ secreted by the organism. The $CaCO_3$ is extracted by the organism from sea water. Certain corals live in solitude, in which case a single polyp produces a single skeleton, but more commonly a compound coral structure is associated with a colony of polyps. Colonies grow by repeated division of its member polyps by asexual budding. As new polyps develop, old ones die, and over time enormous masses of carbonate rock are formed. Corals also reproduce sexually during brief periods of the year. The resulting coral larvae have limited swimming ability and they are largely dispersed and settled by ocean currents, limited by current circulation patterns and the duration of the planktonic phase. When coral larvae find a suitable substrate, they attach themselves to it and metamorphose into a juvenile polyp. Coral larvae establish most successfully on a firm, rocky substrate because they cannot initiate reef development where sediments are mobile.

An important aspect of the ecology of reef-building corals is their symbiotic relationship with small single-celled algae, known as **zooxanthellae**, which inhabit the endodermal tissue of the coral polyp. The zooxanthellae benefit from this relationship by being able to photosynthesise in a protected, nutrient-rich microenvironment. The zooxanthellae also give the coral colonies their distinctive bright colours. In return, the zooxanthellae remove metabolic waste, provide the coral with nutrients and also influence the rate of calcification (or skeletal deposition) as the colony grows.

10.7.2 Environmental controls on coral distribution and growth

The presence/absence and abundance of coral species are strongly controlled by global and local environmental conditions. The main environmental variables affecting zooxanthellae and therefore coral survivorship are sea water temperature and light availability (Figure 10.18). Locally, water turbidity (suspended sediment content), salinity, wave action and nutrient levels can also be important. These environmental factors are discussed below.

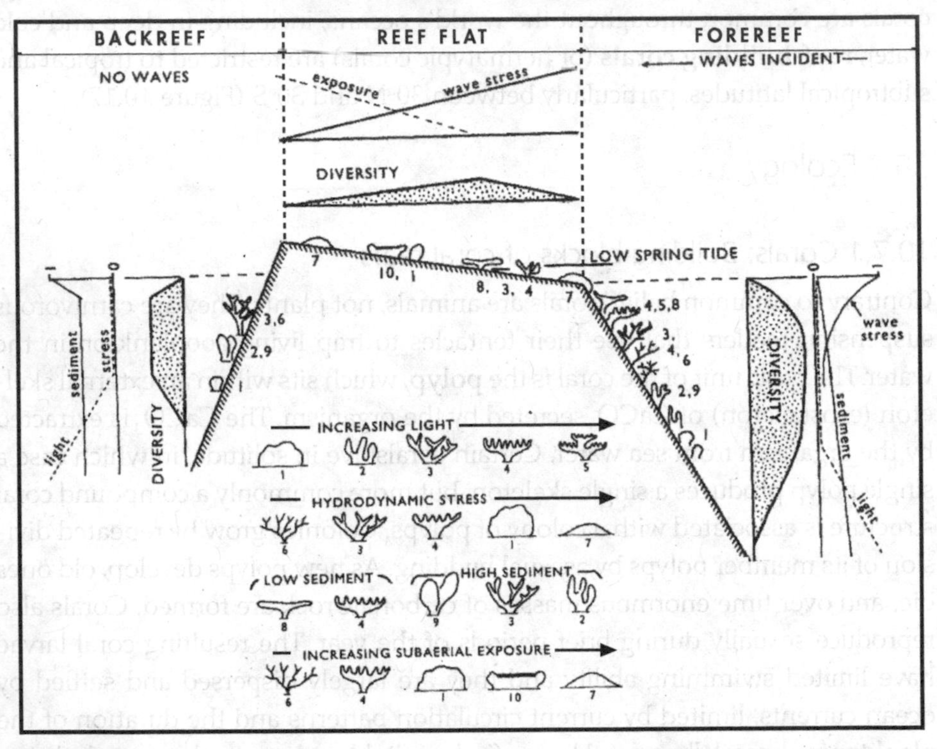

Figure 10.18 Variations in coral community forms and diversity across a reef in response to changing environmental conditions. (From Chappell, 1980.) (Copyright © 1980 Macmillan Magazines, reproduced with permission.)

Sea temperature controls the global distribution of coral reefs and abundance of reef-building corals. The lowest temperature that reef-building corals are able to withstand is 18°C, and the limit of coral reef occurrence lies very close to the 20°C sea surface isotherm for the coldest month, and so corals are therefore restricted to the tropics and subtropics. Coral growth is also limited by high temperature, and many reef builders will not survive very long where water temperatures are in excess of 34°C. Coral reefs occur in two main provinces (Figure 10.17):

- The **Indo-Pacific province** is centred on Indonesia and the Philippines and characterised by high biodiversity (80 genera, 500 species).
- The **Atlantic province** is located mainly in the Caribbean and characterised by a lower biodiversity (20 genera, 65 species).

Within each province, the environmental control of sea water temperature is evident. First, coral species richness decreases with increase in latitude, which results from decreasing water temperatures away from the equator (see Section 11.3). Second, coral reefs are generally less well-developed along the eastern coasts of oceans (i.e. west coast of continents), which is due to the active upwelling of cool water (colder than 18°C) along these coasts (see Section 11.3). The difference in species richness between the Indo-Pacific and Atlantic provinces is unrelated to temperature, however, and may be due to Holocene sea-level history, wave conditions and tidal range. There is also a smaller area of shallow coastal water in the Caribbean compared to the Indo-Pacific.

Short-term rises in water temperature can severely stress corals, causing them to expel their zooxanthellae. Without zooxanthellae, corals lose their pigment and take on a whitened appearance. This phenomenon is known as **coral bleaching**. A moderate bleaching event will reduce coral calcification rates and reproductive ability but may not necessarily cause widespread mortality. Severe bleaching, however, may cause widespread coral mortality and damage to the reef system. Increased frequency and severity of bleaching events due to increased ocean temperatures is a likely effect of global climate change (Box 10.1).

Box 10.1 Coral bleaching: The impact of global warming?

Global warming impacts on marine ecosystems are very varied but widespread (Fischlin *et al.*, 2007). Changes to the marine environment that have an impact on corals include sea surface temperature (SST), pH, dissolved oxygen content and water quality. Sea surface temperature rise is a key issue affecting the sustainability of coral ecosystems because it promotes **coral bleaching**. The causes and processes of coral bleaching are complex and include the concentration and type of zooxanthellae present and thermal stress to zoothanthellae and algae. Light levels, light wavelength, water optical quality (clarity), suspended sediment concentration, sea-level variations and rate of SST increase are also important. Furthermore, coral reef response is affected by local factors including reef geometry, age, heterogeneity and ecosystem dynamics. Cantin *et al.* (2010) show that the growth rate of the massive-reef species *Diploastrea heliopora* reaches a maximum at a SST of 30.25°C. Above this value, the growth rate declines rapidly (i.e. coral bleaching occurs), with a 5 per cent loss for every 0.2°C increase in SST. The precise temperature threshold and growth rate decrease varies between different coral species.

During El Niño events there is a short-term (few months duration) increase in SST in equatorial seas, which leads directly to episodes of coral bleaching. A study by Smith *et al.* (2008) of coral reefs in the eastern Indian

Ocean shows that different coral species respond differently to El Niño SST warming. During the strong El Niño event of 1997–98 hard and soft corals both decreased by over 80 per cent and algae doubled in abundance. Cover by individual coral families decreased by –3 to –37 per cent. There were also differences between coral families in how quickly the decrease took place, and how quickly they recovered afterwards. After six years, the hard coral species had recovered by 40 per cent but there was no change in the cover of soft corals. The community structure had been changed completely as a result of the 1997–98 El Niño.

Many studies are now considering the impact of global warming on coral reefs (Fischlin *et al.*, 2007). This is because the long-term effects on reef biodiversity, community structure and regeneration are still not known with certainty. Coral bleaching events could increase by 80 per cent if atmospheric CO_2 values reach double pre-industrial values. Higher CO_2 values cause an increase in ocean water acidity which in turn causes a reduction in calcification rate which affects coral growth rate and the strength of the coral exoskeleton. Other factors affecting future coral health include dissolved oxygen content within the water column, ocean storminess and waves, and river runoff.

Coral growth is also controlled by light because the symbiotic zooxanthellae depend on adequate light levels for photosynthesis. Light levels change dramatically with depth, so reef-building corals prefer shallow and clear water. For example, in Jamaica, light intensity at 25 m depth in clear water is only 1 per cent of surface illumination (Viles and Spencer, 1995). There are also significant changes in light wavelength with depth. The decrease in light level causes a reduction in coral diversity at depths as shallow as 10 m. Although reef-building coral may be found up to depths of 100 m, the maximum depth at which reefs are being built is rarely more than 50 m.

Coral growth is constrained by water level. Above approximately mean low water springs, zooxanthellae are susceptible to increased exposure where direct sunlight can elevate temperature and winds can cause desiccation. The control of water level is clearly seen on reef flats where **microatolls** are commonly found. These are small (less than 6 m in diameter) individual sub-circular coral colonies with living sides and dead surfaces. The elevation of the dead surface is directly related to atmospheric exposure (i.e. low tide level), and fossil (raised) microatolls have been used for determining past sea levels (Woodroffe and McLean, 1990).

Locally, water turbidity and salinity can be important controls on reef growth.

Turbidity refers to the presence of fine sediment suspended in the water and affects corals in two ways. First, high turbidity reduces water clarity and light penetration, thereby inhibiting coral growth rates and maximum coral depth. Second, high turbidity is associated with high sedimentation rates, and sediment can 'smother' corals. Corals can survive where salinity ranges from 27 to 40 ppt, but coral growth is restricted outside this range. The combination of high turbidity and low salinity is often why coral reefs are absent near river mouths. Conversely, corals located some distance from river mouths can record variations in river discharge through variations in coral properties. For example, fluorescent bands within corals of the Great Barrier Reef, Australia, contain humic and fulvic compounds from the Burdekin River of north Queensland, and record variations in river discharge over the last few centuries (Isdale *et al.*, 1998).

Exposure to wave action is another factor of coral growth. One would perhaps expect energetic waves to limit coral growth due to the physical force of breaking waves. However, the opposite is the case. Vigorous water mixing due to energetic waves prevents the deposition of suspended sediments, preserves a balance between oxygen and carbon dioxide within the water, and brings zooplankton to within reach of the reef community. As a result, coral reefs on exposed, windward sides of tropical islands are generally better developed than on more sheltered leeward sides. Major tropical storms, however, can significantly damage coral reefs.

Nutrient levels are also important. Coral reef communities thrive when nutrient levels are low. **Eutrophication** (nutrient enrichment) may result in the replacement of (hard) coral reef communities with (soft) macroalgae communities. Such a shift from hard to soft communities may result from sewage disposal in the lagoons of coral atolls, or on corals adjacent to river mouths that drain from agricultural land or urban areas. Of particular concern is the fact that re-establishment of coral reef communities may require far higher levels of water quality with respect to nutrients, salinity, turbidity and light than is required to maintain an established reef community.

10.7.3 Generalised cross-section through a coral reef

A large number of reef-building coral species exist and can assume a variety of forms depending on environmental and genetic factors. Ten generalised coral growth forms are indicated in Figure 10.18. These coral forms show a distinct zonation across a reef depending primarily on water depth (light level), wave action, turbidity and atmospheric exposure (Chappell, 1980). Three principal zones can be identified: (1) the forereef, which is the outer or seaward portion of the reef; (2) the reef flat, which is sub-horizontal and sometimes exposed during spring low tide; and (3) the backreef, which is the inner or landward portion of the reef. These zones can be seen clearly across coral islands (Figure 10.19a).

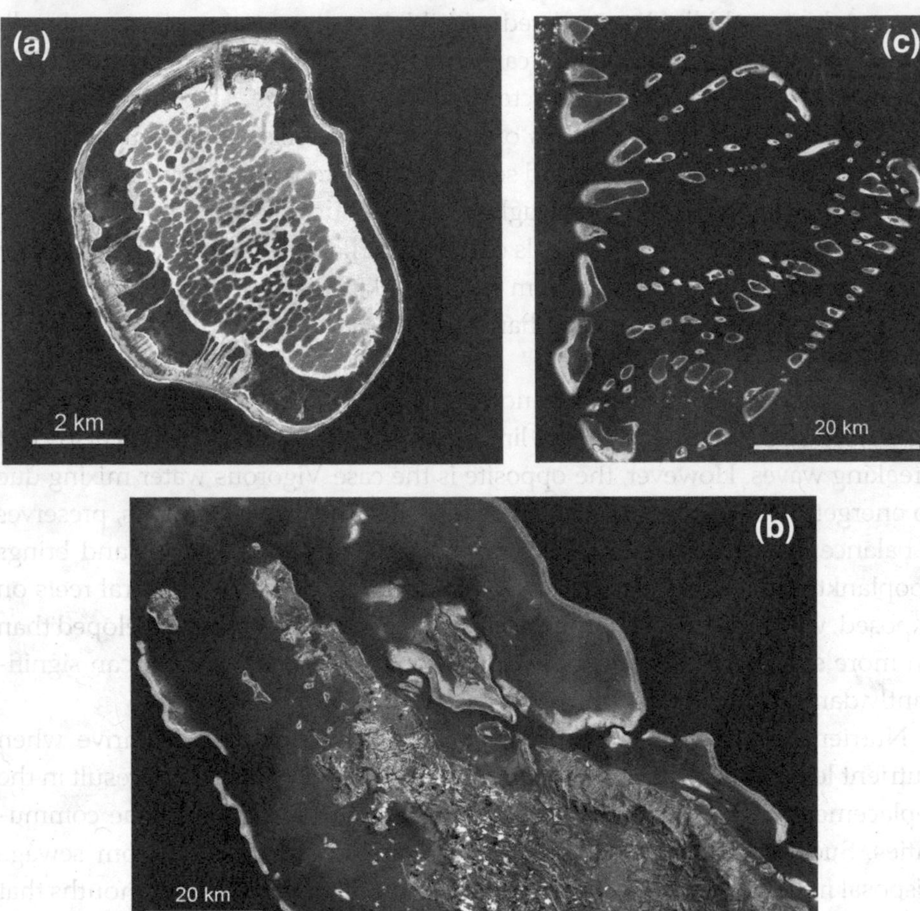

Figure 10.19 (a) Satellite view of Mataiva Atoll, part of the Tuamotu archipelago, French Polynesia, South Pacific. (Copyright © 2010 NASA, reproduced with permission.) (b) Satellite view of the barrier reef of New Caledonia, Coral Sea, South Pacific. (Copyright © 2008 NASA, reproduced with permission.) (c) Satellite view of North and South Malosmadulu Atolls of the Maldive Islands, Indian Ocean. (Copyright © 2005 NASA, reproduced with permission.)

The **forereef** is generally very steep with gradients of 30–40°. The lengths of these slopes range from several metres (on fringing reefs) to several km (on oceanic coral atolls). In all cases, the forereef is exposed to the highest wave energy. Optimal coral growth conditions prevail here due to intense water mixing. Species diversity is high with a large variety of growth forms present. Coralline algae can build low ridges on outer, windward edges of the forereef exposed to wave action. These algal ridges can stand up to 1 m above low tidal level and provide protection of the reef flat behind.

The **reef flat** lies to the lee of the algal ridge and generally forms the widest part of a coral reef. It is sub-horizontal in morphology and varies in width from less than 100 m to several km. On the reef flat, wave action is limited and exposure to light during low tide imposes significant environmental stress such that the reef flat has low species diversity. The reef flat consists of a pavement cemented by calcareous algae and incorporating detrital elements from the forereef. The reef flat may also have thin, mobile sheets of sand, gravel or coral rubble which can sometimes form a significant beach.

The morphology of the **backreef** located behind the reef flat depends on the type of reef under consideration. On barrier reefs and coral atolls, a lagoon is found behind the reef flat (Figure 10.19a), whereas on fringing reefs, the backreef merges with the land. Wave action on the backreef is insignificant, resulting in limited water mixing and sediment deposition. Fresh water and sediments from streams draining the hinterland provide additional environmental stress. Coral growth conditions are therefore sub-optimal on the backreef and species diversity is low.

10.8 Reef growth

The growth rate of individual corals ranges from 1–20 cm yr^{-1} depending on the species in question. Consequently, the rate of **coral growth** varies across a reef system according to its ecological zonation. However, coral growth does not equate to **reef platform growth** which occurs on longer spatial and temporal scales. The distinction between the growth of individual corals and reef platforms is critical for understanding the geomorphic development of reefs.

It is a common perception that tropical reefs are composed of corals preserved in their growth position. This perception is misleading because many reef corals comprise a minor proportion of the reef matrix. Other marine organisms can also secrete calcium carbonate, and these can add to overall reef architecture. The basic framework of a reef is thus a porous structure, provided by corals and coralline algae, cemented and sealed by other encrusting/cementing organisms (calcareous algae, sponges, bryozoans) and infilled by fragments from other calcareous organisms (foraminifera, molluscs, echinoids, calcareous plants).

Constructional activity on the reef is always offset by reef organisms that destroy carbonate substrates. Destructive processes include physical (through mechanical wave stress) and biological breakdown. Bioerosion processes (borers and grazers) effectively limit reef framework construction and produce sediments on reef fronts and tops. As a consequence of the reworking and cementation of carbonate material, rates of reef platform growth are much lower than those of individual corals.

Coral reef communities have high gross primary productivity, despite the fact that surrounding waters have low concentrations of key nutrients, specifically phosphate and nitrate (Viles and Spencer, 1995). The most important consequence of reef metabolic processes is **calcification**. Field studies show typical calcification rates of 10 kg of $CaCO_3$ m^{-2} yr^{-1} for coral thickets, 4 kg of $CaCO_3$ m^{-2} yr^{-1} on typical Indo-Pacific reef flats, and 0.8 kg of $CaCO_3$ m^{-2} yr^{-1} on lagoon floor and rubble substrates (Kinsey, 1985). The production of calcium carbonate can be converted into a potential vertical reef growth rate. Since the density of calcium carbonate is 2.9 g cm^{-3} and assuming an average porosity of reef sediment is 50 per cent, a calcification rate of 10 kg m^{-2} yr^{-1} corresponds to a potential upward growth of 7 mm yr^{-1}. A comparison of reef growth rates for different tectonic settings and reef types suggests a wide range of reef responses over the last 8000 years, with reef growth rates of around 1–10 mm yr^{-1} (Figure 10.20).

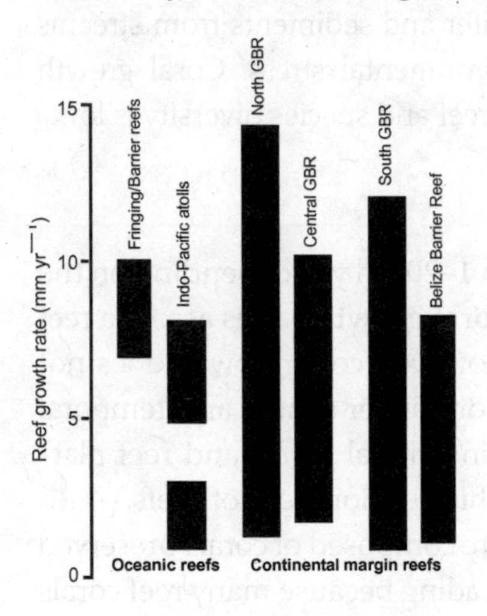

Figure 10.20 Rates of Holocene reef growth based on radiometrically-dated core material. GBR refers to Great Barrier Reef. (Modified from Spencer, 1994.)

10.9 Types of coral reefs and their formation

Coral reefs occur in two major oceanographic settings:

- **Shelf reefs** are found on the continental shelf in water depths less than *c.* 200 m and usually have foundations of continental crust. They are usually located closely to the mainland or islands and may be significantly affected by fresh river water and sediments. Typical shelf reefs include the Great Barrier Reef (Australia) and along the coast of Belize (central America).
- **Oceanic reefs** may rise several km from the ocean floor. They have volcanic foundations and experience limited influence from terrestrial processes. Examples include Pacific island reefs such as Mayotte Island, Society Islands and New Caledonia.

Historically, reef environments have been subdivided into fringing reefs, barrier reefs and coral atolls. Although this classification is not all-inclusive, it is discussed below. The scheme is least suitable for shelf reefs, which are more variable than oceanic reefs in their morphology and ecology.

Fringing reefs extend out from mainland or island coasts, so are narrow where the submarine slope is steep, and wide where it is gentle (Figure 10.21a). Fringing reefs are usually thin and most comprise veneers of Holocene coral overlying platforms of non-reef origin. The forereef of a fringing reef is usually steep with active and abundant coral growth. The reef flat is often strewn with sand and rubble, and coral growth is relatively inactive. A fringing reef is often breached by passages located opposite stream mouths where the inflow of fresh water and sediment inhibit coral growth. Some fringing reefs, especially in east Africa, have a narrow depression, termed a **boat channel**, between the reef and the land (Guilcher, 1988). The depression, which may be 100–200 m wide and

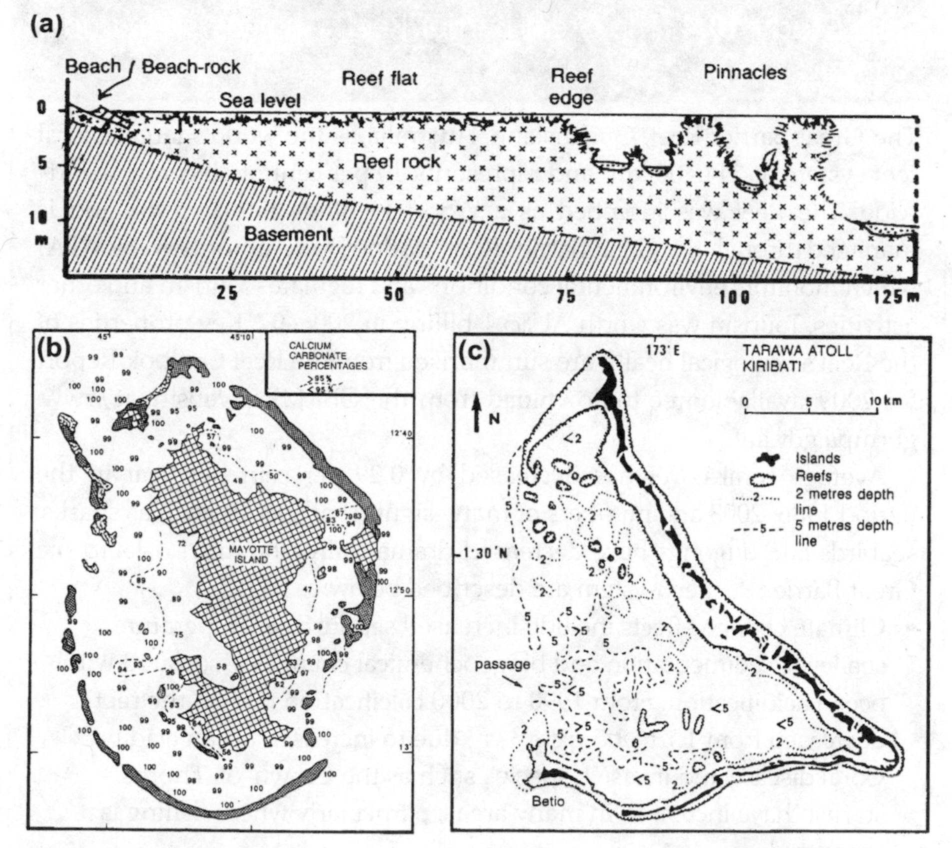

Figure 10.21 (a) Fringing reef with incipient boat channel, Aqaba harbour, Jordan, Red Sea. (b) Barrier reef of Mayotte Island, Indian Ocean. (c) Tarawa Atoll, Kiribati, Micronesia. (From Guilcher, 1988.) (Copyright © 1988 John Wiley & Sons, reproduced with permission.)

up to 3 m deep, develops where sediment from the land inhibits coral growth.

Barrier reefs are narrow, elongate structures separated from mainland or island coasts by lagoons that are generally less than 30 m deep (Figure 10.21b). The barrier reef morphology of New Caledonia is shown in Figure 10.19b. Barrier reefs may be continuous or separated by passages of varying width and depth which allow for ingress of ocean water and sediment into the lagoon. The nature of lagoon sediments can indicate the relative contributions of sediment supply from the land (siliclastic material) and from corals and other marine organisms (carbonate material). Studies of the proportion of $CaCO_3$ in lagoonal sediments in the Great Barrier Reef show that terrigenous sedimentation dominates the inner lagoon, with marine sedimentation dominant in the outer lagoon. The impact of climate change and management on the Great Barrier Reef ecosystem are discussed in Case Study 10.4.

Case Study 10.4 The Great Barrier Reef: An ecosystem under change

The Great Barrier Reef, Queensland, Australia, is the world's largest coral reef system at 344,400 km^2 and represents 10 per cent of all corals worldwide. The Reef was inscribed as a World Heritage Site in 1981 and is managed through the Great Barrier Reef Marine Park Authority (GBRMPA) which monitors environmental conditions and regulates tourism and other activities. Tourism was worth AU$5.1 billion in 2006–07. Key properties of the Reef's ecological health are summarised from the Reef Outlook Report for 2009, available free to download from the GBRMPA website at www.gbrmpa.gov.au/

Average coral cover has decreased by 0.29 per cent per year in the period 1986–2008 and numbers of many significant species such as sharks, seabirds and dugongs have decreased dramatically. The main risks to the Great Barrier Reef ecosystem are described below.

- Climate change effects include increased sea surface temperature, sea level, sedimentation and biogeochemical changes associated with ocean acidification. From 1990 to 2000 calcification rate on the reef decreased from 1.7 to 1.5 g cm^{-2} yr^{-1} due to increased ocean acidity. Coral disease and invasive species such as the Crown-of-Thorns starfish have increased in many areas, particularly where fishing is permitted.
- Nearshore sediment, pollutant and nutrient inputs from rivers are causing eutrophication and algal growth that have a negative impact

on water quality and coral health.

- Development of the Queensland coasts includes draining and modifying wetlands and mangroves, increased coastal population, freshwater usage, and activities such as mining.
- Legal and illegal fishing and poaching can impact on species of conservation concern and through removing top predators such as sharks.

Monitoring Great Barrier Reef ecosystems shows that reef habitats are capable of recovering from short-term disturbances, but the future capacity for reef recovery (i.e. reef resilience) will be likely reduced because environmental disturbances are predicted to increase in their frequency and severity. The Great Barrier Reef ecosystem is graded as 'high risk' and the outlook for the ecosystem in future is graded as 'poor'.

Coral atolls are reefs that surround one or several central lagoons (Figure 10.21c). There are 425 coral atolls in the world, most of which are in the Indian and Pacific Oceans, with only a few in the Caribbean (Stoddart, 1965). Atolls can have circular, elliptical or horse-shoe shapes and vary greatly in size. The large atolls of the Maldives, for example, are in excess of 75 km wide, but most atolls are about 10 km wide (Figure 10.19c). Small atolls, termed **faros** or **atollons**, may be present on the margin of larger atolls. Atolls are generally asymmetric in plan view, being often wider and better developed on the windward side, while outlets to the ocean are generally on the leeward side. The lagoon floor is a smooth depositional surface from which pinnacles and ridges of live coral may protrude. Small platform-like reefs, known as **patch reefs**, may also be present in the lagoon. In contrast with sediments in the lagoons of barrier reefs, sediments in atoll lagoons are entirely calcareous. Sand and gravel beaches and islands may be formed where sediment is washed in through channels or over the bordering reef on to the surface of reef flats. The morphodynamics of these reef and islands is discussed in Section 10.11.

10.9.1 Darwin's model of coral reef formation

During the voyage of HMS *Beagle* from 1832–36, Charles Darwin realised that reef-building corals only live in relatively shallow water (Darwin, 1842). He inferred that the vertical thickness of oceanic barrier and coral atolls is far greater than the maximum depth at which these animals live, which can only be explained by the progressive subsidence of the platform on which the reefs had built.

Darwin's subsidence theory describes the stages of development of the three main types of coral reef (Figure 10.22). A fringing reef surrounding a slowly sinking island grows upwards, inwards and outwards as sea level rises relative to the land. As the land behind the reef subsides, the environmental conditions in the lagoon, characterised by high sedimentation rates, fresh water runoff and limited water mixing, are not favourable for coral reef growth and result in the death of reef-building organisms. A lagoon then forms. The width and depth of the lagoon increase as subsidence proceeds, and the fringing reef develops into a barrier reef. If the barrier reef encircles the island and the island itself sinks below sea level a coral atoll results. Should the rate of subsidence exceed the rate of upward growth of coral, a drowned atoll may result.

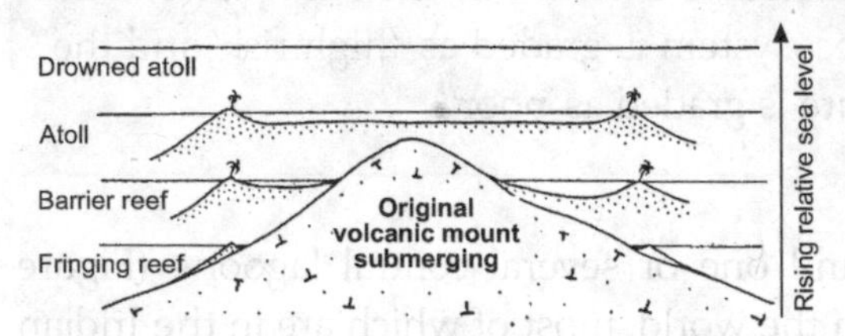

Figure 10.22 Darwin's theory of atoll formation.

When coupled with **plate tectonics**, Darwin's subsidence theory can explain the occurrence and evolution of many linear island chains in the Pacific, such as the Hawaiian and Society Islands. Figure 10.23 shows a model for coral atoll development in the Pacific. The Pacific plate migrates in a west-northwesterly direction, away from the mid-oceanic ridge, at a rate of 80–150 mm yr^{1} while at the same time subsiding at a rate of 0.02–0.03 mm yr^{-1}. Irregular volcanic activity associated with a fixed hot spot results in the formation of a chain of volcanic islands. Coral reefs form on these slowly migrating and subsiding islands and, during their travel, fringing reefs are gradually transformed into barrier reefs, then atolls. As the atolls continue to migrate towards higher latitudes, the reduction in water temperature causes a decrease in coral growth rates. Eventually, vertical reef growth can no longer keep up with the subsiding substrate and the coral atoll drowns, resulting in a **guyot**.

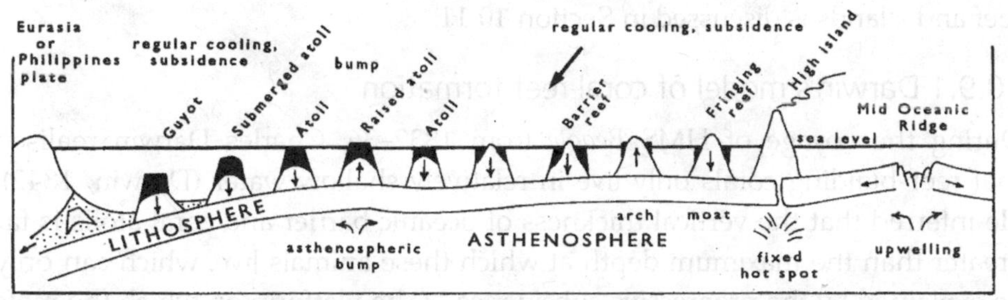

Figure 10.23 Model of reef evolution on the Pacific Plate. (From Guilcher, 1988; modified from Scott and Rotondo, 1983.) (Copyright © 1988 John Wiley & Sons, reproduced with permission.)

Cores through coral platforms and seismic data have now conclusively demonstrated the general validity of Darwin's subsidence theory. For example, drilling on Eniwetak atoll in the Marshall Islands (Pacific Ocean) demonstrates that this atoll consists of more than 1250 m of shallow-water coral limestone on top of a volcano that rises 3.2 km from the ocean floor (Ladd *et al.*, 1967).

10.10 Reef morphology and Quaternary sea-level changes

Darwin's theory adequately explains the formation of oceanic atolls over time scales of millions of years where the typical rate of island subsidence is 0.01–0.1 mm yr^{-1}. Superimposed upon this gradual subsidence of the reef basement have been sea-level fluctuations of *c.* 100–130 m over the Quaternary (Chapter 2) at rates of around 1–10 mm yr^{-1} and which exerted a strong control on reef growth and coral island shape over millennial time scales. Present sea-level rise in the Pacific of 2 mm yr^{-1} is also an important factor for contemporary coral systems, discussed below. Over the Quaternary, periods of subaerial exposure during glacial lowstands followed by drowning and reef re-establishment during interglacial highstands are thought to be the overriding control on the morphology of contemporary coral reefs.

10.10.1 Effect of subaerial weathering during glacial lowstands

The geologist Reginald Daly (1871–1957) was the first to be concerned with the effect of relative sea-level changes on coral reef development (Daly, 1915). According to his **glacial control theory**, modern coral reefs started forming during periods of low Pleistocene sea level, such that the shape of contemporary coral reef systems is largely inherited. For example, many deep passages through barrier reefs and coral atolls are remnants of Pleistocene river valleys that drained from the emerged lagoon into a lowered sea (Guilcher, 1988). Coral reefs emerged above sea level during glacial lowstands have been subjected to **karst processes** by solution of calcium carbonate. In the **antecedent karst theory** put forward by Purdy (1974), modern coral reefs are thin accretions over older reefs that have been strongly modified by subaerial karst processes (Figure 10.24). Purdy proposed that the shape of many, if not most, reefs is karst-induced. Rainfall solution is most rapid towards the interior of steep-sided reef platforms, and inhibited at the edges due to rapid runoff. Karst-eroded surfaces, with raised rims and central solution depressions, are re-colonised when atolls are drowned by rising sea level. Barrier reefs may also be karst-induced, where the lagoon is the result of enhanced limestone solution compared to the seaward edge of the reef.

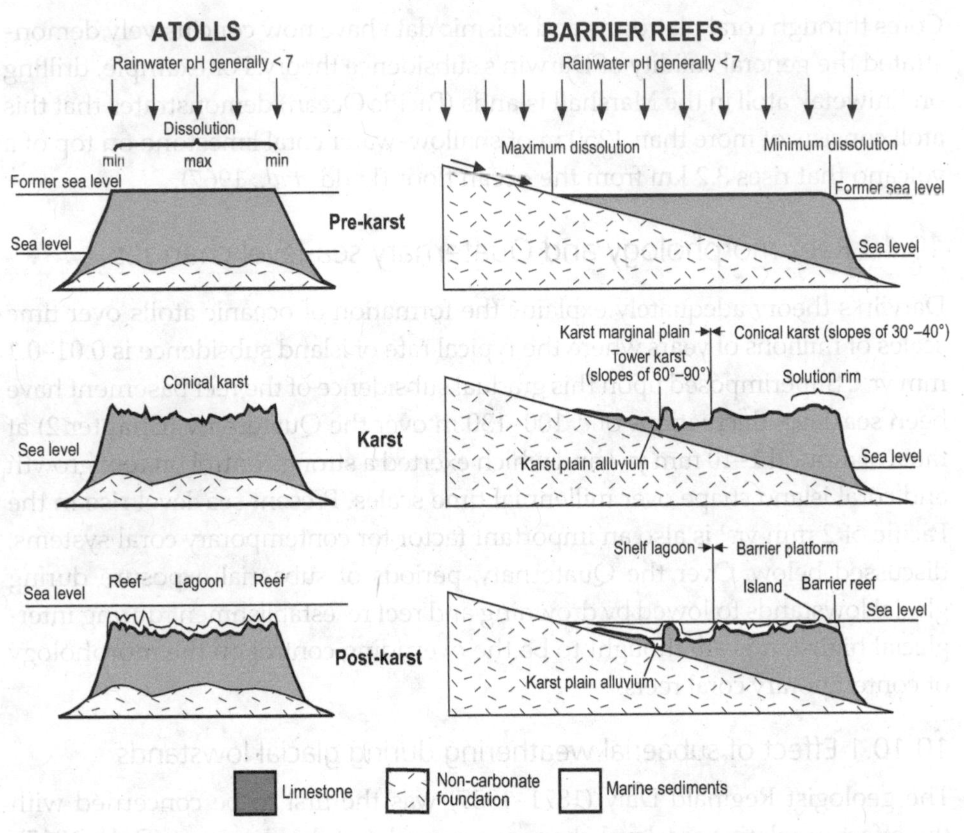

Figure 10.24 The evolution of atolls and barrier reefs according to the antecedent karst theory. (Modified from Purdy, 1974.)

It is difficult to determine the importance of karst processes in shaping modern coral reefs. Although karst processes have operated on emerged coral reefs during glacial periods, the subaerial weathering rate of limestone surfaces of less than 0.5 mm yr^{-1} means that there might not have been sufficient time for prominent karst landforms to have developed. On the other hand, evidence of karstification can be found in many reef environments (Guilcher, 1988), suggesting that it is least locally significant.

10.10.2 Coral reef growth strategies during sea-level rise

Changing relative and/or absolute sea levels are important factors in coral reef development. Of particular interest is the ability of vertical reef growth to keep pace with rising sea level. Studies on Caribbean and tropical Atlantic reefs have shown how these reefs have responded to sea-level rise. Neuman and Macintyre (1985) define three types of reef strategies according to their response to Holocene sea-level rise:

- **Keep-up reefs** grow upwards fast enough to keep pace with rising sea level and are able to maintain shallow, frame-building communities throughout the sea-level rise.
- **Catch-up reefs** begin as shallow water reefs and become deeper as the rate of sea-level rise exceeds the vertical accretion rate. When the rate of sea-level rise decreases, reefs catch-up with sea level. The stratigraphy of catch-up reefs has an upward-deepening sequence that develops when rising sea level leaves the reef surface behind, followed by an upward-shallowing sequence which forms when the reef surface catches-up.
- **Give-up reefs** fall behind rising sea level and are eventually drowned. Their stratigraphy consists of an upward-deepening sequence followed by a deep-water facies.

Differences in growth strategy depend on the rate of sea-level rise, the time lag for recolonisation of the inundated reef surface, and accommodation space for vertical reef growth. The duration of sea-level rise and the depth below which reef growth ceases determines whether a coral reef has the opportunity to catch-up. Spencer (1994) suggests the give-up/catch-up threshold lies at a rate of sea-level rise of 20 mm yr^{-1} and the catch-up/keep-up threshold at 8–10 mm yr^{-1}. The most common growth strategy of the Great Barrier Reef during the Holocene was the catch-up mode (Hopley, 1994). Initially, reef growth could not keep pace with rapidly rising sea level during the early Holocene, but most reefs caught up with sea level over a few thousands of years after the mid-Holocene. Many Caribbean shelf reef systems, however, exhibited give-up behaviour during the Holocene (Macintyre, 1988). The demise of these reefs is generally attributed to a reduction in vertical growth rates due to a decrease in water quality after flooding of the continental shelf at the start of the Holocene under high turbidity and nutrient level conditions.

The reef growth strategy has important implications for the geomorphic development of reefs and associated sedimentary deposits. While reefs grow vertically, the majority of calcified material is retained in the reef framework. However, once reefs attain their maximum vertical extent, excess calcification is shed from the reef as detrital sediments which become subsequently available for reef island formation and lagoon infilling.

10.10.3 Present sea-level change and coral reef response

The response of coral reefs to present sea-level rise depends primarily on whether vertical reef growth can keep up with rising sea level. Reef accretion rates range from 1–10 mm yr^{-1}, the lower rate of which is broadly similar to the rate of global sea-level rise of 3.1±0.4 mm yr^{-1} (Figure 2.11). A rate of 10 mm yr^{-1} is commonly taken as the maximum sustained vertical reef accretion rate

(Buddemeier and Smith, 1988). Coral reefs should generally be able to keep up with the predicted rate of sea-level rise, provided that other factors such as increased water temperature and damaging anthropogenic influences are not acting simultaneously (Fischlin *et al.*, 2007).

Coral atolls and reef islands are considered particularly vulnerable to sea-level rise, due to their low-lying position. The response of these environments to sea-level rise depends on the balance between reef growth, sediment supply from reef flat to island, and the rate of sea-level rise. Recent reviews on coral islands have emphasised their variability and resilience (e.g. Bradbury and Seymour, 2009). Despite this, coral islands are among the most sensitive environments to long-term sea-level rise and climate change. This is especially the case where these effects are superimposed on destructive short-term effects such as El Niño-induced warming, hurricanes, damaging human activities and declining environmental quality. The impacts of El Niño events on Pacific coral reefs are examined in Box 10.2.

Box 10.2 Coral responses to El Niño events

Changes to sea surface temperature (SST) and the hydrological cycle (rainfall volume) during El Niño events in the central Pacific are strongly imprinted on coral reef systems through variations in the oxygen and carbon isotopic properties in the coral's calcium carbonate skeleton. The basis behind isotopic ratios of oxygen was explained in Box 2.2. El Niño events in the central Pacific region are associated with positive temperature and positive precipitation anomalies. These are linked together largely because higher SST leads to higher rates of evaporation and in turn to higher precipitation. Oxygen isotope variability from coral reef records on Christmas (Kiritimati) Island in the central Pacific shows a clear relationship to SST and precipitation (Figure 10.25). During El Niño events here, precipitation increases by 4–6 times average values (Woodroffe and Gagan, 2000). Higher rates of evaporation lead to the precipitation containing less of the heavier $\delta^{18}O$ isotope relative to the lighter $\delta^{16}O$ isotope. The variability in isotope ratios is strongly correlated with SST ($r = 0.80$) and both show a temperature variability of around 5 °C between El Niño and non-El Niño conditions. These records come from *Porites* coral from a microatoll on a reef flat on Christmas Island where shallow, warm water is flushed over the reef crest and on to the microatoll. High-resolution (monthly) records were obtained by drilling and slicing through the coral bands at around 1 mm intervals.

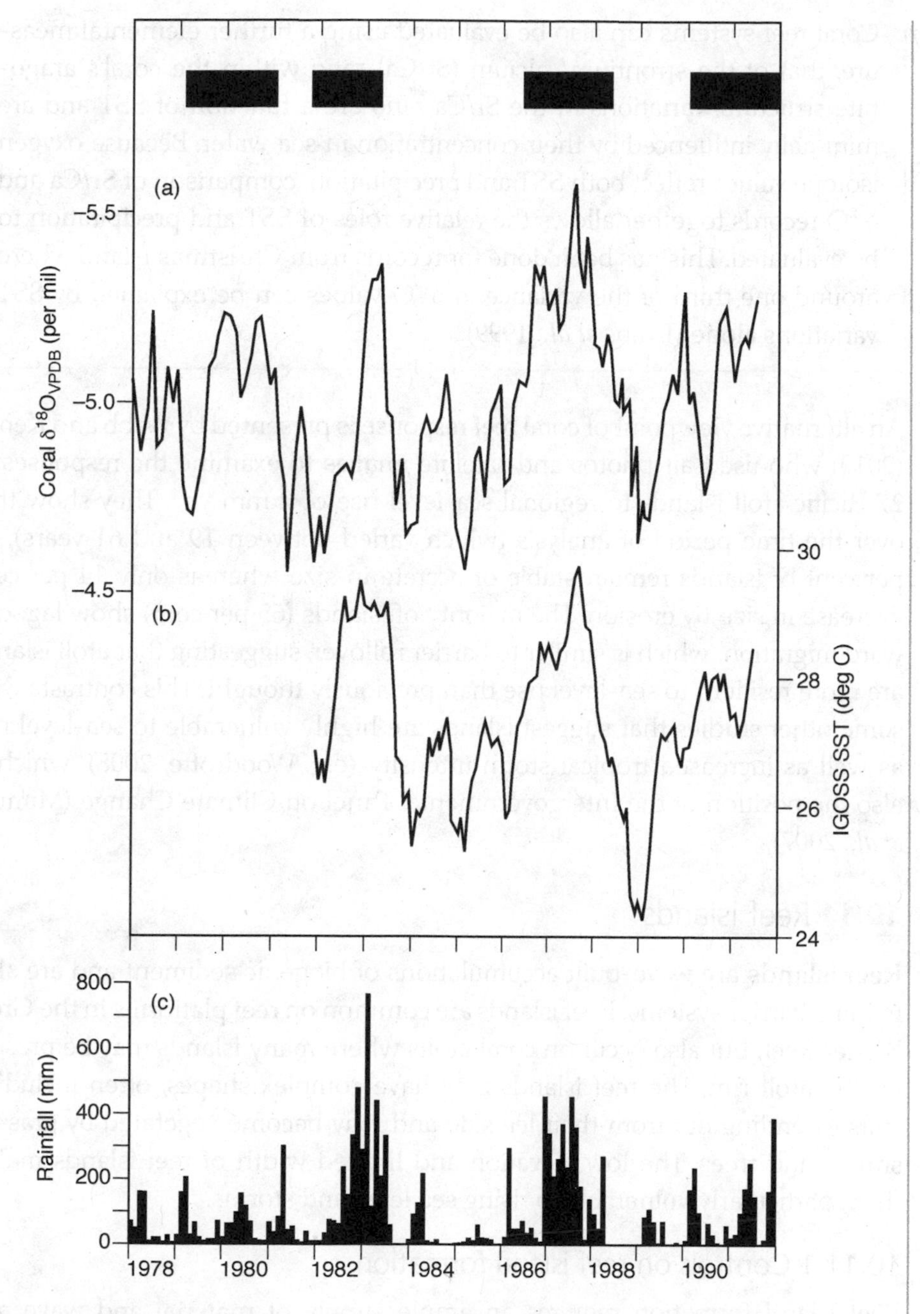

Figure 10.25 (a) Time-series of $\delta^{18}O$ measurements through modern microatoll in Christmas Island, central Pacific. Periods with warm El Niño conditions are indicated by shaded blocks, (b) SST record, (c) monthly rainfall for Christmas Island 1978–91. (From Woodroffe and Gagan, 2000.) (Copyright © 2000 American Geophysical Union, reproduced with permission.)

Coral reef systems can also be evaluated using a further elemental measure, that of the strontium/calcium (Sr/Ca) ratio within the coral's aragonite structure. Variations in the Sr/Ca ratio are a function of SST and are minimally influenced by their concentration in sea water. Because oxygen isotope values reflect both SST and precipitation, comparison of Sr/Ca and $\delta^{18}O$ records together allows the relative roles of SST and precipitation to be evaluated. This has been done for records from Christmas Island where around one third of the variance in $\delta^{18}O$ values can be explained by SST variations alone (Evans *et al.*, 1999).

An alternative viewpoint of coral reef response is presented by Webb and Kench (2010) who used air photos and satellite images to examine the responses of 27 Pacific atoll islands to regional sea-level rise of 2 mm yr^{-1}. They show that over the time period of analysis (which varied between 19 and 61 years), 86 per cent of islands remain stable or accrete in size whereas only 14 per cent decrease in size by erosion. The majority of islands (65 per cent) show lagoonward migration, which is similar to barrier rollover, suggesting that atoll islands are more resilient to sea-level rise than previously thought. This contrasts with some other studies that suggest islands are highly vulnerable to sea-level rise as well as increased tropical storm intensity (e.g. Woodroffe, 2008), which is also the position of the Intergovernmental Panel on Climate Change (Mimura *et al.*, 2007).

10.11 Reef islands

Reef islands are wave-built accumulations of biogenic sediment and are akin to mini-barrier systems. Reef islands are common on reef platforms in the Great Barrier Reef, but also occur on coral atolls where many islands may be present on the atoll rim. The reef islands may have complex shapes, often including spits extending out from their lee side and may become vegetated by grasses, shrubs and trees. The low elevation and limited width of reef islands makes them particularly vulnerable to rising sea level and storms.

10.11.1 Controls on reef island formation

Reef island formation requires an ample supply of material and wave and current patterns that concentrate sediment on the reef. The biogenic sediments that make up reef islands are derived from erosion of coral reefs by waves and the activities of borers and grazers. Both wave and bioerosion produce large quantities of loose reefal material, ranging from mud to boulders. Part of this material will move down the forereef (coarse material) or be washed away (fine material). However, a large amount will be thrown onto the reef flat by waves, in particular during storms, eventually forming reef islands.

The main processes responsible for sediment transport and reef island formation on reef flats, are gravity waves and net currents (wind-, wave- and tide-driven). Similar to sediment transport in the nearshore of wave-dominated shorefaces, wave-induced bed shear stresses entrain and mobilise sediments on reef flats, while net currents move the sediment and determine transport pathways. The amount of incident wave energy that can propagate across the reef surface controls the size and quantity of the transported sediments. Wave height across the reef surface is governed by the interaction between the tidal water level and the elevation of the reef flat. At low tide, when a large proportion of the reef flat may be emerged, incident waves break on the reef crest and only a limited amount of wave energy is transmitted across the flat. In contrast, at high tide, particularly during storms and/or spring tides, water may be 2–3 m deep on the reef surface, allowing a greater amount of wave energy to propagate. Nelson (1994) found that the maximum value of the ratio of wave height to water depth (H/h) never exceeds 0.55 on reef platforms. Thus, wave height on the reef flat is strongly controlled by the local water depth. As a result, tidal modulation of the wave height is apparent, with larger waves at high tide and low waves at low tide (Figure 10.26). Tidal currents dominate water circulation within lagoons and passages. Wave breaking at the reef crest may force water flow and biogenic sediments from the reef flat into the lagoon.

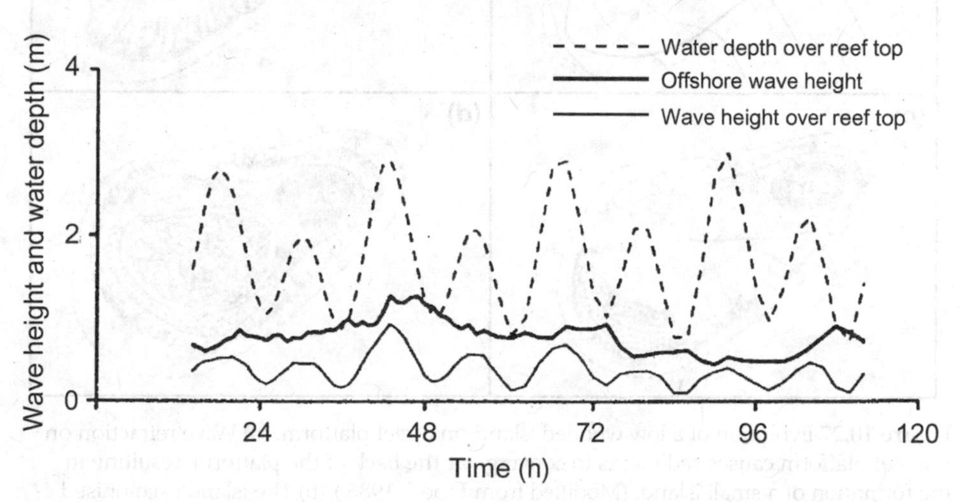

Figure 10.26 Time series of wave and tide data measured on the reef flat of John Brewer Reef, Australia. The data show that waves on the reef flat are higher during high tide than low tide conditions. (Modified from Nelson, 1994.)

10.11.2 Models of reef island deposition

Islands formed of reef debris can be divided into two types:

- **Motus** are long, narrow islands found mainly on the windward side of coral atolls. Their location is controlled by the diminished capacity of currents to transport sediment across the reef and as a consequence they are found in linear chains that parallel the reef rim. The morphology of motus is characterised by ridges comprising gravel to cobble-size material at the oceanside and sand ridges at the lagoonside. Coarse material on the outer part of motus is transported across the reef flat under storm conditions, while sand-size material is deposited either through bypassing of sediment in passages between islands, or by locally generated wind waves in the lagoon.
- **Cays** are smaller than motus and consist of sand and gravel, rather than gravel and cobbles. They are generally found on the leeward side of coral atolls, where lower incident wave energy means that only smaller-sized sediment can be transported and deposited.

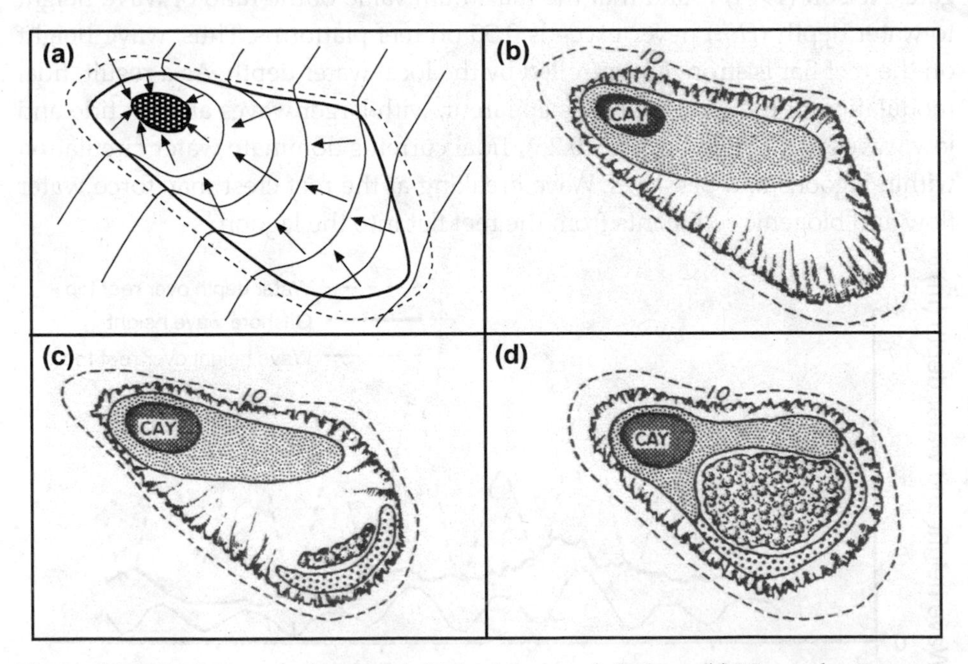

Figure 10.27 Evolution of a low wooded island on a reef platform. (a) Wave refraction on the reef platform causes sediments to converge at the back of the platform, resulting in the formation of a small island. (Modified from Flood, 1986.) (b) The island is colonised by vegetation. (c) A shingle ridge develops at the windward side of the reef platform. (d) Mangroves start colonising the area between the shingle ridge and the sandy cay. (From Bird, 2000.) (Copyright © 2000 John Wiley & Sons, reproduced with permission.)

The occurrence of finer-grained cays and coarser-grained motus is strongly linked to the incident wave climate: cays are primarily found in low-energy wave settings and motus associated with high-energy atolls (Bayliss-Smith, 1988). The distinction between cays and motus is by no means sharp.

Islands that form on individual reef platforms, rather than coral atolls, commonly have similar shapes to the reef platform upon which they are deposited (McKoy *et al.*, 2010). Flood (1986) accounted for the formation of such islands through wave refraction around reefs. For example, on circular reefs, waves refracting around the reef platform converge on a focal point on the reef top where the sediments are deposited. Over time, continued deposition allows such islands to build radially towards the reef edge. Figure 10.27 shows such development of a reef island due to wave refraction, and illustrates the further development of the island into a **low wooded island**, a type of reef island typically found in the Great Barrier Reef, Australia.

The formation of reef islands postdates the development of the reef flat on which they form, but little is known about the pattern and mode of sediment accretion on reef islands or their chronology of development. Woodroffe *et al.* (1999) have indicated a number of ways in which reef islands on atolls may have developed over the last 3000–4000 years, depending on whether they have developed gradually or episodically, and whether they are still accreting. Since that date, different parts of the reef system show different autogenic patterns of evolution, with the outer reef growing upwards but the island shoreline itself responding seasonally to variations in storminess (Kench *et al.*, 2005). In addition to such evolutionary behaviour, the pattern of sediment deposition on reef islands can also vary greatly. Reef islands may have developed from a central core, but can also have developed by oceanward, lagoonward or vertical accretion.

10.11.3 Reef island morphodynamics

Reef island size, shape and position on reef platforms are sensitive to changes in waves, storminess and sea level. Bayliss-Smith (1988) developed a model of reef island behaviour based on the size of material that comprises the islands and whether islands are found in storm or non-storm environments (Figure 10.28). The magnitude and frequency of storms play a vital role in controlling reef island morphology:

- Extreme, catastrophic storms (cyclones or hurricanes) are a mechanism for coarse sediment production, but such events will also cause shoreline erosion and strip islands of fine sediment during overwashing. Cays are particularly vulnerable to storm erosion. Tsunami can also lead to shoreface erosion and landward transport of coastal debris on all types of coastal islands, but Kench *et al.* (2006) argue that atolls are resilient to tsunami forcing.

- More frequent, lower-magnitude storms can also erode cays. However, the more energetic wave conditions can cause onshore transport of coral rubble on the reef flat, so as to build up eroded motus beaches and shorelines. The accretion of storm debris onto old shorelines can lead to a net enlargement of islands, both in width and, through littoral drift and spit development, in length.
- The net effect of small storms and calm weather conditions is erosive on motus. This is because wave action will remove sediment from the islands without compensating replenishment from new coral rubble. Motus will therefore shrink at least on their seaward coasts, but coral growth on the forereef will build up a reservoir of coarse sediment which will eventually become available when the next extreme storm strikes. On cays, the effect of small storms and calm weather conditions is island growth through the influx of sand-size material (McKoy *et al.*, 2010).

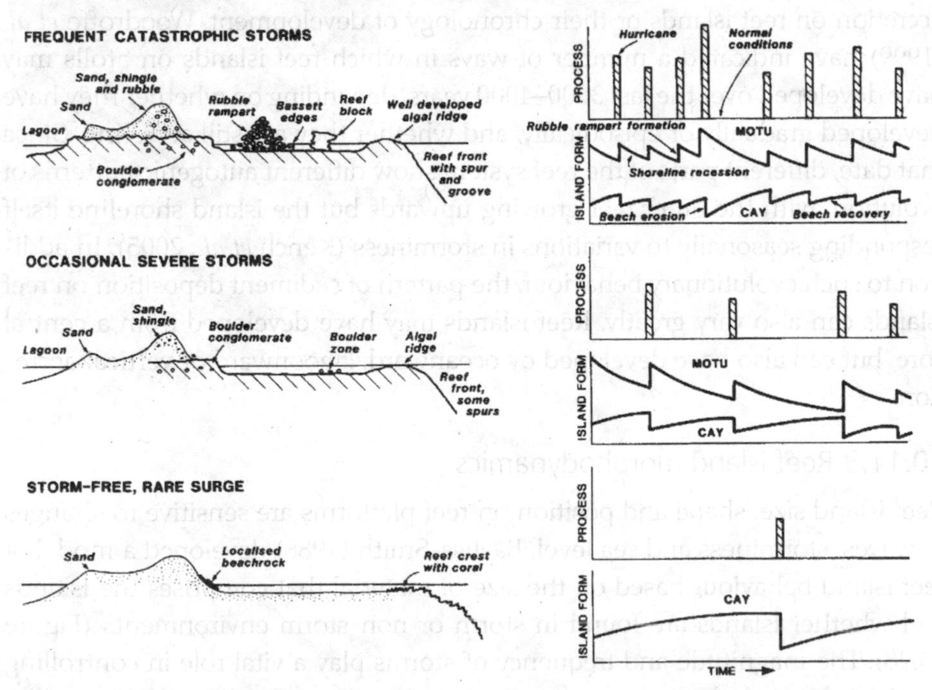

Figure 10.28 The morphology of reefs and reef islands in areas of different storm occurrence, and the response of form to process. Storm events are indicated by bars in the process diagram. Rapid change during storms is followed by redistribution of coarse material on motus, and slow recovery on cays between storms. (From McLean and Woodroffe, 1994; modified from Bayliss-Smith, 1988.) (Copyright © 1994 Cambridge University Press, reproduced with permission.)

Due to their low elevation, reef islands are generally considered vulnerable to sea-level rise, but little is known about their possible responses. Kench and Cowell (2000) carried out numerical experiments and indicated that the morphological response may range from the standard Bruun response, to barrier migration. The most unstable and therefore most vulnerable islands have low-lying margins and are so narrow that the entire island can undergo washover. The most stable and therefore least vulnerable islands are wide with higher elevation margins. In this case, washover is limited and recession of the seaward ridge results in island narrowing. These islands are also expected to undergo vertical aggradation to keep pace with sea-level rise. This occurs because washover carries sediments from the beach onto the island. The production and supply of biogenic material is also important for reef island development. Secondary effects of sea-level rise are also likely to be significant, and include changes in littoral drift caused by changes to wave patterns by increased water depths over the reefs, and directional shifts in regional wave regime.

SUMMARY

- Rocky coasts are dominantly erosional, affected by mechanical wave erosion, weathering, bioerosion and mass movements. The characteristic landforms are cliffs and shore platforms, both of which can be significantly affected by inheritance.
- The morphology of coastal cliffs depends primarily on the rate of downslope transport of material by mass movements and the ability of marine processes to remove this debris. Rock type (lithology and structure) and sea-level history are also important factors. Shore platforms are sloping or sub-horizontal intertidal rock surfaces that form in front of an eroding coastal cliff. Platform lowering is accomplished by wave erosion and/or weathering.
- Coral reefs represent the *in situ* production and deposition of calcium carbonate ($CaCO_3$) in shallow tropical and subtropical oceans. Coral growth is controlled by water temperature and light but local factors of water turbidity, salinity, wave action and nutrient levels are also important.
- Oceanic reefs develop on a slowly subsiding land surface. They start as fringing reefs along the margins of volcanic islands and slowly develop into barrier reefs and then coral atolls as the land surface continues to fall. Reef morphology is strongly influenced by antecedent conditions, including Quaternary sea-level changes and karstic weathering. Coral reef response to sea-level rise falls into three categories: (1) keep-up reefs grow upwards fast enough to keep pace with rising sea level; (2) catch-up reefs are initially left behind by rising sea levels, but manage to catch-up when the rate of

sea-level rise slows down; and (3) give-up reefs fall behind rising sea level and are eventually drowned.

Reflective questions

These questions are designed to test your comprehension of material covered in this chapter. Suggested answers to these questions can be found on this book's website.

10a. Outline the geological controls on the different mass movement processes that affect cliffs.

10b. Why are high magnitude events such as tsunami and 'mega-storms' significant along rocky coasts?

10c. Describe the relationship between atmospheric CO_2, ocean acidification and coral calcification.

10d. Describe how waves influence the morphology of coral islands.

Further reading

Guilcher, A., 1988. *Coral Reef Geomorphology*. Wiley, Chichester. (This is the only text available that deals specifically with coral reef geomorphology.)

Sheppard, C.R.C., **Davy**, S.K. and **Pilling**, G.S., 2009. *The Biology of Coral Reefs*. Oxford University Press, Oxford. (A recent text on coral reef ecosystems.)

Sunamura, T., 1992. *Geomorphology of Rocky Coasts*. Wiley, New York. (This text contains equations linking process to form and quantifying wave forces and rock resistance.)

Trenhaile, A.S., 1987. *The Geomorphology of Rock Coasts*. Oxford University Press, Oxford. (This text is a must for anyone interested in learning more about rocky coasts.)

CHAPTER 11

COASTS AND CLIMATE

AIMS

Global climate is a major forcing factor of coastal processes and geomorphology. This chapter discusses how global climate regime affects the coast, focusing on spatial (in arctic, paraglacial and tropical regions), and temporal patterns (related to the Little Ice Age, North Atlantic Oscillation, El Niño–Southern Oscillation and monsoons).

11.1 Introduction

We have already seen that coastal processes are related to variations in sea level (Chapter 2), tides (Chapter 3), waves (Chapter 4), sediment properties (Chapter 5) and winds and storms (Chapters 9 and 10). In this chapter we focus on spatial and temporal patterns of global climate regimes and their impacts on coastal processes and geomorphology. Examining the effects of climate on coasts allows us to determine to what extent coastal behaviour is due to global/regional-scale climate, *versus* other factors including sediment supply and human activity.

Understanding of climate–coastal systems' relationships is important in the light of ongoing anthropogenic climate change or 'global warming'. The **Intergovernmental Panel on Climate Change** (IPCC) identifies the physical and ecological impacts of global warming on coastal and marine environments. These include changes in coastal geomorphology, storm surges, flood heights, waves, coral reefs, coastal wetlands, marine ecosystems and marine fisheries (Rosenzweig *et al.*, 2007). The IPCC shows schematically relationships between coastal morphodynamic behaviour, climate forcing and human activity (Figure 11.1). It is important to note that climate forcing usually operates over large spatial and temporal scales, but its precise impacts are mediated by local physical geography of the environment it operates on, including land surface geology, topography and biosphere, coastline geometry and coastal type, and human activity (the 'societal sub-system').

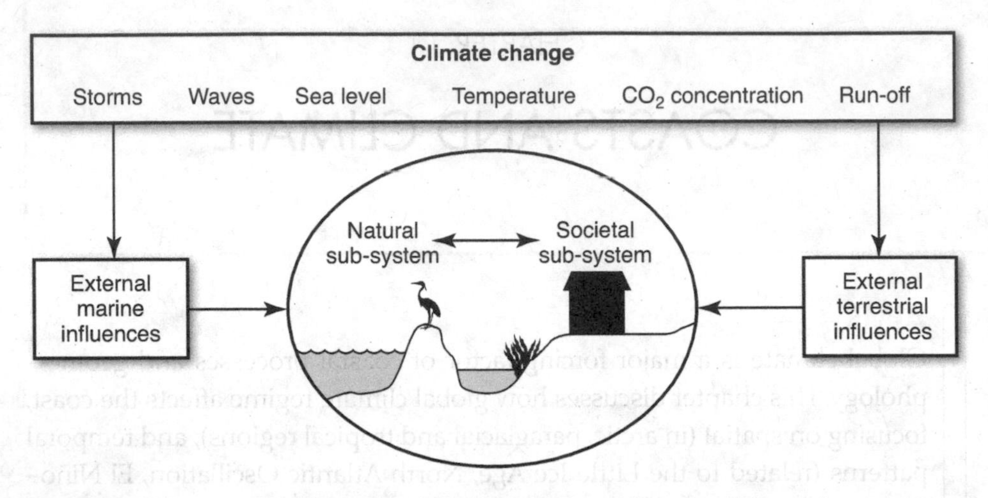

Figure 11.1 Illustration of the major climatic components that contribute to the morphodynamic behaviour of coastal systems. (From Nicholls *et al.*, 2007.) (Copyright © 2007 Cambridge University Press, reproduced with permission.)

11.2 Detection, attribution and quasi-equilibrium in systems

A key problem in correctly identifying climate impacts is **detection and attribution**. This refers to detecting statistically-significant changes to the climate system that can be attributed to a specific forcing mechanism. In coastal systems, detection and attribution refers to identifying a change to coastal morphology that can be specifically attributed to an individual event such as a storm. However, in most coastal systems factors such as coastal management, sediment supply and vegetation are also contributing forcing factors. Key assumptions are that coastal morphology starts out at quasi-equilibrium, which is disturbed by the forcing event; that the system returns to quasi-equilibrium following the event; and that measurement of coastal morphology captures this equilibrium state. However, coastal systems are constantly changing and so we cannot easily determine when it is at equilibrium and when it is not. A final assumption is that the system responds in phase with (i.e. at the same time as) the forcing mechanism and that the response time is very short. In reality, however, response time depends on the nature of the forcing and the system being forced, including its scale. Milankovitch forcing will take longer to drive a system than one being driven by a single storm. Global or regional systems will take longer to respond than local systems. This also has implications for measurement of the system: When has the storm finished driving the system? Is it after one hour, or one day, or when storm waves dissipate, or when wind speed drops? These are important considerations for determining whether a system has finished responding to a specific forcing event.

11.3 Development of zonal climates

The biologist Wladimir Köppen (1846–1940) noted that, across continental areas, ecosystems and therefore climate types form an approximately east–west (termed zonal) pattern, parallel to lines of latitude. This zonal climate pattern is due mainly to variations in incoming solar radiation, termed **insolation**, which is the heat energy received on the Earth's surface from the Sun. Insolation values on the ground surface vary spatially and temporally. Insolation decreases towards the poles because the radiation flux develops a more oblique trajectory through the atmosphere with increasing latitude, so the insolation has a greater chance of being absorbed or reflected rather than reach the ground surface. Seasonal patterns of insolation are also zonal (Figure 11.2a). At the equator, interannual variability of insolation is low, in other words summer and winter insolation values are very similar (Figure 11.2b). With increasing latitude seasonality becomes more pronounced. Above the Arctic Circle (poleward of 66.5° north and south) summer insolation is high, whereas during the polar winters insolation is zero.

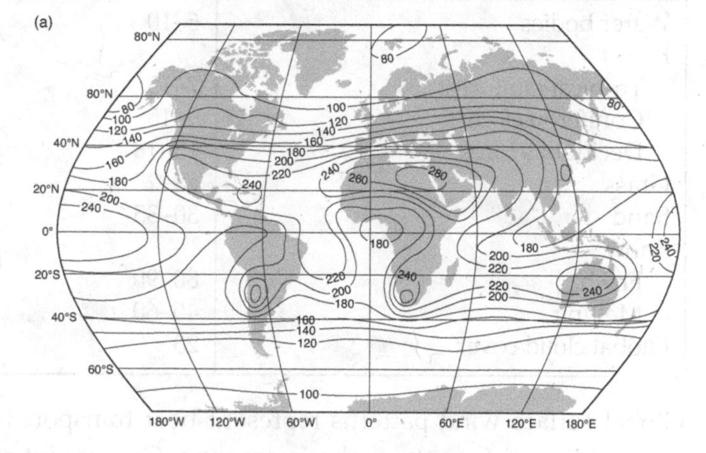

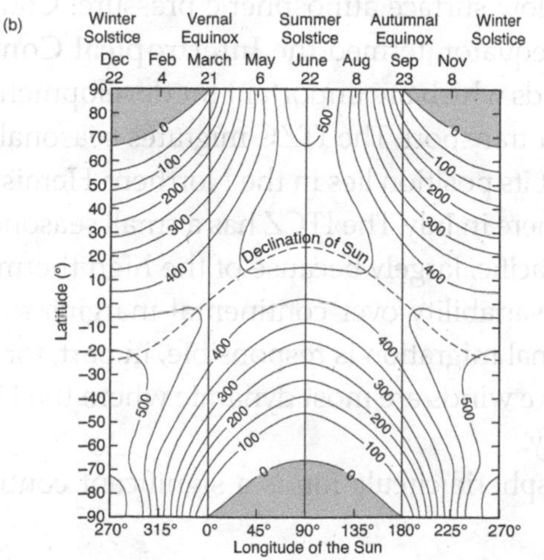

Figure 11.2 (a) Distribution of solar radiation in W m^{-2} yr^{-1} received at the ground surface. (From Barry and Chorley, 2003.) (Copyright © 2003 Routledge Education, reproduced with permission.) (b) Distribution of solar radiation in W m^{-2} yr^{-1} arriving on the ground surface by time of year (horizontal axis) and latitude (vertical axis). (From Wells, 1997.) (Copyright © 1997 John Wiley & Sons, reproduced with permission.)

Although insolation varies substantially over different time and space scales, its effect on the heat budget of individual regions is strongly moderated by the reflective qualities of the land or ocean surface, termed its albedo (Table 11.1). **Albedo** refers to the proportion of insolation that is reflected back into space rather than being absorbed at the surface. A low albedo, typical of dark surfaces, means that the surface absorbs insolation and warms up, whereas a high value means the surface reflects (or re-radiates) insolation and so remains cold. Water bodies have a low albedo where the Sun is overhead (Table 11.1), but a higher albedo where the Sun is at an oblique angle. Land surfaces generally have lower albedo than water surfaces at the same latitude. Land surfaces therefore heat up and cool down more quickly than adjacent water bodies, leading to the formation of land and sea breezes (Box 11.1).

Table 11.1 Albedo values of various surfaces. (From Barry and Chorley, 2003.)

Surface type	Albedo (%)
Water bodies	6–10
Forest	
Tropical rainforest	7–15
Coniferous forest	9–15
Deciduous forest	15–18
Grass	18–25
Sand	30–35
Snow	
Fresh	80–90
Melting	40–60
Global cloud cover	23

Global surface wind patterns represent heat transport from areas of high to areas of low surface atmospheric pressure. Convergent surface air flow at the thermal equator, termed the **Intertropical Convergence Zone** (ITCZ), forms tradewinds which are important for development of swell waves and longshore sediment transport. The ITZC migrates seasonally into the winter hemisphere, such that its position lies in the Northern Hemisphere in January and Southern Hemisphere in July. The ITCZ has a small seasonal variation in position over the central Pacific, largely because of the high thermal capacity of the ocean. It has greatest variability over continental margins such as in southeast Asia where its seasonal migration is responsible, in part, for the southeast Asian monsoon. Nearshore winds are most dynamic where the ITCZ experiences most seasonal variability.

Atmospheric circulation is a significant control on wave climate because it

Box 11.1 Formation of land and sea breezes

Land and sea breezes are common along many coasts where land surfaces heat up and cool down more quickly than the adjacent sea. This means that land surfaces attain a higher temperature than the sea surface by early–mid-afternoon, but cool down quickly in the evening, when insolation heating stops. The sea's greater heat capacity means that, by the evening, the sea surface temperature often exceeds the land temperature. These diurnal temperature contrasts (> 5°C) between land and sea drive changes in the atmospheric pressure gradient that lead, in turn, to sea breezes developed during the daytime and land breezes developed in the evening, most commonly in summer months. During the daytime, air rises over the warm land mass and cooler air is drawn from the sea, setting up an onshore (sea to land) **sea breeze**. These commonly reach a maximum by mid-afternoon and can penetrate tens of kilometres inland. The sea breeze weakens when the land starts to cool down. Sea breezes along the coast of Perth, Western Australia, have wind speeds up to 10 m s^{-1} (Masselink and Pattiaratchi, 1998). Obliquely-incident sea breeze waves increase wave height, decrease wave period, and increase longshore flow velocity and sediment transport. A **land breeze** takes place during late afternoon–early evening when the land starts to cool and a reversed temperature gradient leads to the breeze becoming offshore (land to sea). Land breezes are usually weaker than sea breezes, are often warm and dry, and funnelled downslope as katabatic winds, such as along the west coast of the Americas and South Island, New Zealand.

determines wind strength and direction that in turn control development of swell waves. Changes in atmospheric circulation can therefore lead to significant coastal impacts. Zonal wind patterns can identify those coasts along which onshore or offshore winds are most dominant. There is least variability in the areas dominated by tradewinds, and greatest variability at the boundaries of atmospheric circulation cells in the subtropics and midlatitudes where cyclones are generated. It tends to be at these boundaries where wave regimes are most sensitive to atmospheric forcing, so we can infer that coastal processes along subtropical and midlatitude coasts are most strongly affected by changes in atmospheric circulation patterns.

Insolation heating of ocean surfaces results in more complex circulation patterns because of the greater density of water than air, variations in ocean water depths, and the coupled relationship between water temperature, salinity

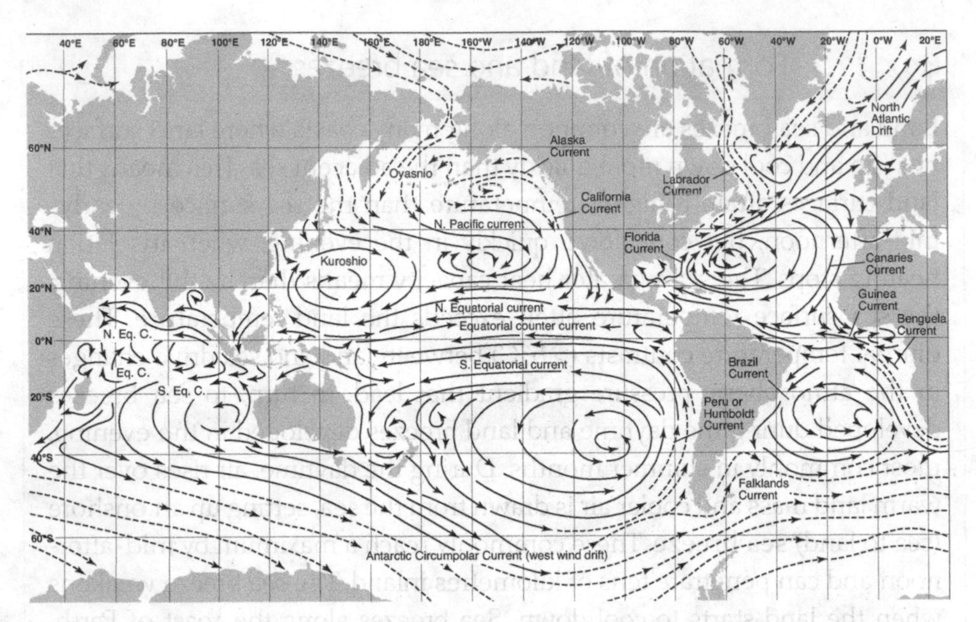

Figure 11.3 Pattern of surface ocean currents. Cool currents are shown by dashed arrows and warm currents by solid arrows. (From Open University, 1998.) (Copyright © 1998 Butterworth Heinemann, reproduced with permission.)

and density properties. Over the open ocean, patterns of sea surface temperatures (SSTs) largely mirror zonal patterns of insolation and show similar strong temperature gradients. Arctic SSTs are around 0°C and tropical SSTs are up to 28°C. Water temperatures at these latitudes are significant because they determine the likelihood of formation of sea ice and tropical cyclones, respectively. Global patterns of salinity also have a zonal pattern, with salinity generally decreasing towards the poles. At low latitudes, high SSTs lead to high rates of surface water evaporation and so high surface salinity. Higher latitude areas have a greater input of fresh water from rivers and by seasonal sea ice melt, which helps dilute the saline marine water.

Large-scale ocean circulation patterns that result from insolation heating (Figure 11.3) are similar to wind patterns but have less variability because of water's greater density. However, they are more important in regulating global climate because oceans have around 30 times the heat capacity of the atmosphere. In order for ocean surface currents to be generated, winds have to blow from a consistent direction and over a prolonged period (several hours or longer). This means that a long ocean fetch can generate both swell waves typical of a fully arisen sea (Section 4.3) and ocean surface currents. Zonal surface ocean currents are observed at the equator and around the Arctic and Southern oceans. Connecting these zonal currents are large-scale **gyres** that circulate as a result of Coriolis and pressure gradient forces. Due to the anticlockwise rotation of the Earth, when viewed from above, the sea surface is higher on western than eastern sides of large ocean basins. As a result, the pressure gradient force moves the water mass down-gradient from west to east. As the water mass starts moving (i.e. when it has a velocity of greater than zero), the Coriolis force deflects the water mass from its path, forming large-scale circulation cells.

Boundary currents within gyres impact on coastal and nearshore processes. Western boundary currents such as the North Atlantic Drift, Brazil Current and Kuroshio Current track along the western edges of gyres and are warm-water currents. Higher water surface levels in the western than eastern sides of the gyres lead to steeper water surface slopes, so these currents are generally deep, fast and of high volume. They are not associated with coastal upwelling. Eastern boundary currents such as the Canary Current, Humboldt Current, Benguela Current and California Current are found along the eastern edges of ocean basins. These currents are much shallower, slower and of smaller magnitude, but are associated with coastal upwelling. **Upwelling** is the process by which cooler, denser waters are brought upwards in the water column, often aided by offshore winds and low-pressure atmospheric cells. Upwelling tends to keep sediment plumes closer to the coast and inhibits the offshore transport of fines. The upwelling water is nutrient-rich and undersaturated with respect to carbon dioxide (CO_2). High nutrient availability on the surface helps drive phytoplankton growth, a critical part of the marine foodchain, and helps draw down atmospheric CO_2 into the ocean. Reduced upwelling off the west coast of central South America during El Niño events causes a major reduction in phytoplankton availability for anchovy fish species and then sea birds. The opposite process of **downwelling** occurs where warmer surface waters lose heat to the atmosphere, become cooler, denser and start to sink. This is the major mechanism by which North Atlantic Deep Water is formed, by density sinking of North Atlantic Drift surface waters.

Variations in wind-driven wave height and direction are important in beach morphodynamics, especially on long-fetch, sandy coasts such as in Australia. For example, Short and Trembanis (2004) use 26 years of monthly beach profiles at Narrabeen beach, New South Wales, to show how the beach changes in response to changes in wave climate. They show that both erosion–accretion cycles and beach rotation take place. Across this time series, variation in beach width is systematic, not random: when the northern part of the beach is particularly wide, the southern part is particularly narrow and *vice versa*, with a correlation coefficient of −0.73. On a gravelly beach at Slapton Sands (Devon, southern England), southerly storms cause intertidal erosion, supratidal deposition and net sediment loss, whereas easterly storms cause intertidal deposition, supratidal erosion and net sediment gain (Ruiz de Alegria-Arzaburu and Masselink, 2010). Variations in storm wave direction cause changes in the direction of longshore transport, causing the beach to change in width by more than 10 m over time scales of a few weeks. The **beach rotation** seen in these examples shows that, at these locations and over these time scales, the coastal sediment cells are just about closed, with sediment merely redistributed from one end of the beach to the other. It also shows that the beach system responds dynamically to external forcing by which it can maintain quasi-equilibrium. The degree of beach rotation is therefore a measure of the extent of wave-forcing of the system, and the extent to which the system 'bounces back' when the forcing changes.

11.4 Zonal climates, coastal processes and geomorphology

So far, we have considered the main properties of large-scale atmospheric and oceanic circulation. A useful way of considering their effects on coastal processes and geomorphology is to examine different latitudinal zones. We have already seen that zonal climate is controlled by zonal patterns of insolation so we can hypothesise that different coastlines within single latitudinal zones will be similar in morphology, processes and morphodynamic controls. We can test this hypothesis by considering the coastlines of arctic (> 65°), paraglacial and temperate (40–65°) and tropical (< 40°) areas. These latitudinal bands are somewhat arbitrary but their coastlines have many properties in common that are linked to their climatic regime, including temperature, precipitation, winds and waves.

11.4.1 Arctic coasts

These areas include northern Russia (including Siberia), northern Canada (including Canadian arctic islands and Hudson Bay), Alaska, northern Scandinavia and coastal Antarctica. These areas have low mean annual air and ground temperatures, low annual precipitation, strong katabatic (down-slope) winds which are commonly directed offshore, sea ice and permafrost, recent or present glaciers, and rocky and high relief, or tundra/permafrost and low relief coastlines.

Permafrost forms where the ground is frozen (below 0°C) for longer than two years. **Permafrost** most readily affects poorly-drained peatlands (tundra) which have high organic and water content. Freezing of the ground surface takes place as a result of the negative winter heat balance where low insolation and high albedo lead to low land surface temperatures. Freezing takes place from the surface downwards, penetrating deeper over time. Permafrost increases the mass strength and cohesion of tundra materials and decreases their permeability. As a result, coastal permafrost can stabilise the coastline

Figure 11.4 Oblique view of thermokarst lakes, Québec, Canada. (Photo: F. Camels, from André, 2009.) (Copyright © 2009 Geological Society of London, reproduced with permission.)

and lead to lower than expected rates of coastline retreat. The coastline can also be protected by the presence of shorefast (perennial) and seasonal sea ice. Seasonal **sea ice** cover develops in the autumn–spring period when there is low insolation and low SSTs (below –1.75°C). The sea ice buffers the effects of storm waves and reduces fetch length.

Over the last decades, arctic areas are warming more quickly than the global average. This phenomenon, termed **arctic amplification**, is due to a decrease in highly-reflective snow and sea ice cover, which reduces albedo and allows arctic areas to warm more quickly. For example, permafrost on the Mackenzie River delta, arctic Canada, has warmed by 2.5°C since the 1970s (Burn and Kokelj, 2009). There is a close relationship between permafrost warming and coastal erosion. For example, along the Beaufort Sea (northern Canada) the area affected by erosion increased by 160 per cent in the period 1952–2000, in particular along north- and west-facing coasts exposed to wave attack (Lantuit and Pollard, 2008). Erosion rates averaged at 13.6 m yr^{-1} in the period 2002–07 (Jones *et al.*, 2009). Increased erosion is caused by a combination of warming permafrost (reducing coastline mass strength) and decreased sea ice duration

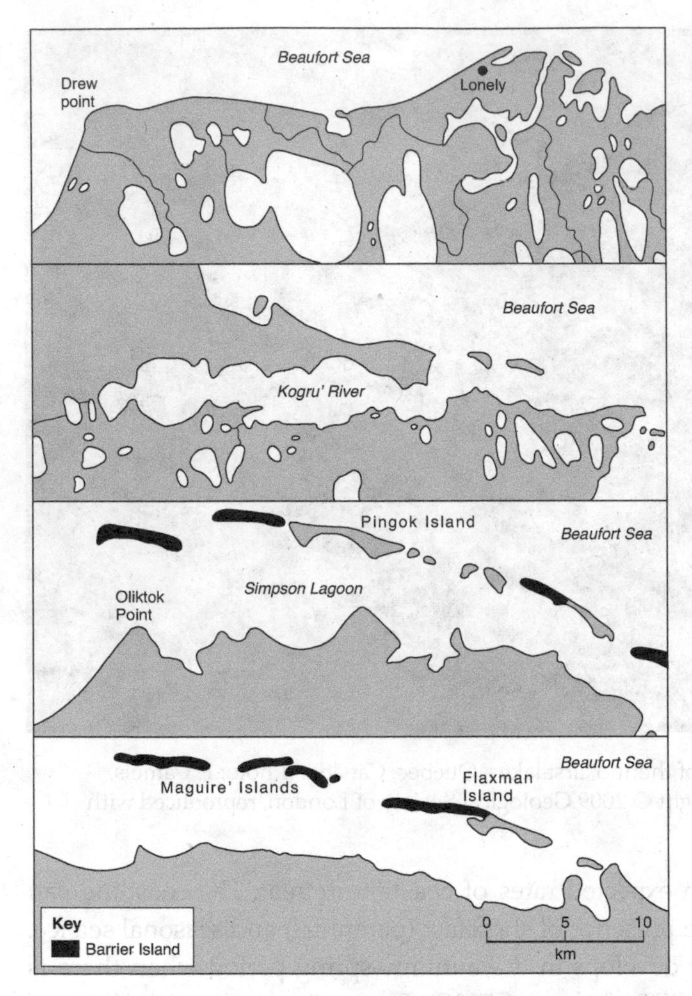

Figure 11.5 Map views of different locations around the Beaufort Sea, illustrating the variety of coastal forms as permafrost degrades. (From Hill *et al.*, 1994.) (Copyright © 1994 Cambridge University Press, reproduced with permission.)

and thickness. This increases coastal vulnerability to wave attack, where waves can develop over a longer fetch. Coastal erosion takes place by wave undercutting of the permafrost margin, leading to retrogressive thaw slumps, landslides and mass flows.

Permafrost warming also leads to the formation and enlargement of **taliks**, which are unfrozen 'holes' in the permafrost. Taliks are often filled with water, forming thermokarst lakes (Figure 11.4). Coastal erosion of taliks is very rapid, leading to high sediment supply and residual sediment mounds with spits, barrier islands and lagoons (Figure 11.5).

11.4.2 Paraglacial and temperate coasts

These areas include the densely populated lowlands and coasts of Europe and North America, southern South America and New Zealand. These areas have strong cyclonic circulation, onshore storm winds and waves and are, even today, influenced by late Pleistocene glaciers which directly eroded or deposited sediment along these coasts, or were located nearby. The relationship between glaciation, landscape geomorphology and sediment supply is described by the concept of paraglaciation. **Paraglacial coasts** are 'those on or adjacent to formerly ice-covered terrain, where glacially excavated landforms or glaciogenic sediments have a recognizable influence on the character and evolution of the coast and nearshore deposits' (Forbes and Syvitski, 1994, p.376). Paraglacial processes include glacial, proglacial, periglacial and slope processes that work more energetically as a result of the former presence of glaciers. Glacio-isostatic variations in land level and sea level are also paraglacial.

A useful way to understand the development of paraglacial coasts is to consider how sediment supply varies in paraglacial environments. Church and Ryder (1972) examined changes in fluvial sediment yield in Baffin Island and British Columbia, Canada, and showed that sediment yield is highest immediately following ice retreat, decreasing exponentially over time. Highest sediment supply to paraglacial coasts occurred around 14,000–8000 years BP, when sea-level change was particularly rapid. Many of the geomorphological elements making up paraglacial coasts were emplaced at this time, and represent part of the **geological inheritance** of such coasts.

Paraglacial coasts include: (1) trailing-edge, low-relief coasts that are sediment dominated (e.g. western Europe, eastern North America) and (2) leading edge, rocky, high-relief and sediment-poor coasts with inland mountains and a narrow coastal fringe (e.g. western North and South America, New Zealand).

Subdued, trailing-edge coasts were affected by continental ice sheets during the last glaciation. Large quantities of glacial sediment were transported and deposited beneath and in front of the ice. These glaciers also extended across the surrounding continental shelves which, due to eustatic sea-level fall, were dry land. Evidence for this comes from the presence on the shelf and coast of subglacial bedforms such as **drumlins** composed of glacial till. Drumlins are common along the coasts of western Ireland (Clew Bay, Galway Bay), northeast USA (Boston Harbor, Massachusetts), and northwest USA (Puget Sound, Washington State). In till cliffs in eastern Ireland, Greenwood and Orford (2008) show that coastal erosion rates are extremely variable (mean of 76±49 mm yr^{-1}) but that 86 per cent of erosion takes place in winter. Carter and Orford (1988) presented a model for beach development as a result of erosion of drumlin islands (Figure 11.6). The model shows that as the till is eroded and sediment accumulates at the cliff base, longshore and tidal processes redistribute this sediment. Finer sediments are transported preferentially, resulting in a coarse sediment lag at the foot of the retreating drumlin cliff (Figure 11.7). Carter

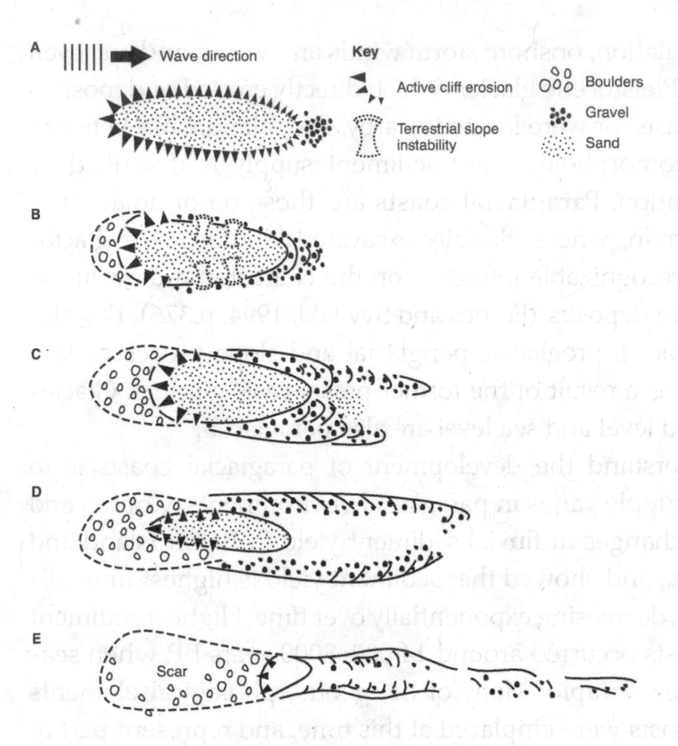

Figure 11.6 Sequence of development of beach forms as a result of drumlin island erosion, and deposition of this sediment in a down-drift direction. (From Carter and Orford, 1988.) (Copyright © 1988 American Geographical Society, reproduced with permission.)

and Orford (1988) identify this event sequence as an important mechanism by which gravel and boulder barriers, spits and tombolos are formed along Atlantic Canada (Nova Scotia, New Brunswick and Newfoundland). Globally, the largest areas of coastal sand dunes, estuaries, barrier islands and sandy beaches have been developed by reworking of glacial sediments.

Paraglacial coasts affected by valley glaciers have deeply incised U-shaped valleys which are flanked by high-relief bedrock uplands. Typical examples include British Columbia (western Canada), Patagonia (southern South America), north Norway and the Southern Alps (South Island, New Zealand).

Figure 11.7 Photo of the till cliff and fronting boulder beach at White Strand, Galway Bay, western Ireland. (Photo: J. Knight.)

Similar high-relief, rocky coasts are also found in non-glaciated areas such as Japan. In glaciated areas, valley glaciers develop and spread radially outwards from small cirque glaciers in mountains or from larger ice caps. Present-day ice caps include Vatnajökull (Iceland), Jostedalsbreen (Norway), Austfonna (Svalbard), the Juneau icefield (Alaska) and Patagonia.

Many coastal U-shaped valleys are significantly overdeepened, with valley floors located below present sea level and the drowned valley forming a **fjord**. Fjords should be distinguished from drowned V-shaped river valleys (rias or fjards). Fjords and the geometry of glaciated coasts are very strongly controlled by the orientation of bedrock structures. Fjord coasts have straight, steep bedrock cliffs with a very narrow and sediment-starved shoreface. As a result, these coasts are generally insensitive to tidal range or sea-level change. Steep and unstable rock slopes, active periglacial weathering, rockfalls and landslides cause high rates of cold-climate mass wastage, termed **solifluction**. In western Norway, stratified solifluction deposits record episodes of enhanced sediment supply under cooler climatic conditions (Blikra and Nemec, 1998).

11.4.3 Tropical coasts

Equatorial and subequatorial tropical coasts include those in central Africa, Indian Ocean, southeast Asia, Indonesia, northern Australia, the Caribbean and northern South America. These areas have high values but low variability of precipitation, humidity and mean annual air temperature. Due to high precipitation, large rivers with well-developed coastal plains and deltas are common (e.g. Orinoco, Amazon, Niger, Indus, Ganges–Brahmaputra, Mekong). These coasts are strongly affected by El Niño (interannual; discussed below) and monsoon (seasonal) climate cycles.

Coastal dynamics during the summer monsoon are substantially modified by the behaviour of incoming rivers, including river mouth position, river discharge and sediment load. Studies of the Ganges–Brahmaputra river system show that although it is largely driven by tectonic uplift of the Tibetan Plateau (which itself controls the dynamics of the south and southeast Asia monsoon) increased summer monsoon rainfall dramatically increases sediment yield from this region. Over time, the creation of relief by tectonic uplift during the Pleistocene has led to increased weathering, river downcutting and high sediment yield. This sediment nourishes the rivers' distributary delta and Bengal Fan, located in the Bay of Bengal. The Ganges–Brahmaputra river system transports around 1×10^8 t yr^{-1} of suspended sediment (8 per cent of the global total) to the Bengal Fan, of which around 21 per cent is deposited on the subaqueous delta and 29 per cent on the distal fan (Wasson, 2003). During the mid-Holocene, however, sediment deposition on the floodplain was much higher (with less sediment

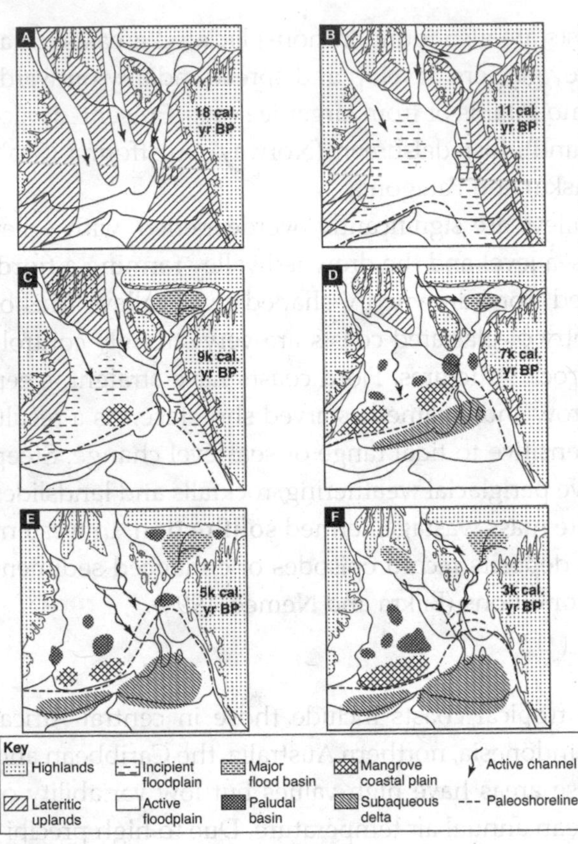

Figure 11.8 Maps showing the late Pleistocene development of the Ganges–Brahmaputra delta. (A) Lowstand system with incised alluvial valleys. (B) Initial phase of delta growth with rising sea level. (C) Aggradation of subaerial delta during rapid sea-level rise. (D) Decelerating sea-level rise, progradation of coastal plain and initial growth of subaqueous delta. (E) Progradation of subaerial and subaqueous deltas. (F) Delta similar to modern system. (From Goodbred and Kuehl, 2000.) (Copyright © 2000 Elsevier Science, reproduced with permission.)

reaching the Bengal Fan) compared to present (Figure 11.8), suggesting that there are cycles of aggradation and erosion. Changes on the Tibetan Plateau by deforestation, soil erosion and climate change are leading to problems on the rivers' floodplain, including floods, sedimentation and crop and infrastructure damage. This is particularly important in Bangladesh where sea-level rise, coastal erosion, cyclones and floods are a big threat. Mangroves are critical to the maintenance and protection of these coastlines (Case Study 11.1).

High SSTs in tropical regions mean that deep low pressure cells called **tropical storms** are commonly formed. Intense tropical storms (>10^4 km hr^{-1}), variously termed tropical cyclones, hurricanes or typhoons, are formed mainly during the summer with SSTs above a critical threshold of 26.5°C. High SSTs lead to high rates of evaporation, high atmospheric humidity, and rapidly rising air parcels which cause intense convective circulation and thunderstorms. Tropical storms are generated over the open ocean within the limits of the ITCZ and steered westwards by tradewinds. When these storms make landfall, storm surges brought about by the deep low pressure conditions, strong winds and large waves often result in coastal flooding and intense damage. The example of Hurricane Katrina (August 2005) is described in Case Study 3.2.

Case Study 11.1 Mangroves and tropical coastal change

Mangroves are common along tropical and subtropical coasts, especially along the margins of estuaries and tidally-influenced rivers. They help reduce erosion by waves and tides through their stabilising roots, and by encouraging sediment deposition by decreasing water velocities. Mangroves are also important ecosystems, helping maintain coastal fisheries and other plant species. Although mangroves can prograde readily in response to sea-level rise, they are vulnerable to changes in storm frequency and magnitude, salinity and inundation regime. McLeod and Salm (2006) discuss the major climatic and anthropogenic factors affecting mangrove survival, including increases in temperature, CO_2, precipitation, storms and sea level. They identify management strategies that can improve mangrove survival, including establishing greenbelts, buffer zones and corridors, and habitat restoration.

Saenger and Siddiqi (1993) describe a programme of mangrove afforestation in Bangladesh whose coastline is highly vulnerable to sea-level rise and cyclone impacts. In the period 1965–90, 120,000 ha of mangroves were planted, including on reclaimed areas, and have provided a range of benefits including coastal protection, local employment, forestry products and increased biodiversity.

11.5 Climate variability over decadal to seasonal time scales

Climatic variability over decadal to seasonal time scales is important in coastal morphodynamics because it is relevant for coastal policy, management, monitoring and planning. Four examples of climatic variability on different time scales are the Little Ice Age, North Atlantic Oscillation, El Niño–Southern Oscillation and monsoons.

11.5.1 Little Ice Age

The Little Ice Age (LIA) refers to the period *c.* AD 1550–1850 in which there was climatic deterioration, particularly in Europe. The principal causes of the **Little Ice Age** are examined in Box 11.2. During the LIA as a whole, temperatures decreased in Europe by 1.5–2.5°C, with increased precipitation, increased storminess and sea level 10–30 cm lower (Lamb, 1982). These climate changes resulted from a reorganisation of synoptic atmospheric circulation. Climate throughout the period of the LIA was not uniform, however, and there was decadal-scale variability in temperature and other properties.

Box 11.2 Causes of the Little Ice Age

The **Little Ice Age** was not a period of uniform climate but was characterised by decadal-scale variations in temperature and storminess (Lamb, 1982). Much of this variability was driven by changes in climate forcing. The Little Ice Age period was one of low solar activity. A common historic measure of solar activity is sunspot number. Sunspots are dark, cool patches on the Sun's surface, easily observed using telescopes, and historical records of sunspot numbers have been maintained for hundreds of years. Sunspot numbers vary cyclically over 11-year (Schwab) and 22-year (Hale) time scales and are associated with a decrease in insolation of around 0.1 per cent. While this is relatively small, the cumulative effects of low sunspot numbers could be significant, because it can also include atmospheric and land surface feedbacks. Low sunspot numbers were observed during several phases of the Little Ice Age, including the Maunder Minimum (*c.* 1645–1715) and Dalton Minimum (*c.* 1795–1825) (Figure 11.9). Higher atmospheric concentrations of reflective volcanic dust may have also contributed to cooling. For example, the large eruption of Tambora, Indonesia, in 1815 was followed in Europe by 'the year without a summer' in 1816.

Evidence for the impacts of the Little Ice Age comes from a range of geomorphological, historical and documentary records across Europe, including episodes of coastal flooding, winter sea and river ice formation, enhanced coastal storminess leading to active sand dune migration, and coastal erosion. Winter sea ice cover in Canada, Greenland and Iceland increased significantly during the Little Ice Age, and was a contributory factor to settlement abandonment in Greenland.

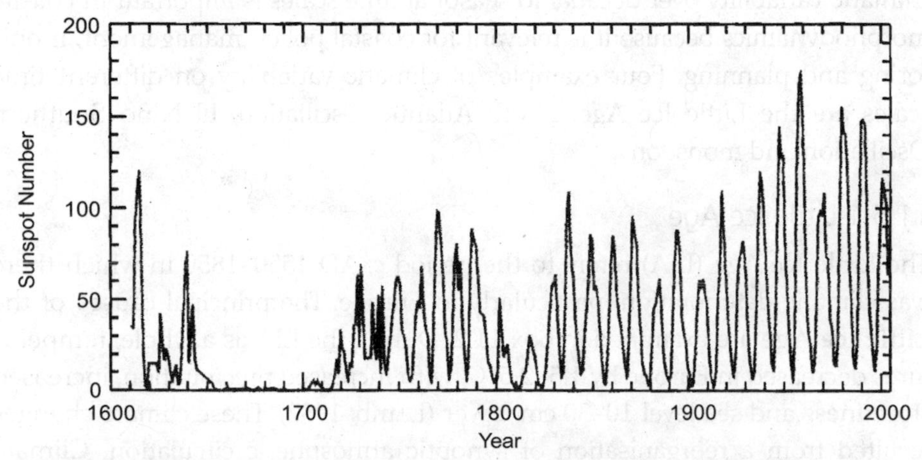

Figure 11.9 Sunspot numbers between AD 1600–2000, showing the Maunder Minimum (*c.* 1645–1715) and Dalton Minimum periods (*c.* 1795–1825). (From Wallace and Hobbs, 2006.) (Copyright © 2006 Academic Press, reproduced with permission.)

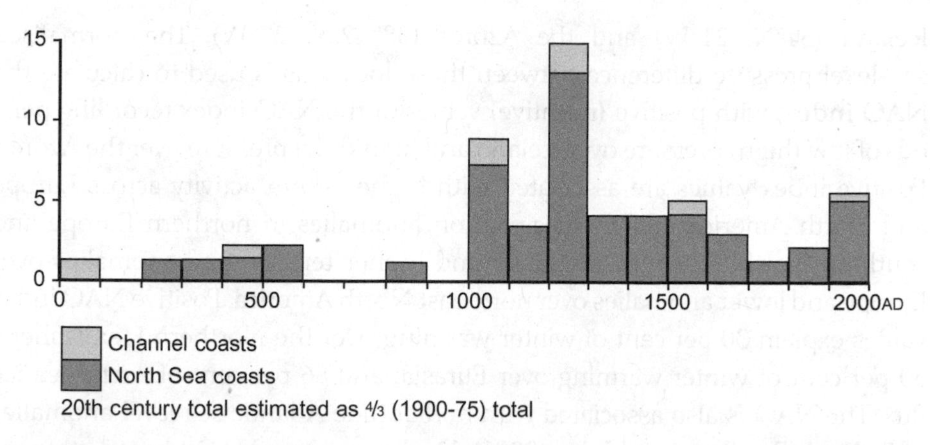

Figure 11.10 Number of sea floods per century on North Sea and English Channel coasts. (From Lamb, 1982.) (Copyright © 1982 Methuen, reproduced with permission.)

Prior to instrumental climate records, many storms producing strong winds and large waves were recorded in diaries, annals, letters, account books and other documents. These are a valuable resource for reconstructing the timing, magnitude and impacts of past events. The climatologist Hubert Lamb (1913–97) used historical documentary evidence from the North Sea region in order to identify patterns of past coastal storms. He calculated a **Storm Severity Index** based on

$$SVI = V_{max} A_{max} D \quad (11.1)$$

where V_{max} is the storm's maximum surface wind speed, A_{max} is the maximum area affected by the storm, and D is storm duration in hours. Lamb showed that storms and associated sea floods reached a maximum frequency and magnitude during the LIA (Figure 11.10). Climate impacts on coasts during the LIA can be evaluated using geomorphological and sedimentary evidence. Enhanced storminess and wave intensity during the LIA, in addition to cooler temperatures that suppressed plant growth, resulted in coastal sand dune instability and rapid dune migration. This took place in particular along exposed Atlantic and North Sea coasts including eastern England, Denmark and Ireland. However, it is unclear whether dune development was caused by a change in storminess, sea level, sediment supply, or a combination of these and other factors.

11.5.2 North Atlantic Oscillation

On multiannual time scales, cyclic variations in high and low atmospheric cells include the Antarctic, Arctic and North Atlantic Oscillations. The **North Atlantic Oscillation** (NAO) refers to an alternation of atmospheric mass between

Iceland (64°N, 21°W) and the Azores (38°42′N, 27°W). The normalised sea-level pressure difference between these locations is used to calculate the **NAO index**, with positive (negative) values of the NAO index recording periods of low (high) pressure over Iceland and high (low) pressure over the Azores. Positive index values are associated with higher storm activity across Europe and North America; higher precipitation anomalies in northern Europe and southern North America/Caribbean; and higher temperature anomalies over Europe and lower anomalies over northeast North America. Positive NAO index values explain 30 per cent of winter warming over the Northern Hemisphere, 50 per cent of winter warming over Eurasia, and 86 per cent of Arctic sea ice flux. The NAO is also associated with midlatitude Atlantic sea level anomalies of 3–8 cm (Esselborn and Eden, 2001). During positive NAO phases, stronger westerly winds and waves into northern Europe are associated with periods of enhanced coastal change. For example, shifts in the position of ebb channels and channel outlets in northwest Ireland estuaries have a close correspondence to positive index phases (Burningham, 2005). Dated periods of historical coastal sand dune activity in Portugal coincide with negative index phases, consistent with the southern Europe location of strong storms during this phase (Clarke and Rendell, 2006).

11.5.3 El Niño–Southern Oscillation

El Niño–Southern Oscillation (ENSO) refers to the large-scale reorganisation of atmosphere and ocean circulation in the equatorial Pacific, operating on a quasi-cycle of 3–7 years and driven largely by changes in SSTs. Usually, cool water from the north-going Humboldt Current upwells near the coast of Peru in the eastern equatorial Pacific. A warm SST region is located in the western Pacific. This SST gradient leads to strong westerly tradewinds which promote further upwelling. El Niño events lead to reorganisation of this circulation pattern. The state of ENSO is measured by the **Southern Oscillation Index**, defined as the normalised air pressure difference between Darwin (12°S, 150°W) and Tahiti (12°S, 131°E). Negative values of the index, where Darwin has higher pressure and Tahiti lower pressure, correspond to the El Niño phase, and positive values to the non-El Niño phase, commonly termed La Niña. El Niño events have a duration of 12–18 months and generally start in anomalous warming of eastern Pacific surface waters as upwelling decreases significantly. This warm pool spreads westward into the central Pacific by slackening tradewinds. The locations of high and low air pressure cells, associated with dry and wet weather conditions respectively, also migrate with the warm and cool water pools. The climatological properties of El Niño events are described in more detail by Cane (2005).

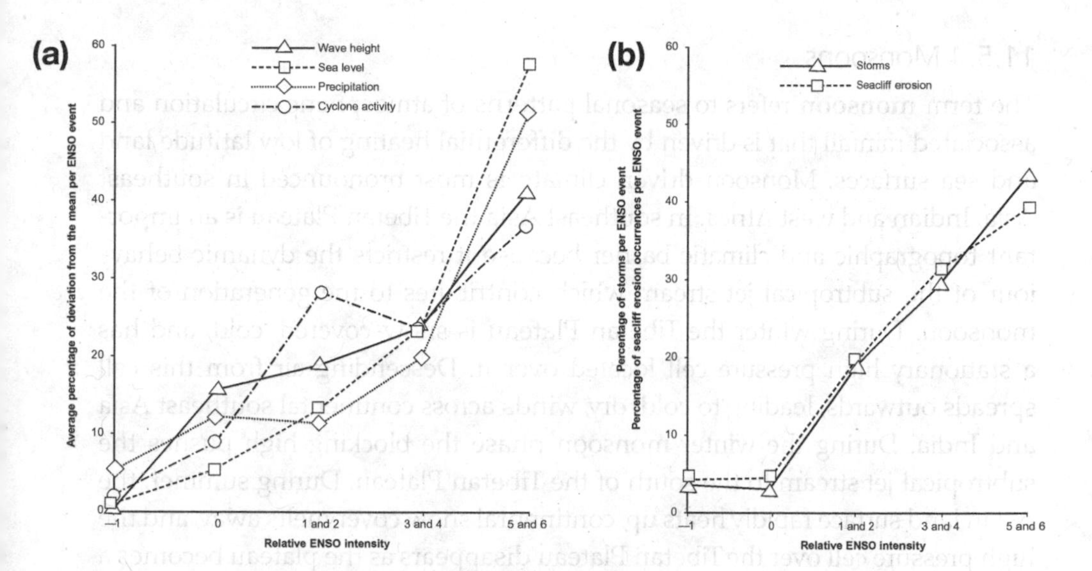

Figure 11.11 Variation in climatic factors (a) and coastal effects (b) as a result of ENSO intensity, 1910–95, where value of –1 is low and 6 is high. (From Storlazzi and Griggs, 2000.) (Copyright © 2000 Geological Society of America, reproduced with permission.)

Climatological changes during **El Niño events** have significant implications for coastal processes. Reorganisation of SSTs, surface current strength and surface winds means that surrounding coasts experience temperature and precipitation anomalies that affect river discharge and sediment supply, wave height, wave direction and storminess. Storlazzi and Griggs (2000) describe relationships between El Niño intensity and physical and climatic properties that affect the coastline of southern California, western USA. They find that precipitation, sea-level height and wave height increase nonlinearly with increased El Niño intensity (Figure 11.11). During El Niño winter conditions a low pressure cell is located over the North Pacific and high pressure cell over northern Canada, causing west to east storm tracks to the California coast. This causes increased precipitation, sea level and wave height which are the major causes of coastal erosion and are statistically correlated ($p< 0.01$) with strong El Niño events. Similar reorganisation of high and low pressure cells during El Niño events also causes enhanced erosion and localised reversal of longshore sediment transport along the coast of Florida (Hepner and Davis, 2004). El Niño events lead to temperature and precipitation anomalies worldwide, termed **teleconnections**. For example, changes in snow accumulation rate and oxygen isotope ratios in Andean glaciers reflect changes in SST anomalies in the equatorial Atlantic where this precipitation is derived. Atlantic SST anomalies lag the Pacific by 6–8 months and are smaller in magnitude (Hastenrath, 2006).

11.5.4 Monsoons

The term **monsoon** refers to seasonal patterns of atmospheric circulation and associated rainfall that is driven by the differential heating of low latitude land and sea surfaces. Monsoon-driven climate is most pronounced in southeast Asia, Indian and west Africa. In southeast Asia the Tibetan Plateau is an important topographic and climatic barrier because it restricts the dynamic behaviour of the subtropical jet stream which contributes to the generation of the monsoon. During winter the Tibetan Plateau is snow-covered, cold, and has a stationary high pressure cell located over it. Descending air from this cell spreads outwards, leading to cold, dry, winds across continental southeast Asia and India. During the winter monsoon phase the blocking high pushes the subtropical jet stream to the south of the Tibetan Plateau. During summer, the Asian land surface rapidly heats up, continental snow cover melts away, and the high pressure cell over the Tibetan Plateau disappears as the plateau becomes a heat source with ascending air. At this time the subtropical jet stream shifts in position from the south to the north of the Tibetan Plateau, drawing in behind it warm, moist air from the South China Sea and Bay of Bengal, leading to summer monsoon rains over southeast Asia and India.

Monsoon-dominated coasts therefore show seasonal patterns of offshore (winter) and onshore winds (summer), high waves, high river sediment input and changes to longshore transport. In eastern India, the summer monsoon leads to thinner, steeper beaches and the removal of berms and bars, which recover during the winter. This leads to typical 'summer' and 'winter' beach profiles. Longshore transport direction can also change, depending on coastal orientation. In Karala, southwest India, beach profiles show that twice as much erosion takes place during the (onshore) summer monsoon than the amount of accretion that takes place during the offshore winter monsoon, meaning that many beaches are in long-term sediment deficit. At the mouth of the Mekong River in Vietnam, the strong offshore winter monsoon leads to waves from the northeast and sediment transport dominantly to the southwest (Tamura *et al.*, 2010). During the summer monsoon the situation is reversed, in addition to a dramatic increase in river discharge and sediment load. Mud and very fine sand is deposited near the river mouth during the summer monsoon and reworked southwards by stronger waves during the winter monsoon. The interplay between summer and winter monsoon seasons determines the asymmetric behaviour of coastal sediments and landforms at the Mekong River mouth.

SUMMARY

- Climate is an important control on coastal processes because it leads to variability in sea level, wind and waves. Incoming solar radiation drives heating of land and sea surfaces which, in turn, generate zonal patterns of atmosphere and ocean circulation.
- Arctic coasts are affected by the presence of permafrost and sea ice. Temperate, paraglacial coasts have a glacial inheritance that gives rise to high sediment supply along low-relief coasts and low sediment supply along rocky coasts. Tropical coasts are strongly affected by monsoons.
- Cycles of climatic variability take place over decadal (North Atlantic Oscillation), interannual (ENSO) and seasonal (monsoons) time scales. Variations in wind and wave conditions by these events can cause cycles of change along many mid and low latitude coasts, particularly to sandy beaches and estuaries.

Reflective questions

These questions are designed to test your comprehension of material covered in this chapter. Suggested answers to these questions can be found on this book's website.

11a. Explain how increased temperature can affect sea level, wind and waves.

11b. Why do variations in wind and waves lead to beach rotation?

11c. Describe the seasonal changes in albedo that take place in (i) high latitude, and (ii) low latitude locations.

11d. Outline how variations in temperature and precipitation inland can lead to changes on tropical coasts.

Further reading

Barry, R.G. and **Chorley**, R.J., 2003. *Atmosphere, Weather and Climate* (8th edn). Routledge, London. (This is a classic textbook in climatology and provides an excellent grounding in the topic.)

Lamb, H.H., 1982. *Climate, History and the Modern World*. Methuen, London. (This is a good summary of climate impacts during the Little Ice Age.)

Nicholls, R.J., **Wong**, P.P., **Burkett**, V.R., **Codignotto**, J.O., **Hay**, J.E., **McLean**, R.F., **Ragoonaden**, S. and **Woodroffe**, C.D., 2007. 'Coastal systems and low-lying areas.' In: Parry, M.L., Canziani, O.F., Palutikof, J.P., van der Linden, P.J. and Hanson, C.E. (eds) *Climate Change 2007: Impacts, Adaptation and Vulnerability. Contribution of Working Group II to the Fourth Assessment Report of the Intergovernmental Panel on Climate Change*. Cambridge University Press, Cambridge, UK, 315–356. (This chapter of the IPCC report addresses the impacts of global warming on coastal systems.)

Wells, N., 1997, *The Atmosphere and Ocean, a physical introduction*. Wiley, Chichester. (This is a good textbook that focuses on the physical workings of the climate system.)

CHAPTER 12

FUTURE COASTS

AIMS

This chapter considers some of the wider issues that affect coastal processes, geomorphology and management, including human activity and the use of coastal resources. Coasts are vulnerable to the effects of climate change, in particular sea-level rise, which poses challenges for the sustainability and management of coasts over the next decades to centuries.

12.1 Introduction

Throughout this book we have considered the major processes and landforms of the coastal zone that operate over different spatial and temporal scales and within the context of coastal systems. We have also considered the ongoing role of climate change and human activity in modifying these processes, landforms and systems. In this final chapter we consider wider issues of coastal resources, sustainability and management, because these factors are intimately related to the development of coasts over the next decades to centuries.

12.1.1 Climate change and the diversity of coastal impacts

The IPCC Fourth Assessment Report (2007) considers climate projections to the year 2100 based on different CO_2 emissions scenarios. Although these modelled outcomes are numerically well-constrained, climate conditions so far in the future are uncertain and have high error margins (see Bindoff *et al.* (2007) for results affecting the coastal zone). Also unknown are the demographic, economic, social, political, cultural, transport and energy conditions of the 2100 world. These properties are significant because they are closely connected to how the coastal zone is used, managed, engineered and valued at regional and local scales. Regional climate scenarios set the scene for coastal responses to future climate. For example, regional scenarios produced by the United Kingdom Climate Impacts Programme 2009 (UKCP09, see www.ukcip.org.uk/) for the decades 2020s, 2050s and 2080s can be used to help identify regional coastal impacts at these time points, and therefore help in formulating policy and management strategies to minimise negative impacts. Some of the major effects of future climate change on coastal environments are listed in Table 12.1.

Table 12.1 Summary of the likely effects of ongoing global warming on coastal environments. Note that no scale is implied, so some of these effects will be global and others local, and their timings and/or time lags will be different.

Climatic changes	Likely effects
Increased sea surface temperature	Increased water stratification, eutrophication, decreased coastal upwelling, coral bleaching, decreased dissolved oxygen content in water, ecosystems put under thermal stress, increase in cyclone/hurricane intensity
Increase in storminess	Increased onshore wind strength, increased storm intensity and frequency
Increased wave height	Increased coastal erosion, nearshore pollutant dispersal
Increased sea level	Coastal flooding, inundation of small islands, coastal erosion
Decreased sea ice cover	Increased arctic coastal erosion
Increased atmospheric CO_2	Ocean acidification
Change in ENSO/ monsoon/hurricane intensity	Implications for hazard management and mitigation
Increased frequency/ magnitude of coastal hazards	Increased coastal vulnerability and likely decreased resilience
Increased permafrost temperatures	Increased surface runoff, coastal hazards, increased rate of arctic coastal erosion
Decreased snow and ice cover	Change in timing of peak river discharge, coastal sediment supply
Decreased river sediment transport	Sediment starvation at coast, increased coastal erosion
Changes in nearshore wind climate	Implications for renewable energy production
Changes in coastal climate regime	Implications for development and sustainability of tourist activities, increased groundwater extraction, salinisation of coastal aquifers

Table 12.1 continued

Increased coastal erosion	Implications for coastal landscape, resource and cultural management
Changes in coastal eco-systems	Implications for viability of aquaculture, biodiversity
Coastal wetland drainage and desiccation	Methane degassing
Increase in invasive species	Ecosystem and biodiversity loss
Increased coastal erosion	Need for increased coastal management, engineering, infrastructure

Table 12.1 shows that while many of these impacts are wide-ranging and cover different spatial and temporal scales, they are strongly determined by local topography, geology, land surface processes (including river dynamics and mass movements) and the capacity of human systems to manage or adapt to change effectively. In particular, there are diverse and unpredictable impacts of rivers and ecosystems along coasts. The following examples briefly illustrate some of the complex coastal responses to climate change.

Changes in precipitation in the coastal hinterland will have a first order control on river discharge bringing freshwater and sediment to the coast. Increased seasonality of precipitation is likely to result in increasing seasonal behaviour of rivers, with low discharges and warm and stratified water present during summers, in particular within estuaries. Additionally, changes in temperature and precipitation throughout the river catchment may influence land-use type, and nitrate and phosphorus runoff. Increased variability in river discharge is also likely to lead to increased need for river flood management, which decreases sediment transport through river systems and, consequently, decreases coastal sediment supply. Beach narrowing and steepening increases their vulnerability to sea-level change or enhanced erosion (Taylor *et al.*, 2004). Sea-level rise leads to increased coastal erosion and '**coastal squeeze**', whereby coastal landforms are restricted in their capacity for landward migration, thereby reducing the breadth of the coastal plain as sea level rises. As a result, coastal morphological and ecological systems occupy a narrower coastal strip, making them more vulnerable to future change. The process of coastal squeeze is particularly rapid on shallowly-sloping coasts associated with tidal flats and salt marsh, but also occurs along rocky coasts fronted by shore platforms.

Increased land and shallow-water temperatures (and lengthened growing season) will in many cases increase the productivity of coastal ecosystems, which is of particular importance for aquaculture. This advantage is likely to be offset, however, by decreased dissolved oxygen content within the water column and bottom sediments, and increased level of disease by overwintering pathogens and invasive species. Higher nutrient loadings can also lead to eutrophication, pollutant dispersal and colliform/algal problems in estuaries and nearshore environments. In turn, these primary producers can impact on marine invertebrates and molluscs, native sea birds and waders, and migratory bird species. It should be noted that many coastal sites are areas of high biodiversity and/or endemism. In summary, future climate change has the capacity to impact on different types of coastlines and in diverse and sometimes unanticipated ways. This sets the scene for examining our relationship to the coast and its resources, including coastal zone management.

12.1.2 Human activity and the coast

Coastlines generally have high population totals, high population density, high real-estate value, and large cities that are located near sea level and/or on deltas. There are many different estimates of coastal populations: presently, almost 40 per cent of the world's population live within 100 km of the coast (Cohen *et al.*, 1997), but up to 75 per cent could be living within 60 km of the coast by 2020 (Edgren, 1993). In the conterminous USA, 3 per cent of its population is at risk from a 1 per cent annual chance (100-year) coastal flood (Crowell *et al.*, 2010).

Case Study 12.1 Protecting coastal cities: Venice and London

The island city of Venice, Italy, is both subsiding as a result of sediment compaction and groundwater extraction, and is vulnerable to storm surges through the Adriatic Sea which are funnelled into and trapped within its surrounding lagoon (Ravera, 2000). As a result, variations in storm surge frequency and sea surface elevation are key elements of flood hazard and erosion risk for Venice, which is inscribed as a UNESCO World Heritage Site. An engineering scheme called MOSE (Modulo Sperimentale Elettromeccanico, or Electromechanical Experimental Module) is under development to help protect Venice from future storm surges. This project, which has a cost of €4.7 billion, involves constructing a total of 78 floodgates across three inlets into the lagoon, which can be raised vertically from their seafloor position to prevent sea water ingress during exceptional high tide or storm surge conditions greater than 1.10 m above normal water levels. The scheme is expected to be completed by 2014.

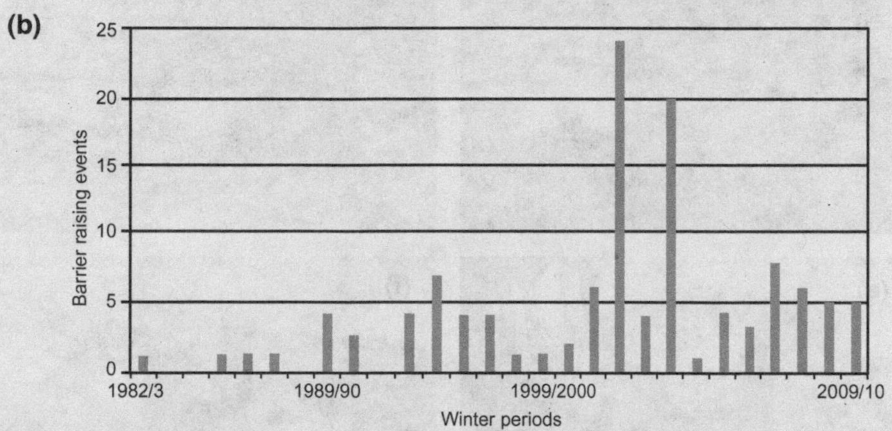

Figure 12.1 (a) Photo of the Thames Barrier. (Photo: H. Burningham.) (b) Graph showing the number of times the Thames Barrier has been raised, 1982–2010. (Based on data from Environment Agency.)

A similar problem of storm surges, sea-level rise and subsidence affects London, where storm surges can be funnelled into the River Thames estuary. The Thames Barrier, which is 520 m long and was completed in 1982 at a cost of £534 million, is located east of the city and comprises tilting gates that can be lowered into position to prevent upstream flooding. The barrier defences have been deployed much more frequently than anticipated – to 2011, 78 times for tidal surges and 41 times for river floods (Figure 12.1). Models also suggest that the surge elevation for which the barrier was designed to withstand – a 1-in-1000 year event – will be a 1-in-200 year event by the 2050s.

For the same coastal flood scenario, 260 km^2 of the metropolitan area of New York City would be inundated (Colle *et al.*, 2008), and such floods would affect 10 per cent of the population of the world's largest coastal cities, with assets to the value of US$3000 billion at risk (2005 figures) (Hanson *et al.*, 2011). Coastal

Figure 12.2 Photos of coastal archaeological features in Ireland. (a) Organic-stained and shell-rich occupation horizon within dune sands. (b) Buried shipwreck. (c) Kelp grid. (d) One arm of an intertidal fish trap. (e) Buried prow of a logboat. (f) World War II pillbox now found on a beach as a result of dune erosion. (Photos: J. Knight.)

cities are therefore vulnerable to a combination of sea-level rise and storm surges (Case Study 12.1).

Coastal locations, particularly near river mouths, were the first to be occupied by humans in postglacial times. These locations provided resources (food, water, transport, etc.) that were exploited by early settlers. Today, the coastal zone also has a range of uses, including tourism, residential, recreation, industrial/commercial, agriculture, waste disposal, aquaculture/fishing, nature reserves, military/strategic and coastal defence uses. The **archaeology** of coastal landscapes is therefore a preserved record of relationships between coastal geomorphology and human activity over long time periods. Evidence for past human activity in the coastal zone includes those features shown in Figure 12.2. In the midlatitudes such as the UK, Ireland, USA and South Africa, the coastal archaeological record commonly extends back to the Mesolithic period (in Europe, around 7000 years BP). Over this time frame, changes in coastline geography (by coastal erosion, deposition and sea-level change) mean that some archaeological features are now found raised or submerged, or at threat from coastal erosion. For example, Holocene sea-level rise in the central North Sea progressively drowned a palaeo-landscape, informally termed 'Doggerland'. Geophysical data and artefacts recovered by trawlers show how this landscape and associated human activity changed over time (Coles, 2000). Archaeological features in the present intertidal zone are particularly vulnerable to sea-level change, coastal erosion and human pressure. Conservation of cultural heritage is a key issue in coastal zone management.

12.2 Coastal resources and sustainability

Human activity makes use of the physical, ecological and scenic properties and attributes of the coastal zone, hence these can be termed **coastal resources**. Sustainability refers to the extent to which resources can be utilised over time without depleting the net resource base for future generations, but increased coastal population has meant that many resources are being rapidly depleted. For example, **groundwater extraction** from coastal aquifers is a significant problem in areas underlain by sediments or sedimentary rocks (see Section 2.3.3). Net groundwater depletion, where groundwater recharge rate is less than the extraction rate, leads to relative sea-level rise as the land surface subsides. Oil and gas extraction also leads to subsidence. Desalinisation plants, producing fresh water from sea water, now supply some developed coastal cities.

Offshore **aggregate extraction** is a significant activity on sandy paraglacial shelves and many nearshore environments. Aggregate is used for construction and engineering and often, in coastal settings, is associated with dredging within harbours and barrier island inlets. These activities are important in deepening of the nearshore and redistributing sediments of different grain sizes, which can lead to changes in nearshore wave climate, in particular an increase in wave energy. A recent volume on offshore aggregate resources highlights interactions between sea bed topography, landforms, waves and tides (Van Lancker *et al.*, 2010). Aggregate extraction can also disturb the sea bed, with implications for benthic habitats. Aggregate, heavy mineral and rare-earth extraction from beaches, known as sand mining, was a common practice but has declined since the 1970s, largely because it reduces beach width and can make cliffs more vulnerable to wave attack.

Coastal ecosystems are a key component of coastal sustainability with respect to their biodiversity and role in carbon cycling. Coastal sand dunes, estuaries, wetlands and the intertidal zone are areas of high **biodiversity**. Coastal biodiversity can be maintained or enhanced through the use of protected areas or by maintaining the geomorphological integrity and connectivity of coastal landscapes (Defeo *et al.*, 2009). Wetlands and the nearshore zone affected by upwelling and phytoplankton blooms are areas of high **gross primary productivity** (GPP) and as such are important in the global carbon cycle because they draw down atmospheric CO_2 as the phytoplankton grow. Due to coastal squeeze and sea-level rise, the capacity for carbon uptake by wetlands and the intertidal zone is decreasing, irrespective of any increases in GPP as a result of increased temperature and CO_2. Methane, an important contributor to global warming, is released by degassing from coastal wetlands by organic decomposition. For example, Magenheimer *et al.* (1996) found that salinity and water table position explained 29 per cent of the variance of the methane flux from salt marshes, and biomass and water table position explained 63 per cent of the variance of the CO_2 flux. Variations in wetland and intertidal environments therefore affect the workings of the global carbon cycle.

Human activity, in particular tourism, puts pressure on coastal landscapes, for example through reclamation of estuaries and salt marsh, construction of sea walls and groynes, groundwater extraction, trampling on sand dunes and development of coastal infrastructure including car and caravan parks. Many of these activities come into conflict as a result of coastal changes caused by climate (van Vuren *et al.*, 2004). **Sustainable coastal management** that marries understanding of physical systems with consideration of socioeconomic changes along coasts is an important theme in coastal science (Cooper and McKenna, 2008). An important aspect of sustainability is the economic and social **sustainability of coastal communities**. Despite being, in some places, expensive and desirable places to live, many coastal towns have low social mobility, seasonally-high unemployment, high poverty and other measures of deprivation including health, education and housing and low **social justice**. Involvement of local communities in decision-making, economic and social planning and policy, which also includes developing coastal resources, is an important part of sustainable economic regeneration. Glavovic (2008) argues that participatory action by coastal communities in planning and policy can help make them more resilient to coastal change by erosion and storm hazards, and to build communities that are 'sea-worthy'. Here, resilience refers to the ability of coastal communities to recover from adverse impacts of forcing, such as sea-level rise or extreme storms. It is often the most vulnerable people in coastal communities that are most at risk when these communities come under stress by coastal hazards. For example, 85 per cent of fatalities from Hurricane Katrina (described in Case Study 3.2) were over 51 years of age, despite this age group making up only 25 per cent of the pre-hurricane population (Jonkman *et al.*, 2009).

A further aspect of coastal sustainability relates to **renewable energy** which utilises coastal wind, tide and wave resources (see Coley (2008) for discussion of renewable energy and climate change). Problems with wind power include its variability, with low wind speeds in particular associated with high pressure cells during winter, which means that wind energy production can be unreliable. The visual impact of wind turbines within sight of the coast is a key area of contention with respect to coastal planning (Devine-Wright, 2005). With respect to waves, the highest annual average wave power (of around 20–70 kW m^{-1}) is found in the midlatitudes where strong storms and swell waves are dominant. Lower wave power in equatorial regions is often compensated for by consistency of wave properties as a result of tradewinds. Wind and wave structures on the sea bed can cause scouring, increased sediment transport and therefore changes to sea bed morphology. Although tidal currents are very site-specific, they have high predictability. Inside tidal barrages, however, tidal range and energy can be reduced, with impacts on intertidal areas.

12.3 Valuing the coast

12.3.1 Values and identity

The long relationship between human activity/occupation and the coastal zone means that coastal landforms, landscapes and scenery have a perceived value beyond their intrinsic and functional values. Indeed, it is often the cultural and historical associations of coastal settlements and distinctive coastal landforms that have greatest resonance with respect to imbuing values, identity and a sense of place. As many coastal environments are used by different people and for different purposes, many studies by social scientists have focused on why and how different user groups value coastal resources in different ways and shape their own values and identities. Studies have included examining the roles of territoriality and segregation, race, age, gender, display, masculinity, political expression, freedom and resistance (e.g. Preston-Whyte, 2001). Particular cultural associations and expressions of **identity** have also focused on activities such as surfing (Lanagan, 2002). Bondi Beach is closely associated with Australian national identity (Huntsman, 2001). Changes to coastal environments often cause these values and identities to be renegotiated. For example, in Louisiana (USA), Burley *et al.* (2007) show that residents in areas undergoing coastal erosion have a heightened **sense of place** and sense of attachment, because the erosion highlights the fragility and uniqueness of their environment. Conversely, coastal management and restoration can leave residents feeling isolated and alienated from their environment when they are not involved in developing and deploying policy and management. This is why local stakeholders are increasingly being involved with coastal management initiatives.

The value of coasts for recreation and leisure purposes has been an impetus for developing coastal resources nearer to population centres or as holiday resorts. For example, in many large cities seasonal or permanent beaches are often built in urban squares or along river margins. These **urban beaches**, such as in London, Paris, Amsterdam, Berlin and Toronto, commonly have a mix of sand, rock and water features with areas for recreation, seating and sunbathing. These 'designer' beaches do not attempt to mimic coastal landforms, but rather to provide an area for urban relaxation. Coasts developed for recreation purposes are also common in economically-developed areas (Case Study 12.2) and along the shores of inland lakes.

Case Study 12.2 Artificial islands: Coastal engineering in Dubai

Highly engineered and artificial coastal environments are commonly found where tourism is an important economic activity, and where infrastructure and engineering costs are perceived to be outweighed by economic and other benefits. However, it could be argued that this is a short-term view that does not consider wider environmental impacts of coastal development, including long-term changes to coastal sediment cells.

For example, development of tourism infrastructure in the port city of Dubai (United Arab Emirates) has focused around the construction of offshore island complexes or archipelagos into the Persian Gulf, with hotels, restaurants, shopping malls, spas and other entertainment and leisure facilities. Three archipelagos have been built to date: two palm-shaped islands (Palm Jumeirah and Palm Jebel Ali, with a third one, Palm Deirah, planned) and one resembling a map of the world (The World Islands). These artificial islands were constructed by deposition of dredged and imported sediment onto the sea bed (10 m deep) which were then stabilised by rock armour and sandy beach frontages. The first of these, Palm Jumeirah (Figure 12.3), was constructed in 2001–09 at a total estimated cost of US$12.3 billion. However, there are a number of environmental problems associated with island development. Subsidence has taken place as a result of sediment consolidation

Figure 12.3 Part of the port city of Dubai (United Arab Emirates) showing the artificial tourist islands of Palm Jumeirah in the Persian Gulf. (Image copyright © 2010 NASA, reproduced with permission.)

and sea bed disturbance. The presence of breakwaters and islands has influenced tidal flow patterns and resulted in sedimentation and low-oxygen conditions in water bodies between the islands. There are also infrastructure problems related to transport and services. These represent significant challenges for the sustainability of such island developments.

12.3.2 Methods of valuing the coast

One of the ways in which to evaluate and rank different coastal stretches or environments with respect to one another is to assign a numerical value to a checklist of coastal attributes. Ergin *et al.* (2006) describe a 26-attribute scheme for evaluating **coastal scenery**, including both physical and human attributes (Table 12.2). Based on this scheme, they identify five coastal classes, from Class

Table 12.2 List of coastal scenic attributes considered in the scheme of Ergin *et al.* (2006). These attributes are given a relative integer score of 1 (low) to 5 (high).

Physical environment		Human environment
Cliff height	Dunes	Disturbance factor
Cliff slope	Valley	Litter
Special features	Landform	Sewage
Beach type	Tides	Non-built environment
Beach width	Landscape features	Built environment
Beach colour	Vistas	Access type
Shore slope	Water colour	Skyline
Shore extent	Vegetation cover	Utilities
Shore roughness	Seaweed	

1 (very attractive) to Class 5 (very unattractive). This type of scenic evaluation builds from a seminal paper (in 1969) by the geomorphologist Luna B. Leopold (1915–2006) who used 46 variables to calculate a 'uniqueness factor' for fluvial landscapes that in turn can calculate 'landscape scale' and 'landscape interest'. Identifying which landscape elements make landscapes either unique or common is important in coastal management. For example, Nordstrom (1990) describes how management practices can be used to replicate natural coastal landforms and so enhance a coast's intrinsic value.

Another method of valuing the coast is **contingent valuation**, also known as willingness-to-pay. The contingent valuation method aims to put a 'price' on personal and subjective benefits that users receive or experience from environmental services, for example, the peacefulness of walking on a deserted beach, or spectacle of storm waves breaking over rocks. The method surveys users' opinions of how much they would be willing to pay in order to receive a specific environmental service, and has been often used to gauge public preference for particular coastal management strategies (e.g. Blakemore and Williams, 2008).

A related technique employs **cost-benefit analysis**, whereby alternative management or engineering schemes are compared economically by calculating the capital cost of undertaking and maintaining the works, divided by the benefits derived from these works for the time period over which these schemes will operate. Cost-benefit analysis is a useful means by which to identify the most cost-effective strategies, and is commonly used in coastal management (Carter, 1988). Changes in recurrence interval of storms or flood events may mean that management or engineering strategies that were previously too expensive now become economically viable. Similarly, reduction in land prices and/or increase in coastal protection costs may mean that coastal sections that were previously protected may in the future be left to erode. Cost-benefit analysis works best when considering the economic viability of coastal engineering structures such as sea walls and groynes. The impact of coastal engineering on tourism value or increased water quality is less easy to calculate by cost-benefit analysis, and contingent valuation is more useful in these cases.

12.4 Managing coastal change

The ongoing and accelerating rise in eustatic sea level (see Section 2.4.1) is of major concern to coastal nations as it can result in enhanced coastal erosion and flooding of coastal wetlands. Likely impacts on human activities in the coastal zone will include disrupted transport routes and loss of agricultural land in addition to an increase in coastal hazards. Here we use the example of coastal and human response to sea-level rise to illustrate some key management issues associated with future coastal change.

12.4.1 Coastal impacts of sea-level rise

One way of assessing the potential risk of rising sea levels is by **coastal inundation maps**, which indicate areas threatened by sea-level rise (Figure 12.4). Such maps are generated by projecting a predicted sea-level rise (for example 2 m rise over 200 years) onto a contour map and shading those areas that would be flooded by such a rise. Although this visual analysis is appealing, it has a weak scientific basis since there are significant height errors involved, depending on spatial scale, and feedbacks are not considered. However, inundation maps are often used uncritically by planners and insurance companies to identify areas at flood risk.

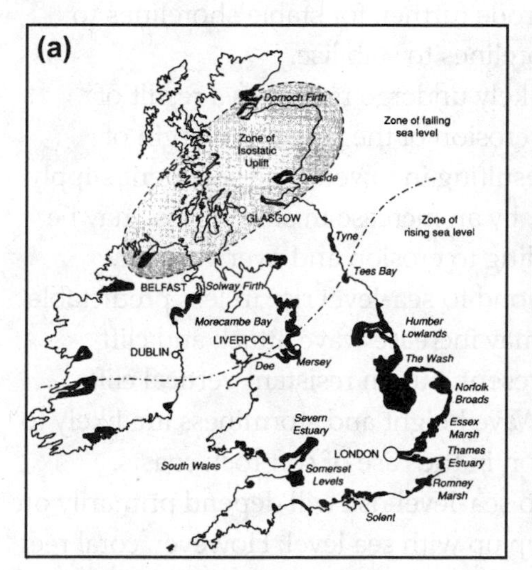

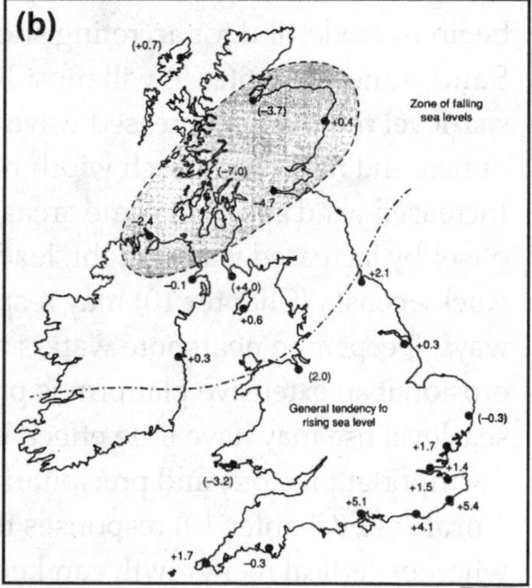

Figure 12.4 Map of the British Isles showing: (a) areas of isostatic uplift (grey shading) and those threatened by sea-level rise (black shading) and (b) rates of relative sea-level rise in mm per year. (From Briggs *et al.*, 1997; modified from Boorman *et al.*, 1989.) (Copyright © 1997 Taylor & Francis Books, reproduced with permission.)

Geomorphological responses to sea-level rise have been considered in detail throughout this book and to summarise these are:

- **Deltas** (Chapter 6) are dependent on fluvial sediment supply to build up the delta surface and prograde the delta front. Most deltas worldwide are experiencing decreased sediment supply due to catchment engineering, and so are threatened by coastal erosion. Sea-level rise and delta subsidence will likely mean that deltas will continue to erode at increasing rates and delta hinterlands will experience increased flood events.

- **Estuaries** (Chapter 7) tend to widen and deepen as sea level rises, and if tides penetrate farther upstream, tidal range may increase and tidal currents may increase in strength. Salt marsh and mangrove environments (and to a lesser extent tidal flats) grow through vertical accretion of sediment on the wetland surface and progradation at the edges. The response of these environments to sea-level rise is affected by organic and inorganic sediment supply, and depends on the relative rates of accretion and sea-level rise.
- **Beaches and barriers** (Chapter 8) generally retreat as sea level rises, either through erosion, migration or overstepping of the barrier which is dependent on the substrate gradient. Barriers can also prograde under transgressive conditions if there is a large enough sediment supply, so this supply rate is a critical control on beach and barrier response to sea-level rise. Bird (1993) argues that with future sea-level rise there will be a tendency for eroding shorelines to erode further, for stable shorelines to begin to erode, and for accreting shorelines to stabilise.
- **Sand dunes** (Chapter 9) will most likely undergo retreat as a result of sea-level rise due to increased wave erosion of the seaward margin of dunes, and narrowed beach width resulting in lower dune sediment supply. Increased wind speed in some areas, by an increase in storminess, may be offset by increased wave height, leading to erosion and overwashing.
- **Rocky coasts** (Chapter 10) may respond to sea-level rise in less predictable ways. Deepening nearshore waters may increase wave attack and cliff erosion if an extensive platform is present, but on resistant vertical cliffs, sea-level rise may have little effect. Wave height and storminess are likely to be important factors, and precipitation in the case of soft rock coasts.
- **Coral reef** (Chapter 10) responses to sea-level rise will depend primarily on whether vertical reef growth can keep up with sea level. However, coral reef ecosystems are also sensitive to increased sea surface temperature (causing coral bleaching) which will decrease their viability. Hurricane effects are also important for coral reefs and islands.

12.4.2 Coastal vulnerability

As apparent from the previous section, geomorphological responses to sea-level rise are very diverse, depending on the physical properties of the coastal stretch under consideration and the nature of the forcing factors, which also include human activity. An important concept in the management of the effects of these forcing factors is **vulnerability**, which refers to the sensitivity of the system to be affected by variations in forcing, such as by sea-level rise along coasts.

We have already seen that different types of coastline have different levels of sensitivity to external forcing. A useful way to consider coastal response to forcing is to consider the relationship between the coast's vulnerability and resilience. Vulnerability is defined above, whereas **resilience** is a measure of the system's capacity to respond to the consequences of disturbance by forcing. Resilience is highest where the coast has high resistance. **Resistance** describes the ability of a coastal system to avoid disturbance. High vulnerability tends to be associated with high coastal impacts (i.e. coastal change takes place very readily) and low capacity of the coastal system to withstand these impacts (Figure 12.5). In the context of coastal management, McFadden *et al.* (2007) argue that

$$V = I - A \qquad \textbf{(12.1)}$$

where V is vulnerability, I is the coastal impacts and A is the effect of adaptation by the mitigating effects of coastal management. This means that effective management can make the coastline more resilient to change, thereby reducing its vulnerability. At the same time, poor management can make coasts less resilient to change and hence more vulnerable. An example of the latter can be to allow housing developments on the coastal floodplain.

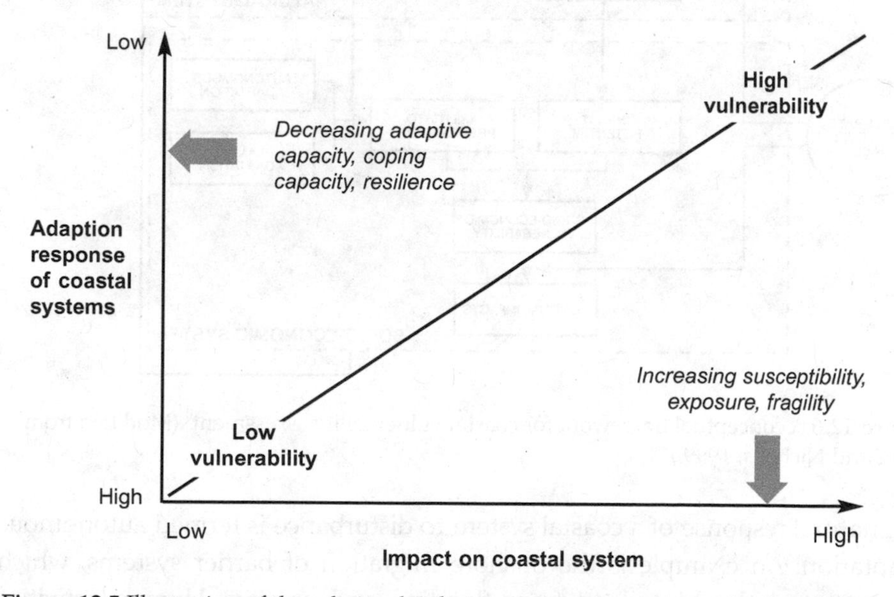

Figure 12.5 Illustration of the relationship between coastal systems and vulnerability. (Redrawn from McFadden *et al.*, 2007.) (Copyright © 2007 Elsevier Science, reproduced with permission.)

Klein and Nicholls (1999) make a distinction between **natural vulnerability** and **socioeconomic vulnerability** to sea-level rise (Figure 12.6). Analysis of coastal vulnerability starts with a notion of the natural system's susceptibility to the biophysical effects of sea-level rise by looking at the coastal system's potential to be affected by sea-level rise. For example, a subsiding delta is more susceptible to the adverse effects of sea-level rise than a rocky coast. The natural capacity of the coastal system to deal with the impacts of sea-level rise – its resilience and resistance – is also important in determining the natural vulnerability. Many natural features contribute to coastal resilience by providing ecological buffers (coral reefs, salt marshes and mangrove forests) and morphological protection (sand and gravel beaches, barriers and coastal dunes).

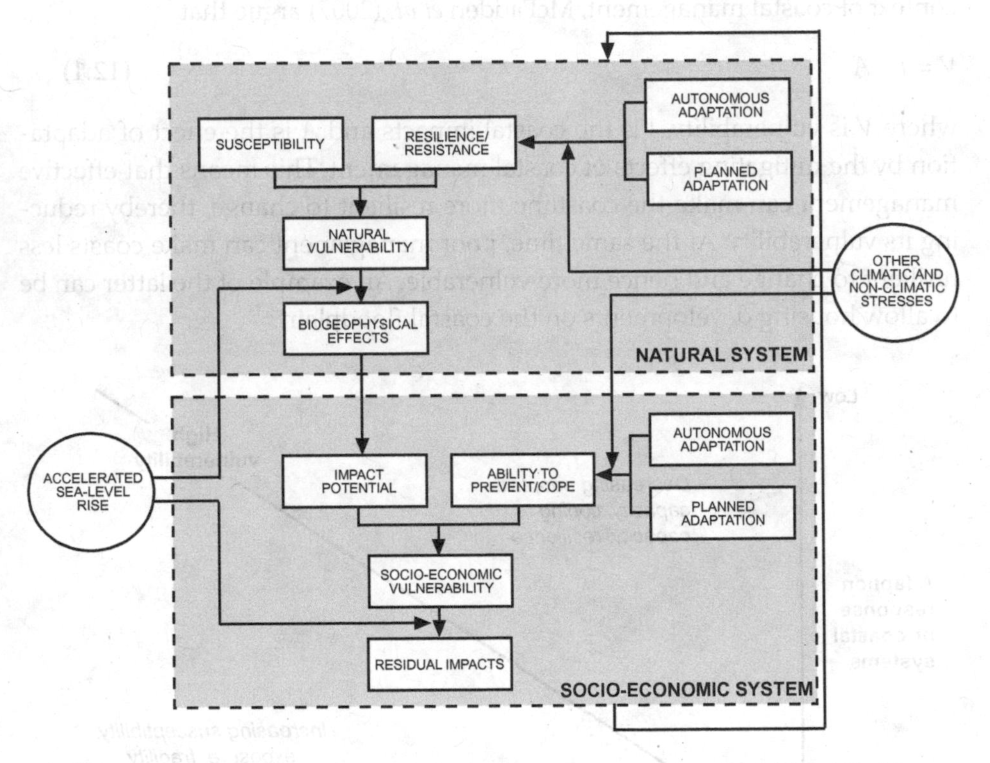

Figure 12.6 A conceptual framework for coastal vulnerability assessment. (Modified from Klein and Nicholls, 1999.)

The natural response of a coastal system to disturbance is termed autonomous adaptation. An example is the onshore migration of barrier systems, which helps protect the hinterland from flooding and erosion. Human activities often reduce resilience and resistance, thereby increasing natural vulnerability. However, planned adaptation can reduce natural vulnerability by enhancing the system's resilience and resistance. An example of planned adaptation is managed retreat, which involves allowing the coastline to recede to a new line of defence to restore natural coastal processes and systems, such as mud flats, salt marsh areas and dunes. Managed retreat helps to reduce coastal vulnerability by increasing coastal resilience.

12.4.3 Managing sea-level rise

Management strategies are adaptive responses that can help reduce coastal vulnerability to sea-level rise and other forcing factors. There are four principal management strategies that have been used to cope with coastal erosion due to sea-level rise:

- **Do nothing** – This option is only viable if the coastline under question is undeveloped and nothing is at stake by giving up the land to coastal erosion.
- **Managed retreat** – This option, also referred to as managed realignment, may involve the relocation of coastal communities and/or industry which therefore minimises future risk and protection costs. However, social and economic costs associated with relocation and compensation are potentially high and this option requires a strong governmental role with supportive legislation.
- **Accommodation** – This allows continued occupancy and use of vulnerable coastal areas by adapting to, rather than protecting fully against, adverse impacts. It means learning to live with sea-level rise and coastal flooding. Accommodation options include elevating buildings, enhancing storm and flood warning systems, and modifying drainage. The accommodation option can also involve changing activities, such as changing farming practices to suit the new environment, or accepting the risks of inundation and increasing insurance premiums. The accommodation option requires high levels of organisation and community participation.
- **Protection** – This option is also referred to as **hold-the-line** and involves physically protecting the coast through hard engineering structures or beach nourishment. Protection has social, economic and political advantages because assets and investments are safeguarded while economic activity can continue largely unhindered. Protection is the most expensive option to implement and maintain, and is only economically justifiable if the land to be protected is of high value.

The first three strategies are based on the premise that increased land losses and coastal flooding will be allowed to occur and that some coastal functions and values will be changed or lost. These strategies can help to maintain the dynamic nature of the coast and allow it to adjust to rising sea levels naturally. It is beneficial to allow as many coastal regions to retreat as naturally as possible because erosion will liberate sediments, which may lessen the impact of sea-level rise elsewhere. Hence, the first three options are most sustainable from a geomorphological point of view (although not necessarily from a socio-economic perspective). Which management option is most suitable on any given coast depends more on socioeconomic and political factors than on physical factors. For example, Tol *et al.* (2008) examine the adaptive capacity of different European countries to sea-level rise. They argue that, irrespective of coastline type, policy responses vary considerably between countries, which means that the impacts of future sea-level rise will be more strongly determined by coastal management policy rather than coastal processes and geomorphology.

12.4.4 Coastal management policy for the twenty-first century

Coastal management covers three broad areas (Carter, 1988):

- Policy relates to the political and administrative framework through which coastal management is regulated, principally through legislation and education.
- Planning is the process of allocation of environmental, ecological, social or economic resources. Planning may be negative in that it discourages development, or positive in that it encourages it.
- Practice covers the techniques needed for implementation of planning decisions, or for undertaking restorative or remedial works. Practices range from building coastal protection structures to planting marram grass.

The three areas are closely related, and developments in one area tend to inform developments in the other two.

In the past, coastal management was often concerned with single issues that could be dealt with by a single authority. This is no longer the case due to increased complexity of coastal management issues and the spatial and temporal scales at which they operate. **Integrated coastal management** (ICM) refers to the management approach that considers the range of scientific and human issues affecting a coastal stretch, different coastal users (including administrative authorities, environmental organisations, industry and other interest groups), available legislation, engineering and expertise, and which effectively integrates policy, planning and practice. As such, ICM is a dynamic process in which a coordinated strategy is developed and implemented for the allocation of resources to achieve the conservation and sustainable use of the coastal zone. Geographically, ICM embraces upland watersheds, the shoreline (beaches, dunes and wetlands), estuaries and nearshore waters, and requires integration in the following areas:

- programmes for economic development, environmental quality development and land use
- programmes for sectors such as food production (including agriculture and fishing), energy, transportation, water resources, waste disposal and tourism
- tasks of coastal management (planning, analysis, implementation, operation, maintenance, monitoring and evaluation) performed continuously over time

- responsibilities for management tasks among levels of government (international, national, regional, state/provincial and local) and between the public and private sectors
- available resources for management (i.e. personnel, funds, materials, equipment).

The ICM concept is now universally applied to coastal management problems in both developed and developing countries, and is identified by the IPCC as a key tool for dealing with future sea-level rise.

Typical output from an ICM strategy covers areas of policy, planning and practice and in both statutory and non-statutory ways. For example, in the Netherlands, a Dynamic Preservation strategy was adopted in 1991 with the aim of maintaining the coastline at its 1990 position, as measured by changes to the coastal sediment budget between the coastal dune foot (+3 m above sea level) and –5 m below sea level. Based on this strategy, any coastal erosion taking place would be compensated for by nourishment. In 2000 a Coastal Foundation Zone strategy was adopted, which extended this sediment budget approach to the area between the dune foot and –20 m below sea level. Non-statutory initiatives tend to be more locally based and are not legally binding, although they are advisory. They are generally in the form of **coastal** (also called **shoreline**) **management plans**, although not all management plans are non-statutory. Coastal management plans can chart out a course for the future development of a coastal stretch and/or assist in resolving current management problems. They take into consideration the physical and human environments, wave, tide and sediment patterns, and the trajectory of coastal change over short (0–20 year) to long (50–100 year) time scales.

More recently, ICM now has a greater emphasis on the sustainability of both the physical environment and management practices, particularly the effects on climate change, so the term **sustainable coastal management** (SCM) is now more commonly used. Central to SCM is the use of sediment budgets as a diagnostic tool to measure the 'sustainability' of coastal landforms, such as the size of a beach and the effectiveness of management strategies (e.g. Brommer and Bochev-van der Burgh, 2009). The adaptive capacity of the coastline to respond to forcing is also central to SCM. For example, if the coastal stretch exhibits high resilience then it is likely to give rise to greater sustainability. Understanding of coastal responses to climate change is therefore one of the most pressing issues in order to develop sustainable coastal management strategies for the twenty-first century and beyond.

SUMMARY

- Future global climate changes provide a context for evaluating how coasts will change in the future, in particular their responses to sea-level rise. These coastal changes are relevant for human activity which is disproportionately concentrated in the coastal zone and which is therefore vulnerable to changes in sea level, coastal erosion, coastal squeeze, storminess, land subsidence, river discharge, aquifer contamination and other factors.
- The sustainability of coastal resources is based on the rate at which they are depleted relative to the rate at which they are enhanced or maintained. Human activity is closely associated with the depletion of many coastal physical resources, but coasts are also valued for their scenic and aesthetic qualities. Conservation of both coastal physical and scenic resources is an important priority for coastal management.
- Sea-level rise will impact different coastal environments in different ways, most commonly through enhanced erosion and changes in sediment supply which result in increased coastal vulnerability. Management responses to sea-level rise are the most important means by which coastal vulnerability can be decreased.

Reflective questions

These questions are designed to test your comprehension of material covered in this chapter. Suggested answers to these questions can be found on this book's website.

12a. Outline the reasons why future climate change is likely to result in many different responses rather than a single coastal response.

12b. Which coastal resources are likely to be sustainable over decadal time scales, and why?

12c. Critically consider Ergin *et al.*'s (2006) checklist of attributes used for evaluating coastal scenery (Table 12.2). What other factors could be included?

12d. How can the adaptive capacity of the coastal zone be increased with respect to ongoing climate change?

Further reading

Huntsman, L., 2001. *Sand in Our Souls: The beach in Australian history and culture*. Melbourne University Press, Melbourne, Australia. (An interesting and entertaining book that describes the relationship between culture and coasts.)

Jones, A.L. and **Phillips**, M., 2010. *Disappearing Destinations – Climate change and future challenges for coastal tourism.* CABI Publications, London. (This book has case studies that address the impacts of climate change on human use of the coastal zone.)

McFadden, L., **Nicholls**, R.J. and **Penning-Rowsell**, E., (eds) 2007. *Managing Coastal Vulnerability*. Elsevier, Amsterdam. (This is a very useful textbook with case studies on vulnerability as applied to coastal management.)

Monmonier, M., 2008. *Coast Lines: How mapmakers frame the world and chart environmental change*. University of Chicago Press, Chicago. (This book describes the history of coastal cartography and how environmental change has shaped the coast.)

REFERENCES

Aagaard, T., **Davidson-Arnott**, R., **Greenwood**, B. and **Nielsen**, J., 2004. 'Sediment supply from shoreface to dunes: linking sediment transport measurements and long-term morphological evolution.' *Geomorphology*, 60, 205–224.

Aagaard, T., **Kroon**, A., **Greenwood**, B. and **Hughes**, M.G., 2010. 'Observations of offshore bar decay: Sediment budgets and the role of lower shoreface processes.' *Continental Shelf Research*, 30, 1497–1510.

Aagaard, T., **Kroon**, A., **Hughes**, M.G. and **Greenwood**, B., 2008. 'Field observations of nearshore bar formation.' *Earth Surface Processes and Landforms*, 33, 1021–1032.

Aagaard, T. and **Masselink**, G., 1999. 'Chapter 4: The surf zone.' In: A.D. Short (ed) *Beach Morphodynamics*. Wiley and Sons, London, 72–118.

Allen, G.P., **Salomon**, J.C., **Bassoulet**, P., **Du Penhoat**, Y. and **Du Grandpré**, C., 1980. 'Effects of tides on mixing and suspended sediment transport in macrotidal estuaries.' *Sedimentary Geology*, 26, 69–90.

Allen, J.R.L., 1984. *Sedimentary Structures: Their Character and Physical Basis*. Elsevier, Amsterdam.

Allen, J.R.L., 1985. *Principles of Physical Sedimentology*. George Allen and Unwin, London.

Allen, J.R.L., 1994. 'Fundamental properties of fluids and their relation to sediment transport.' In: K. Pye (ed) *Sediment Transport and Depositional Processes*. Blackwell Scientific Publications, Oxford, 25–60.

Allen, P.A., 1997. *Earth Surface Processes*. Blackwell Science, Oxford.

Allison, R.J., (ed) 1992. *The Coastal Landforms of West Dorset*. Geologists' Association Guide no. 47, Geologists' Association, London.

André, M.-F., 2009. 'From climatic to global change geomorphology: contemporary shifts in periglacial geomorphology.' In: J. Knight and S. Harrison (eds) *Periglacial and Paraglacial Processes and Environments*. Geological Society of London Special Publications, 320, 5–28.

Arens, S.M., 1994. *Aeolian Processes in the Dutch Foredunes*. PhD Thesis, University of Amsterdam.

Argus, D.F. and **Peltier**, W.R., 2010. 'Constraining models of postglacial rebound using space geodesy: a detailed assessment of model ICE-5G (VM2) and its relatives.' *Geophysical Journal International*, 181, 697–723.

Augustinus, P.G.E.F., 1989. 'Cheniers and chenier plains: A general introduction.' *Marine Geology*, 90, 219–229.

Austin, M.J., **Scott**, T.M., **Brown**, J.W., **Brown**, J.A., **MacMahan**, J., **Masselink**, G. and **Russell**, P.E., 2010. 'Temporal observations of rip current circulation on a macro-tidal beach.' *Continental Shelf Research*, 30, 1149–1165.

Avoine, J., 1981. *L'Estuare de la Seine: Sediments et Dynamique Sedimentaire*. These, Docteur de Specialite. Université de Caen, France.

Bagnold, R.A., 1941. *The Physics of Blown Sand and Desert Dunes*. Methuen, London.

Bagnold, R.A., 1963. 'Mechanics of marine sedimentation.' In: M.N. Hill (ed) *The Sea*, Vol. 3. Wiley-Interscience, New York, 507–528.

Bagnold, R.A., 1966. *An Approach to the Sediment Transport Problem from General Physics.* Professional Paper no. 422-I, United States Geological Survey.

Bailey, S.D. and **Bristow**, C.S., 2004. 'Migration of parabolic dunes at Aberffraw, Anglesey, north Wales.' *Geomorphology*, 59, 165–174.

Barry, R.G. and **Chorley**, R.J., 2003. *Atmosphere, Weather and Climate* (8th edn). Routledge, London.

Bartrum, J.A., 1916. 'High water rock platforms: A phase of shoreline erosion.' *Transactions New Zealand Institution*, 48, 132–134.

Bascom, W.H., 1953. 'Characteristics of natural beaches.' *Proceedings 4th International Conference on Coastal Engineering*. ASCE, 163–180.

Bascom, W.H., 1980. *Waves and Beaches*. Anchor Books, Garden City.

Bascom, W.N., 1951. 'The relationship between sand size and beach-face slope.' *Transactions, American Geophysical Union*, 32, 866–874.

Bates, C.C., 1953. 'Rational theory of delta formation.' *Bulletin of the American Association of Petroleum Geologists*, 37, 2119–2161.

Battjes, J.A., 1974. 'Surf similarity.' *Proceedings 14th International Conference on Coastal Engineering*. ASCE, 466–480.

Bayliss-Smith, T.P., 1988. 'The role of hurricanes in the development of reef islands, Ontong Java Atoll, Solomon Islands.' *Geographical Journal*, 154, 377–391.

Bayram, A., **Larson**, M. and **Hanson**, H., 2007. 'A new formula for the total longshore sediment transport rate.' *Coastal Engineering*, 54, 700–710.

Been, K., and **Sills**, G.C., 1981. 'Self weight consolidation of soft soils: An experimental and theoretical study.' *Geotechnique*, 31, 519–535.

Belderson, R.H., **Pingree**, R.D. and **Griffiths**, D.K., 1986. 'Low sea level tidal origin of Celtic sea sand banks – evidence from numerical modelling of M2 tidal streams.' *Marine Geology*, 73, 99–108.

Berger, A.L., 1992. 'Astronomical theory of paleoclimates and the last glacial-interglacial cycle.' *Quaternary Science Reviews*, 11, 571–582.

Bindoff, N.L., **Willebrand**, J., **Artale**, V., **Cazenave**, A., **Gregory**, J., **Gulev**, S., **Hanawa**, K., **Le Quéré**, C., **Levitus**, S., **Nojiri**, Y., **Shum**, C.K., **Talley**, L.D. and **Unnikrishnan**, A., 2007. 'Observations: Oceanic Climate Change and Sea Level.' In: S. Solomon, D. Qin, M. Manning, Z. Chen, M. Marquis, K.B. Averyt, M. Tignor and H.L Miller (eds) *Climate Change 2007: The Physical Science Basis.* Contribution of Working Group I to the Fourth Assessment Report of the Intergovernmental Panel on Climate Change. Cambridge University Press, Cambridge, 386–432.

Bird, E.C.F., 1993. *Submerging coasts: The Effects of a Rising Sea Level on Coastal Environments*. Wiley and Sons, Chichester.

Bird, E.C.F., 2000. *Coastal Geomorphology: An Introduction*. Wiley and Sons, Chichester.

Blakemore, F. and **Williams**, A., 2008. 'British tourists' valuation of a Turkish beach using contingent valuation and travel cost methods.' *Journal of Coastal Research*, 24, 1469–1480.

Blanchon, P. and **Shaw**, J., 1995. 'Reef drowning during the last deglaciation: Evidence for catastrophic sea-level rise and ice-sheet collapse.' *Geology*, 23, 4–8.

Blikra, L.H. and **Nemec**, W., 1998. 'Postglacial colluvium in western Norway: depositional processes, facies and palaeoclimatic record.' *Sedimentology*, 45, 909–959.

Blum, M.D. and **Roberts**, H.H., 2009. 'Drowning of the Mississippi Delta due to insufficient sediment supply and global sea-level rise.' *Nature Geoscience*, 2, 488–491.

Boggs, S., 1995. *Principles of Sedimentology and Stratigraphy* (2nd edn). Prentice Hall, New Jersey.

Boorman, L.A., **Goss-Custard**, J.D. and **McGrorty**, S., 1989. *Climate Change, Rising Sea Level and the British Coast.* London, HMSO.

Bowen, A.J., **Inman**, D.L. and **Simmons**, V.P., 1968. 'Wave 'setdown' and setup.' *Journal of Geophysical Research*, 73, 2569–2577.

Bradbury, R.H. and **Seymour**, R.M., 2009. 'Coral reef science and the new commons.' *Coral Reefs*, 28, 831–837.

Bretschneider, C.L., 1952. 'The generation and decay of wind waves in deep water.' *Transactions American Geophysical Union*, 33, 381–389.

Briggs, D., **Smithson**, P., **Addison**, K. and **Atkinson**, K., 1997. *Fundamentals of the Physical Environment* (2nd edn). Routledge, London.

Brommer, M.B. and **Bochev-van der Burgh**, L.M., 2009. 'Sustainable Coastal Zone Management: A concept for forecasting long-term and large-scale coastal evolution.' *Journal of Coastal Research*, 25, 181–188.

Brunsden, D. and **Goudie**, A., 1997. *Classic Landforms of the West Dorset Coast*. Geographical Association, Sheffield.

Bruun, P., 1954. *Coastal Erosion and the Development of Beach Profiles*. US Army Beach Erosion Board, Technical Memorandum, 44.

Bruun, P., 1962. 'Sea level rise as a cause of shore erosion.' *Journal of Waterway, Port, Coastal and Ocean Engineering*, ASCE, 88, 117–130.

Bryant, E., 2001. *Tsunami: The underrated hazard*. Cambridge University Press, Cambridge.

Buddemeier, R.W. and **Smith**, S.V., 1988. 'Coral reef growth in an era of rapidly rising sea level: Predictions and suggestions for long-term research.' *Coral Reefs*, 7, 51–56.

Burley, D., **Jenkins**, P., **Laska**, S. and **Davis**, T., 2007. 'Place attachment and environmental change in coastal Louisiana.' *Organization & Environment*, 20, 347–366.

Burn, C.R. and **Kokelj**, S.V., 2009. 'The environment and permafrost of the Mackenzie Delta area.' *Permafrost and Periglacial Processes*, 20, 83–105.

Burningham, H., 2005. 'Morphodynamic behaviour of a high-energy coastal inlet: Loughros Beg, Donegal, Ireland.' In: D.M. FitzGerald and J. Knight (eds) *High Resolution Morphodynamics and Sedimentary Evolution of Estuaries*. Springer, New York, 215–243.

Burningham, H., 2008. 'Contrasting geomorphic response to structural control: The Loughros estuaries, northwest Ireland.' *Geomorphology*, 97, 300–320.

Buscombe, D. and **Masselink**, G., 2006. 'Concepts in gravel beach dynamics.' *Earth-Science Reviews*, 79, 33–52.

Cane, M., 2005. 'The evolution of El Niño, past and future.' *Earth and Planetary Science Letters*, 230, 227–240.

Cantin, N.E., **Cohen**, A.L., **Karnauskas**, K.B., **Tarrant**, A.M. and **McCorkle**, D.C., 2010. 'Ocean warming slows coral growth in the central Red Sea.' *Science*, 329, 322–325.

Carr, A.P., 1971. 'Experiments on longshore transport and sorting of pebbles: Chesil Beach, England.' *Journal of Sedimentary Petrology*, 41, 1084–1104.

Carr, A.P., **Gleason**, R. and **King**, A., 1970. 'Significance of pebble size and shape in sorting by waves.' *Sedimentary Geology*, 4, 89–101.

Carter, R.W.G., 1988. *Coastal Environments*. Academic Press, London.

Carter, R.W.G., **Nordstrom**, K.F. and **Psuty**, N.P., 1990. 'The study of coastal dunes.' In: K.F Nordstrom, N.P. Psuty and R.W.G. Carter (eds) *Coastal Dunes: Form and Process*. John Wiley & Sons, London, 1–14.

Carter, R.W.G. and **Orford**, J.D., 1988. 'Conceptual model of coarse clastic barrier formation from multiple sediment sources.' *The Geographical Review*, 78, 221–239.

Carter, R.W.G. and **Wilson**, P., 1990. 'The geomorphological, ecological and pedological development of coastal foredunes at Magilligan Point, Northern Ireland.' In: K.F. Nordstrom, N. Psuty and R.W.G. Carter (eds) *Coastal Dunes: Form and Process*. Wiley, London, 129–157.

Carter, R.W.G. and **Wilson**, P., 1991. 'Chronology and geomorphology of the Irish dunes.'

In: M.B. Quigley (ed) *A guide to the sand dunes of Ireland*. Dublin: European Union for Dune Conservation and Coastal Management, 18–41.

Carter, R.W.G. and **Woodroffe** C.D., 1994. 'Coastal evolution: An introduction.' In: R.W.G. Carter and C.D. Woodroffe (eds) *Coastal Evolution*. Cambridge University Press, Cambridge, 1–31.

Cartwright, D.E., 1999. *Tides: A Scientific History*. Cambridge University Press, Cambridge.

Castelle, B., **Turner**, I.L., **Bertin**, X. and **Tomlinson**, R., 2009. 'Beach nourishments at Coolangatta Bay over the period 1987–2005: Impacts and lessons.' *Coastal Engineering*, 56, 940–950.

CERC, 1984. *Shore Protection Manual* (4th edn). Coastal Engineering Research Center, Waterway Experiment Station, Corps of Engineers, Vicksburg.

Chalikov, D., 2009. 'Freak waves: their occurrence and probability.' *Physics of Fluids*, 21, 076602, doi:10.1063/1.3175713.

Chandler, J.H. and **Brunsden**, D., 1995. 'Steady state behaviour of the Black Ven mudslide: The application of archival analytical photogrammetry to studies of landform change.' *Earth Surface Processes and Landforms*, 20, 255–275.

Chapman, D.M., 1990. 'Aeolian sand transport – an optimized model.' *Earth Surface Processes and Landforms*, 15, 751–760.

Chappell, J., 1980. 'Coral morphology, diversity and reef growth.' *Nature*, 286, 249–252.

Chappell, J. and **Shackleton**, N.J., 1986. 'Oxygen isotopes and sea level.' *Nature*, 324, 137–140.

Chelli, A., **Pappalardo**, M., **Llopis**, I.A. and **Federici**, P.R., 2010. 'The relative influence of lithology and weathering in shaping shore platforms along the coastline of the Gulf of La Spezia (NW Italy) as revealed by rock strength.' *Geomorphology*, 118, 93–104.

Chepil, W.S., 1945. 'Dynamics of wind erosion. 3. The transport capacity of the wind.' *Soil Science*, 60, 475–480.

Cherniawsky, J.Y., **Foreman**, M.G.G., **Kang**, S.K., **Scharroo**, R. and **Eert**, A.J., 2010. '18.6-year lunar nodal tides from altimeter data.' *Continental Shelf Research*, 30, 575–587.

Church, M. and **Ryder**, J.M., 1972. 'Paraglacial sedimentation: a consideration of fluvial processes conditioned by glaciation.' *Bulletin of the Geological Society of America*, 83, 3059–3071.

Clarke, J.A., **Farrell**, W.E. and **Peltier**, W.R., 1978. 'Global changes in postglacial sea level: A numerical calculation.' *Quaternary Research*, 9, 265–278.

Clarke, M.L. and **Rendell**, H.M., 2006. 'The effects of storminess, sand supply and the North Atlantic Oscillation on sand invasion and coastal dune accretion in western Portugal.' *The Holocene*, 16, 341–355.

Clemmensen, L.B., **Andreasen**, F., **Heinemeier**, J. and **Murray**, A., 2001. 'A Holocene coastal aeolian system, Vejers, Denmark: landscape evolution and sequence stratigraphy.' *Terra Nova*, 13, 129–134.

Coco, G., **Huntley**, D.A. and **O'Hare**, T.J., 2001. 'Regularity and randomness in the formation of beach cusps.' *Marine Geology*, 178, 1–9.

Coco, G. and **Murray**, A.B., 2007. 'Patterns in the sand: From forcing templates to self-organization.' *Geomorphology*, 91, 271–290.

Coco, G., **O'Hare**, T.J. and **Huntley**, D.A., 2000. 'Investigation of a self-organisation model for beach cusp formation and development.' *Journal of Geophysical Research*, 105, 21991–22002.

Cohen, J.E., **Small**, C., **Mellinger**, A., **Gallup**, J. and **Sachs**, J., 1997. 'Estimates of coastal populations.' *Science*, 278, 1211–1212.

Colella, A. and **Prior**, D.B. (eds), 1990. *Coarse-grained Deltas*. IAS Special Publication no 10, Blackwell Science, Oxford.

Coleman, J.M., 1982. *Deltas: Processes of Deposition and Models for Exploration*. International Human Resources Development Corporation, Boston.

Coleman, J.M., 1988. 'Dynamic changes and processes in the Mississippi River delta.' *Geological Society of America Bulletin*, 100, 999–1015.

Coleman, J.M., **Roberts**, H.H. and **Stone**, G.W., 1998. 'Mississippi River delta: an overview.' *Journal of Coastal Research*, 14, 698–716.

Coles, B.J., 2000. 'Doggerland: the cultural dynamics of a shifting coastline.' In: K. Pye and J.R.L. Allen (eds) *Coastal and Estuarine Environments: sedimentology, geomorphology and geoarchaeology*. Geological Society of London Special Publications, 175, 393–401.

Coley, D.A., 2008. *Energy and Climate Change – creating a sustainable future*. John Wiley & Sons, London.

Colle, B.A., **Buonaiuto**, F., **Bowman**, M.J., **Wilson**, R.E., **Flood**, R., **Hunter**, R., **Mintz**, A. and **Hill**, D., 2008. 'New York City's vulnerability to coastal flooding.' *Bulletin of the American Meteorological Society*, 89, 829–841.

Cooper, G.R., 2007. Black Ven, Dorset. In: R.G. Cooper (ed) *Mass Movements in Great Britain*. Geological Conservation Review Series, No 33, Joint Nature Conservation Committee, Peterborough, 223–244.

Cooper, J.A.G. and **McKenna**, J., 2008. 'Working with natural processes: the challenge for coastal protection strategies.' *Geographical Journal*, 174, 315–331.

Cooper, J.A.G. and **Pilkey**, O.H., 2004. 'Sea-level rise and shoreline retreat: time to abandon the Bruun Rule.' *Global and Planetary Change*, 43, 157–171.

Cooper, S.R. and **Brush**, G.S., 1993. 'A 2,500-year history of anoxia and eutrophication in Chesapeake Bay.' *Estuaries*, 16, 617–626.

Corey, A.T., 1949. *Influence of shape on the fall velocity of sand grains*. Unpublished MS thesis, A&M College, Colorado.

Correggiari, A., **Cattaneo**, A. and **Trincardi**, F., 2005. 'The modern Po Delta system: Lobe switching and asymmetric prodelta growth.' *Marine Geology*, 222–223, 49–74.

Coughenour, C.L., **Archer**, A.W. and **Lacovara**, K.J., 2009. 'Tides, tidalites, and secular changes in the Earth–Moon system.' *Earth-Science Reviews*, 97, 59–79.

Cowell, P.J. and **Thom**, B.G., 1994. 'Morphodynamics of coastal evolution.' In: R.W.G. Carter and C.D. Woodroffe (eds) *Coastal Evolution*. Cambridge University Press, Cambridge, 33–86.

Cowell, P.J., **Hanslow**, D.J. and **Meleo**, J.F., 1999. 'The shoreface.' In: A.D. Short (ed) *Handbook of Beach and Shoreface Morphodynamics*. Wiley and Sons, Chichester, 39–71.

Crowell, M., **Coulton**, K., **Johnson**, C., **Westcott**, J., **Bellomo**, D., **Edelman**, S. and **Hirsch**, E., 2010. 'An estimate of the US population living in 100-year coastal flood hazard areas.' *Journal of Coastal Research*, 26, 201–211.

Dalrymple, R.W., **Zaitlin**, B.A., and **Boyd**, R., 1992. 'Estuarine facies models: Conceptual basis and stratigraphic implications.' *Journal of Sedimentary Petrology*, 62, 1130–1146.

Daly, R.A., 1915. 'The glacial-control theory of coral reefs.' *American Academy of the Arts and Science*, 51, 155–251.

Dana, J.D., 1849. *Geology*. Putnam, New York, Report US Exploration expedition 1838–1842, 10, 35–38.

Darwin, C.R., 1842. 'The structure and distribution of coral reefs. Being the first part of the geology of the voyage of the Beagle, under the command of Capt. Fitzroy, R.N. during the years 1832 to 1836.' Smith Elder and Co., London.

Davidson-Arnott, R., 2010. *Introduction to Coastal Processes and Geomorphology*. Cambridge University Press, Cambridge.

Davidson-Arnott, R.G.D., **Yang**, Y., **Ollerhead**, J., **Hesp**, P.A. and **Walker**, I.J., 2008. 'The effects of surface moisture on aeolian sediment transport threshold and mass flux on a

beach.' *Earth Surface Processes and Landforms*, 33, 55–74.
Davies, J.L., 1980. *Geographical Variation in Coastal Development* (2nd edn). Longman, New York.
Davis, R.A. and **Hayes**, M.O., 1984. 'What is a wave-dominated coast?' *Marine Geology*, 60, 313–329.
Dean, R.G., 1977. *Equilibrium Beach Profiles: US Atlantic and Gulf Coasts*. Ocean Engineering Technical Report, 12.
Dean, R.G., 1987. 'Coastal sediment processes: Toward engineering solutions.' *Coastal Sediments' 87*, ASCE, 1–24.
Dean, R.G., 1991. 'Equilibrium beach profiles: Characteristics and applications.' *Journal of Coastal Research*, 7, 53–84.
Defant, A., 1958. *Ebb and Flow*. The University of Michigan Press.
Defeo, O., **McLachlan**, A., **Schoeman**, D.S., **Schlacher**, T.A., **Dugan**, J., **Jones**, A., **Lastra**, M. and **Scapini**, F., 2009. Threats to Sandy Beach Ecosystems: a review. *Estuarine, Coastal and Shelf Science*, 81, 1–12.
Defant, A., 1961. *Physical Oceanography*. Volume 2, Pergamon Press.
De Lange, W.P. and **Moon**, V.G., 2005. 'Estimating long-term cliff recession rates from shore platform widths.' *Engineering Geology*, 80, 292–301.
Devine-Wright, P., 2005. 'Beyond NIMBYism: towards an integrated framework for understanding public perceptions of wind energy.' *Wind Energy*, 8, 125–139.
Dickson, M.E., **Walkden**, M.J.A. and **Hall**, J.W., 2007. 'Systemic impacts of climate change on an eroding coastal region over the twenty-first century.' *Climatic Change*, 84, 141–166.
Dyer, K.R., 1986. *Coastal and Estuarine Sediment Dynamics*. Wiley and Sons, Chichester.
Dyer, K.R., 1994. 'Estuarine sediment transport and deposition.' In: K. Pye (ed) *Sediment Transport and Depositional Processes*. Blackwell Scientific Publications, Oxford, 25–60.
Dyer, K.R., 1998. *Estuaries: A Physical Introduction*. Wiley and Sons, Chichester.
Edgren, G., 1993. 'Expected economic and demographic developments in coastal zones world-wide.' *World Coast '93*, National Institute for Coastal and Marine Management, Coastal Zone Management Centre, Noordwijk, The Netherlands, 367–370.
Eisma, D., 1993. *Suspended Matter in the Aquatic Environment*. Springer-Verlag, Berlin.
Elliott, T., 1986. 'Deltas.' In: H.G. Reading (ed) *Sedimentary Environments and Facies*. Blackwell Science, Oxford, 113–154.
Ellis, J.T., **Li**, B., **Farrell**, E.J. and **Sherman**, D.J., 2009. 'Protocols for characterizing aeolian mass-flux profiles.' *Aeolian Research*, 1, 19–26.
Emery, K.O. and **Kuhn**, G.G., 1982. 'Sea cliffs: Their processes, profiles and classification.' *Geological Society of America Bulletin*, 93, 644–654.
EPICA community members, 2004. 'Eight glacial cycles from an Antarctic ice core.' *Nature*, 429, 623–628.
Ergin, A., **Williams**, A.T. and **Micallef**, A., 2006. 'Coastal scenery: appreciation and evaluation.' *Journal of Coastal Research*, 22, 958–964.
Eronen, M., 1983. 'Late Weichselian and Holocene shore displacement in Finland.' In: D.E. Smith and A.G. Dawson (eds) *Shorelines and Isostasy*. Academic Press, London, 183–207.
Esselborn, S. and **Eden**, C., 2001. 'Sea surface height changes in the North Atlantic Ocean related to the North Atlantic Oscillation.' *Geophysical Research Letters*, 28, 3473–3476.
Evans, M.N., **Fairbanks**, R.G. and **Rubenstone**, J.L., 1999. 'The thermal oceanographic signal of El Niño reconstructed from a Kiritimati Island coral.' *Journal of Geophysical Research*, 104 (C6), 13409–13421.
Fairbanks, R.G., 1989. 'A 17,000-year glacio-eustatic sea level record: Influence of glacial melting rates on the Younger Dryas event and deep-ocean circulation.' *Nature*, 342,

637–642.

Fairbridge, R.W., 1983. 'Isostasy and eustasy.' In: D.E. Smith and A.G. Dawson (eds) *Shorelines and Isostasy*. Academic Press, London, 3–25.

Farris, A.S. and **List**, J.H., 2007. 'Shoreline change as a proxy for subaerial beach volume change.' *Journal of Coastal Research*, 23, 740–748.

Fenton, J.D. and **McKee**, W.D., 1990. 'On calculating the lengths of water waves.' *Coastal Engineering*, 14, 499–513.

Finkl, C.W., 2004. 'Coastal classification: systematic approaches to consider in the development of a comprehensive scheme.' *Journal of Coastal Research*, 20, 166–213.

Fischlin, A., **Midgley**, G.F., **Price**, J.T., **Leemans**, R., **Gopal**, B., **Turley**, C., **Rounsevell**, M.D.A., **Dube**, O.P., **Tarazona**, J. and **Velichko**, A.A., 2007. 'Ecosystems, their properties, goods, and services.' In: M.L. Parry, O.F. Caniani, J.P. Palutikof, P.J. van der Linden and C.E. Hanson (eds) *Climate Change 2007: Impacts, Adaptation and Vulnerability.* Contribution of Working Group II to the Fourth Assessment Report of the Intergovernmental Panel on Climate Change. Cambridge University Press, Cambridge, 211–272.

Flood, P.G., 1986. 'Sensitivity of coral cays to climatic variations, southern Great Barrier Reef.' *Coral Reefs*, 5, 13–18.

Folk, R.L. and **Ward**, W., 1957. 'Brazos River bar: A study in the significance of grain size parameters.' *Journal of Sedimentary Petrology*, 27, 3–26.

Forbes, D.L. and **Syvitski**, J.P.M., 1994. 'Paraglacial coasts.' In: R.W.G. Carter and C.D. Woodroffe (eds) *Coastal Evolution*. Cambridge University Press, Cambridge, 373–424.

Francis, J.R.D., 1973. 'Experiments on the motion of solitary grains along the bed of a water stream.' *Philosophical Transactions of the Royal Society, London Series A*, 332, 443–471.

French, J.R., **Clifford**, N.J. and **Spencer**, T., 1993. 'High frequency flow and suspended sediment measurements in a tidal wetland channel.' In: N.J. Clifford, J.R. French and J. Hardisty (eds) *Turbulence: perspectives on flow and sediment transport*. Wiley and Sons, Chichester, 249–278.

French, J.R., **Spencer**, T., **Murray**, A.L. and **Arnold**, A.S., 1995. 'Geostatistical analysis of sediment deposition in two small tidal wetlands, Norfolk, UK.' *Journal of Coastal Research*, 11, 295–570.

Friedman, G.M., 1961. 'Distinction of dune, beach, and river sands from their textural characteristics.' *Journal of Sedimentary Petrology*, 31, 514–529.

Friedrichs, C.T. and **Aubrey**, D.G., 1988. 'Non-linear tidal distortion in shallow well-mixed estuaries: A synthesis.' *Estuarine, Coastal and Shelf Science*, 27, 521–545.

Fritz, H.M., **Blount**, C., **Sokoloski**, R., **Singleton**, J., **Fuggle**, A., **McAdoo**, B.G., **Moore**, A., **Grass**, C. and **Tate**, B., 2007. 'Hurricane Katrina storm surge distribution and field observations on the Mississippi Barrier Islands.' *Estuarine, Coastal and Shelf Science*, 74, 12–20.

Frouin, M., **Sebag**, D., **Laignel**, B., **Ogier**, S., **Verrecchia**, E.P. and **Durand**, A., 2006. 'Tidal rhythmites of the Marais Vernier Seine estuary, France and their implications for relative sea-level.' *Marine Geology*, 235, 165–175.

Fu, L.L., **Vazquez**, J. and **Parke**, M.E., 1987. 'Seasonal variability of the Gulf Stream from satellite altimetry.' *Journal of Geophysical Research*, 92, 749–754.

Galloway, W.E., 1975. 'Process framework for describing the morphologic and stratigraphic evolution of deltaic depositional systems.' In: M.L. Broussard (ed) *Deltas, Models for Exploration*. Houston Geological Society, Houston, 87–98.

Galvin, C.J., 1968. 'Breaker type classification on three laboratory beaches.' *Journal of Geophysical Research*, 73, 3651–3659.

Garel, E., **Pinto**, L., **Santos**, A. and **Ferreira**, Ó., 2009. 'Tidal and river discharge forcing upon water and sediment circulation at a rock-bound estuary (Guadiana estuary, Portugal).' *Estuarine, Coastal and Shelf Science*, 84, 269–281.

Gehrels, R., 2010. 'Sea-level changes since the Last Glacial Maximum: an appraisal of the IPCC Fourth Assessment Report.' *Journal of Quaternary Science*, 25, 26–38.

George, D.A. and **Hill**, P.S., 2008. 'Wave climate, sediment supply and the depth of the sand–mud transition: A global survey.' *Marine Geology*, 254, 121–128.

Glavovic, B.C., 2008. 'Sustainable coastal communities in the age of coastal storms: Reconceptualising coastal planning as 'new' naval architecture.' *Journal of Coastal Conservation*, 12, 125–134.

Gómez-Pujola, L., **Orfila**, A., **Canellasa**, B., **Alvarez-Ellacuria**, A., **Méndez**, F.J., **Medina**, R. and **Tintoré**, J., 2007. 'Morphodynamic classification of sandy beaches in low energetic marine environment.' *Marine Geology*, 242, 235–246.

Goodbred Jr., S.L. and **Kuehl**, S.A., 2000. 'The significance of large sediment supply, active tectonism, and eustasy on margin sequence development: Late Quaternary stratigraphy and evolution of the Ganges–Brahmaputra delta.' *Sedimentary Geology*, 133, 227–248.

Gornitz, V., 1993. 'Mean sea level changes in the recent past.' In: R.A. Warrick, E.M. Barrow and T.M.L. Wigley (eds) *Climate and Sea Level Change: Observations, Projections and Implications*. Cambridge University Press, Cambridge, 25–44.

Gornitz, V., 1995. 'Sea level rise: A review of recent past and near-future trends.' *Earth Surface Processes and Landforms*, 20, 7–20.

Goudie, A., 1990. *Geomorphological Techniques* (2nd edn). Unwin Hyman, London.

Goudie, A.S., 2006. 'The Schmidt Hammer in geomorphological research.' *Progress in Physical Geography*, 30, 703–718.

Greenwood, R.O. and **Orford**, J.D., 2008. 'Temporal patterns and processes of retreat of drumlin coastal cliffs – Strangford Lough, Northern Ireland.' *Geomorphology*, 94, 153–169.

Griggs, G.B. and **Trenhaile**, A.S., 1994. 'Coastal cliffs and platforms.' In: R.W.G. Carter and C.D. Woodroffe (eds) *Coastal Evolution*. Cambridge University Press, Cambridge, 425–450.

Guilcher, A., 1988. *Coral Reef Geomorphology*. Wiley and Sons, Chichester.

Guza, R.T. and **Inman**, D., 1975. 'Edge waves and beach cusps.' *Journal of Geophysical Research*, 80, 2997–3012.

Guza, R.T. and **Thornton**, E.B., 1985. 'Observations of surf beat.' *Journal of Geophysical Research*, 90, 3161–3172.

Hallermeier, R.J., 1981. 'A profile zonation for seasonal sand beaches from wave climate.' *Coastal Engineering*, 4, 253–277.

Hallock, P., 2005. 'Global change and modern coral reefs: New opportunities to understand shallow-water carbonate depositional systems.' *Sedimentary Geology*, 175, 19–33.

Hansom, J.D., 1988. *Coasts*. Cambridge University Press, Cambridge.

Hansom, J.D., **Barltrop**, N.D.P. and **Hall**, A.M., 2008. 'Modelling the processes of cliff-top erosion and deposition under extreme storm waves.' *Marine Geology*, 253, 36–50.

Hansom, J.D. and **Hall**, A.M., 2009. 'Magnitude and frequency of extra-tropical North Atlantic cyclones: A chronology from cliff-top storm deposits.' *Quaternary International*, 195, 42–52.

Hanson, J.L., **Tracy**, B.A., **Tolman**, H.L. and **Scott**, R.D., 2009. 'Pacific hindcast performance of three numerical wave models.' *Journal of Atmospheric and Oceanic Technology*, 26, 1614–1633.

Hanson, S., **Nicholls**, R., **Ranger**, N., **Hallegatte**, S., **Corfee-Morlot**, J., **Herweijer**, C. and **Chateau**, J., 2011. 'A global ranking of port cities with high exposure to climate extremes.' *Climatic Change*, 104, 89–111.

Harris, P.T., **Hughes**, M.G., **Baker**, E.K., **Dalrymple**, R.W. and **Keene**, J.B., 2004. 'Sediment transport in distributary channels and its export to the pro-deltaic environment in a tidally-dominated delta: Fly River, Papua New Guinea.' *Continental Shelf Research*, 24, 2431–2454.

Haslett, S.K., 2000. *Coastal Systems*. Routledge, London.

Hesp, P.A., **Davidson-Arnott**, R., **Walker**, I.J. and **Ollerhead**, J., 2005. 'Flow dynamics over a foredune at Prince Edward Island, Canada.' *Geomorphology*, 65, 71–84.

Hasselmann, K., **Ross**, D.B., **Muller**, P. and **Sell**, W., 1976. 'A parametric wave prediction model.' *Journal of Physical Oceanography*, 6, 200–228.

Hastenrath, S., 2006. 'Circulation and teleconnection mechanisms of Northeast Brazil droughts.' *Progress in Oceanography*, 70 (2–4), 407–415.

Hayes, M.O. and **Boothroyd**, J.C., 1969. 'Storms as modifying agents in the coastal environment.' In: M.O. Hayes (ed) *Coastal Environments: NE Massachusetts*. Department of Geology, University of Massachusetts, Amherst, 290–315.

Hepner, T.L. and **Davis**, R.A., Jr., 2004. 'Effect of El Niño (1997–98) on beaches of the peninsular Gulf Coast of Florida.' *Journal of Coastal Research*, 20, 776–791.

Hesp, P., 1999. 'The beach, backshore and beyond.' In: A.D. Short (ed) *Handbook of Beach and Shoreface Dynamics*. Wiley and Sons, Chichester, 145–169.

Hesp, P.A. and **Short**, A.D., 1999. 'Barrier morphodynamics.' In: A.D. Short (ed) *Handbook of Beach and Shoreface Morphodynamics*. Wiley and Sons, Chichester, 307–333.

Hijma, M.P. and **Cohen**, K.M., 2010. 'Timing and magnitude of the sea-level jump preluding the 8200 yr event.' *Geology*, 38, 275–278.

Hill, P.R., **Barnes**, P.W., **Héquette**, A. and **Ruz**, M.-H., 1994. 'Arctic coastal plain shorelines.' In: R.W.G. Carter and C.D. Woodroffe (eds) *Coastal Evolution*. Cambridge University Press, Cambridge, 341–372.

Holman, R.A. and **Bowen**, A.J., 1982. 'Bars, bumps, and holes: Models for the generation of complex beach topography.' *Journal of Geophysical Research*, 87, 457–468.

Holman, R.A., 1983. 'Edge wave sand the configuration of the shoreline.' In: P.D. Komar (ed) *CRC Handbook of Coastal Processes and Erosion*. CRC Press, Boca Raton, 21–33.

Hopley, D., 1994. 'Continental shelf reef systems.' In: R.W.G. Carter and C.D. Woodroffe (eds) *Coastal Evolution*. Cambridge University Press, Cambridge, 303–340.

Horikawa, K., **Hotta**, S., **Kubota**, S. and **Katori**, S., 1983. 'Field measurements of blown sand transport rate by trench trap.' *Coastal Engineering Japan*, 26, 100–120.

Houser, C. and **Hamilton**, S., 2009. 'Sensitivity of post-hurricane beach and dune recovery to event frequency.' *Earth Surface Processes and Landforms*, 34, 613–628.

Houston, J.A., **Edmundson**, S.E. and **Rooney**, P.J., 2001. *Coastal Dune Management: Shared Experience of European Conservation Practice*. Liverpool University Press, Liverpool.

Hsu, S.A., 1971. 'Wind stress criteria in eolian sand transport.' *Journal of Geophysical Research*, 76, 8684–8686.

Hsu, S.A., 1977. 'Boundary layer meteorological research in the coastal zone.' In: H.J. Walker (ed) *Geoscience and Man*. School of Geoscience, Louisiana State University, Baton Rouge, LA, 18, 99–111.

Hsu, K.J., 2004. *Physics of Sedimentology: textbook and reference*. Springer, Berlin.

Hsu, J.R.C., **Silvester**, R. and **Xia**, Y.M., 1987. 'New characteristics of equilibrium shaped bays.' *Proceedings 8th International Conference on Coastal Engineering*, ASCE, 140–144.

Hughes, M.G. and **Cowell**, P.J., 1987. 'Adjustment of reflective beaches to waves.' *Journal of Coastal Research*, 3, 153–167.

Hughes, M.G. and **Turner**, I., 1999. 'The beachface.' In: A.D. Short (ed) *Handbook of Beach and Shoreface Morphodynamics*. Wiley and Sons, Chichester, 119–144.

Hughes, M.G., **Harris**, P.T. and **Hubble**, T.C.T., 1998. 'Dynamics of the turbidity maximum zone in a micro-tidal estuary: Hawkesbury River, Australia.' *Sedimentology*, 45, 397–410.

Hughes, M.G., **Masselink**, G. and **Brander**, R.W., 1997. 'Flow velocity and sediment transport in the swash zone of a steep beach.' *Marine Geology*, 138, 91–103.

Huntsman, L., 2001. *Sand in Our Souls: The beach in Australian history and culture*. Melbourne University Press, Melbourne, Australia.

Hutchinson, J.N., 1973. 'The response of London Clay cliffs to differing rates of toe erosion.' *Geologia Applicata e Idrogeologia*, 8, 221–239.

Inman, D.L. and **Brush**, B.M., 1973. 'The coastal challenge.' *Science*, 181, 20–32.

Inman, D.L. and **Nordstrom**, K.F., 1971. 'On the tectonic and morphologic classification of coasts.' *Journal of Geology*, 79, 1–21.

Isdale, P.J., **Stewart**, B.J., **Tickle**, K.S. and **Lough**, J.M., 1998. 'Palaeohydrological variation in a tropical river catchment: a reconstruction using fluorescent bands in corals of the Great Barrier Reef, Australia.' *The Holocene*, 8, 1–8.

Isla, F.I., 2008. 'ENSO-dominated estuaries of Buenos Aires: The interannual transfer of water from Western to Eastern South America.' *Global and Planetary Change*, 64, 69–75.

Jennings, J.N., 1967. 'Cliff-top dunes.' *Australian Geographical Studies*, 5, 40–49.

Johnson, J.W., 1919. *Shore Processes and Shoreline Development*. Wiley, New York. [facsimile edition: Hafner, New York (1965).]

Jones, A.L. and **Phillips**, M., 2010. *Disappearing Destinations – Climate change and future challenges for coastal tourism*. CABI Publications, London.

Jones, B.M., **Arp**, C.D., **Jorgenson**, M.T., **Hinkel**, K.M., **Schmutz**, J.A. and **Flint**, P.L., 2009. 'Increase in the rate and uniformity of coastline erosion in Arctic Alaska.' *Geophysical Research Letters*, 36, L03503, doi:10.1029/2008GL036205.

Jonkman, S.N., **Maaskant**, B., **Boyd**, E. and **Levitan**, M.L., 2009. 'Loss of life caused by the flooding of New Orleans after Hurricane Katrina: analysis of the relationship between flood characteristics and mortality.' *Risk Analysis*, 29, 676–698.

Kadib, A.A., 1965. 'A function for sand movement by wind.' *University of California Hydraulics Engineering Laboratory Report HEL*, 2–8, Berkeley.

Kaiser, M.F.M. and **Frihy**, O.E., 2009. 'Validity of the equilibrium beach profiles: Nile Delta Coastal Zone, Egypt.' *Geomorphology*, 107, 25–31.

Kawamura, R., 1951. 'Study of sand movement by wind.' *University of California Hydraulics Engineering Laboratory Report HEL*, 2–8, Berkeley.

Kelletat, D.H., 1995. 'Atlas of Coastal Geomorphology and Zonality.' *Journal of Coastal Research*, Special Issue 13, 286pp.

Kench, P.S. and **Cowell**, P., 2000. 'The morphological response of atoll islands to sea-level rise. Part 2: Application of the modified Shoreface Translation Model.' *Journal of Coastal Research*, Special Issue 34, 645–656.

Kench, P.S., **McLean**, R.F., **Brander**, R.W., **Nichol**, S.L., **Smithers**, S.G., **Ford**, M.R., **Parnell**, K.E. and **Aslam**, M., 2006. 'Geological effects of tsunami on mid-ocean atoll islands: The Maldives before and after the Sumatran tsunami.' *Geology*, 34, 177–180.

Kench, P.S., **McLean**, R.F. and **Nichol**, S.L., 2005. 'New model of reef-island evolution: Maldives, Indian Ocean.' *Geology*, 33, 145–148.

Khalifa, M.A., **El Ganainy**, M.A. and **Nasr**, R.I., 2009. 'Statistical and uncertainty analysis of longshore sediment transport evaluations for the Egyptian northern coast: a case study application.' *Journal of Coastal Research*, 25, 1002–1014.

King, C.A.M., 1972. *Beaches and Coasts*. Edward Arnold, London.

King, D.M., **Cooper**, N.J., **Morfett**, J.C. and **Pope**, D.J., 2000. 'Application of offshore breakwaters to the UK: A case study at Elmer Beach.' *Journal of Coastal Research*, 16, 172–187.

Kinsey, D.W., 1985. 'Metabolism, calcification and carbon production: 1. Systems level studies.' *Proceedings 5th International Coral Reef Symposium*, 505–526.

Kinsman, B., 1984. *Wind Waves* (2nd edn). Dover Publications, New York.

Kirby, R., 1988.'High concentration suspension (fluid mud) layers in estuaries.' In: J. Dronkers and W. van Leussen (eds) *Physical Processes in Estuaries*. Springer-Verlag, 463–487.

Klein, R.J.T. and **Nicholls**, R.J., 1999.'Assessment of coastal vulnerability to climate change.' *Ambio*, 28, 182–187.

Kleinhans, M.G. and **Grasmeijer**, B.T., 2006.'Bed load transport on the shoreface by currents and waves.' *Coastal Engineering*, 53, 983–996.

Knight, J., 2005. 'Controls on the formation of coastal ventifacts.' *Geomorphology*, 64, 243–254.

Knight, J. and **Burningham**, H., 2001. 'Formation of bedrock-cut ventifacts and late Holocene coastal zone evolution, County Donegal, Ireland.' *Journal of Geology*, 109, 647–660.

Knight, J. and **Burningham**, H., 2011.'Boulder dynamics on an Atlantic-facing rock coastline, northwest Ireland.' *Marine Geology*, 283, 56–65.

Knight, J., **Burningham**, H. and **Barrett-Mold**, C., 2009. 'The geomorphology and controls on development of a boulder-strewn rock platform, NW Ireland.' *Journal of Coastal Research*, Special Issue, 56, 1646–1650.

Knight, J., **Orford**, J.D., **Wilson**, P. and **Braley**, S.M., 2002. 'Assessment of temporal changes in coastal sand dune environments using the log-hyperbolic grain size method.' *Sedimentology*, 49, 1229–1252.

Komar, P.D., 1976. *Beach Processes and Sedimentation*. Prentice-Hall, New Jersey.

Komar, P.D., 1998. *Beach Processes and Sedimentation* (2nd edn). Prentice Hall, New Jersey.

Komar, P.D. and **Inman**, D.I., 1970.'Longshore sand transport on beaches.' *Journal of Geophysical Research*, 75, 5514–5527.

Kostaschuk, R., **Best**, J., **Villard**, P., **Peakall**, J. and **Franklin**, M., 2005.'Measuring flow velocity and sediment transport with an acoustic Doppler current profiler.' *Geomorphology*, 68, 25–37.

Kubicki, A., 2008.'Large and very large subaqueous dunes on the continental shelf off southern Vietnam, South China Sea.' *Geo-Marine Letters*, 28, 229–238.

Kuenen, P.H., 1948.'The formation of beach cusps.' *Journal of Geology*, 56, 34–40.

Kuhnle, R.A. and **Wren**, D.G., 2009.'Size of suspended sediment over dunes.' *Journal of Geophysical Research*, 114, F02020, doi:10.1029/2008JF001200.

Kvale, E.P., 2006.'The origin of neap-spring tidal cycles.' *Marine Geology*, 235, 5–18.

Kvale, E.P., **Cutright**, J., **Bilodeau**, D., **Archer**, A., **Johnson**, H.R. and **Pickett**, B., 1995. 'Analysis of modern tides and implications for ancient tidalites.' *Continental Shelf Research*, 15, 1921–1943.

Ladd, H.S., **Tracey**, J.I. and **Gross**, M.G., 1967.'Drilling on Midway Atoll, Hawaii.' *Science*, 156, 1088–1094.

Lamb, H.H., 1982. *Climate, History and the Modern World*. Methuen, London.

Lambeck, K., 1993.'Glacial rebound and sea-level change: An example of a relationship between mantle and surface processes.' *Tectonophysics*, 223, 15–37.

Lambeck, K. and **Chappell**, J., 2001.'Sea level change through the Last Glacial Cycle.' *Science*, 292, 679–686.

Lanagan, D., 2002.'Surfing in the third millennium: commodifying the visual argot.' *The Australian Journal of Anthropology*, 13, 283–291.

Lantuit, H. and **Pollard**, W.H., 2008.'Fifty years of coastal erosion and retrogressive thaw slump activity on Herschel Island, southern Beaufort Sea, Yukon Territory, Canada.' *Geomorphology*, 95, 84–102.

Lane, A., 2004.'Bathymetric evolution of the Mersey Estuary, UK, 1906–1997: causes and effects.' *Estuarine, Coastal and Shelf Science*, 59, 249–263.

Larson, M. and **Sunamura**, T., 1993. 'Laboratory experiment on flow characteristics at a beach step.' *Journal of Sedimentary Petrology*, 63, 495–500.

Lawson, N.V. and **Abernathy**, C.L., 1975. 'Long-term wave statistics off Botany Bay.' *Proceedings 2nd Australian Conference on Coastal and Ocean Engineering*, 167–176.

Leeder, M., 1999. *Sedimentology and Sedimentary Basins: From Turbulence to Tectonics*. Blackwell Scientific Publications, Oxford.

Leopold, L.B., 1969. 'Landscape esthetics.' *Natural History*, 78, 36–45.

Lettau, K. and **Lettau**, H., 1977. 'Experimental and micrometeorological field studies of dune migration.' In: K. Lettau and H. Lettau (eds) *Exploring the World's Driest Climate*. University of Wisconsin-Madison, IES Report 101, 110–147.

Levin N., 2011. 'Climate-driven changes in tropical cyclone intensity shape dune activity on Earth's largest sand island.' *Geomorphology*, 125, 239–2522.

Lewis, D.W. and **McConchie**, D., 1994. *Analytical Sedimentology*. Chapman & Hall, New York.

Li, M. and **Zhong**, L., 2009. 'Flood–ebb and spring–neap variations of mixing, stratification and circulation in Chesapeake Bay.' *Continental Shelf Research*, 29, 4–14.

Longuet-Higgins, M.S. and **Stewart**, R.W., 1962. 'Radiation stresses and mass transport in gravity waves with applications to surf beat.' *Journal of Fluid Mechanics*, 13, 481–504.

Longuet-Higgins, M.S. and **Stewart**, R.W., 1964. 'Radiation stresses in water waves: A physical discussion with applications.' *Deep-Sea Research*, 11, 529–562.

Lotfy, M.F. and **Frihy**, O.E., 1993. 'Sediment balance in the nearshore zone of the Nile delta coast, Egypt.' *Journal of Coastal Research*, 9, 654–662.

Lüthi, D., **Le Floch**, M., **Bereiter**, B., **Blunier**, T., **Barnola**, J.-M., **Siegenthaler**, U., **Raynaud**, D., **Jouzel**, J., **Fischer**, H., **Kawamura**, K. and **Stocker**, T.F., 2008. 'High-resolution carbon dioxide concentration record 650,000–800,000 years before present.' *Nature*, 453, 379–382.

Lynch, K., **Jackson**, D.W.T. and **Cooper**, J.A.G., 2010. 'Coastal foredune topography as a control on secondary airflow regimes under offshore winds.' *Earth Surface Processes and Landforms*, 35, 344–353.

Macintyre, I.G., 1988. 'Modern coral reefs of the western Atlantic: New geological perspective.' *American Association of Petroleum Geologists' Bulletin*, 72, 1360–1369.

MacMahan, J.H., **Thornton**, E.B. and **Reniers**, A.J.H.M., 2006. 'Rip current review.' *Coastal Engineering*, 53, 191–208.

Magenheimer, J.F., **Moore**, T.R., **Chmura**, G.L. and **Daoust**, R.J., 1996. 'Methane and carbon dioxide flux from a macrotidal salt marsh, Bay of Fundy, New Brunswick.' *Estuaries*, 19, 139–145.

Mandelbrot, B., 1967. 'How long is the coast of Britain? Statistical self-similarity and fractional dimension.' *Science*, 156, 636–638.

Masetti, R., **Fagherazzi**, S. and **Montanari**, A., 2008. 'Application of a barrier island translation model to the millennial-scale evolution of Sand Key, Florida.' *Continental Shelf Research*, 28, 1116–1126.

Masselink, G. and **Anthony**, E., 2001. 'Location and size of intertidal bars on macrotidal ridge and runnel beaches.' *Earth Surface Processes and Landforms*, 26, 759–774.

Masselink, G. and **Pattiaratchi**, C.B., 1998. 'The effects of sea breeze on beach morphology, surf zone hydrodynamics and sediment resuspension.' *Marine Geology*, 146, 115–135.

Masselink, G. and **Short**, A.D., 1993. 'The effect of tide range on beach morphodynamics and morphology: A conceptual beach model.' *Journal of Coastal Research*, 9, 785–800.

Masselink, G. and **Turner**, I., 1999. 'The effect of tides on beach morphodynamics.' In: A.D. Short (ed) *Handbook of Beach and Shoreface Morphodynamics*. Wiley and Sons,

Chichester, 204–229.

Mateo, Z.R.P. and **Siringan**, F.P., 2007. 'Tectonic control of high-frequency Holocene delta switching and fluvial migration in Lingayen Gulf bayhead, northwestern Philippines.' *Journal of Coastal Research*, 23, 182–194.

Matsukura, Y. and **Matsuoka**, M., 1991. 'Rates of tafoni weathering on uplifted shore platforms in Nojima-Zaki, Boso Peninsula, Japan.' *Earth Surface Processes and Landforms*, 16, 51–56.

Maun, M.A., 2009. *The Biology of Coastal Sand Dunes*. Oxford University Press, Oxford.

May, J.P. and **Tanner**, W.F., 1973. 'The littoral power gradient and shoreline changes.' In: D.R. Coates (ed) *Coastal Geomorphology*. New York State University Press, New York, 43–61.

McCook, L.J., **Jompa**, J. and **Diaz-Pulido**, G., 2001. 'Competition between coral and algae on coral reefs: a review of evidence and mechanisms.' *Coral Reefs*, 19, 400–417.

McFadden, L., **Nicholls**, R.J. and **Penning-Rowsell**, E., (eds) 2007. *Managing Coastal Vulnerability*. Elsevier, Amsterdam.

McKoy, H., **Kennedy**, D.M. and **Kench**, P.S., 2010. 'Sand cay evolution on reef platforms, Mamanuca Islands, Fiji.' *Marine Geology*, 269, 61–73.

McLean, R.F. and **Woodroffe**, C.D., 1994. 'Coral atolls.' In: R.W.G. Carter and C.D. Woodroffe (eds) *Coastal Evolution*. Cambridge University Press, Cambridge, 267–302.

McLeod, E. and **Salm**, R.V., 2006. *Managing Mangroves for Resilience to Climate Change*. IUCN, Gland, Switzerland. 64pp.

Meehl, G.A., **Stocker**, T.F., **Collins**, W.D., **Friedlingstein**, P., **Gaye**, A.T., **Gregory**, J.M., **Kitoh**, A., **Knutti**, R., **Murphy**, J.M., **Noda**, A., **Raper**, S.C.B., **Watterson**, I.G., **Weaver**, A. and **Zhao**, Z.-C., 2007. 'Global Climate Projections.' In: S. Solomon, D. Qin, M. Manning, Z. Chen, M. Marquis, K.B. Averyt, M. Tignor and H.L. Miller (eds) *Climate Change 2007: The Physical Science Basis*. Contribution of Working Group I to the Fourth Assessment Report of the Intergovernmental Panel on Climate Change. Cambridge University Press, Cambridge, 747–845.

Meeker, L.D. and **Mayewski**, P.A., 2002. 'A 1400-year high-resolution record of atmospheric circulation over the North Atlantic and Asia.' *The Holocene*, 12, 257–266.

Mehta, A.J., 1989. 'On estuarine cohesive sediment suspension behaviour.' *Journal of Geophysical Research*, 94, 14303–14314.

Meyer-Peter, E. and **Muller**, R., 1948. 'Formulas for bedload transport.' *Proceedings 2nd Congress of International Association of Hydraulic Research*, Stockholm, 39–64.

Milliman, J.D. and **Meade**, R.H., 1983. 'World-wide delivery of river sediments to the oceans.' *Journal of Geology*, 91, 1–21.

Milne, G.A. and **Mitrovica**, J.X., 2008. 'Searching for eustasy in deglacial sea-level histories.' *Quaternary Science Reviews*, 27, 2292–2302.

Monmonier, M., 2008. *Coast Lines: How mapmakers frame the world and chart environmental change*. University of Chicago Press, Chicago.

Moore, L.J., **List**, J.H., **Williams**, S.J. and **Stolper**, D., 2010. 'Complexities in barrier island response to sea level rise: Insights from numerical model experiments.' North Carolina Outer Banks. *Journal of Geophysical Research*, 115, F03004, doi:10.1029/2009JF001299.

Moore, R.D., **Wolf**, J., **Souza**, A.J. and **Flint**, S.S., 2009. 'Morphological evolution of the Dee Estuary, Eastern Irish Sea, UK: A tidal asymmetry approach.' *Geomorphology*, 103, 588–596.

Mörner, N-A., 1987. 'Models of global sea level changes.' In: M.J. Tooley and I. Shennan (eds) *Sea Level Changes*. Blackwell, Oxford, 332–355.

Morton, R.A., **Clifton**, H.E., **Buster**, N.A., **Peterson**, R.L. and **Gelfenbaum**, G., 2007. 'Forcing of large-scale cycles of coastal change at the entrance to Willapa Bay,

Washington.' *Marine Geology*, 246, 24–41.

Munk, W.H. and **Traylor**, M.A., 1947. 'Refraction of ocean waves: A process linking underwater topography to beach erosion.' *Journal of Geology*, 55, 1–26.

Nandasena, N.A.K., **Paris**, R. and **Tanaka**, N., 2011. 'Reassessment of hydrodynamic equations: Minimum flow velocity to initiate boulder transport by high energy events (storms, tsunamis).' *Marine Geology*, 281, 70–84.

Naylor, L.A. and **Stephenson**, W.J., 2010. 'On the role of discontinuities in mediating shore platform erosion.' *Geomorphology*, 114, 89–100.

Naylor, L.A., **Stephenson**, W.J. and **Trenhaile**, A.S., 2010. 'Rock coast geomorphology: Recent advances and future research directions.' *Geomorphology*, 114, 3–11.

Nelson, R.C., 1994. 'Depth limited design wave heights in very flat regions.' *Coastal Engineering*, 23, 43–59.

Nemec, W., 1990. 'Aspects of sediment movement on steep delta slopes.' In: A. Colella and D.B. Prior (eds) *Coarse-grained Deltas*. IAS Special Publication no 10, Blackwell Science, Oxford, 29–73.

Neuman, A.C. and **Macintyre**, I.G., 1985. 'Reef response to sea level rise: Keep-up, catch-up or give-up.' *Proceedings 5th International Coral Reef Symposium*, 105–110.

Nicholls, R.J., **Wong**, P.P., **Burkett**, V.R., **Codignotto**, J.O., **Hay**, J.E., **McLean**, R.F., **Ragoonaden**, S. and **Woodroffe**, C.D., 2007. 'Coastal systems and low-lying areas.' In: M.L. Parry, O.F. Canziani, J.P. Palutikof, P.J. van der Linden and C.E. Hanson (eds) *Climate Change 2007: Impacts, Adaptation and Vulnerability*. Contribution of Working Group II to the Fourth Assessment Report of the Intergovernmental Panel on Climate Change. Cambridge University Press, Cambridge, UK, 315–356.

Nickling, W.G., 1978. 'Eolian sediment transport during dust storms: Slims River valley, Yukon Territory.' *Canadian Journal of Earth Sciences*, 15, 1069–1084.

Niedoroda, A.W. and **Swift**, D.J.P., 1991. 'Shoreface processes.' In: J.B. Herbich (ed) *Handbook of Coastal and Ocean Engineering, Volume 2*. Gulf Publishing, Houston, 736–770.

Nielsen, P., 1992. *Coastal Bottom Boundary Layers and Sediment Transport*. World Scientific, Singapore.

Nordstrom, K.F., 1980. 'Cyclic and seasonal beach response: A comparison of oceanside and bayside beaches.' *Physical Geography*, 1, 177–196.

Nordstrom, K.F., 1990. 'The concept of intrinsic value and depositional coastal landforms.' *Geographical Review*, 80, 68–81.

Nordstrom, K.F., 1994. 'Developed coasts.' In: R.W.G. Carter and C.D. Woodroffe (eds) *Coastal Evolution*. Cambridge University Press, Cambridge, 477–509.

Nordstrom, K.F., **Psuty**, N. and **Carter**, R.W.G. (eds), 1990. *Coastal Dunes: Form and Process*. Wiley, London.

Nott, J., 2003. 'Waves, coastal boulder deposits and the importance of the pre-transport setting.' *Earth and Planetary Science Letters*, 210, 269–276.

Open University, 1994. *Waves, Tides and Shallow-Water Processes*. Pergamon Press, Oxford.

Open University, 1998. *Ocean Circulation*. Butterworth-Heinemann, Oxford.

Open University, 2000. *Waves, Tides and Shallow-Water Processes*. Butterworth-Heineman, Oxford.

Orford, J.D., **Carter**, R.W.G. and **Jennings**, S.C., 1991. 'Coarse clastic barrier environments: Evolution and implications for Quaternary sea level interpretation.' *Quaternary International*, 9, 87–104.

Orford, J.D., **Carter**, R.W.G. and **Jennings**, S.C., 1996. 'Control domains and morphological phases in gravel-dominated coastal barriers of Nova Scotia.' *Journal of Coastal Research*, 12, 589–604.

Orford, J.D., **Wilson**, P., **Wintle**, A.G., **Knight**, J. and **Braley**, S., 2000. 'Holocene coastal dune initiation in Northumberland and Norfolk, eastern UK: climate and sea-level changes as possible forcing agents for dune initiation.' In: I. Shennan and J. Andrews (eds) *Holocene land-ocean interaction and environmental change around the North Sea*. Geological Society Special Publication, 166, 197–217.

Patsch, K. and **Griggs**, G., 2008. 'A sand budget for the Santa Barbara Littoral Cell, California.' *Marine Geology*, 252, 50–61.

Peterson, C.D., **Stock**, E., **Hart**, R., **Percy**, D., **Hostetler**, S.W. and **Knott**, J.R., 2010. 'Holocene coastal dune fields used as indicators of net littoral transport: west coast, USA.' *Geomorphology*, 116, 115–134.

Pethick, J., 1984. *An Introduction to Coastal Geomorphology*. Edward Arnold, London.

Pettijohn, F.J., **Potter**, P.E. and **Siever**, R., 1987. *Sand and Sandstone*. Springer-Verlag, New York.

Pilkey, O.H., **Cooper**, J.A.G. and **Lewis**, D.A., 2009. 'Global distribution and geomorphology of fetch-limited barrier islands.' *Journal of Coastal Research*, 25, 819–837.

Pinet, P.R., 2000. *Invitation to Oceanography* (2nd edn). Jones and Bartlett Publishers, Sudbury.

Pirazzoli, P.A., 1989. 'Recent sea-level changes in the North Atlantic.' In: D.B. Scott, P.A. Pirazzoli and C.A. Honig (eds) *Late Quaternary Sea-level correlation and Applications*. Kluwer, Dordrecht, NATO ASI Series C, vol. 256, 153–167.

Pirazzoli, P.A., 1996. *Sea-Level Changes: The Last 20,000 Years*. Wiley and Sons, Chichester.

Pond, S. and **Pickard**, G.L., 1983. *Introductory Dynamical Oceanography* (2nd edn). Pergamon, Oxford.

Pond, S. and **Pickard**, G.L., 1995. *Introductory Dynamical Oceanography* (2nd revised edn). Butterworth-Heinemann, Oxford.

Pope, J. and **Dean**, J.L., 1986. 'Development of design criteria for segmented breakwaters.' *Proceedings 20th International Conference on Coastal Engineering*, ASCE, 2144–2158.

Porter-Smith, R., **Harris**, P.T., **Andersen**, O.B., **Coleman**, R., **Greenslade**, D. and **Jenkins**, C.J., 2004. 'Classification of the Australian continental shelf based on predicted sediment threshold exceedance from tidal currents and swell waves.' *Marine Geology*, 211, 1–20.

Posamentier, H.W. and **Vail**, P.R., 1988. 'Eustatic controls on clastic deposition II – Sequence and systems tract models.' In: C.K. Wilgus, B.S. Hastings, C.G.StC. Kendall, H.W. Posamentier, C.A. Ross and J.C. Van Wagoner (eds) *Sea-Level Changes: An Integrated Approach*. SEPM Special Publication 42, 125–154.

Postma, G., 1990. 'Depositional architecture and facies of river and fan deltas: A synthesis.' In: A. Colella and D.B. Prior (eds) *Coarse-grained Deltas*. IAS Special Publication no 10, Blackwell Science, Oxford, 13–28.

Potter, I.C., **Chuwen**, B.M., **Hoeksema**, S.D. and **Elliott**, M., 2010. 'The concept of an estuary: A definition that incorporates systems which can become closed to the ocean and hypersaline.' *Estuarine, Coastal and Shelf Science*, 87, 497–500.

Powers, M.C., 1953. 'A new roundness scale for sedimentary particles.' *Journal of Sedimentary Petrology*, 23, 117–119.

Prandle, D., 2009. *Estuaries: Dynamics, Mixing, Sedimentation and Morphology*. Cambridge University Press, Cambridge.

Preston-Whyte, R., 2001. 'Constructed leisure spaces: the seaside at Durban.' *Annals of Tourism Research*. 28, 581–596.

Psuty, N.P., 1992. 'Spatial variation in coastal foredune development.' In: R.W.G. Carter, T.G.F. Curtis and M.J. Sheehy-Skeffington (eds) *Coastal Dunes*. Proceedings 3rd European Dune Congress, Balkema, Rotterdam, 3–13.

Pugh, D.T., 1987. *Tides, Surges and Mean Sea-Level*. Wiley and Sons, Chichester.

Purdy, E.G., 1974.'Reef configurations: Cause and effect.' In: L.F. Laporte (ed) *Reefs in Time and Space*. Society of Economic Paleontologists and Mineralogists, Special Publication 18, 9–76.

Pye, K., 1994.'Properties of sediment particles.' In: K. Pye (ed) *Sediment Transport and Depositional Processes*. Blackwell Scientific Publications, Oxford, 1–24.

Pye, K. and **Blott**, S.J., 2008.'Decadal-scale variation in dune erosion and accretion rates: An investigation of the significance of changing storm tide frequency and magnitude on the Sefton coast, UK.' *Geomorphology*, 102, 652–666.

Ravera, O., 2000.'The Lagoon of Venice: the result of both natural factors and human influence.' *Journal of Limnology*, 59, 19–30.

Reading, H.G. and **Collinson**, J.D., 1996.'Clastic coasts.' In: H.G. Reading (ed) *Sedimentary Environments: Processes, Facies and Stratigraphy*. Blackwell Science, Oxford, 154–231.

Reineck, H.-E. and **Singh**, I.B., 1980. *Depositional Sedimentary Environments: With Reference to Terrigenous Clastics*. Springer-Verlag, Berlin.

Reinson, G.E., 1984.'Barrier island and associated strand-plain systems.' In: R.G. Walker (ed) *Facies Models* (2nd edn). Geoscience Canada Reprint Series 1, Geological Association of Canada, 119–140.

Reis, A.H. and **Gama**, C., 2010.'Sand size versus beachface slope – an explanation based on the Constructal Law.' *Geomorphology*, 114, 276–283.

Rignot, E., **Velicogna**, I., **van den Broeke**, M.R., **Monaghan**, A. and **Lenaerts**, J., 2011. 'Acceleration of the contribution of the Greenland and Antarctic ice sheets to sea level rise.' *Geophysical Research Letters*, 38, L05503, doi:10.1029/2011GL046583.

Robertson, W., **Zhang**, K., **Finkl**, C.W. and **Whitman**, D., 2008.'Hydrodynamic and geologic influence of event-dependent depth of closure along the South Florida Atlantic Coast.' *Marine Geology*, 252, 156–165.

Robinson, L.A., 1977.'Marine erosive processes at the cliff foot.' *Marine Geology*, 23, 257–271.

Rodrigue, L.B., **Curole**, G.P., **Lee**, D.M. and **Dearmond**, D.A., 2008. *Operations, Maintenance, and Monitoring Report for Isles Dernieres Restoration, Phase 0, Trinity Island (TE-24) Project*.'Coastal Protection and Restoration Authority of Louisiana, Office of Coastal Protection and Restoration, Thibodaux, Louisiana', 26 pp.

Rosenzweig, C., **Casassa**, G., **Karoly**, D.J., **Imeson**, A., **Liu**, C., **Menzel**, A., **Rawlins**, S., **Root**, T.L., **Seguin**, B. and **Tryjanowski**, P., 2007.'Assessment of observed changes and responses in natural and managed systems.' In: M.L. Parry, O.F. Canziani, J.P. Palutikof, P.J. van der Linden and C.E. Hanson (eds) *Climate Change 2007: Impacts, Adaptation and Vulnerability*. Contribution of Working Group II to the Fourth Assessment Report of the Intergovernmental Panel on Climate Change. Cambridge University Press, Cambridge, UK, 79–131.

Rossi, A., **Massei**, N., **Laignel**, B., **Sebag**, D. and **Copard**, Y., 2009.'The response of the Mississippi River to climate fluctuations and reservoir construction as indicated by wavelet analysis of streamflow and suspended-sediment load, 1950–1975.' *Journal of Hydrology*, 377, 237–244.

Roy, P.S., **Cowell**, P.J., **Ferland**, M.A. and **Thom**, B.G., 1994.'Wave-dominated coasts.' In: R.W.G. Carter and C.D. Woodroffe (eds) *Coastal Evolution*. Cambridge University Press, Cambridge, 121–186.

Roy, P.S., **Thom**, B.G. and **Wright**, L.D., 1980.'Holocene sequences on an embayed high-energy coast: An evolutionary model.' *Sedimentary Geology*, 26, 1–19.

Ruddiman, W.F., 2006.'Orbital changes and climate.' *Quaternary Science Reviews*, 25, 3092–3112.

Ruessink, B.G., **Pape**, L. and **Turner**, I.L., 2009.'Daily to interannual cross-shore sandbar migration: observations from a multiple sandbar system.' *Continental Shelf Research*, 29, 1663–1677.

Ruiz de Alegria-Arzaburu, A. and **Masselink**, G., 2010. 'Storm response and beach rotation on a gravel beach, Slapton Sands, UK.' *Marine Geology*, 278, 77–99.

Saenger, P. and **Siddiqi**, N.A., 1993. 'Land from the sea: the mangrove afforestation program of Bangladesh.' *Ocean & Coastal Management*, 20, 23–39.

Saito, Y., **Yang**, Z. and **Hori**, K., 2001. 'The Huanghe (Yellow River) and Changjiang (Yangtze River) deltas: a review on their characteristics, evolution and sediment discharge during the Holocene.' *Geomorphology*, 41, 219–231.

Salisbury, R.D., 1892. 'Overwash plains and valley trains.' *Geological Survey of New Jersey, Annual Report of the State Geologist*, 7, 96–114.

Sallenger, A.H. and **Holman**, R.T., 1985. 'Wave energy saturation on a natural beach with variable slope.' *Journal of Geophysical Research*, 90, 11939–11944.

Sanderson, P.G. and **Eliot**, I., 1996. 'Shoreline salients, cuspate forelands and tombolos on the coast of Western Australia.' *Journal of Coastal Research*, 12, 761–773.

Saye, S.E., **Pye**, K. and **Clemmensen**, L.B., 2006. 'Development of a cliff-top dune indicated by particle size and geochemical characteristics: Rubjerg Knude, Denmark.' *Sedimentology*, 53, 1–21.

Saye, S.E., **van der Wal**, D., **Pye**, K. and **Blott**, S.J., 2005. 'Beach–dune morphological relationships and erosion/accretion: An investigation at five sites in England and Wales using LIDAR data.' *Geomorphology*, 72, 128–155.

Scheffers, A., **Scheffers**, S., **Kelletat**, D. and **Browne**, T., 2009. 'Wave-emplaced coarse debris and megaclasts in Ireland and Scotland: boulder transport in a high-energy littoral environment.' *Journal of Geology*, 117, 553–573.

Schoonees, J.S., **Theron**, A.K. and **Bevis**, D., 2006. 'Shoreline accretion and sand transport at groynes inside the Port of Richard Bay.' *Coastal Engineering*, 53, 1045–1058.

Schwing, F.B., **Murphree**, T., **deWitt**, L. and **Green**, P.M., 2002. 'The evolution of oceanic and atmospheric anomalies in the northeast Pacific during the El Nino and La Nina events of 1995–2001.' *Progress in Oceanography*, 54, 459–491.

Scott, G.A.J. and **Rotondo**, G.M., 1983. 'A model to explain the differences between Pacific plate island atoll types.' *Coral Reefs*, 1, 139–150.

Sella, G.F., **Stein**, S., **Dixon**, T.H., **Craymer**, M., **James**, T.S., **Mazzotti**, S. and **Dokka**, R.K., 2007. 'Observation of glacial isostatic adjustment in "stable" North America with GPS.' *Geophysical Research Letters*, 34, L02306, doi:10.1029/2006GL027081.

Shepard, F.P., 1963. *Submarine Geology* (2nd edn). Harper & Row, New York.

Sheppard, C.R.C., **Davy**, S.K. and **Pilling**, G.S., 2009. *The Biology of Coral Reefs*. Oxford University Press, Oxford.

Sherman, D.J. and **Bauer**, B.O., 1993. 'Dynamics of beach-dune systems.' *Progress in Physical Geography*, 17, 413–447.

Sherman, D.J., **Jackson**, D.W.T., **Namikas**, S.L. and **Wang**, J., 1998. 'Wind-blown sand on beaches: An evaluation of models.' *Geomorphology*, 22, 113–133.

Shields, A., 1936. *Anwendung der Ähnlichkeits-Mechanik und der Turbulenz-forschung auf die Geschiebebewegung*. Preussische Versuchsanstalt für Wasserbau und Schiffbau, 26.

Short, A.D., 1985. 'Rip current type, spacing and persistence, Narrabeen Beach, Australia.' *Marine Geology*, 65, 47–71.

Short, A.D. (ed), 1999. *Handbook of Beach and Shoreface Morphodynamics*, Wiley and Sons, Chichester.

Short, A.D., 2006. 'Australian beach systems – nature and distribution.' *Journal of Coastal Research*, 22, 11–27.

Short, A.D., 2010. 'Sediment transport around Australia – sources, mechanisms, rates, and

barrier forms.' *Journal of Coastal Research*, 26, 395–402.
Short, A.D. and **Hesp**, P.A., 1982. 'Wave, beach and dune interactions in southeastern Australia.' *Marine Geology*, 48, 259–284.
Short, A.D. and **Trembanis**, A.C., 2004. 'Decadal scale patterns in beach oscillation and rotation Narrabeen Beach, Australia – time series, PCA and wavelet analysis.' *Journal of Coastal Research*, 20, 523–532.
Sleath, J.F.A., 1984. *Sea Bed Mechanics*. Wiley and Sons, New York.
Smith, D.E., **Fretwell**, P.T., **Cullingford**, R.A. and **Firth**, C.R., 2006. 'Towards improved empirical isobase models of Holocene land uplift for mainland Scotland, UK.' *Philosophical Transactions of the Royal Society of London, Series A*, 364, 949–972.
Smith, D.E., **Shi**, S., **Cullingford**, R.A., **Dawson**, A.G., **Dawson**, S., **Firth**, C.R., **Foster**, I.D.L., **Fretwell**, P.T., **Haggart**, B.A., **Holloway**, L.K. and **Long**, D., 2004. 'The Holocene Storegga Slide tsunami in the United Kingdom.' *Quaternary Science Reviews*, 23, 2291–2321.
Smith, L.D., **Gilmour**, J.P. and **Heyward**, A.J., 2008. 'Resiliance of coral communities on an isolated system of reefs following catastrophic mass-bleaching.' *Coral Reefs*, 27, 197–205.
Soulsby, R.L., 1983. 'The bottom boundary layer of shelf seas.' In: B. Johns (ed) *Physical Oceanography of Coastal and Shelf Seas*. Elsevier, Amsterdam, 189–266.
Soulsby, R.L., 1997. *Dynamics of Marine Sands: A Manual for Practical Applications*. Thomas Telford, London.
Southard, J.B. and **Boguchwal**, L.A., 1990. 'Bed configurations in steady unidirectional water flows. Part 2: Synthesis of flume data.' *Journal of Sedimentary Petrology*, 60, 658–679.
Spencer, T., 1994. 'Tropical coral islands – an uncertain future?' In: N. Roberts (ed) *The Changing Global Environment*. Blackwell, Oxford, 190–209.
Spencer, T. and **Viles**, H.A., 2002. 'Bioconstruction, bioerosion and disturbance on coral reefs and rocky carbonate coasts.' *Geomorphology*, 48, 23–50.
Stanley, D.J. and **Chen**, Z., 1993. 'Yangtze delta, eastern China: I. Geometry and subsidence of Holocene depocenter.' *Marine Geology*, 112, 1–11.
Stephenson, W.J. and **Kirk**, R.M., 1996. 'Measuring erosion rates using the micro-erosion meter: 20 years of data from shore platforms, Kaikoura Peninsula, South Island, New Zealand.' *Marine Geology*, 131, 209–218.
Stephenson, W.J. and **Kirk**, R.M., 2000. 'Development of shore platforms on Kaikoura Peninsula, South Island, New Zealand, Part II: The role of subaerial weathering.' *Geomorphology*, 32, 43–56.
Stockdon, H.F., **Holman**, R.A., **Howd**, P.A. and **Sallenger**, A.H., 2006. 'Empirical parameterization of setup, swash, and runup.' *Coastal Engineering*, 53, 573–588.
Stoddart, D.R., 1965. 'The shape of atolls.' *Marine Geology*, 3, 369–383.
Stone, G.W. and **Donley**, J.C. (eds), 1998. 'The world deltas symposium: A tribute to James Plummer Morgan (1919–1995). Special Thematic Section' *Journal of Coastal Research*, 14, 695–916.
Storlazzi, C.D. and **Griggs**, G.B., 2000. 'Influence of El Niño–Southern Oscillation (ENSO) events on the evolution of central California's shoreline.' *GSA Bulletin*, 112, 236–249.
Sunamura, T. and **Mizuno**, O., 1987. 'A study on depositional shoreline forms behind an island.' *Annual Report Institute Geosciences*, University of Tsukuba, Tsukuba, No. 13, 71–73.
Sunamura, T., 1992. *Geomorphology of Rocky Coasts*. Wiley and Sons, New York.
Svendsen, I.A., 1984. 'Mass flux and undertow in the surf zone.' *Coastal Engineering*, 8, 347–365.
Sverdrup, H.U. and **Munk**, W.H., 1946. 'Theoretical and empirical relations in forecasting

breakers and surf.' *Transactions American Geophysical Union*, 27, 828–836.
Switzer, A.D. and **Burston**, J.M., 2010.'Competing mechanisms for boulder deposition on the southeast Australian coast.' *Geomorphology*, 114, 42–54.
Takeda, I. and **Sunamura**, T., 1982.'Formation and height of berms.' *Transactions, Japanese Geomorphological Union*, 3, 145–157.
Tamura, T., **Horaguchi**, K., **Saito**, Y., **Nguyen**, V.L., **Tateishi**, M., **Ta**, T.K.O., **Nanayama**, F. and **Watanabe**, K., 2010.'Monsoon-influenced variations in morphology and sediment of a mesotidal beach on the Mekong River delta coast.' *Geomorphology*, 116, 11–23.
Thom, B.G. and **Hall**, W., 1991.'Behaviour of beach profiles during accretion and erosion dominated periods.' *Earth Surface Processes and Landforms*, 16, 113–127.
Thornton, E.B. and **Guza**, R.T., 1982.'Energy saturation and phase speeds measured on a natural beach.' *Journal of Geophysical Research*, 87, 9499–9508.
Thurman, H.V. and **Burton**, E.A., 2001. *Introductory Oceanography* (9th edn). Prentice Hall, New Jersey.
Tol, R.S.J., **Klein**, R.J.T. and **Nicholls**, R.J., 2008.'Towards successful adaptation to sea-level rise along Europe's coasts.' *Journal of Coastal Research*, 24, 432–442.
Trenhaile, A.S., 1987. *The Geomorphology of Rock Coasts*. Oxford University Press, Oxford.
Trenhaile, A.S., 1997. *Coastal Dynamics and Landforms*. Oxford University Press, Oxford.
Trenhaile, A.S., 1999.'The width of shore platforms in Britain, Canada and Japan.' *Journal of Coastal Research*, 15, 355–364.
Trenhaile, A.S. and **Mercan**, D.W., 1984.'Frost weathering and the saturation of coastal rocks.' *Earth Surface Processes and Landforms*, 9, 321–331.
Tucker, M.E., 1995. *Sedimentary Petrology: An Introduction to the Origin of Sedimentary Rocks*. Blackwell Science, Oxford.
Turcotte, D.L., 2007.'Self-organized complexity in geomorphology: Observations and models.' *Geomorphology*, 91, 302–310.
Turner, I., 1993.'Water table outcropping on macro-tidal beaches: A simulation model.' *Marine Geology*, 115, 227–238.
Turner, I.L., 1995.'Simulating the influence of groundwater seepage on sediment transported by the sweep of the swash zone across macro-tidal beaches.' *Marine Geology*, 125, 153–174.
Udo, K., **Kuriyama**, Y. and **Jackson**, D.W.T., 2008.'Observations of wind-blown sand under various meteorological conditions at a beach.' *Journal of Geophysical Research*, 113, F04008, doi:10.1029/2007JF000936.
Uncles, R.J., **Bale**, A.J., **Stephens**, J.A., **Frickers**, P.E. and **Harris**, C., 2010.'Observations of floc sizes in a muddy estuary.' *Estuarine, Coastal and Shelf Science*, 87, 186–196.
Uncles, R.J., **Barton**, M.L. and **Stephens**, J.A., 1994.'Seasonal variability of fine-sediment concentrations in the turbidity maximum region of the Tamar estuary.' *Estuarine, Coastal and Shelf Science*, 38, 19–39.
Van Dijk, M., **Postma**, G. and **Kleinhans**, M.G., 2009.'Autocyclic behaviour of fan deltas: an analogue experimental study.' *Sedimentology*, 56, 1569–1589.
van Dyke, M., 1982. *An Album of Fluid Motion*. The Parabolic Press, Stanford.
van Houwelingen, S., **Masselink**, G. and **Bullard**, J., 2008.'Dynamics of multiple intertidal bars over semidiurnal and lunar tidal cycles, North Lincolnshire, England.' *Earth Surface Processes and Landforms*, 33, 1473–1490.
Van Lancker, V., **Bonne**, W., **Uriarte**, A. and **Collins**, M. (eds), 2010. EUMARSAND: European Marine Sand and Gravel Resources. *Journal of Coastal Research*, Special Issue 51, 1–226.
van Vuren, S., **Kok**, M. and **Jorissen**, R.E., 2004.'Coastal defense and societal activities

in the coastal zone: compatible or conflicting interests?' *Journal of Coastal Research*, 20, 550–561.

Viles, H. and **Spencer**, T., 1995. *Coastal Problems*. Edward Arnold, London.

Villard, P.V. and **Osborne**, P.D., 2002. 'Visualisation of wave-induced suspension patterns over two-dimensional bedforms.' *Sedimentology*, 49, 363–378.

Walkden, M.J.A. and **Hall**, J.W., 2005. 'A predictive Mesoscale model of the erosion and profile development of soft rock shores.' *Coastal Engineering*, 52, 535–563.

Wallace, J.M. and **Hobbs**, P.V., 2006. *Atmospheric Science – an introductory survey* (2nd edn). Academic Press, Amsterdam.

Wang, P., **Kirby**, J.H., **Haber**, J.D., **Horwitz**, M.H., **Knorr**, P.O. and **Krock**, J.R., 2006. 'Morphological and sedimentological impacts of Hurricane Ivan and immediate poststorm beach recovery along the northwestern Florida barrier-island coasts.' *Journal of Coastal Research*, 22, 1382–1402.

Ward, W.T., 2006. 'Coastal dunes and strandplains in southeast Queensland: sequence and chronology.' *Australian Journal of Earth Sciences*, 53, 363–373.

Wasson, R.J., 2003. 'A sediment budget for the Ganga–Brahmaputra catchment.' *Current Science*, 84, 1041–1047.

Webb, A.P. and **Kench**, P.S., 2010. 'The dynamic response of reef islands to sea-level rise: Evidence from multi-decadal analysis of island change in the Central Pacific.' *Global and Planetary Change*, 72, 234–246.

Wells, N., 1997. *The Atmosphere and Ocean, a physical introduction*. Wiley, Chichester.

Wemelsfelder, P.J., 1953. 'The disaster in the Netherlands caused by the storm flood of February 1, 1953.' *Proceedings of the 4th Coastal Engineering Conference*, ASCE, 256–271.

Wiegel, R.L., 1964. *Oceanographical Engineering*. Prentice-Hall, Englewood Cliffs.

Wiggs, G.F.S., **Baird**, A.J. and **Atherton**, R.J., 2004. 'The dynamics effects of moisture on the entrainment and transport of sand by wind.' *Geomorphology*, 59, 13–30.

Wilson, P., 1987. 'Soil formation on coastal beach and dune ridge sands at Magilligan Point Nature Reserve, Co. Londonderry.' *Irish Geography*, 20, 43–49.

Wilson, P., **Orford**, J.D., **Knight**, J., **Braley**, S. and **Wintle**, A.G., 2001. 'Late-Holocene (post-4000 years BP) coastal dune development in Northumberland, northeast England.' *The Holocene*, 11, 215–229.

Woodroffe, C.D., 2002. *Coasts: form, process and evolution*. Cambridge University Press, Cambridge.

Woodroffe, C.D., 2008. 'Reef-island topography and the vulnerability of atolls to sea-level rise.' *Global and Planetary Change*, 62, 77–96.

Woodroffe, C.D. and **Gagan**, M.K., 2000. 'Coral microatolls from the central Pacific record late Holocene El Niño.' *Geophysical Research Letters*, 27, 1511–1514.

Woodroffe, C.D. and **McLean**, R., 1990. 'Microatolls and recent sea level change on coral atolls.' *Nature*, 344, 531–534.

Woodroffe, C.D., **McLean**, R.F., **Smithers**, S.G. and **Lawson**, E.M., 1999. 'Atoll reef-island formation and response to sea-level change: West Island, Cocos (Keeling) Islands.' *Marine Geology*, 160, 85–104.

Wright, L.D., 1976. 'Morphodynamics of a wave-dominated river mouth.' *Proceedings 15th Coastal Engineering Conference*, ASCE, 1721–1737.

Wright, L.D., 1977. 'Sediment transport and deposition at river mouths: A synthesis.' *Geological Society of America Bulletin*, 88, 857–868.

Wright, L.D., 1989. 'Benthic boundary layers of estuarine and coastal environments.' *Reviews in Aquatic Sciences*, 1, 75.

Wright, L.D. and **Short**, A.D., 1984. 'Morphodynamic variability of surf zones and beaches: A synthesis.' *Marine Geology*, 56, 93–118.

Wright, L.D. and **Thom**, B.G., 1977.'Coastal depositional landforms: A morphodynamic approach.' *Progress in Physical Geography*, 1, 412–459.

Wright, L.D., **Coleman**, J.M. and **Thom**, B.G., 1973.'Processes of channel development in a high-tide-range environment: Cambridge Gulf-Ord River delta, Western Australia.' *Journal of Geology*, 81, 15–41.

Wright, L.D., **Yang**, Z.S., **Bornhold**, B.D., **Keller**, G.H., **Prior**, D.B. and **Wiseman**, W.J., 1986.'Hyperpycnal plumes and plume fronts over the Huanghe (Yellow River) delta front.' *Geo-Marine Letters*, 6, 97–105.

Yasso, W.E., 1965.'Plan geometry of headland-bay beaches.' *Marine Geology*, 73, 702–714.

Young, I.R. and **Holland**, G.J., 1996. *Atlas of Oceans, Wind and Wave Climate*. Elsevier, Oxford.

Yokoyama, Y., **Lambeck**, K., **De Deckker**, P., **Johnston**, P. and **Fifeld**, L.K., 2000.'Timing of the Last Glacial Maximum from observed sea-level minima.' *Nature*, 406, 713–716.

Zhang, W., **Wang**, Y. and **Lee**, S.J., 2008.'Simultaneous PIV and PTV measurements of wind and sand particle velocities.' *Experiments in Fluids*, 45, 241–256.

Zingg, A.W., 1953.'Wind tunnel studies of the movement of sedimentary material.' *Proceedings of the Fifth Hydraulics Conference. State University of Iowa, Studies in Engineering Bulletin*, 34, 111–135.

Zingg, T. 1935. Beitrag zur Schotteranalyse. *Schweizerische Mineralogische und Petrographische Mitteilungen*, 15, 39–140.

INDEX